CELL BIOLOGY

A Comprehensive Treatise

Volume 2

The Structure and Replication of Genetic Material

CONTRIBUTORS

Friedrich Bonhoeffer
Jonathan O. Carlson
Cedric I. Davern
Ruth M. Hall
Roger Hand
Leland H. Johnston
Burke H. Judd
R. F. Kimball
Julie Korenberg
Anthony W. Linnane
W. J. Peacock
David E. Pettijohn
Hans Ris
Peter Symmons

CELL BIOLOGY

A Comprehensive Treatise

Volume 2

The Structure and Replication of Genetic Material

Edited by

DAVID M. PRESCOTT

LESTER GOLDSTEIN

Department of Molecular, Cellular and Developmental Biology
University of Colorado
Boulder, Colorado

Academic Press *New York San Francisco London 1979*
A Subsidiary of Harcourt Brace Jovanovich, Publishers

ACADEMIC PRESS, INC.
111 Fifth Avenue, New York, New York 10003

United Kingdom Edition published by
ACADEMIC PRESS, INC. (LONDON) LTD.
24/28 Oval Road, London NW1 7DX

Library of Congress Cataloging in Publication Data
Main entry under title:

The Structure and replication of genetic material.

(Cell biology, a comprehensive treatise ; v. 2)
Includes bibliographies and index.
1. Chromosomes. 2. Chromosome replication.
3. Deoxyribonucleic acid repair. I. Prescott,
David M., Date II. Goldstein, Lester.
III. Series.
QH574.C43 vol. 2 [QH600] 574.8'7s [574.8'732] 78–10457
ISBN 0–12–289502–9 (v. 2)

PRINTED IN THE UNITED STATES OF AMERICA

79 80 81 82 9 8 7 6 5 4 3 2 1

Contents

7 Strandedness of Chromosomes and Segregation of Replication Products

W. J. Peacock

8 Eukaryotic Chromosome Replication and Its Regulation

Roger Hand

9 DNA Repair and Its Relationship to Mutagenesis, Carcinogenesis, and Cell Death

R. F. Kimball

List of Contributors

Numbers in parentheses indicate the pages on which the authors' contributions begin.

Friedrich Bonhoeffer (59), Max-Planck-Institut für Virusforschung, 74 Tübingen, Spemannstr. 35, West Germany

Jonathan O. Carlson (1), Department of Biophysics and Genetics, University of Colorado Medical Center, Denver, Colorado 80262

Cedric I. Davern* (131), Department of Biology and Department of Microbiology, University of Utah, Salt Lake City, Utah 84132

Ruth M. Hall (171), Department of Biochemistry, Monash University, Clayton, Victoria 3168, Australia

Roger Hand (389), Department of Microbiology, McGill University, Montreal, Quebec H3A 2B4, Canada

Leland H. Johnston (59), Division of Microbiology, National Institute for Medical Research, The Ridgeway, Mill Hill, London NW7 1AA, England

Burke H. Judd (223), The University of Texas at Austin, Department of Zoology, Austin, Texas 78712

R. F. Kimball (439), Biology Division, Oak Ridge National Laboratory, Oak Ridge, Tennessee 37830

Julie Korenberg† (267), Department of Zoology, University of Wisconsin, Madison, Wisconsin 53705

Anthony W. Linnane (171), Department of Biochemistry, Monash University, Clayton, Victoria 3168, Australia

W. J. Peacock‡ (363), Division of Plant Industry, Commonwealth Scientific and Industrial Research Organization, Canberra, ACT, Australia

David E. Pettijohn (1), Department of Biophysics and Genetics, University of Colorado Medical Center, Denver, Colorado 80262

Hans Ris (267), Department of Zoology, University of Wisconsin, Madison, Wisconsin 53705

Peter Symmons (59), Max-Planck-Institut für Virusforschung, 74 Tübingen, Spemannstr. 35, West Germany

* Present address: University of Utah, Salt Lake City, Utah 84112.

† Present address: University of Miami School of Medicine, Miami, Florida.

‡ Present address: Division of Plant Industry, Commonwealth and Scientific Industrial Research Organization, P.O. Box 1600, Canberra City, ACT 2601, Australia.

Preface

Volume 1 of this treatise dealt with the genetic mechanisms of cells. A logical extension of this topic is the consideration of genetics at the molecular level, and this volume, therefore, deals with the structure and replication of the genetic material both in the nucleus (including bacterial and viral nucleoids) and in cytoplasmic organelles. Volumes 3 and 4 will be concerned with genetic expression, covering transcription and translation, respectively. These four volumes will complete the first part of the treatise and establish a basis on which to deal with broad questions of cell structure and function, such as cell reproduction, differentiation, and cell–cell interactions, in subsequent volumes.

Continuing the objective presented in the Preface to Volume 1, we have planned this and succeeding volumes to serve as primary sources of fundamental knowledge for graduate students, investigators working in peripheral areas, and for anyone else in need of information on some particular phase of cell biology. Thus we asked authors to write chapters emphasizing reasonably well established facts and concepts, but not to attempt the more traditional up-to-the-minute reviews that investigators working in specialized fields count on. A measure of the maturity of cell biology also became evident from the fact that it has been a relatively simple matter to construct each volume around a single, unified theme.

David M. Prescott
Lester Goldstein

Titles of Other Volumes

1

Chemical, Physical, and Genetic Structure of Prokaryotic Chromosomes

David E. Pettijohn and Jonathan O. Carlson

ISBN 0-12-289502-9

I. INTRODUCTION

The word chromosome (colored body) was coined to describe the darkly staining bodies that condense from components of the eukaryotic nucleus as cells begin mitosis. We now know that these structures consist of a very long DNA molecule (one per chromatid) and various other molecules that are bound to the DNA. Some of the latter adducts are probably involved in arranging the structure of the packaged DNA and in organizing its replication, transcription, recombination, and other genetic processes. In a prokaryotic cell, the genomic DNA is also condensed into an easily recognized structure, although its shape and size are not as regular as in a eukaryotic chromosome. In this chapter these prokaryotic structures will be referred to as chromosomes. By analogy with the terminology for eukaryotic chromosomes, molecules bound to the condensed prokaryotic DNA will be considered part of the chromosome. The purpose of this chapter is to summarize the current state of knowledge of the structure of prokaryotic chromosomes.

The chromosome can be described at several levels of organization. At one extreme there have been extensive studies of the intracellular morphology of bacterial chromosomes using various microscopy techniques. At the other extreme, nucleotide sequences of limited regions of the chromosomal DNA have been worked out to define the actual chemical structure of parts of the chromosome. In between these two levels of study, other investigations have defined intermediate levels of chromosomal organization, using genetic mapping techniques, biochemical dissection of isolated chromosomes, fine-structure microscopy, and physicochemical approaches. In this chapter, we shall review these developments. The order of discussion will be such that the gross structure is presented first, after which progressively finer levels of organization will be introduced, finally extending down to DNA sequence studies.

By analogy with protein structure, the structure of DNA in a chromosome can be described from different points of reference. For example, genetic mapping or DNA sequencing provides a one-dimensional representation of the chromosomal DNA, analogous in some sense to the primary structure of an enzyme. The organization of the three-dimensional structure of DNA in chromosomes, which has some analogy to tertiary structure organization in enzymes, is just beginning to be elucidated. The factors and structures that determine the positions of folds in the double-helical DNA and organize superhelical turns as yet are not well defined. A complete description of three-dimensional chromosome structure obviously will have implications for the mechanisms of DNA recombination, replication, transcription, and their regulation. While the latter subjects are covered in other chapters of these volumes, the subjects of these chapters necessarily overlap with the present discussion. When more is learned about the three-dimensional structure of the chromosome, one would expect even greater complementarity among these subjects.

Chromosome structure has been studied in many different prokaryotic systems. Space does not allow a survey of these; rather, we shall concentrate on the best-understood chromosomes of a few biological systems. Specifically emphasized will be the chromosome of *Escherichia coli* and the chromosomes of λ and T4 phages. Review of many other interesting systems (for example, bacterial spores) will be restricted or omitted, because comparatively less is known about their chromosomes. In addition, this chapter is not intended to be comprehensive. In an area as intensively studied as prokaryotic chromosomes, it has been necessary to limit the discussion to certain subjects and approaches.

II. GROSS ORGANIZATION OF PROKARYOTIC CHROMOSOMES

A. Structure of the Bacterial Nucleoid *in Vivo*

The bacterial chromosome (or nucleoid) can be visualized in the cell by a variety of different microscopy techniques (for reviews, see Ryter, 1968; Pettijohn, 1976). In all cases, the nucleoid is seen to occupy only a portion of the intracellular volume; however, no nuclear membrane or other structural components segregates it from the cytoplasm. Apparently, interactions within the structure define its state of condensation. The size and shape of the nucleoids vary when cells are grown in various media or under different conditions (see Figure 1). When cells of *E. coli* are grown in rich media, the apparent "surfaces" of the nucleoids appear more uneven and convoluted than those seen in cells growing slowly in minimal media (Ryter and Chang, 1975). When protein synthesis or RNA synthesis is inhibited, the nucleoids can change in size and shape (Kellenberger *et al.*, 1958; Daneao-Moore and Higgins, 1972; Dworsky and Schaechter, 1973). It seems that the structure of the nucleoid is dynamic in the sense that at least its gross organization is dependent on the physiological state of the cell. Also, the gross appearance of the nucleoids varies among the different bacterial strains. For example, in *Bacillus subtilis* the nucleoids tend to be smaller and more compact than in *E. coli*.

It has been suggested that the intracellular nucleoids visualized by microscopy represent only the tightly condensed genetically inactive DNA of the chromosome (Ryter and Chang, 1975) that perhaps is analogous to heterochromatin. DNA sequences containing actively transcribed genes

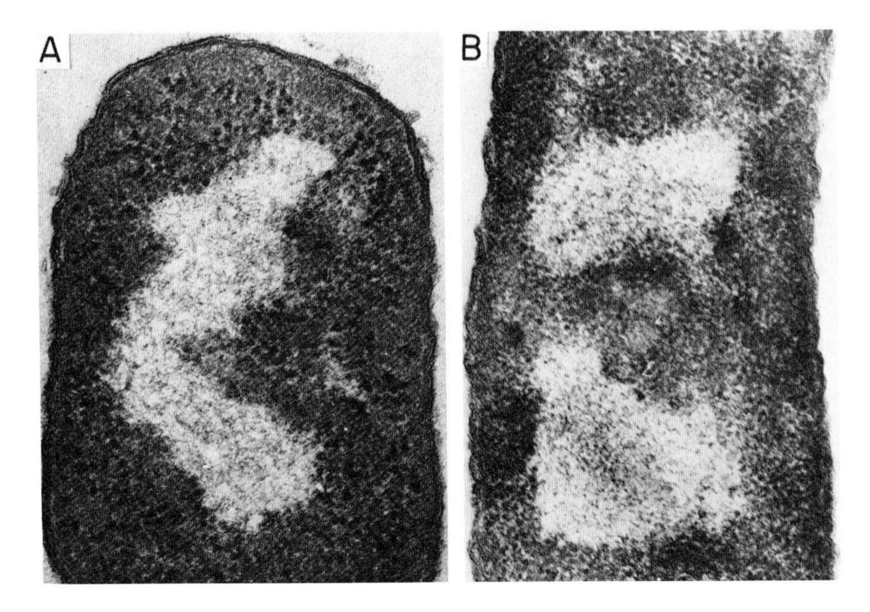

Fig. 1. Electron micrographs of thin sections of *E. coli* cells. (A) Bacterial nucleoid visualized in a cell grown in complex medium (× 90,000). (B) Bacterial nucleoid seen in a cell grown in synthetic medium (× 80,000). (From A. Ryter and A. Chang, 1975, with permission.)

may be "looped out" of the nucleoid into the cytoplasm, where they are available for interaction with RNA polymerase and ribosomes. In prokaryotes, the transcription of genes and translation of mRNA are tightly coupled and occur simultaneously (Miller *et al.*, 1970). The looping out of active genes would avoid the necessity of ribosomes functioning within the tightly packed structure of the visible nucleoid. DNA loops in the cytoplasm are not seen by conventional electron microscopy of cellular thin sections; but this is not surprising since the dense background in the cytoplasm would preclude the observation of a strand of the double helix. High-resolution autoradiographic methods have been used to examine the intracellular distribution of DNA with respect to the nucleoid surface (Ryter and Chang, 1975). The proportion of autoradiographic grains from ^{3}H-labeled DNA found external to the visible nucleoid in rapidly growing cells, in which genes are more transcriptionally active, was compared to the proportion found external in slowly growing cells. The distribution of grains was more external to the nucleoid in the more rapidly growing cells. This favors the idea that DNA loops form outside the visible nucleoid *in vivo*.

In unsynchronized growing cells, the DNA's of the nucleoids exist in various stages of partial replication. As will be described in Chapter 3, the number of DNA replication forks per chromosome increases as the division rate of the cells increases in richer media. As a consequence, the DNA content per nucleoid varies in the same organism growing at different rates. Indeed, the amount of DNA per cell can vary from one genome equivalent in very slow growing cells to greater than five genome equivalents in cells growing in very rich media (Kubitschek and Freedman, 1971). This occurs not only because the DNA content per nucleoid is greater but also because rapidly growing cells frequently contain more than one nucleoid. It is unclear whether the multiple nucleoids that appear to be physically independent (see, for example, Figure 1) are really free of interconnections prior to cell division. This point becomes important in the studies of isolated nucleoids that will be described below.

Detailed electron microscopic observations of thin sections of bacteria have demonstrated the existence of sites of attachment of the chromosome to the cellular envelope (for reviews, see Ryter, 1968; Leibowitz and Schaechter, 1975). As will be discussed in Chapter 3, these sites are believed to be responsible for the segregation of daughter chromosomes during cell division. They also may constitute the so-called replisome sites where DNA replication occurs on the bacterial membrane.

B. Gross Structure and Composition of Isolated Nucleoids

During the early 1970's, methods were developed for isolating the condensed DNA of *E. coli* in a structure that resembles the intracellular bacterial nucleoid (for review, see Pettijohn, 1976). The average size and DNA content of the isolated chromosomes, visualized in aqueous suspension, are similar to their values for the chromosome still in the cell (Hecht *et al.*, 1975). The gross stability of these prokaryotic chromosomes seems to differ markedly from that of eukaryotic metaphase chromosomes or chromatin. Whereas a compact state of the DNA in eukaryotic chromosomes and chromatin can be maintained (at least to some extent) in a variety of solvents and at various ionic strengths, the DNA of prokaryotic chromosomes readily unfolds when cells are opened in solutions containing only weak nonionic detergents, buffers, or concentrations of monovalent cations less than about 0.8 *M*. This may reflect an unusual sensitivity of the chromosomal proteins to the weak detergents normally used to disrupt cells (Dworsky, 1975a). It may also occur because the factors that organize the condensed state of the bacterial DNA are more weakly attached to the DNA. For example, no major classes of proteins analogous to histones, which maintain strong associations with

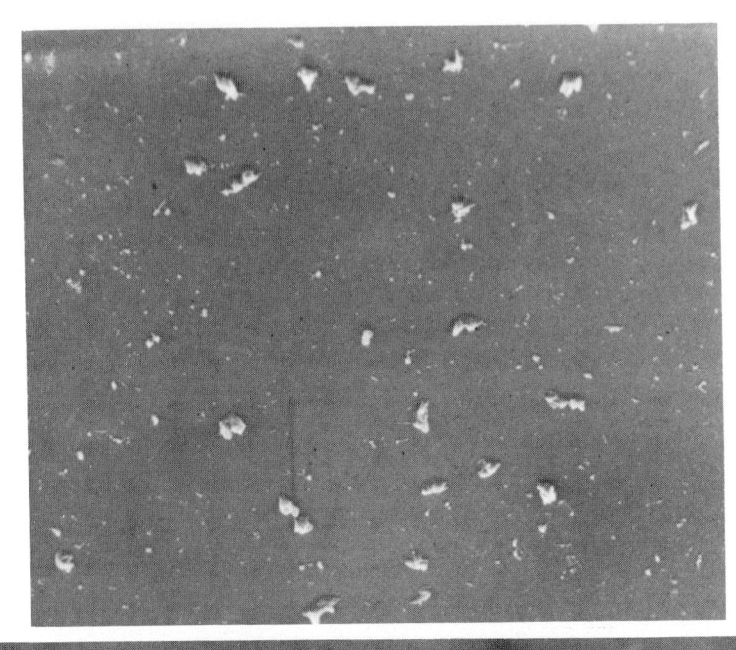

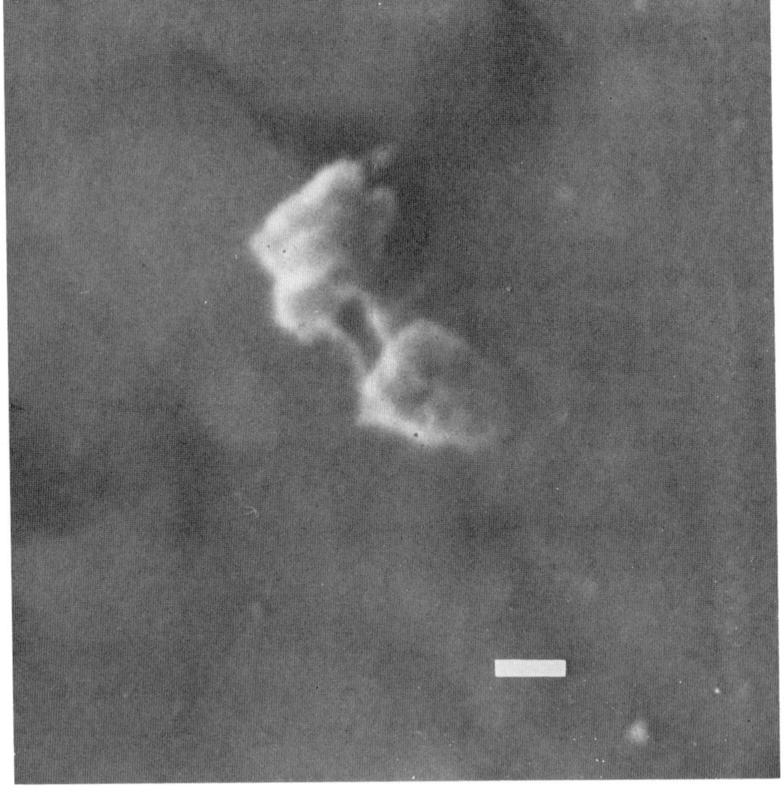

DNA over a very wide range of solvents, are found in prokaryotes. The known proteins of *E. coli* that resemble histones chemically are more readily dissociated from the DNA (these will be discussed later).

Three essentially different approaches have been developed for isolating the *E. coli* chromosome while maintaining the DNA in a condensed state: (1) high concentrations of monovalent counter-ions (about 1.0 *M*) are introduced to stabilize the DNA during lysis and during purification (Stonington and Pettijohn, 1971); (2) low concentrations of counter-ions are employed, but the concentrations of detergents used to disrupt the cell envelope are reduced to a minimum (Dworsky, 1975a,b); (3) low salt concentrations are used, but multivalent counter-ions such as spermidine or Mg^{2+} are introduced to stabilize the chromosome (Kornberg *et al.*, 1974). The stabilizing effect of the added counter-ions is believed to occur because there is shielding of charge repulsions in the closely packed RNA and DNA of the chromosome. Use of the latter two methods has not yet enabled separation of the chromosome from associations with the cell envelope. As a result, major components of the isolated system are envelope proteins, and it has been difficult to determine which of the more minor molecules are bound only to the cell membranes and which are part of the chromosome. Nucleoids isolated by the first method can be prepared either as membrane-associated chromosomes (Pettijohn *et al.*, 1973; Worcel and Burgi, 1974; Ryder and Smith, 1974) or as chromosomes that are either free of all, or contain only very minor amounts of the cell envelope (Pettijohn, 1976). These will be referred to here as membrane-free nucleoids. Because of their comparative simplicity, membrane-free nucleoids prepared by the first method have been used in most of the physicochemical studies of chromosome structure. However, the studies that have been accomplished with chromosomes prepared by the other methods as yet do not indicate any differences in the DNA structure.

Shown in Figure 2 are isolated *E. coli* nucleoids visualized by scanning electron microscopy. Similar to what is seen *in vivo*, the sizes and shapes of the chromosomes are not constant but appear to vary somewhat. As implied above, this is partly because their DNA contents vary, since the chromosomes differ in their state of replication. The shape variations may also reflect that dynamic state of the organization of the DNA which was

---

Fig. 2. Isolated bacterial chromosomes visualized by scanning electron microscopy. Membrane-free nucleoids isolated by the high-salt method (Giorno *et al.*, 1975) were adsorbed onto fragments of glass coverslips. After critical point drying the chromosomes were fixed, shadowed with gold, and examined in the scanning electron microscope. Top: general field showing many chromosomes (× 5000). Bottom: one selected chromosome (× 5000). Bar represents 0.2 μm.

also indicated by their morphology *in vivo*. As previously noted (Hecht *et al.*, 1975), many chromosomes have a doubletlike organization within which the chromosome appears to have two halves (see Figure 2). This may occur because the multiple nucleoids in a cell that appear to be separate structures are actually interconnected, so that they remain associated after isolation. Supporting evidence comes from the observation that the average DNA content per isolated nucleoid is similar to the average DNA content per cell, even when isolated from populations in which many cells contain more than one nucleoid. Whether these attachments are attributable to incomplete segregation of DNA, a remnant piece of the cell envelope, or some other specialized structure is unknown.

The nucleoids are usually isolated by centrifugation through sucrose gradients. This technique has also been used frequently to study the state of folding of the DNA (see Figure 3 below). The nucleoids sediment in a band that is broader than that for particles of homogeneous size; this is to be expected, since they appear to vary over a limited size range when examined by any of the microscopy techniques. Centrifugation in sucrose gradients can separate fractions of narrower ranges of sedimentation coefficient (Worcel and Burgi, 1972). Also, there is evidence that chromosomes isolated from cells that have been aligned in their DNA replication cycle have homogeneous sedimentation coefficients (Worcel and Burgi, 1974). This would be expected if the limited heterogeneity in sedimentation coefficient were primarily due to chromosomes in different stages of partial replication.

Chromosomes isolated by the high-salt method have been analyzed to determine their molecular composition. The results are summarized in Table I. The nucleoids contain a significant amount of RNA, most (and perhaps all) of which is nascent RNA chains attached to the chromosome via associated RNA polymerase molecules (Pettijohn *et al.*, 1970a; Pettijohn, 1976). It has been demonstrated that the endogenous RNA polymerase molecules are distributed on the genes of the isolated chromosome at sites similar to their sites *in vivo* (Pettijohn *et al.*, 1970b). Electrophoretic analyses of the proteins of the isolated "membrane-free" nucleoid showed that the bound proteins are mostly RNA polymerase molecules. Large amounts of the major envelope proteins are bound to the membrane-associated nucleoid. Trace amounts of these envelope proteins also accompany the membrane-free nucleoid. Few of the other minor proteins of the isolated nucleoids have yet been identified or their functions determined. A more detailed description of the molecular components of the isolated nucleoids may be found in a recent review (Pettijohn, 1976). Preliminary analyses of nucleoids isolated by Dworsky's procedure (low ionic strength, low salt) have also been obtained

TABLE I

Properties of Nucleoids Isolated from *Escherichia coli* (30 min Generation Time)

	Membrane-free nucleoid	Membrane-associated nucleoid[a]
DNA content (weight fraction)[b]	0.6	~0.4
RNA content (weight fraction)[b]	0.3	0.15–0.35
Protein content (weight fraction)[b]	0.05–0.1	~0.4
Lipid content (fraction of total labeled lipid)[c]	<1%	~20%
Sedimentation rate (weight average)[d]	1900 S°	5500 S°
Buoyant density in CsCl	1.69 ± 0.02 gm/cm³	1.46 ± 0.02 gm/cm³
DNA mass per singlet nucleoid	$9 \pm 1.0 \times 10^{-9}$ μg	—
DNA mass per doublet nucleoid	$15 \pm 2 \times 10^{-9}$ μg	—
Genome equivalents per singlet	2.2	—
Genome equivalents per doublet	3.6	—
Fraction of total cellular membrane bound to nucleoids	<0.01	~0.2
Bound proteins		
RNA polymerase	Yes	Yes
Envelope proteins	Little or none	Yes

[a] Membrane-associated nucleoids can have variable amounts of membrane; therefore, these properties may vary.
[b] Weight fraction means fraction of total nucleoid dry weight.
[c] Lipids labeled with [^{14}C]oleic acid.
[d] Corrected for rotor speed effect indicated by S°.

(Dworsky, 1976; D. Pettijohn, unpublished results). Most of the nucleoid proteins observed in gel electrophoresis are the same as those obtained with membrane-associated nucleoids isolated by the high-salt methods; however, a few high molecular proteins may be present in an enriched amount in nucleoids prepared by this method (Dworsky, 1976).

C. Gross Structure of Bacteriophage Chromosomes

Bacteriophage DNA can be grossly organized into at least two distinct structures. When condensed in phage heads, the DNA has a packaging density (defined as the ratio of the DNA mass to the inner head volume) greater than 0.6 gm/cm³, which approaches the density of crystalline DNA (about 1.8 gm/cm³). In this tightly packaged state, the DNA is genetically inactive, since DNA replication and transcription do not occur in a phage head. In many cases, after infection the genetically active phage DNA appears in "nucleoid-like" structures within the cell (for example, phage T4; Kellenberger *et al.*, 1958). The packaging density of the DNA in these structures is much less than that inside the phage heads and may be somewhat less than that of the *E. coli* nucleoid (Hamilton and Pettijohn, 1976). The more relaxed state of condensation of the intracellular phage DNA may be required to maximize its interaction with the enzymes responsible for the genetic expression and replication of the phage genome.

The intracellular DNA's of several phages have been isolated in condensed conformations (for examples, see Frankel, 1968; Huberman, 1968; Altman and Lerman, 1970; Serwer, 1974; Curtis and Alberts, 1976; Hamilton and Pettijohn, 1976). The topological organization of the DNA in these structures is not clear, but it is known that each of the bodies contains 30–200 phage equivalents of DNA. At least in the cases of T4 and T7 phages the molecular interactions that maintain this condensed association are insensitive to strong ionic detergents, proteases, and RNase. Curtis and Alberts (1976) have shown that DNA fragments of one-quarter to two genome lengths are released from the T4 DNA complex when it is incubated with a single-strand-specific DNA endonuclease. The rate of release of the fragments is first-order and oligomeric intermediates are not seen, suggesting that each DNA fragment is held at roughly genome length intervals by some "core material." The core is apparently resistant to the above lytic enzymes and protein denaturants. DNA in this complex is the precursor to the DNA in progeny phage. Genomic lengths of the DNA are encapsulated from this complex into preformed heads in phages T4 and λ (Laemmli *et al.*, 1974; Kaiser *et al.*, 1975). There is evidence that during packaging the DNA becomes associated with the capsid at specific sites, and in T4 a specific internal protein (P22) is cleaved while the DNA is being packaged (Laemmli and Favre, 1973). This cleavage liberates a very acidic internal protein that apparently causes the T4 DNA to condense (Laemmli, 1975). The purified acidic protein can cause purified T4 DNA to collapse into particles having dimensions similar to that of T4 phage heads. Analogous condensations of T4 DNA can be obtained in solutions containing poly(ethylene oxide) or polylysine. This conformational transition (the ψ transition), first discovered by Lerman (see Lerman, 1973), occurs when repulsive interactions between the added polymer and the DNA result in the DNA assuming a conformation that minimizes the interaction with the added polymer. It is proposed that this interaction facilitates the entry of the DNA into the phage head. This transition may play a role in interconverting the phage DNA into its two different states of packaging. Similar polymer-induced transitions in DNA may also be involved in maintaining condensed states of DNA in other chromosomes (Lerman, 1973).

When T4 DNA is maximally condensed in the ψ state, it becomes sensitive to hydrolysis by DNA endonucleases that are specific for single-stranded DNA. The DNA is cut at sites separated by 200–400 base pairs (Laemmli, 1975), which is about the periodicity of DNA folds required to package the DNA into particles of the observed size. This could mean that a bend or kink in the DNA at each fold exposes a DNA strand in a conformation that makes it susceptible to the single-strand-specific endonuclease. Earlier evidence suggested that DNA packaged in phage heads may be distorted such that a fraction of the base-pairing or base-stacking interactions is disrupted (for example, see Tikchonenko *et al.*, 1971). The DNA in phage heads reacts with a variety of chemical agents that normally do not react with double-helical DNA but which react readily with single-stranded DNA. Studies of this kind suggest that there are changes in the secondary structure of packaged DNA. There are molecular models which show how kinks can be positioned in a double helix as it is folded in chromosomes (Crick and Klug, 1975). Thus, modifications in the double-helical structure may occur when DNA is tightly packaged.

D. Interactions of Phage and Plasmid DNA's with Bacterial Chromosomes

Plasmid DNA that is not covalently attached to the host bacterial chromosome (nonintegrated) can still cosegregate with the bacterial chromosome (Hohn and Korn, 1969). This may occur because of some mutual association with common element(s) of the cell envelope (Jacob *et al.*, 1966). On the other hand, some recent evidence indicates that there may be a noncovalent physical association between the plasmid DNA and the bacterial chromosome (Kline and Miller, 1975). When nucleoids are isolated from cells containing a nonintegrated F sex factor plasmid, the closed circular plasmid DNA cosediments with the nucleoid. The extent of the association between the two chromosomes is independent of whether the nucleoids are isolated in a membrane-free or membrane-associated form. By contrast, closed circular DNA of the plasmid colicin E1 (Col E1), which is known to be distributed randomly in the *E. coli* cytoplasm, is associated to a much lesser extent with the isolated bacterial nucleoid. However, even with Col E1, one or two copies of the closed circular plasmid DNA (less than 10% of the total Col E1) are associated with the isolated bacterial chromosome. These findings appear to apply generally to a number of different plasmid systems (Kline *et al.*, 1976). The different plasmids seem to be either F-like or Col E1-like. The degree to which the plasmid DNA is associated with the bacterial chromosome is determined by the genotype of the plasmid, and not by the size of its DNA. This argues against the possibility that the association is due primarily to a trapping phenomenon, which would be expected to be related to the size of the plasmid DNA. Furthermore, it has been observed that many of the plasmids that are Col E1-like, and not showing extensive association with the isolated bacterial chromosome, also segregate well into minicells, which lack a bacterial chromosome. By contrast, the plasmids that segregate poorly into minicells are those that are F-like and extensively associated with the isolated bacterial chromosome. It has been proposed that the nonintegrated plasmid–chromosome complexes may arise when the circular plasmid DNA forms an added loop or domain of supercoiling in the bacterial chromosome (see Section III,B below).

Infection of *E. coli* with bacteriophage T4 has a profound effect on the bacterial chromosome (for examples, see Luria and Human, 1950; Kellenberger *et al.,* 1958). A series of distinct transformations in the gross chromosomal structure take place. The first observable change occurs within 2–3 minutes after infection, when the nucleoids undergo "nuclear disruption." During this process, the host DNA moves from a largely central location to juxtaposition with the cell membrane. Mutants (*ndd* mutants) defective in this process have been isolated (Snustad and Conroy, 1974); these are normal with respect to phage growth and development.

Later in T4 infection, the bacterial DNA becomes dispersed throughout the cytoplasm. When isolated from these cells (or from cells somewhat earlier in infection) by the high-salt methods normally used to preserve chromosome structure, the bacterial DNA is unfolded (Snustad *et al.,* 1974). The DNA unfolding is independent of nuclear disruption, since it also occurs in cells infected with the *ndd* mutants. Estimates of the number of single- or double-strand breaks in the host DNA also indicate that the unfolded DNA or the DNA after nuclear disruption is essentially intact. Thus, the effects on chromosome structure are primarily on interactions stabilizing or organizing the chromosome, not on the covalent integrity of the DNA. Later in infection, T4-specific nucleases *do* degrade the host DNA. This breakdown also occurs in the cells infected with *ndd* mutants, showing that nuclear disruption is not essential for the hydrolysis of the host DNA. There is evidence supporting the hypothesis that nuclear disruption occurs because new sites of attachment of the host DNA to the cell envelope are introduced after T4 infection. These sites, which are under control of T4 gene D2b, are altered or inactivated in the *ndd* mutant.

The T4-induced unfolding of the host bacterial DNA requires the product of the T4 *unf* gene. Mutations (unf^-) that are deficient in DNA unfolding (Snustad *et al.*, 1976) have been isolated. The product of the *unf* gene operates independently of the gene products for host DNA degradation or nuclear disruption, since all three phenomena can proceed in the absence of one or both of the other. T4 *alc* mutants are also unable to unfold the *E. coli* chromosome (Sirotkin *et al.*, 1977). The product of the *alc* gene is required not only for host DNA unfolding, but is also necessary for the transcription of late T4 genes. The *alc* gene product is a T4-coded RNA polymerase subunit that interacts with host RNA polymerase to modify its transcriptional properties. It was proposed that the modification of RNA polymerase eliminates the ability of polymerase to transcribe DNA containing cytosine. Since T4 DNA normally contains hydroxymethylcytosine instead of cytosine, the *alc* gene product probably acts to shut off the transcription of some of the genes in the host genome and thereby favors transcription of T4 DNA. The *alc* gene maps very close to the *unf* gene, and it is possible that they are the same. The reason the *alc* and *unf* gene products unfold the bacterial chromosome may be that they inhibit the synthesis of some host RNA species (Sirotkin *et al.*, 1977). As mentioned below, it has been demonstrated that nascent RNA chains whose synthesis is inhibited by rifampicin are necessary for the stability of the isolated bacterial chromosome. Further research will be necessary to settle the significance of *alc* and *unf* gene action.

III. LONG-RANGE ORGANIZATION OF DNA IN CHROMOSOMES

A. DNA Circularity and Supercoiling

The DNA in the nonreplicating chromosome of *E. coli* is a single circular double-helical molecule of about 2.7×10^9 daltons or 4.1×10^3 kb, having a circumference of approximately 1.3 mm. The circular configuration of the *E. coli* DNA molecule is not unusual among prokaryotes, and indeed, circular DNA's have been isolated from a large variety of organisms. The vast majority, if not all, of the extrachromosomal, freely replicating plasmids, such as F factors, drug resistance factors, and colicins which are associated with bacteria are circular, and many, but not all of the bacteriophage DNA's are either circular or become circular during some portion of their life cycle. It is not clear what advantage, if any, the circular configuration carries; however, since circular DNA has no ends, the DNA would be resistant to degradation by exonucleases. It also seems possible that replication of the last few nucleotides in a sequence may require nucleotides beyond the last nucleotide to be added to a new chain, and this relationship is obtained by a circle (Chapter 3). The formation of circles also provides the possibility of a topological constraint on the helical-winding number, a constraint which in turn leads to the possibility of DNA supercoiling. This idea will be discussed next.

If all the phosphodiester bonds in the sugar–phosphate backbones of both strands of a circular DNA molecule are intact, the DNA molecule is termed a covalently closed circle. Covalently closed circular DNA has a unique property that distinguishes it from linear or open circular (nicked in one strand) DNA; namely, the two strands of the double helix are not free to vary in the total number of turns they make about each other. This property is called topological constraint and is defined for any particular DNA molecule by the topological winding number (α), which is the number of times one strand crosses over the other strand if the molecule is constrained to lie in a plane. As long as the two strands of the DNA are covalently continuous, α must remain constant. α may be expressed as the sum of the numbers of two different kinds of turns, $\alpha = \beta + \tau$. β is the number of secondary structure turns, that is, the number of times the two strands wind around each other to form the double helical structure of DNA. τ is the number of tertiary or superhelical turns that are due to the winding of the double helix around itself or around a superhelical axis. In an aqueous environment, in which most experiments are conducted, DNA assumes its thermodynamically most stable secondary structure, which has one double-helical turn for every ten base pairs. Thus, under these conditions β equals the number of base pairs in the DNA molecule divided by ten. If, in a covalently closed circular DNA molecule, β does not equal α, the DNA is trapped in a stressed high-energy configuration, which causes the winding of the double helix about itself to form superhelical turns. The number of superhelical turns is equal to the difference between α and β: $\tau = \alpha - \beta$.

If α is greater than β, the DNA strands will be overwound with respect to the thermodynamically most favorable structure and the superhelical turns will be positive. If α is less than β, the strands will be underwound, and the superhelical turns will be negative. Superhelical turns can be either interwound (formed by winding two regions of a double helix about each other) or toroidal (formed by higher–order coiling of a double helix about a superhelical axis without winding two double helices around each other) (see model in Figure 4 for illustration of toroidal superhelices). Both toroidal and interwound superhelices can exist in equilibrium in a DNA molecule; they have opposite handedness so that left-handed toroidal helices are equivalent to right-handed interwound superhelices.

Since two DNA molecules of different molecular weights will have a different number of τ turns for the same degree of twisting, it is useful to define an intrinsic quantity σ, the superhelical density, as the number of superhelical turns for every ten base pairs. Thus, two molecules that have the same superhelical density have the same degree of twisting regardless of their molecular weights. It is interesting to note that, with few exceptions, superhelical DNA's isolated from a variety of organisms, both prokaryotic and eukaryotic, have about the same superhelical densities. This superhelical density is about one negative superhelical turn per 200 bp's. The ubiquity of this superhelical density suggests that a common DNA secondary and/or tertiary structure exists inside all cells.

B. Domains of Supercoiling in Isolated Bacterial Chromosomes

The DNA in isolated nucleoids is supercoiled with about the same superhelical density as found in other naturally occurring, closed circular DNA's. The evidence for this comes both from physicochemical studies (Worcel and Burgi, 1972), as well as from electron microscopic observations (Delius and Worcel, 1973). The sedimentation rate of isolated nucleoids varies when different amounts of ethidium bromide (EB) are intercalated in the DNA. At a critical concentration of EB (2.0 μg/ml), the sedimentation coefficient of the chromosome is minimal, and at higher or lower concentrations of EB the coefficient is increased. This kind of biphasic transition is typical of negatively supercoiled DNA's, which are more compact at EB concentrations above and below a critical concentration. σ is negative at zero or low concentrations of EB, positive at high concentrations, and zero at the critical concentration.

A nick introduced with DNase in one strand of a DNA double helix provides a swivel that permits rotation of one strand around the other. Thus, it becomes possible to change the topological winding number (α), in a nicked circular DNA. This has the effect of relieving strain due to underwinding of a double helix, and therefore the supercoiling is relieved. Ordinarily, a supercoiled circular DNA molecule requires only one nick to relieve all the supercoiling (Vinograd *et al.*, 1965); however, many separate nicks in the isolated *E. coli* chromosome are required to relieve all of its supercoiling (Worcel and Burgi, 1972). This suggested that there are restraints on the unwinding of the double helix when it is packaged in nucleoid form. A swivel that relieves supercoiling in one region of the DNA cannot also relieve supercoiling in another region, presumably because the rotations of one strand about the other cannot be propagated into the adjacent region. This indicates that the packaged DNA is organized into a series of domains of supercoiling, each of which is segregated by separate topological constraints. A domain of supercoiling is defined as a region of a DNA double helix bounded by two sites that restrict the rotation of the double helix.

There is an alternative explanation for why multiple swivels are required to relax all the supercoiling in the nucleoid. One might imagine that the unwinding of supercoils about a single nick in a very large DNA molecule is slow, and the measurements of the amount of supercoiling were made too soon after introducing a nick to allow complete relaxation; however, relaxation goes more quickly when multiple nicks are introduced. The requirement for multiple swivels, then, would be only a kinetic requirement existing because of the experimental design. Several lines of evidence argue against this. First, it is clear from other studies to be described below that there *are* restraints on the nucleoid DNA that define the positions of DNA folds. If these molecular interactions were sufficiently strong, they would also be expected to restrict rotation of the double helix. Also, when a few nicks (10–30 per genome equivalent of DNA) are introduced with γ-irradiation, the nucleoids have sedimentation properties of partially relaxed DNA, and the remnant supercoiling has the same superhelical density as in the initial unnicked chromosome (Lydersen and Pettijohn, 1977). This is to be expected if the partial relaxation occurs because some domains have lost all supercoiling and others have lost none. A partial relaxation due to progressive kinetic unwinding would be more likely to reduce the average superhelical density. Moreover, the amount of partial relaxation does not decline detectably when the nicked chromosomes are held for different times prior to analysis, as would be expected if the unwinding had occurred with slow kinetics.

1. Interactions Stabilizing the Isolated Bacterial Chromosome

When isolated nucleoids are reacted with agents that dissociate or denature the nucleoid proteins, the DNA unfolds, indicating that certain proteins are essential for the stability of the isolated chromosome (Pettijohn *et al.*, 1973; Drlica and Worcel, 1975). Likewise, when the nucleoids are incubated with RNase, the DNA is unfolded, suggesting that some of the nucleoid RNA components are also essential for the condensed structure of the DNA (Stonington and Pettijohn, 1971). The unfolding transitions are manifested by changes in sedimentation coefficient (see Figure 3), viscosity (Drlica and Worcel, 1975), rotor speed dependency of sedimentation coefficient (Hecht *et al.*, 1977), and microscopic observation (Hecht, 1976) of the nucleoid DNA. As the nucleoid RNA is more extensively hydrolyzed, the DNA becomes progressively more unfolded (Figure 3), suggesting that many RNA molecules are involved in stabilizing the isolated chromosome. When RNA hydrolysis is complete, the DNA

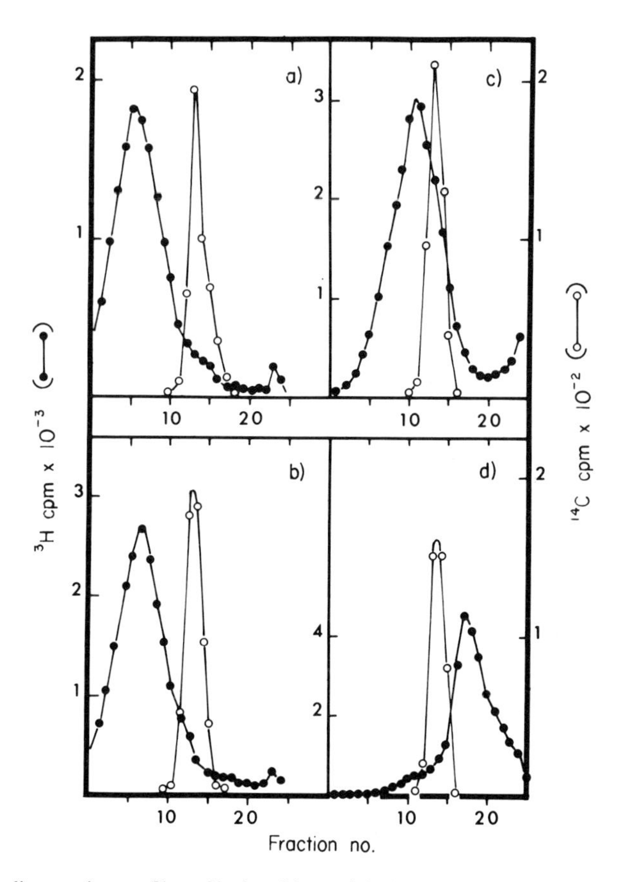

Fig. 3. Sedimentation profiles of isolated bacterial chromosomes with the packaged DNA unfolded to various extents. Isolated bacterial nucleoids with ^3H-labeled DNA were mixed with ^{14}C-labeled T4 phage that served as a sedimentation marker (1025 S). The nucleoids were reacted with different concentrations of RNase to partially unfold the DNA and sedimented on sucrose gradients. (a) Incubated with zero RNase; (b, c, and d) incubated with increasing concentrations of RNase.

appears to be completely or nearly completely unfolded (Hecht *et al.*, 1977), and it has lost all detectable supercoiling (Drlica and Worcel, 1975). The essential RNA molecules appear to be nascent RNA molecules bound to the chromosome, since growth of the cells with rifampicin for periods just long enough to eliminate nascent RNA molecules leads to DNA unfolding during isolation (Pettijohn and Hecht, 1973; Dworsky and Schaechter, 1973). Recent studies indicate that a complete unfolding of the nucleoid DNA occurs when nucleoids are isolated from the cells grown with rifampicin (D. Pettijohn, unpublished result). Thus, complete

DNA unfolding seems to be obtained when either the nucleoid RNA or the nucleoid proteins is dissociated from the DNA. One way in which this could occur is if the essential proteins of the isolated chromosome are those that stabilize the association of the essential nascent RNA molecules. A model depicting this is given in Figure 4. This model of the isolated chromosome also brings together the earlier findings concerning the organization of the domains of supercoiling (see above, and particularly Worcel and Burgi, 1972).

The model indicates that an RNA molecule bound to two or more separate sites on the isolated chromosome determines the position of a DNA fold and also restrains the rotation of the double helix at the sites of attachment. The rotation of the DNA is restricted by the maximum possible coiling of the RNA about the DNA. The RNA molecule holds together two regions of the double helix that are far apart on the linear chromosome map. Many such interactions involving different RNA molecules would segregate the DNA into a series of domains. The DNA in each

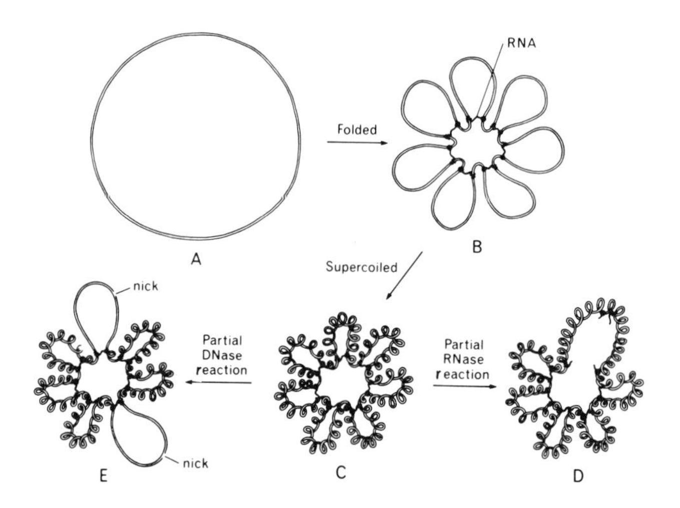

Fig. 4. A model of the conformational organization of DNA in an isolated bacterial chromosome. (A) The nonreplicating, circular completely unfolded DNA. (B) DNA containing folds restrained by RNA–DNA interactions; seven domains of supercoiling are shown; the actual number is 100 ± 30 per genome equivalent of DNA. (C) The nucleoid DNA folded and supercoiled; left-handed toroidal supercoils are drawn, as first proposed by Worcel and Burgi (1972); these also may convert into interwound supercoils after isolation (see Figure 5). (D) Partially unfolded DNA of the nucleoid obtained after hydrolyzing some of the stabilizing RNA; adjacent domains coalesce, resulting in fewer domains with more DNA per domain. (E) Chromosome having two single-stranded DNA breaks; these are in different domains and only the supercoiling in those domains is relaxed. Certain elements of this model were first proposed by Worcel and Burgi (1972) (From Pettijohn and Hecht, 1973.)

domain is supercoiled with a superhelical density of 0.05 or 1 superhelical turn per 200 base pairs. It is assumed that the essential RNA chains are nascent molecules attached at their 3′-ends to the DNA by an RNA polymerase molecule. The other attachment site may involve an RNA–DNA hybrid (Hecht and Pettijohn, 1976; Lydersen and Pettijohn, 1977). This type of organization would be dynamic in that the particular RNA–DNA sites would be in a constant state of flux.

The number of domains of supercoiling in the isolated nucleoid can be estimated from the number of single-strand breaks required to relax the supercoiling. Assuming that all the supercoiling in the chromosome is relaxed when each domain has at least one nick, Poisson statistics can be applied. The number estimated from reaction with DNase is 12–100 domains per genome equivalent of DNA (Worcel and Burgi, 1972; Pettijohn and Hecht, 1973). This is a rough estimate because of difficulties in measuring precisely the number of breaks and also because it is not clear whether DNase introduces random nicks in the DNA. Recent measurements using gamma irradiation to nick the DNA have provided a more accurate estimate of 100 ± 30 domains per genome equivalent (Lydersen and Pettijohn, 1977). Since singlet nucleoids in this study contained an average of about 2.2 genome equivalents of DNA, there would be on average about 200 domains per nucleoid.

The number of domains of supercoiling declines when chromosomes are partially unfolded with RNase (Pettijohn and Hecht, 1973). That is, the number of single-strand breaks required to relax all the supercoiling is reduced when nucleoid DNA is unfolded partially. Yet in the absence of added nicks, no detectable loss of supercoiling occurs in the partially unfolded DNA. These findings are consistent with the model shown in Figure 4D. Adjacent domains in nucleoids partially unfolded with RNase can coalesce to form larger domains. If there are no nicks in the chromosome, supercoils cannot unwind in these coalesced domains, so there is no detectable loss of supercoiling. Yet a single nick will now relax all of the DNA in this expanded domain, explaining the reduction in the number of nicks required for complete relaxation.

The model described in Figure 4 accounts for all of the experimental results described here concerning the structure of the isolated nucleoid. It should be emphasized, however, that the relationship of this structure to the chromosome *in vivo* remains to be established. There is no certainty that the interactions that stabilize the isolated chromosome are identical to those that function in the cell. The methods for isolating the chromosome do not preserve the association of all the chromosomal proteins with the DNA. In particular, the histonelike protein HU, which has been discovered in *E. coli* (Rouvière-Yanif and Gros, 1975), is apparently dissociated from the chromosome by the high salt concentrations used during isolation. It is not known whether other DNA-bound proteins, which may have a role in chromosome structure, are also absent. As will be described below, such chromosomal proteins may have a role in organizing toroidal superhelical spacings in the chromosomal DNA.

2. *Electron Microscopic Observations of Spread Isolated Nucleoids*

The isolated bacterial nucleoid spread with a monolayer of cytochrome *c* has been visualized by electron microscopy (Figure 5). In this form, the

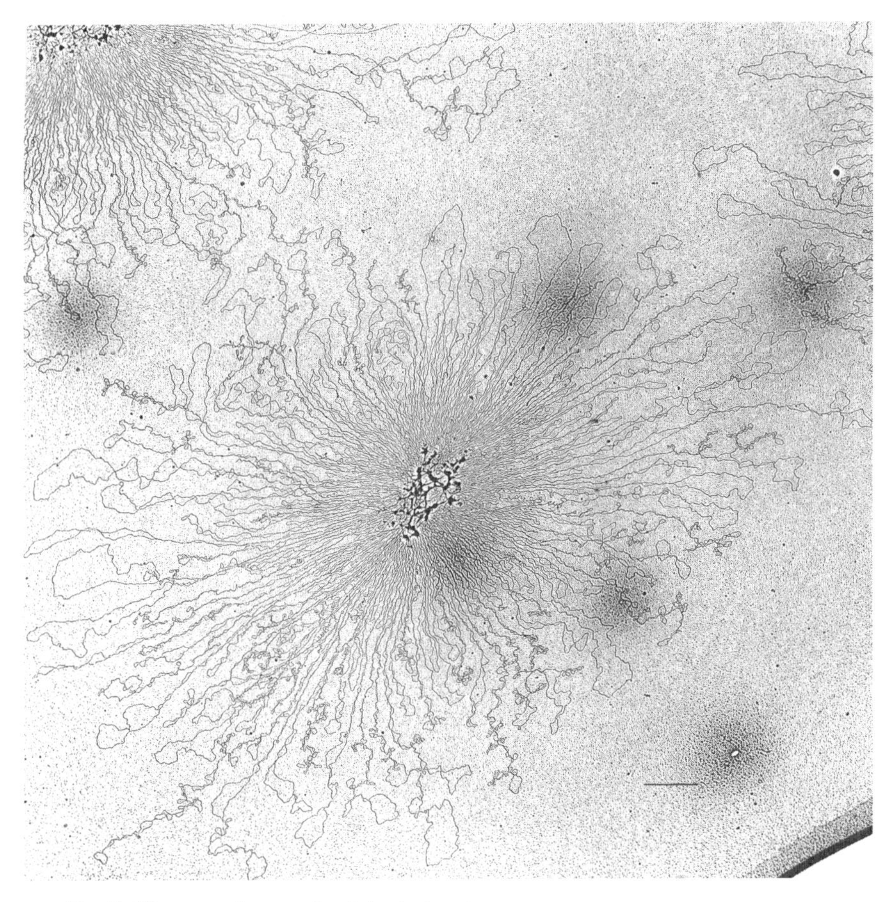

Fig. 5. Electron micrograph of the isolated "membrane-free" nucleoid. The nucleoids were spread with cytochrome *c*. Bar represents 1 μm. [Micrograph courtesy of Dr. Ruth Kavenoff; see Kavenoff and Bowen (1976).]

maximum dimension of the structure is 5- to 20-fold greater than in the cell. Strikingly regular arrays of loops can be seen, and many of the loops appear to be twisted with interwound superhelices (Delius and Worcel, 1973; Kavenoff and Ryder, 1976; Kavenoff and Bowen, 1976). The loops seem to occur in limited size ranges. The number of loops per nucleoid can be estimated; these range from 65 to 130 per membrane-associated nucleoid and from 98 to 194 per "membrane-free" nucleoid. The fact that more loops are seen in the latter chromosome may indicate that some of the loops have remained condensed in the central region of the membrane-associated nucleoid. The average number of loops agrees reasonably well with the estimates of the average number of domains of supercoiling, i.e., 220 (see above). Clearly, most of the DNA of the chromosome is in loops, since tracings indicated about 5×10^9 daltons of DNA per nucleoid or approximately two genome equivalents. This agrees with earlier estimates of the average DNA content per singlet nucleoid.

The dense material at the center of the spread chromosome appears to be mostly RNA. When the chromosomes on the grids are exposed to RNase, the dense material is greatly depleted. Although this procedure results in no detectable unfolding of the DNA, one would expect the DNA conformation to be relatively fixed when it is bound on the cytochrome *c* monolayers.

Most of all of the supercoils seen in electron micrographs of the chromosome appear to be interwound supercoils. This may explain at least in part why the nucleoid DNA appears to be so spread out. Toroidal superhelices greatly compress the length of a DNA double helix, while interwound superhelices in a DNA loop do not greatly reduce the length of the loop. Toroidal superhelices that exist in the cell may be converted into interwound twists when the nucleoid is isolated or when it is spread for electron microscopy. This could occur if some chromosomal components (histonelike proteins, for example) that hold toroidal supercoils are dissociated from the DNA during isolation (Griffith, 1976).

3. Properties of Membrane-Associated Chromosomes

Bacterial nucleoids isolated with associated membrane by the high-salt methods have many properties similar to those of the membrane-free nucleoid. The DNA is supercoiled with the same superhelical density, and unfolding of the DNA occurs after deproteinization or after degradation of the nucleoid RNA. The primary differences in properties are attributable to the mass or physical properties of the associated cell envelope (see Table I). For example, the sedimentation coefficient is greater, and the buoyant density is less; likewise, the lipid and protein contents are much greater. The additional protein is primarily derived from the major envelope proteins.

In an effort to identify the molecular interactions that maintain the association of chromosome to envelope, cross-linkage of bromouridine-labeled DNA to cellular envelope has been induced by uv irradiation of membrane-associated nucleoids. Two specific envelope proteins are selectively crosslinked to the DNA, one of 80,000, the other of 50,000 daltons (Worcel and Portalier 1976). A protein similar to the larger one has also been isolated in a complex with DNA by means of free-flow electrophoresis (Heidrich and Olsen, 1975).

Under some conditions of cellular lysis, the envelope fragments in membrane-associated nucleoids can form vesicles. These then can become entangled in the compact DNA of the chromosome (Meyer *et al.*, 1976). In this state, the envelope fragments *cannot* be solubilized from the DNA by the same procedures used to prepare membrane-free nucleoids, suggesting that nonspecific associations between cell envelope and chromosome may occur during lysis. Caution is indicated in interpreting studies of the association of nucleoid with cell envelope.

The membrane-associated chromosomes isolated in low-salt solvents by Dworsky's procedure have somewhat different properties than those isolated by the other methods (Dworsky, 1975a, 1976). Although superhelical densities are the same, the interactions stabilizing the DNA folding may be altered. For example, the nucleoids can be isolated from cells in which nascent RNA chains have been eliminated by growth with rifampicin; however, unfolding of the DNA still occurs after reaction with RNase. The chromosomes isolated by this method also unfold after exposure to high salt concentrations, including the concentrations used to stabilize the chromosomes isolated by the older methods. The results of this work, as well as other research reviewed in this section, suggest that rearrangements may occur in the molecular interactions of the isolated chromosomes during isolation. Obviously, further research will be required for these to be clarified.

C. Replication and Packaging of Bacteriophage λ DNA

The DNA of the bacteriophage λ undergoes a number of configurational changes during its life cycle. Orderly progression of the DNA through these changes is necessary for reproduction of the phage. The delineation of the sequence and nature of these changes, along with genetic analysis of the genes affecting these changes, has provided much detailed knowledge about the phage's life cycle. The DNA in the bacteriophage particle is condensed into a phage head of about 55 nm in diameter. The DNA extracted from the phage particle is a linear molecule of 30.8×10^6 daltons or 46.5 kb and has a length of about 16 μm (Davidson and Szybalski,

1971). The ends of the DNA molecule have short single-stranded regions, twelve bases long, that are extensions of the 5′-ends. These single-stranded ends are complementary to one another, and the DNA, if left in solution, will form circles (Hershey and Burgi, 1965). End-to-end aggregates also are formed. The proportion of circles and end-to-end aggregates depends on the DNA concentration, and hence, on the relative probabilities of one end colliding with the other end of its own molecule or another molecule (Wang and Davidson, 1966, 1968). The resulting circular DNA contains two nicks 12 base pairs apart in opposite strands where the ends have joined, and can be converted back to linear DNA by raising the temperature enough to denature the short region of double-stranded DNA between the nicks. No nonbase paired gaps are present in the circle, since the nicks can be sealed with DNA ligase to form a covalently continuous circle.

λ DNA is cyclized immediately after injection of the DNA into the bacterium. Cyclization and ligation of the nicks are necessary for the normal growth of the phage. The rate of cyclization in the cell is much greater than can be achieved *in vitro*. The reason for this is not clear, but it has been proposed that the two ends find each other in the cell by a two-dimensional diffusion process perhaps on the cell membrane, rather than by the three-dimensional diffusion process necessary for cyclization of purified DNA in solution (Wang and Davidson, 1968). In any case, cyclization and ligation of the injected DNA are completed quickly, since no linear monomeric λ DNA is found inside cells.

Replication of the DNA begins on these circles and proceeds in both directions from a specific replication origin (*ori;* see Figure 8). Replication generates intermediates, so-called theta structures, analogous to the structure seen by Cairns for the replicating *E. coli* chromosome (Schnös and Inman, 1970) (see Chapter 3). Replication proceeds by this mechanism for a few rounds, generating more covalently closed circles and nicked circles, all of monomer lengths. After this initial stage, the mode of replication switches from the replication of circles to a rolling circle mode of replication. This late mode of replication generates long linear concatameric DNA molecules containing several monomer equivalents of λ DNA arranged tandemly (end-to-end) along its length (Skalka *et al.*, 1972; Stahl *et al.*, 1972). The switch from the early circular mode of replication to the late rolling circle mode of replication is apparently dependent on the function of the γ gene product of λ. This protein somehow inactivates the *rec*BC nuclease of the host, which, in the absence of the γ protein, attacks the linear concatameric molecules but not covalently closed circles, and thereby prevents the progression of replication beyond the early circle mode (Enquist and Skalka, 1973; Greenstein and Skalka, 1975).

The long concatameric DNA produced by rolling circle replication is the substrate for the packaging of monomeric DNA into phage heads. Several lines of evidence from both *in vivo* and *in vitro* experiments allow a fairly detailed reconstruction of the packaging process. Concatameric DNA is necessary for packaging, since phage unable to enter the rolling circle mode of replication do not produce phage particles efficiently. The small amount of DNA that is packaged in this situation is probably the result of recombination between circles to form dimers that can then be packaged (Enquist and Skalka, 1973; Stahl *et al.*, 1972). This is supported by the finding that lysogens of phage deficient in both replication and recombination are unable to package DNA unless two phage genomes are lysogenized in tandem (Freifelder *et al.*, 1974). These experiments imply that monomer circles cannot be packaged and that more than one genome length of DNA is necessary as a packaging substrate. This is further borne out by experiments using *in vitro* packaging systems (Kaiser and Masuda, 1973; Hohn and Hohn, 1974), in which a cell-free extract is made from cells infected with λ. If purified DNA of the proper form is added to this extract, it will be packaged into infectious phage particles. In these systems, circular λ monomers are not packaged, whereas concatameric DNA is (Syvanen, 1975; Hohn, 1975).

In the packaging process, the concatameric DNA interacts with petite λ particles, which are partially assembled heads containing many of the λ head proteins but no DNA (Hendrix and Casjens, 1975; Hohn *et al.*, 1975). This interaction is mediated by the product of λ gene *A*. In the presence of the A protein, spermidine, and ATP, a monomer length of DNA is condensed inside the petite λ protein shell, other head proteins are added to the structure, and the cohesive end sites or *cos* sites are clipped (also mediated by the *A* gene product) to form a filled λ phage head containing a mature sticky-ended DNA molecule (Kaiser *et al.*, 1975). The packaging process is not merely a random process of stuffing a DNA molecule inside the head, since in reaction of tailless λ heads with micrococcal nuclease, only bases on the right sticky end are hydrolyzed (Padmanabhan *et al.*, 1972). This suggests that the right end of the λ DNA is always packaged nearest the tail. Further evidence for polarity in the packaging process comes from a more detailed examination of substrates for the *in vitro* packaging reaction. In this reaction, a λ DNA molecule with another partial λ DNA joined to the left end is packaged far more efficiently than a λ DNA with another partial λ DNA joined to the left end (Syvanen, 1975). *In vivo* evidence for packaging polarity is provided by analysis of phage containing a duplication of the region containing the *cos* site. When the DNA from a preparation of one of these phages is examined by heteroduplex mapping, a heterogeneous population of molecules is observed with some having the duplicated region on the left end, some on the right end, some having duplications on both ends, and some having no duplications at all. However, the great majority of the molecules contain one duplication on the left end. This bias is most easily explained if the packaging process begins at or near the left end of the molecule (still in the concatamer), proceeds toward the right end until the phage head is approximately full, and is completed by cleaving the DNA at the next available *cos* site (Emmons, 1974).

After packaging of the DNA is completed, the preassembled tail is attached to the head, and the infectious phage particle is complete. As can be seen from the above discussion, the configuration of the DNA during the life cycle profoundly influences the outcome of the infective process, and without the progression of the DNA through these configurations, reproduction is not successful.

D. Genetic Organization of Chromosomes

Ever since it was realized that DNA is the cellular component that carries the genetic information from generation to generation, a major thrust of molecular biology has been the determination of the mechanisms involved in the metabolism and function of that polynucleotide, and much research has been directed at determining the way in which the functional units or genes are organized on the chromosomal DNA. Much of this research has involved the bacterium *E. coli,* and as a result, more is known about the organization and function of the *E. coli* genome and its viruses than any other organism.

A number of techniques enable one to determine the position of genes in the chromosomal DNA molecules. We here briefly review a few of these, starting with purely genetic techniques and progressing to more recent physical and chemical techniques that allow very precise mapping of genes.

1. Genetic Methods of Mapping

In general, bacteria and bacteriophage are haploid and possess a single chromosome. However, genetic mapping methods traditionally depend on the introduction of two distinguishable alleles of a particular gene into the same cell and analysis of the progeny. The experimental techniques developed by bacterial geneticists for achieving such diploidy have been cleverly exploited for mapping purposes. Once two DNA molecules containing the two alleles have been introduced into the same cell, they can interact by recombination to form new DNA molecules containing portions of both parental molecules. "Normal" (legitimate) recombination requires fairly extensive homology between the base sequences of the two parent molecules. The recombination event can apparently occur with equal probability (at least to a first approximation) at any position on the DNA molecule, and because of the near-random nature of this process, it can be used to obtain positional information about the genes. If the two parental molecules contain mutations in two different genes, the probability that the recombinant daughter molecule will contain both mutations or neither mutation (that is, the probability that a recombinational event will have occurred in the DNA between the genes) is proportional to a first approximation to the amount of DNA between the genes. By comparing the recombination frequencies between pairs of genes among the progeny of the cell, one may establish the relative distance between the genes on the chromosome, and, if enough recombination frequencies are known, the order of the genes may be established.

The DNA molecules need not recombine but may complement one another by providing functions necessary for the survival of each other among the progeny. This type of interaction is useful in determining whether two mutations lie in the same gene, or whether one allele is dominant over another, or whether a mutation on one DNA molecule affects only itself or other DNA molecules as well.

One of the mechanisms for mapping genes in the *E. coli* chromosome is by determining the time of transfer of the gene from one bacterium to another during bacterial mating or conjugation. Certain strains of *E. coli* carry an autonomous circular plasmid DNA called an F factor. This plasmid is about 62×10^6 daltons or 94.5 kb and carries the genetic information that enables it to transfer itself from an F^+ bacterium to an F^- bacterium by conjugation. Occasionally in these F^+ strains, the F factor becomes covalently integrated into the chromosome of the bacterium and establishes an Hfr (male) strain. Because of its covalent linkage with the chromosome, when the F factor in an Hfr strain transfers itself to an F^- or female strain it is capable of transferring the entire male bacterial chromosome along with it. The transfer starts at a unique site in the integrated F DNA, proceeds with the chromosomal DNA immediately adjacent to the F DNA, continues through the whole chromosome, and ends with the portion of the F factor that was not transferred initially. The entire process takes about 100 minutes. The DNA transferred from the male bacterium to the female bacterium is free to recombine with the DNA of the recipient. If the conjugation process is interrupted during the transfer, for instance by violent agitation of the culture, only the portion of the chromosome transferred up to the time of interruption enters the female. If the male and female bacteria have suitable genetic markers, it is possible to select for recombinant progeny resulting from these matings. By determining the time during the mating that these recombinant bacteria appear, and by comparing these data with similar data for other genetic markers, the relative positions of the genetic markers can be determined. Many different Hfr strains containing the F factor integrated at various sites throughout the bacterial chromosome have been isolated (see Figure 6; Low, 1972). All of these Hfr's are capable of transferring the entire bacterial chromosome during conjugation. This implies that the genetic map is circular, in agreement with observations cited earlier.

More precise data on the location of the individual genes can be obtained by introducing the new DNA into the bacterium with transducing phage. Certain bacteriophage are capable of incorporating host DNA into their virus particles during their growth cycle and transferring this DNA to another bacterium through virus infection. This process is known as transduction, and bacteriophage capable of doing this are known as transducing phage. Transducing phage are characterized by whether they are able to transfer a wide variety of genetic markers (generalized transducing phage), or only one particular marker, or a limited number of genetic markers (specialized transducing phage). One generalized transducing phage that is widely used for mapping studies is bacteriophage P1. When P1 infects a cell, some of the progeny phage particles contain P1-sized (64×10^6 daltons or 97 kb) pieces of host DNA. These pieces of DNA are cut from the host chromosome randomly (at least for practical purposes). The frequency with which two genes are cotransduced, that is, contained in the same phage particle, is inversely related to the distance between the genes; hence, these frequencies can be used to determine the order of the markers. Since the size of DNA in one phage particle is limited to 97 kb, only markers that are separated by less than this amount of DNA can be cotransduced. During conjugation, 97 kb of DNA corresponds to the amount of DNA that can be transferred in about 2.3 minutes.

Other genetic methods that may be used for genetic mapping are reviewed by Bachman *et al.* (1976), and in some of the articles they cite.

2. *Physical and Chemical Methods for Mapping*

The genetic methods of mapping mentioned above are quite valuable in determining the relative order of genes; however, they are less reliable than physical methods for determining the absolute physical distance between genes or the length of DNA contained in a gene. Also, purely genetic methods are not very useful in the mapping of genes for which mutations are difficult to isolate or are not available, such as the ribosomal protein genes. In the last few years, new physical and chemical techniques have been developed whose use in conjunction with genetic techniques makes it possible to map directly the physical positions and dimensions of many genes.

One of the big advances in physical mapping has come about with the development of techniques for visualizing DNA directly by electron microscopy (for review, see Younghusband and Inman, 1974). DNA itself is too thin to be easily observed with the electron microscope. In order to make DNA molecules more easily visible, DNA can be mixed with a basic protein such as cytochrome *c*, which coats the DNA, making it thicker, and hence more easily observable in the electron microscope (Kleinschmidt, 1968). This method allows very precise and reproducible length measurements that are accurate to within about ± 50 base pairs. By mounting DNA under slightly denaturing conditions, such as in the presence of formamide, it is possible to observe single-stranded DNA in an extended form and to obtain reliable length measurements on single-stranded DNA (Inman, 1966; Westmoreland *et al.*, 1969; Davis *et al.*, 1971). These two developments have made it possible to examine sequence relations of a large number of DNA's using the heteroduplex method. In this method, two or more DNA's to be compared are denatured to separate their strands and then placed under renaturing conditions such that heteroduplex DNA molecules containing one strand from each of the original parent DNA's are formed. These heteroduplex molecules contain regions of double-stranded DNA in which the two parents have homologous sequences and loops or bubbles of single-stranded DNA in which the parental sequences are nonhomologous (see, for example, Figure 9). The lengths of the single- and double-stranded regions can be measured, and by comparing the regions of homology and nonhomology with the genetic properties of the parental DNA's, it is possible to determine the physical positions of the genes on the parent molecules. This technique has been valuable in locating genes and in determining the nucleotide sequence relations among related bacteriophages such as the lambdoid phages (Simon *et al.*, 1971; Fiandt *et al.*, 1971), T3 and T7 (Davis and Hyman, 1971), and the T-even phages (Kim and Davidson, 1974), and among various episomes such as F factors, F′ factors, and drug resistance factors (Sharp *et al.*, 1972; Hu *et al.*, 1975a,b; Davidson *et al.*, 1975) and in mapping bacterial genes contained on transducing phages and F′ factors (Ohtsubo *et al.*, 1974a,b,c; Lee *et al.*, 1974; Fiandt *et al.*, 1976).

It is also possible to map genes for RNA molecules such as messenger RNA's transfer RNA's, and ribosomal RNA's by annealing these RNA molecules to single-stranded regions of the heteroduplex DNA (Hyman and Summers, 1972; Deonier *et al.*, 1974; Ohtsubo *et al.*, 1974c). A recent development in physical mapping of RNA molecules is the discovery that under certain conditions the RNA–DNA double helix is more stable than the DNA–DNA double helix. Under these conditions, RNA molecules will displace the complementary DNA strand in double-stranded DNA in the region of homology (Thomas *et al.*, 1976). The resulting structure, called an R loop, appears in the electron microscope as a bubble in the DNA in which one side is an RNA–DNA helix and the other side is a single strand of DNA. Using this technique, it is possible to position precisely the gene coding for any RNA molecule that can be isolated.

By mounting the molecules for electron microscopy under successively stronger denaturing conditions, it is possible to examine the completeness of homology between the two strands of heteroduplex DNA and to estimate the evolutionary "distance" between related DNA molecules (Davis and Hyman, 1971). Using this technique, it is also possible to detect inhomogeneities in base composition along a double-stranded DNA molecule, since AT-rich regions are less stable and denature sooner than GC-rich regions (Inman, 1966, 1967).

In the past, the chemical and physical study of the fine structure of DNA has often been hampered by the inability to obtain precisely defined pieces of DNA of workable length in sufficient quantity and purity to perform the necessary experiments. This problem has been alleviated to a large extent by the discovery of class II restriction enzymes. These enzymes are double-stranded endonucleases that are able to cleave DNA at specific base sequences, usually 3 to 6 base pairs long. To date, close to 90 different restriction enzymes are known, with some 45 different cleavage specificities. The enzymes have been found in a wide variety of bacterial species, but so far only in bacteria (see review by Roberts, 1976). After cleavage of a DNA preparation with a particular restriction enzyme, the resulting DNA fragments can be fractionated according to size by electrophoresis on agarose gels. Restriction fragments can be mapped on a genome by a variety of techniques, such as hybridization to specific RNA's, hybridization to other restriction fragments generated by cleavage of the same DNA by another restriction enzyme, by heteroduplex mapping, by ability to bind specific proteins, or by a number of other techniques (Roberts, 1976). The restriction fragments can be used in *in vitro* transcription–translation systems to identify the protein products coded by the fragment. This approach has been used to map some of the *E. coli* ribosomal protein genes carried on λ transducing phages (Lindahl *et al.*, 1976, 1977).

Recently, two new techniques for DNA sequencing have been developed that allow the sequencing of long stretches of polynucleotides (Sanger and Coulson, 1975; Maxam and Gilbert, 1977). Determination of the DNA sequence allows the ultimate in mapping resolution, since DNA and protein sequences can be compared and the location of the gene can

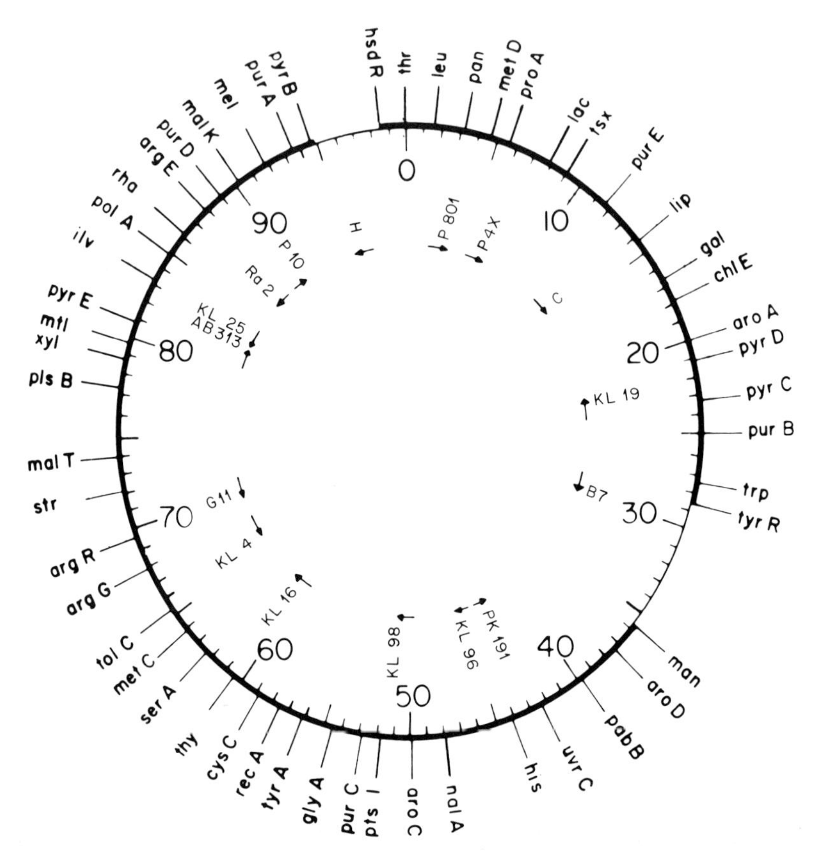

Fig. 6. Circular map of *E. coli* K-12. The large numbers refer to map position in minutes relative to the *thr* locus. The positions of 52 genetic loci are shown outside the circle. The thin portions of the circle represent the only two map intervals that are not spanned by a series of P1 cotransduction linkages. Inside the circle, the leading conjugational transfer regions of a number of Hfr strains are indicated. The replication origin and terminus map near minutes 86 and 32, respectively. (From Bachman *et al.*, 1976, reprinted with permission.)

be determined at the nucleotide level. The power of these techniques is well illustrated by the fact that the entire sequence of the bacteriophage ϕX174 genome has been determined (Sanger *et al.*, 1977). This genome is a single-stranded circular DNA molecule of 5375 bases coding for nine known proteins, but the total molecular weight of the proteins apparently exceeds the coding capacity of the DNA molecule. This dilemma has been resolved by examination of wild-type and mutant DNA sequences in certain genes. Two regions of the DNA molecule were found where two different proteins are coded by the same sequence of DNA, but each is read in two different reading frames for the synthesis of the two proteins (Barrell *et al.,* 1976; Smith *et al.,* 1977).

3. The Genetic Map of E. coli

The genetic map of *E. coli* as determined by the above-mentioned methods (as well as others) can be represented as a continuous unbranched circle having a circumference of 100 minutes (Figure 6), corresponding to the conjugational transfer time. Approximately 650 genes have now been mapped with some precision (Bachman *et al.,* 1976), a number which is 20–30% of that expected for a DNA molecule the size of the *E. coli* chromosome. The circularity of the map coincides nicely with the observed circularity of the DNA, although the circularity of one does not necessarily prove the circularity of the other.

The distribution of the genes on the map shows some interesting features. The known gene loci are not distributed randomly over the map, but rather, are clustered at various locations along the map (Figure 7). Furthermore, the clusters are arranged fairly symmetrically about a point very close to the origin for bidirectional replication of the chromosome (Bachman *et al.,* 1976). The replication terminus is located about 180° away from the replication origin in a long region in which there are no known genes. The genes for the machinery for macromolecular synthesis are mostly within 17 minutes on either side of the replication origin. The significance of this is not clear, but one might speculate that since the genes in this region will be present in more copies per growing cell on the average, this would be advantageous to the cell by allowing faster growth or greater resistance to mutation (Bachman *et al.,* 1976). Another interesting observation is that genes which are related in function have a higher than random probability of being located approximately 90° or 180° away from one another on the map (for example, *pur G, pur I,* and *pur C* map at 53–55 min, while *pur B* maps at 25 min), although there are numerous exceptions. This observation has prompted the proposal of the present genome's evolution from an ancestral structure one-fourth the size of the present one, with two duplications of this genome occurring at some time in the past to produce the present chromosomes (Zipkas and Riley, 1975). The relationship of the distribution of the genes on the map to what is known about the physical structure of the genome is not yet clear.

A clearer understanding of the significance of gene clustering is seen in the organization of operons. Often, the genes needed for the utilization, synthesis, and/or transport of a certain metabolite are grouped closely together on the map and are coordinately expressed. Furthermore, a mutation in one of these genes sometimes will affect the expression of other genes in the group. This type of mutation is called a polar mutation and is

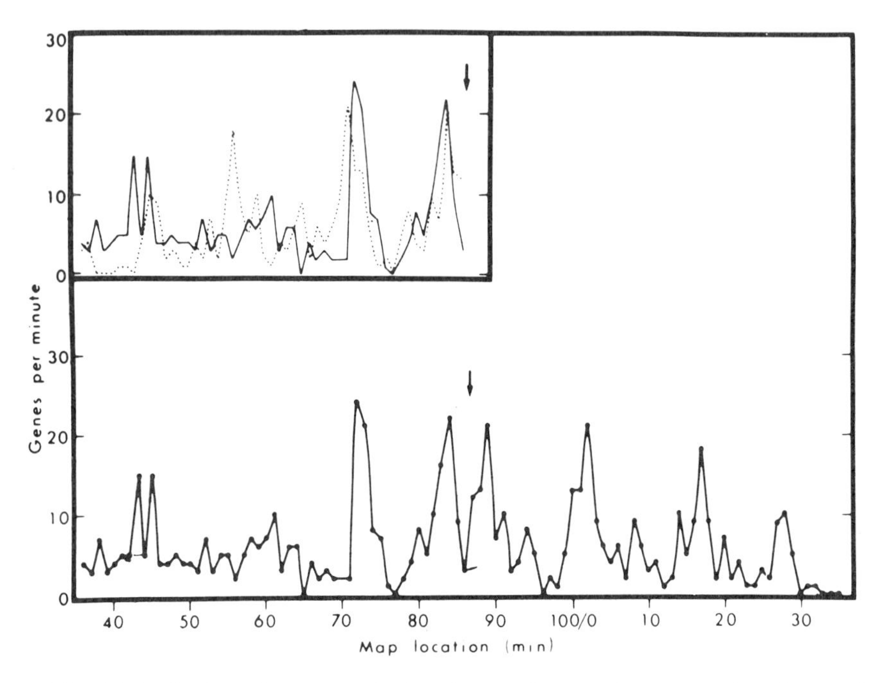

Fig. 7. Distribution of known gene loci on the genetic map. The total number of mapped genes in each one-minute interval of map length is plotted against map location, starting at min 35 and proceeding in a clockwise direction. The arrows placed at min 87 indicate an axis of symmetry for positions of gene clusters on the chromosome. Inset: The continuous line represents the number of genes per minute in the 36–86 min map segment, proceeding clockwise from left to right. The dashed line represents the number of genes per min in the 37–87 min map segment, proceeding counterclockwise left to right. Another possible axis of symmetry lies at min 82. (From Bachman *et al.*, 1976, reprinted with permission.)

believed to result from clustering of the genes in a single transcription unit. Messenger RNA for all the genes in the operon is synthesized as a single polycistronic transcript. Polar mutations affect genes that are transcribed after the gene in which the mutation lies or "downstream" from the mutation with respect to the promotor. However, the mechanism by which polarity affects the downstream genes is not clear. The significance of this is that the grouping of related genes into transcription units or operons allows the expression of several genes to be regulated by a single control site on the DNA and therefore provides a relatively simple mechanism for quantitative control and temporal coordination of a group of related genes. Most operons contain genes that are involved in the metabolism of some specific molecule such as lactose (*lac*), histidine (*his*), tryptophan (*trp*), galactose (*gal*), and arabinose (*ara*). However, an interesting situation exists in the case of a cluster of genes at 72 min that contains genes for several of the proteins of the ribosomal subunits and the α subunit of RNA polymerase (Jaskunas *et al.*, 1975). These genes are apparently present in the same operon, a phenomenon which may imply the coordinate control of the production of machinery for RNA and protein synthesis (Fiandt *et al.*, 1976). In this light, it is also interesting to note that another group of ribosomal protein genes, a set of ribosomal RNA genes, and the genes for the β and β' subunits of RNA polymerase are clustered near 89 min (Lindahl *et al.*, 1975, 1977).

4. The Genetic Map of Bacteriophage Lambda

The genome of bacteriophage λ is a linear DNA molecule of approximately 30.8×10^6 daltons corresponding to about 46.5 kb. Approximately 50 genes have been identified on this DNA molecule, a number which is close to the expected coding capacity for the DNA. The relationship between the arrangement of the genes on the DNA molecule and the life cycle of the phage is perhaps better understood in λ than in any other genetic entity. The map is linear, as is the DNA molecule, and is arranged in functional blocks of genes (Figure 8).

Infection of *E. coli* by λ can have two different outcomes (see review by Herskowitz, 1973). The first, called the lytic response, is devoted to the multiplication of the bacteriophage and results in lysis of the cell with the release of about 100 new phage particles. The second, called the lysogenic response, results in the covalent integration of the λ genome into the *E. coli* chromosome where it can remain stably associated as a prophage for many cell generations. Much about these two responses can be explained by what is known about the transcription of the various λ operons. The proteins involved in the assembly of the phage particle are coded by the left half of the DNA molecule, with the head protein genes (*A–F*) grouped to the left of the tail protein genes (*Z–J*). The function(s) of the *b2* region between the *J* gene and the attachment site (*att*), while not known, is in any event, not essential to phage growth. The attachment site is the point at which the λ molecule integrates into the host chromosome to establish lysogeny. The region from the attachment site through γ contains genes which control the various phage-mediated recombination events, including site-specific recombination (integration) and generalized recombination. The region from γ to *cII* is the control region that regulates the genetic expression of the various operons on the λ genome. The *OP* region controls the replication of the phage DNA. The *Q* gene controls the genetic expression of the late genes of λ, and the genes *S* and *R* on the right end of the molecule code for proteins needed to lyse the cell.

When λ DNA enters the cell, it cyclizes by association of its sticky ends

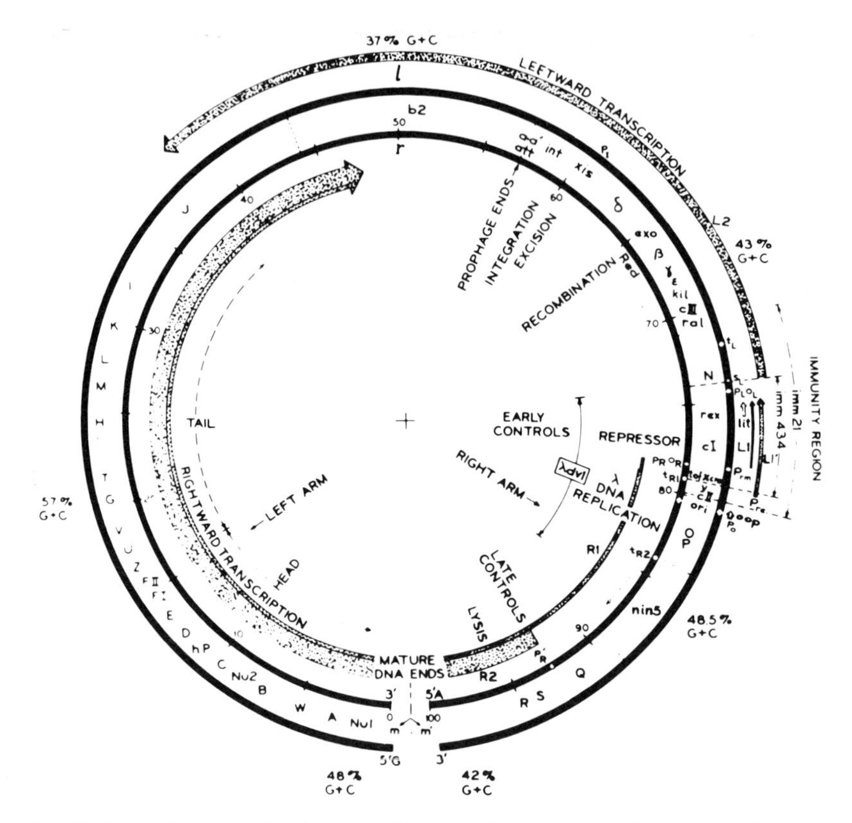

Fig. 8. A genetic and molecular map of bacteriophage lambda. The two complementary strands of the λ DNA molecule (l and r) are represented by the solid lines. The length of the DNA has been divided into %λ units. The GC base composition of various regions of the molecule is indicated outside the circle. The cohesive ends are designated m and m′. The different genes and their positions are shown between the strands. The functions of the various groups of genes are indicated inside the circle. The positions of the various transcription units, their initiation points, and their direction of transcription are indicated by the shaded arrows. (From W. Szybalski, 1974, reprinted with permission.)

and is ligated to form a covalently continuous circle. Expression of the genome begins in the control region with leftward transcription of *N* and rightward transcription of *cro* or *tof* from promotors on either side of the *cI* gene. The *N* gene product in turn stimulates transcription of the genes extending from *cIII* through *int* and the genes from *cII* through *Q*. The decision whether to take the lysogenic or lytic pathway is determined by the interaction of the products of several of these genes, namely, *N, cro, cI, cII,* and *cIII,* either among themselves or with the DNA. The biochemical nature of these interactions is not yet completely understood (Herskowitz, 1973). The lysogenic pathway involves the synthesis of the λ repressor (*cI* gene product), stimulated by *cII* and *cIII* (Reichardt, 1975a) and the integration of the λ genome into the host genome at the λ attachment site, stimulated by *int,* to form the prophage (Gottesman and Weisberg, 1971). After integration, the repressor prevents the expression of all the phage genes except the *cI* gene and thus maintains the lysogenic state. Since λ DNA is a covalently closed circle before integration and the attachment site is in the middle of the λ molecule, the order of the genes on the prophage map is a circular permutation of the order on the DNA molecule isolated from the phage particle. In the lytic pathway, the synthesis of repressor is not sufficiently stimulated to shut down gene expression, but is depressed by the *cro* product. The *N*-stimulated early transcription of the *O* and *P* genes results in the replication of the DNA. Furthermore, *N*-stimulated transcription of the *Q* gene results in the expression of genes needed late in the infection, namely, the lysis genes *R* and *S,* and the head and tail genes *A* through *J*. Thus, the life cycle of λ is controlled by the expression of different functional blocks of genes turned on and off at various times, regulated by the products of genes *N, cro, cI, cII, cIII,* and *Q*.

The order of the genes on λ is similar to the order on the other lambdoid phages, and, in fact, many of these phages share extensive regions of homology in different regions of their genomes, as shown by heteroduplex mapping (Simon *et al.,* 1971; Fiandt *et al.,* 1971). Furthermore, many of these phages can recombine to form viable hybrid phage having some of the genetic properties of each parent. The recombination often involves the substitution of a functional block of genes of one phage for the analogous block of the other (Dove, 1971). This is most strikingly evident in the formation of hybrids between λ and the *Salmonella* phage P22 (Botstein and Herskowitz, 1974). Although these phages grow in different bacteria and have different DNA structure (λ is nonpermuted and has sticky ends, whereas P22 is terminally redundant and circularly permuted), the order of the analogous genes in the two phage genomes is almost identical. Viable hybrids can be constructed between λ and P22 in which the control regions and/or the DNA replication region of P22 is substituted for those regions in λ (Hilliker and Botstein, 1976). This suggests that the lambdoid phages have evolved by divergence within functional blocks of genes, while the functions of the blocks themselves have remained fairly constant. A similar situation probably exists in the case of the P2,-186 family of phage, for T3 and T7, and for the T-even phages.

5. *Movable DNA Segments*

Up to now, our discussion of the genetic organization of prokaryotic genomes has focused primarily on the constant aspects of the organization of genes in the chromosomes. However, there are a number of examples of genes or segments of DNA that can reside and function at a number of locations on the genome. We have already mentioned two examples of this, namely, λ and the F factor. Although both λ and F can exist as autonomous episomes that are not associated with the host chromosomal DNA, both are capable of integrating themselves into the host DNA and remaining stably associated with the host chromosome indefinitely.

The unassociated form of both DNA's is a covalently closed circle, and integration involves recombination between this circle and the host DNA, resulting in the opening of the circle and the insertion of the episome into the host to form a continuous linear structure (Campbell, 1962).

Lambda integrates by a site-specific recombination event mediated by the *int* gene product. The recombination takes place between the λ attachment site on the phage genome and a bacterial attachment site between the *gal* and *bio* genes on the *E. coli* chromosome. While the integration of λ is almost always at this point on the *E. coli* chromosome, if this region is deleted, λ can insert itself at a number of other positions with a reduced efficiency. F also integrates at specific sites on the *E. coli* genome. Heteroduplex studies and genetic studies have shown that F is also organized into functional blocks of genes with one region containing the genes necessary for conjugative transfer of the episome from one bacterium to another, another region responsible for the autonomous replication of the episome, and another region containing a number of different insertion sequences (IS sequences) (Davidson *et al.,* 1975); this organization will be discussed in greater detail later. Heteroduplex mapping studies indicate that the integration of the F factor is the result of a recombination event between one of the IS sequences on the F factor and one of the IS sequences on the host chromosome (Deonier and Davidson, 1976; Davidson *et al.,* 1975; Hu *et al.,* 1975a). The recombination pathway responsible for this is not known.

Although the integration of both λ and F is stable, both are able to reverse the condition to re-form episomes by a process that involves recombination between the two episome–bacterial DNA joints. In λ the process is mediated by the *xis* and *int* gene products. The recombination pathway for the excision of F is not known. Occasionally, the excision process is abnormal, and the recombination takes place at sites other than the episome–bacterial joints. When this happens in λ, bacterial DNA is substituted for part of the phage DNA in the resulting phage particle, and the phage becomes a transducing phage. When abnormal excision takes place with the F factor, the resulting episome also contains bacterial DNA and is termed an F′ factor. The recombination processes responsible for these abnormal excision events have not yet been determined.

Another type of movable DNA segment that was mentioned briefly above with regard to F integration is the IS sequence. IS sequences can be inserted at a large number of sites in a variety of chromosomes, and it is possible that their insertion is completely random throughout the genome. The recombination pathway responsible for their insertion is not known. At present, three IS sequences are known: IS1, IS2, and IS3; these range in size from 700 to 1400 base pairs. They have been found in the *E. coli* chromosome and in other chromosomes such as the F factor and some phage strains (Hu *et al.*, 1975a,b,c; Fiandt *et al.*, 1972; Malamy *et al.*, 1972; Hirsch *et al.*, 1972). Hybridization experiments have provided an estimate of eight copies of IS1 and five copies of IS2 in the *E. coli* chromosome (Saedler and Heiss, 1973). The estimated lifetime of an IS sequence at any location on a chromosome is about 2.5×10^6 generations (Hu *et al.*, 1975c).

The IS sequences have a number of interesting genetic properties. If an IS sequence is inserted into a gene, it stops expression of that gene, and if the gene is in an operon, the mutation is strongly polar. Insertion sequences can be inserted in either orientation (Fiandt *et al.*, 1972), and in the case of IS2 the effect on the expression of downstream genes is dependent on its orientation (Saedler *et al.*, 1974). If IS2 is inserted in one direction, it prevents expression of the downstream genes. However, if IS2 is inserted in the other direction, it causes constitutive expression of the downstream genes. This implies that in one orientation IS2 contains a transcriptional stop signal, and in the other orientation IS2 contains a promotor. Therefore, IS2 can potentially act as a movable control element capable of either promoting or repressing gene expression.

IS sequences also appear to be involved in a number of recombination phenomena. As mentioned above, F factor integration takes place at IS sequences. They apparently are also hot spots for deletion, with one end of the deletion at the IS sequence and the other end at variable positions in the DNA (Reif and Saedler, 1975). They have also been implicated in the formation of certain drug-resistance plasmids. These plasmids are generally organized into two regions. One region, the *RTF* region, is involved in the transfer of the plasmid from one bacterium to another and is identical to the analogous region of the F factor (Sharp *et al.*, 1973). The other region is the *R* determinant region, which confers resistance to various antimicrobial agents to the host bacterium. The junctions between these two regions are IS1 sequences (Hu *et al.*, 1975b; Ptashne and Cohen, 1975). When the plasmids are grown in *Proteus mirabilis* in the absence of drugs, these plasmids dissociate into separate *RTF* and *R* determinant plasmids, and when they are grown in the presence of drugs, the plasmids reassociate to form plasmids containing one *RTF* region and one or more *R* determinant regions (Clowes, 1972; Rownd *et al.*, 1975). It has been proposed that this dissociation and reassociation phenomenon is a result of recombination between the IS sequences at the boundaries of the two regions (Ptashne and Cohen, 1975).

Certain other elements have been shown to be able to move from one position on a chromosome to another and to move between chromosomes (Cohen, 1976). A number of these elements, called transposons, have been shown to insert into a large number of sites in chromosomes and confer to the bacterium resistance to various antibiotics after insertion (Kleckner *et al.*, 1975; Heffron *et al.*, 1975; Berg *et al.*, 1975; Gottesman and Rosner, 1975; Kopecko and Cohen, 1975). Like the IS sequences, they cause strongly polar mutations in the genes into which they are inserted. Electron microscopy of heteroduplex DNA containing these transposons has shown that in some cases the transposon consists of two IS sequences in opposite orientations with the gene(s) for drug resistance in between the two IS sequences (see Figure 9) (Kleckner *et al.*, 1975; Ptashne and Cohen, 1975; Kopecko *et al.*, 1976). Other transposons have shorter inverted repeat sequences at their termini, not yet identified with

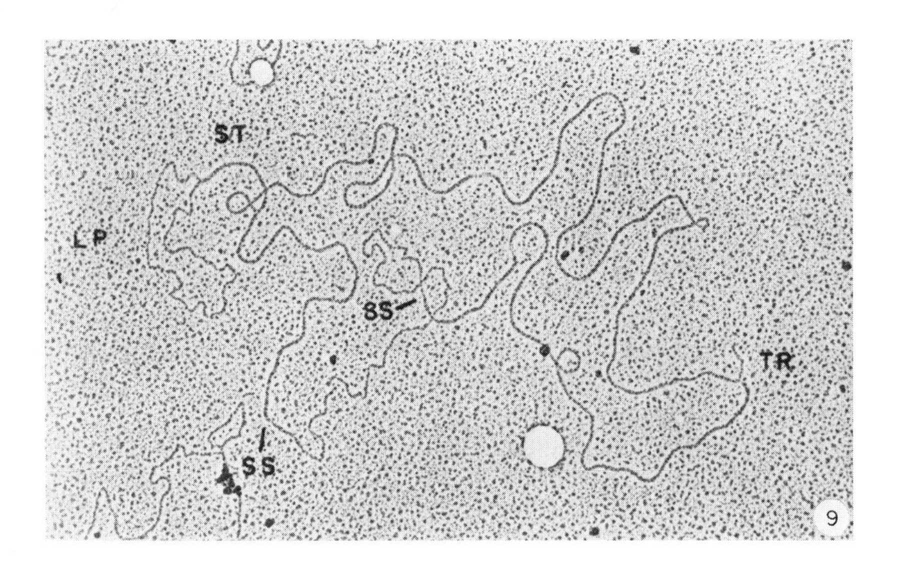

Fig. 9. Electron micrograph of a heteroduplex DNA molecule containing one strand from phage P22 wild type and the other strand from a P22 carrying an inserted transposon that confers resistance to the antibiotic tetracycline. The transposon is visible as a lariatlike structure having a double-stranded stem (ST) and a single-stranded loop (LP). This structure arises because the strand containing the transposon contains the region LP flanked by inverted repeat sequences that renature to form the double-stranded stem. The single-stranded region (SS) and terminal repetition (TR) are due to P22 packaging mechanism. (From Tye *et al.*, 1974, reprinted with permission.)

any known IS sequence (Heffron *et al.*, 1975; Kopecko and Cohen, 1975), and still others contain no detectable inverted repeat structure at all (Gottesman and Rosner, 1975).

The random insertion behavior of IS sequences and transposons is remarkably like that of the bacteriophage mu (Taylor, 1963; Howe and Bade, 1975; Couturier, 1976). Mu is a temperate bacteriophage that is capable of forming lysogens by the insertion of its genome into the host genome. However, unlike λ, mu can insert itself into a large, and possibly random, number of sites in the *E. coli* genome. The integration of mu into a gene causes a polar mutation in that gene, similar to the effect of IS sequences and transposons.

The evolutionary significance of all these movable pieces of DNA is not clear. However, their mobility provides a mechanism for allowing the resistance to various drugs to be spread rapidly throughout a bacterial population (Kopecko *et al.*, 1976; Cohen, 1976). Whether this process can be used for spreading other genetic traits is not known; however, the phenomenon does suggest a mechanism for rapid evolution of genomes and may be a potent adaptive mechanism in bacterial evolution and perhaps evolution in higher organisms as well (Reanney, 1976).

IV. SHORT-RANGE STRUCTURE OF DNA IN CHROMOSOMES

Due to some of the difficulties discussed earlier, the acquisition of primary DNA base sequence data has been slow. However, the sequence data that have been obtained are tantalizing. Sequencing efforts thus far have been concentrated largely on the regions that control the expression of genes. These regions are presumably of particular interest because they do not merely reflect the amino acid sequence of a protein. The control regions of the DNA must possess the structural information governing interactions with the various control proteins, and determination of the base sequence of the DNA is the first step toward elucidating this interaction. Three such control sites were earlier studied in detail genetically, and now the determination of their base sequence has allowed a molecular reconstruction of their function. These sites are the control region for the *lac* operon and the control regions for the early leftward and rightward transcription of λ.

A. *Lac* Control Region

The *lac* operon codes for three proteins which enable the utilization of lactose by the cell. There is a wealth of evidence from experiments, both *in vivo* and *in vitro*, that the expression of the *lac* operon is under negative control by a protein called the *lac* repressor. The *lac* repressor binds to a site on the DNA called the operator and prevents expression of the operon. When lactose or some related compound is added to the repressor–DNA complex, it binds to the repressor, decreases the repressor's affinity for the operator, and thus causes release of the repressor from the DNA. This allows the transcription of the operon by RNA polymerase and expression of the genes. The operon is also under the more general cellular control of the catabolite gene-activator protein (CAP). In the presence of cyclic AMP (cAMP), a cAMP–CAP complex is formed that binds to the DNA and stimulates transcription of the *lac* operon.

Sequence analysis of the DNA of the *lac* operon control region has shown that the binding sites for CAP, RNA polymerase, and repressor, the initiation point for the *lac* mRNA, and the beginning of the first structural gene of the *lac* operon (coding for β-galactosidase) are all contained within about 150 base pairs (bp) of DNA (Dickson *et al.*, 1975). Examination of the topography of this region reveals some interesting features about the molecular aspects of control of the operon (see Figure 10). The binding site for the repressor is contained within a 35-bp sequence of DNA that possesses a twofold symmetry with six mismatched base pairs (Gilbert *et al.*, 1975). This sequence is shown below with the center of symmetry indicated by an arrow and the symmetric base pairs enclosed in boxes.

							↓							
TGTGTG	G	AATTGT	G	A	G	C	G	G	A	T	A	ACAATT	T	CACACA
ACACAC	C	TTAACA	C	T	C	G	C	C	T	A	T	TGTTAA	A	GTGTGT
							↑							

The center of symmetry is about 25 bp from the start of the coding sequence for the β-galactosidase gene. The significance of the twofold symmetry of the operator site is unclear. Repressor binds to the DNA as a tetrameric complex, a phenomenon which suggests that perhaps an axis of symmetry in the subunit construction of the protein might be reflected in an axis of symmetry in its binding site. However, genetic and chemical experiments designed to demonstrate the importance of symmetry to the binding of repressor to DNA thus far have tended to discount the importance of symmetry in this interaction. For example, upon binding of the repressor to the operator sequence several of the bases in the operator show increased or decreased reactivity to the methylating agent dimethyl sulfate compared to the bases in unbound operator DNA. These bases are not symmetrically placed about the center of symmetry, suggesting that the repressor does not bind symmetrically about the center of symmetry

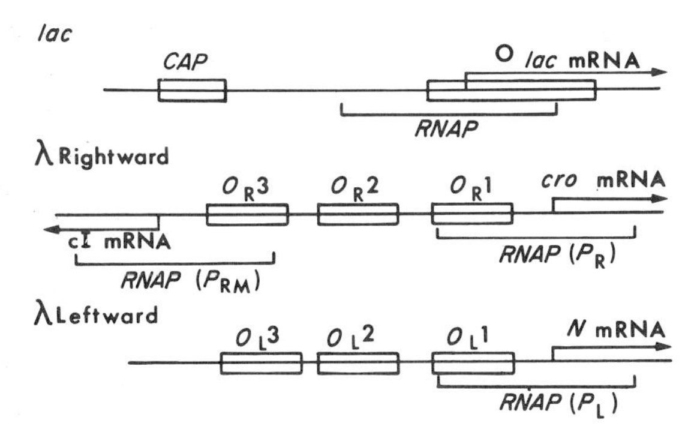

Fig. 10. Topography of the *lac* and λ early control regions. Regions of partial twofold symmetry characterizing the *CAP* binding site and the operators are indicated by closed boxes. The regions of DNA protected against DNase digestion by bound RNA polymerase are indicated by closed brackets. The initiation point for mRNA synthesis and the direction of synthesis is shown with an open-ended bracket with an arrow on the end (from Dickson *et al.*, 1975, and Ptashne *et al.*, 1976).

in the DNA (Gilbert *et al.*, 1976). Also, a chemically synthesized operator sequence having perfect symmetry with no mismatched bases has been inserted into a plasmid DNA. This symmetric operator sequence binds repressor very poorly compared to the native operator sequence when assayed both *in vivo* and *in vitro*. This is contrary to what one might expect if the repressor bound symmetrically about the center of symmetry in the operator (Sadler *et al.*, 1977).

The binding site for RNA polymerase is defined as the DNA protected by bound RNA polymerase from digestion by DNase. This site of 40–45 bp contains the site for initiation of the *lac* mRNA and overlaps the operator region. The center of the RNA polymerase binding site is about 10–12 bp farther from the start of the β-galactosidase gene than the center of symmetry of the operator. The overlap of the binding sites suggests that the repressor prevents transcription of the *lac* operon through steric hindrance, by preventing binding of RNA polymerase to its binding site.

The binding site for the CAP protein is located about 100 bp from the start of the β-galactosidase gene. The binding site has a 14-bp sequence that possesses a twofold symmetry with two mismatched base pairs as shown below.

G T G A G	T	T A	G	C T C A C
C A C T C	A	A T	C	G A G T G

Both mutational evidence and the results of experiments involving protection of bases from chemical modification by bound protein suggest that the CAP protein binds to this site symmetrically. The molecular mechanism for the positive stimulation of transcription is unknown.

Thus, the topography of the control region of the *lac* operon may be summarized as follows (see Figure 10). Centered about 100 bp from the start of the gene for β-galactosidase is the 14-bp region of partial twofold symmetry that characterizes the CAP protein binding site. Moving toward the gene, we next come to the RNA polymerase binding site, which occupies 40–45 bp of DNA and which is centered about 40 bp from the start of the gene. This binding site also contains the site of initiation of the *lac* mRNA, which is 38 bp from the beginning of the gene. Again proceeding toward the gene, 28 bp from the beginning of the gene is the center of the 35 bp region of partial twofold symmetry that contains the *lac* operator. This region also contains the initiation site for *lac* mRNA and overlaps the RNA polymerase binding site by about 25 bp.

B. λ Early Control Regions

As mentioned earlier, the expression of the λ genome is controlled from two sites on either side of the λ *cI* gene, which codes for the λ repressor. Early gene expression comes about as the result of leftward transcription of an operon containing genes *N* through *int,* beginning at a promoter P_L situated to the left of *cI* and rightward transcription of another operon containing genes *cro* through *P,* beginning at a promotor P_R situated to the right of *cI*. Transcription from promotors P_L and P_R is negatively regulated by the binding of λ repressor to operators O_L and O_R, respectively (Herskowitz, 1973). During lysogeny, the λ repressor is synthesized from mRNA transcribed from the *cI* gene between O_LP_L and O_RP_R. This RNA is synthesized by leftward transcription from a promotor P_{RM} located near P_R. Evidence indicates that the λ repressor controls its own synthesis by regulating *cI* mRNA synthesis from P_{RM} both positively and negatively (Reichardt, 1975b; Ptashne *et al.,* 1976; Walz *et al.,* 1976).

Again, these effects can be demonstrated both *in vivo* and *in vitro,* and the determination of the nucleotide sequences of the two control regions, O_LP_L and O_RP_R, contributes greatly to our understanding of the regulation of the expression of λ DNA (Maniatis *et al.,* 1975; Ptashne *et al.,* 1976; Walz and Pirrotta, 1975; Walz *et al.,* 1976). The two regions possess several common features and will be discussed together (see Figure 10). The size of the operator regions protected against DNase digestion by bound repressor depends on the amount of repressor bound to the DNA. Fragments of about 25, 50, and 80 bp are protected depending on the repressor : DNA ratio. This is due to the fact that each operator has three repressor-binding sites, O_L1 O_L2, O_L3, and O_R1, O_R2, O_R3. The relative affinities of repressor for the binding sites are such that O_L1 and O_R1 bind repressor most tightly, O_R2 and O_L2 bind intermediately, and O_R3 and O_L3 bind least tightly. The three binding sites of each operator are arranged such that the strongest binding site is nearest the first structural gene in their respective operon, and the weakest farthest from the first gene. Examination of the nucleotide sequence of the regions shows that each of the six binding sites has a 17-bp region of partial twofold symmetry, and each is at least partially homologous to the other five, as shown below.

```
                     ↓
O_R1    T A T C A C C G C C A G A G G T A
        A T A G T G G C G G T C T C C A T
O_R2    T A A C A C C G T G C G T G T T G
        A T T G T G G C A C G C A C A A C
O_R3    T A T C A C C G C A A G G G A T A
        A T A G T G G C G T T C C C T A A
O_L1    T A T C A C C G C C A G T G G T A
        A T A G T G G C G G T C A C C A T
O_L2    C A A C A C C G C C A G A G A T A
        G T T G T G G C G G T C T C T A T
O_L3    T A T C A C C G C A G A T G G T T
        A T A G T G G C G T C T A C C A A
```

The three 17-bp regions of each operator are separated by three to seven bp.

The RNA polymerase-binding sites, defined as the DNA protected by bound RNA polymerase from DNase digestion, contain the initiation sites for the *cro* and *N* mRNA's, and overlap O_R1 and O_L1, respectively. Again, this suggests that the repressor molecule prohibits the expression of the operons by sterically hindering the binding of RNA polymerase to the promotors.

Thus, the topography of the two control regions of the leftward and rightward early λ operons may be summarized as follows (see Figure 10). Each operator contains three 17-bp regions having partial twofold symmetry centered approximately 16, 40, and 60 bp away from the initiation point for RNA synthesis for the operons. The repressor-binding affinity for the three binding sites decreases with the distance from the initiation point. The RNA polymerase-binding site contains the initiation point and overlaps the region of partial symmetry closest to the initiation point (O_L1 and O_R1).

The initiation point of the *cI* mRNA transcribed from P_{RM} is 18 bp to the left of the 17-bp region of partial symmetry called O_R3. The RNA polymerase-binding site corresponding to P_{RM} contains this initiation point and overlaps O_R3, a pattern which is analogous to the overlaps of P_L and O_L1 and P_R and O_R1. It is known that low levels of repressor in the cell stimulate transcription from P_{RM} and high levels of repressor inhibit transcription from P_{RM} (Reichardt, 1975b). The probable mechanism for this involves binding of the repressor, when present at a low level, to O_R1 (the binding site with the highest affinity for repressor). This prevents transcription of *cro* from P_R and perhaps stimulates transcription of *cI* from P_{RM}. As the cellular level of repressor increases, it binds successively to Q_R2 and then O_R3, further depressing synthesis from P_R and ultimately stopping *cI* RNA synthesis from P_{RM} as well (Ptashne *et al.*, 1976; Walz *et al.*, 1976). In this manner, the repressor regulates its own synthesis as well as stopping gene expression in the rest of the λ genome.

C. Promotors

A number of promotors for a variety of genes have now been sequenced, namely, the *lac* promotor (Dickson *et al.*, 1975), the *gal* promotor (Musso *et al.*, 1977), the *trp* promotor (Bennett *et al.*, 1976), the lambda promotors P_L, P_R, and P_{RM} (Maniatis *et al.*, 1975; Walz and Pirrotto, 1975; Ptashne *et al.*, 1976; Walz *et al.*, 1976), the lambda *oop* promotor (Scherer *et al.*, 1977), the *A2* and *A3* promotors responsible for early gene expression in phage T7, several promotors on phage fd (Schaller *et al.*, 1975; Takanami *et al.*, 1976) and phage φX174 (Sanger *et al.*, 1977), the promotor for an *E. coli* tyrosine tRNA (Sekiya and Khorana, 1974), and a promotor on SV40 DNA used *in vitro* by *E. coli* RNA polymerase (Dhar *et al.*, 1974). Several of these promotors have been isolated as pieces of DNA protected against pancreatic DNase digestion by bound RNA polymerase. All are approximately 40 to 45 bp in length, and all contain the initiation point for RNA synthesis near the center of the protected piece. Many of the promotors contain a common seven base-pair sequence (5′T–A–T–Pu–A–T–G3′) centered nine or ten bp from the RNA initiation point (Pribnow, 1975a,b). It has been suggested that this sequence is responsible for binding RNA polymerase, and that changes in this sequence influence the relative strength of the promotor. In support of this is the fact that a promotor mutation which renders transcription of the *lac* operon independent of the CAP protein has an altered sequence in this region. It is interesting to note that while RNA can be initiated and synthesized from these fragments by RNA polymerase that is bound before DNase digestion, polymerase is unable to form a functional complex and synthesize RNA when added to purified promotor fragments. This implies that some feature needed for the recognition of the promotor by the RNA polymerase is missing from the fragments. This information, at least in the case of the λ promotors, is evidently contained in the sequence extending 20 base pairs farther from the start of the mRNA (Ptashne *et al.*, 1976). Promotor fragments isolated by digestion of polymerase–DNA complexes with a combination of λ exonuclease and S1 nuclease are about 20 bp larger, and are able to bind RNA polymerase and synthesize RNA. Two λ promotor mutations are also contained in this region.

D. Other DNA Sequences That Interact with Specific Proteins

Other DNA sequences that interact with specific proteins include the cohesive end regions of several of the lamboid phages and the P2-186 family of phage (Wu and Taylor, 1971; Padmanabhan and Wu, 1972; Weigel *et al.*, 1973). Some of these sequences show a partial symmetry about an axis; however, the significance of this is not understood.

The sequences recognized by a number of the previously mentioned restriction endonucleases are known (Roberts, 1976). These recognition sites are sequences of three to six base pairs that generally possess at least some degree of symmetry about an axis through the center of the sequence. The endonucleolytic cuts are usually made symmetrically about this axis, however, the positions of the cuts vary from enzyme to enzyme. Some enzymes produce "sticky" 5′-ends, some "sticky" 3′-ends, and some produce ends without single-stranded extensions.

E. Influence of DNA Secondary and Tertiary Structure on Protein–DNA Interactions

From the previous discussion, it is apparent that many of the control proteins that interact with DNA do so with specific base sequences of the DNA. In the attempt to characterize these interactions further, it is of interest to know whether the target base sequences are in a "normal" DNA double helix (A-, B-, C-type helices), or whether they exist in some novel secondary structure that facilitates recognition and binding by the protein. Many recognition sequences (namely, the *CAP* protein site, the *lac* operator, the λ operators, restriction enzyme sites, and the *cos* sites of several sticky-ended phage DNA's) contain at least a partial twofold symmetry, and other regions of inverted symmetry (for example, the regions at the ends of some of the drug-resistance transposons discussed earlier) are known to exist in the DNA of many organisms. This has led to the proposal that perhaps the complementary DNA strands in the symmetric regions separate and form hairpinlike loops by intrastrand base pairing. Thus, the DNA would adopt a so-called cruciform structure with two hairpin loops in opposite strands (Gierer, 1966). However, a number of lines of evidence tend to argue against the formation of such structures. Thermodynamic considerations make the formation of cruciforms energetically less favorable than double helices, because formation of cruciforms requires (1) interruption of the helical stacking interaction at the base of the hairpin, (2) the existence of unpaired regions at the ends of the hairpin loops, and (3) unstable base pairs in the stems resulting from imperfect symmetry of some of the regions. The free energy due to superhelical turns in the DNA could reduce this unfavorable energy requirement, but the superhelical density necessary to overcome the energy barrier involved in the formation of cruciforms is not found experimentally in DNA supercoiled *in vivo* (Wang, 1974). Attempts to find such structures *in vivo* in mouse cells using the crosslinking agent psoralen to trap the DNA in cruciform structures have failed to detect any significant portion of the inverted repeat DNA in cruciform structures, although cruciforms smaller than 200 bp would not have been detected in this study (Cech and Pardue, 1976).

One could argue that the binding of specific proteins could lower the energy barrier to reasonable levels. Studies of *lac* repressor binding to *lac* operator on DNA's of several different superhelical densities indicate that the amount of unwinding of the operator DNA (a region that might be expected to form a cruciform) following repressor binding is much less than would be expected if the operator DNA were converted to a cruciform upon complexing with the repressor (Wang *et al.*, 1974; Bourgeois *et al.*, 1975). Thus, it appears that DNA is rarely if ever in the cruciform configuration, although again it is difficult to rule out cruciform formation by interaction with appropriate proteins.

The presence of superhelical turns in a DNA can have considerable influence on the interaction of some proteins with DNA. Generally, protein–DNA interactions that lead to unwinding of the DNA helix or that occur preferentially on single-stranded DNA increase with the superhelical density of the DNA (Wang, 1974). For example, RNA polymerase is known to unwind the double helix by approximately one turn on binding to DNA (Saucier and Wang, 1972), and both the number of RNA chains initiated per site and the number of sites of initiation increase with superhelical density (Botchan, 1976; Botchan *et al.*, 1973; Wang, 1974). Also, the binding constant of *lac* repressor to *lac* operator increases with the superhelical density of the operator DNA. From this increase in binding, it is possible to infer that the binding of repressor to operator causes an unwinding of the helix of about 90° (Wang *et al.*, 1974; Bourgeois *et al.*, 1975).

The *in vivo* significance of experiments designed to test the effects of tertiary structure on the binding of proteins to DNA is not clear, since the secondary and tertiary structure of DNA inside the cell is not known. However, the fact that superhelical densities of DNA's isolated from a variety of organisms are the same suggests that the same intracellular DNA structure is common to virtually all organisms. Two plausible explanations for the presence of superhelical turns in isolated DNA can be stated in their extreme forms as follows. (1) Superhelical turns reflect a gross difference in solvent environment of the DNA *in vivo* and *in vitro*. The chemical environment of ions, polyamines, and proteins inside the cell is such that the thermodynamically favored secondary and tertiary structure of the DNA is much different from that of isolated DNA in solution. In this case, the stress in the double helix only occurs after the DNA is removed from the intracellular environment. (2) The DNA is actually in a stressed high-energy configuration inside the cell, and superhelical turns are present both inside and outside the cell.

Suggestive experimental evidence supports both viewpoints. In support of the first hypothesis, most of the DNA in eukaryotic cells is complexed with histones. A variety of physical, chemical, and microscopical experiments have shown that a complex of histones containing two of each of four different histone species binds to the DNA at intervals of about 200 base pairs. When histones bind to the DNA, it is compacted, and if histones are bound to a nonsuperhelical, covalently closed circular DNA, and the complex is treated with nicking–closing enzyme and the bound histones are then removed, the resulting closed circular DNA is supercoiled to about the same superhelical density as intracellularly supercoiled DNA (Germond *et al.*, 1975). This suggests that in eukaryotes supercoiling is due to the binding of histones. However, no histones have been found in prokaryotes, although proteins having some histonelike properties have been isolated (Rouvière-Yaniv and Gros, 1975), and electron microscopy has shown that under certain conditions DNA from freshly lysed *E. coli* cells has the beaded appearance characteristic of eukaryotic nucleohistone complexes (Griffith, 1976).

Evidence for the existence of superhelical turns inside the cell comes from several experiments. First, *recA*-mediated recombination of exogenously added DNA in spheroplasts is much more efficient if the added DNA is superhelical than if it is not superhelical, implying that tertiary turns favor recombination (Holloman and Radding, 1976). Second, two drugs, novobiocin and coumermycin, which are known to inhibit DNA replication, are also known to inhibit DNA gyrase, an enzyme which, in the presence of ATP, is able to introduce negative superhelical turns into nonsuperhelical closed DNA (Gellert *et al.*, 1976a,b). When nicked circular DNA is incubated with cell extracts containing gyrase, it is converted to closed superhelical DNA having a superhelical density very close to that of DNA isolated from cells. This reaction is inhibited by novobiocin and coumermycin. If a bacterium that has been lysogenized by λ is infected with λ, the infecting DNA becomes circular and covalently closed and supercoiled to the superhelical density characteristic of intracellular DNA. If this process is performed in the presence of coumermycin, the infecting DNA becomes circular and covalently closed but contains few if any superhelical turns. The implication of these experiments is that superhelical turns are actually present inside the cell and that this is due to strain in the DNA introduced by DNA gyrase and not to the intracellular environment. Furthermore, the strained configuration of the DNA is apparently necessary for certain cellular processes such as recombination and perhaps DNA replication. Two enzymes are known that are potentially capable of regulating the superhelical density inside the cell. DNA gyrase is capable of introducing tertiary turns into the DNA, and the *E. coli* ω protein (a transient DNA swivel) is capable of removing turns from the DNA (Wang, 1971, 1973). The presence of these two enzymes raises the intriguing possibility of regulating cellular processes by varying the tertiary structure of DNA in different regions of the chromosome to stimulate or suppress binding of proteins to the DNA.

REFERENCES

Altman, S., and Lerman, L. (1970). Kinetics and intermediates in the intracellular synthesis of bacteriophage T4 deoxyribonucleic acid. *J. Mol. Biol.* **50,** 235–261.

Bachman, B. J., Low, K. B., and Taylor, A. L. (1976). Recalibrated linkage map of *Escherichia coli* K-12. *Bacteriol. Rev.* **40,** 116–167.

Barrell, B. S., Air, G. M., and Hutchison, C. A., III. (1976). Overlapping genes in bacteriophage φX174. *Nature* (*London*) **264,** 34–41.

Bennett, G. N., Schweingruber, M. E., Brown, K. D., Squires, C., and Yanofsky, C. (1976). Nucleotide sequence of region preceding *trp* mRNA initiation site and its role in promotor and operator function. *Proc. Natl. Acad. Sci. U.S.A.* **73,** 2351–2355.

Berg, D., Davies, J., Allet, B., and Rochaiz, J. D. (1975). Transposition of R factor genes to bacteriophage λ. *Proc. Natl. Acad. Sci. U.S.A.* **72,** 3628–3632.

Botchan, P. (1976). An electron microscopic comparison of transcription on linear and superhelical DNA. *J. Mol. Biol.* **105,** 161–176.

Botchan, P., Wang, J. C., and Echols, H. (1973). Effect of circularity and superhelicity of transcription from bacteriophage λ DNA. *Proc. Natl. Acad. Sci. U.S.A.* **70,** 3077–3081.

Botstein, D., and Herskowitz, I. (1974). Properties of hybrids between *Salmonella* phage P22 and coliphage λ. *Nature* (*London*) **251,** 584–589.

Bourgeois, S., Barkley, M. D., Jobe, A., Sadler, J. R., and Wang, J. C. (1975). Effect of

alterations of *lac* operator DNA on repressor binding. *In* "Protein–Ligand Interactions" (H. Sund and G. Blauer, eds.), pp. 253–267. de Gruyter, Berlin.
Campbell, A. (1962). The episomes. *Adv. Genet.* **11,** 101.
Cech, T. R., and Pardue, M. L. (1976). Electron microscopy of DNA crosslinked with trimethylpsoralen: Test of the secondary structure of eukaryotic inverted repeat sequences. *Proc. Natl. Acad. Sci. U.S.A.* **73,** 2644–2648.
Clowes, R. C. (1972). Molecular structure of bacterial plasmids. *Bacteriol. Rev.* **36,** 361–405.
Cohen, S. N. (1976). Transposable genetic elements and plasmid evolution. *Nature* (*London*) **273,** 731–737.
Couturier, M. (1976). The integration of excision of bacteriophage mu-1. *Cell* **7,** 155–164.
Crick, F., and Klug, A. (1975). Kinky helix. *Nature* (*London*) **255,** 530–532.
Curtis, M., and Alberts, B. (1976). Studies on the structure of intracellular bacteriophage T4 DNA. *J. Mol. Biol.* **102,** 793–816.
Daneao-Moore, L., and Higgins, M. (1972). Morphokinetic reaction of *Streptococcus faecalis* (ATCC 9790) cells to the specific inhibition of macromolecular synthesis: Nucleoid condensation and the inhibition of protein synthesis. *J. Bacteriol.* **109,** 1210–1220.
Davidson, N., and Szybalski, W. (1971). Physical and chemical characteristics of lambda DNA. *In* "The Bacteriophage Lambda" (A. D. Hershey, ed.), pp. 45–77. Cold Spring Harbor Lab., Cold Spring Harbor, New York.
Davidson, N., Deonier, R. C., Hu, S., and Ohtsubo, E. (1975). Electron microscope heteroduplex studies of sequence relations among plasmids of *Escherichia coli* X. Deoxyribonucleic acid sequence organization of F and of F-primes, and the sequences involved in Hfr formation. *In* "Microbiology-1974" (D. Schlessinger, ed.), pp. 56–65. Am. Soc. Microbiol., Washington, D.C.
Davis, R. W., and Hyman, R. W. (1971). A study in evolution: The DNA base sequence homology between coliphages T7 and T3. *J. Mol. Biol.* **62,** 287–301.
Davis, R. W., Simon, M., and Davidson, N. (1971). Electron microscope heteroduplex methods for mapping regions of base sequence homology in nucleic acids. *In* "Methods in Enzymology" (L. Grossman and K. Moldave, eds.), Vol. 21, pp. 413–428. Academic Press, New York.
Delius, H., and Worcel, A. (1973). Electron microscopic visualization of the folded chromosome of *Escherichia coli*. *J. Mol. Biol.* **82,** 107–109.
Deonier, R. C., and Davidson, N. (1976). The sequence organization of the integrated F plasmid in two Hfr strains of *Escherichia coli*. *J. Mol. Biol.* **107,** 207–222.
Deonier, R. C., Ohtsubo, E., Lee, H. J., and Davidson, N. (1974). Electron microscope heteroduplex studies of sequence relations among plasmids of *Escherichia coli*. VII. Mapping the ribosomal RNA genes of plasmid F14. *J. Mol. Biol.* **89,** 618–629.
Dhar, R., Weissman, S. M., Zain, B. S., Pan, J., and Lewis, A. M. (1974). The nucleotide sequence preceding an RNA polymerase initiation site on SV40 DNA. Part 2. The sequence of the early strand transcript. *Nucleic Acids Res.* **1,** 595–613.
Dickson, R. C., Abelson, J., Barnes, W. M., and Reznikoff, W. S. (1975). Genetic regulation: The *lac* control region. *Science* **187,** 27–35.
Dove, W. F. (1971). Biological inferences. *In* "The Bacteriophage Lambda" (A. D. Hershey, ed.), pp. 297–311. Cold Spring Harbor Lab. Cold Spring Harbor, New York.
Drlica, K., and Worcel, A. (1975). Conformational transitions in the *Escherichia coli* chromosome: Analysis by viscometry and sedimentation. *J. Mol. Biol.* **98,** 393–411.
Dworsky, P. (1975a). A mild method for the isolation of folded chromosomes from *Escherichia coli*. *Z. Allg. Mikrobiol.* **15,** 231–242.
Dworsky, P. (1975b). Unfolding of the chromosome of *Escherichia coli* after treatment with rifampicin. *Z. Allg. Mikrobiol.* **15,** 243–247.

Dworsky, P. (1976). Comparative studies on membrane associated folded chromosomes from *E. coli*. *J. Bacteriol.* **126,** 64–71.

Dworsky, P., and Schaechter, M. (1973). Effect of rifampin on the structure and membrane attachment of the nucleoid of *Escherichia coli*. *J. Bacteriol.* **116,** 1364–1374.

Emmons, S. W. (1974). Bacteriophage lambda derivatives carrying two copies of the cohesive end site. *J. Mol. Biol.* **83,** 511–525.

Enquist, L. W., and Skalka, A. (1973). Replication of bacteriophage λ DNA dependent on the function of host and viral genes. I. Interaction of *red, gam,* and *rec*. *J. Mol. Biol.* **75,** 185–212.

Fiandt, M., Hradecna, Z., Lozeron, H. A., and Szybalski, W. (1971). Electron micrographic mapping of deletions, insertions, inversions, and homologies in the DNA's of coliphage lambda and phi 80. *In* "The Bacteriophage Lambda" (A. D. Hershey, ed.), pp. 329–354. Cold Spring Harbor Lab. Cold Spring Harbor, New York.

Fiandt, M., Szybalski, W., and Malamy, M. (1972). Polar mutations in *lac, gal* and phage λ consist of a few IS-DNA sequences inserted with either orientation. *Mol. Gen. Genet.* **119,** 223–231.

Fiandt, M., Szybalski, W., Blattner, F. R., Jaskunas, S. R., Lindahl, L., and Nomura, M. (1976). Organization of ribosomal protein genes in *Escherichia coli*. I. Physical structure of DNA from transducing λ phages carrying genes from the *aroE-str* region. *J. Mol. Biol.* **106,** 817–836.

Frankel, F. (1968). DNA replication after T4 infection. *Cold Spring Harbor Symp. Quant. Biol.* **33,** 485–493.

Freifelder, D., Chud, L., and Levine, E. E. (1974). Requirement for maturation of *Escherichia coli* bacteriophage lambda. *J. Mol. Biol.* **83,** 503–309.

Gellert, M., Mizuuchi, K., O'Dea, M. H., and Nash, H. A. (1976a). DNA gyrase: An enzyme that introduces superhelical turns into DNA. *Proc. Natl. Acad. Sci. U.S.A.* **73,** 3874–3876.

Gellert, M., O'Dea, M. H., Itoh, T., and Tomigawa, J.-I. (1976b). Novobiocin and coumermycin inhibit DNA supercoiling catalyzed by DNA gyrase. *Proc. Natl. Acad. Sci. U.S.A.* **73,** 4474–4478.

Germond, J. E., Hirt, B., Oudet, P., Gross-Bellard, M., and Chambon, P. (1975). Folding of the DNA double helix in chromatin-like structures from Simian Virus 40. *Proc. Natl. Acad. Sci. U.S.A.* **72,** 1843–1847.

Gierer, A. (1966). Model for DNA and protein interactions and the function of the operator. *Nature* (*London*) **212,** 1480–1481.

Gilbert, W., Gralla, J., Majors, J., and Maxam, A. (1975). Lactose operator sequences and the action of *lac* repressor. *In* "Protein–Ligand Interactions" (H. Sund and G. Blauer, eds.), pp. 193–206. de Gruyter, Berlin.

Gilbert, W., Majors, J., and Maxam, A. (1976). How proteins recognize DNA sequences. *In* "Organization and Expression of Chromosomes," Dahlem Workshop Rep. Berlin.

Giorno, R., Hecht, R., and Pettijohn, D. (1975). Analysis by isopycnic centrifugation of isolated nucleoids of *Escherichia coli*. *Nucleic Acids Res.* **2,** 1559–1567.

Gottesman, M. E., and Rosner, J. (1975). Acquisition of a determinant for chloramphenicol resistance by coliphage lambda. *Proc. Natl. Acad. Sci. U.S.A.* **72,** 5041–5045.

Gottesman, M. E., and Weisberg, R. A. (1971). Prophage insertion and excision. *In* "The Bacteriophage Lambda" (A. D. Hershey, ed.), pp. 113–138. Cold Spring Harbor Lab., Cold Spring Harbor, New York.

Greenstein, M., and Skalka, A. (1975). Replication of bacteriophage lambda DNA: *In Vivo* studies of the interaction between the viral *Gamma* protein and the host *rec* BC DNAse. *J. Mol. Biol.* **97,** 543–560.

Griffith, J. (1976). Visualization of prokaryotic DNA in a regularly condensed chromatin-like fiber. *Proc. Natl. Acad. Sci. U.S.A.* **73**, 563–567.

Hecht, R. (1976). *In* "Molecular Mechanism in the Control of Gene Expression" (D. P. Nierlich, J. W. Rutter, and C. F. Fox, eds.), pp. 45–50. Academic Press, New York.

Hamilton, S., and Pettijohn, D. (1976). Properties of condensed bacteriophage T4 DNA isolated from *Escherichia coli* infected with bacteriophage T4. *J. Virol.* **19**, 1012–1027.

Hecht, R., and Pettijohn, D. (1976). Studies of DNA-bound RNA molecules isolated from nucleoids of *Escherichia coli. Nucleic Acids Res.* **3**, 767–788.

Hecht, R., Taggart, R., and Pettijohn, D. (1975). Size and DNA content of purified *E. coli* nucleoids observed by fluorescence microscopy. *Nature (London)* **253**, 60–62.

Hecht, R., Stimpson, D., and Pettijohn, D. (1977). Sedimentation properties of the bacterial chromosome as an isolated nucleoid and as an unfolded DNA fiber: Chromosomal DNA folding measured by rotor speed effects. *J. Mol. Biol.* **111** (in press).

Heffron, F., Rubens, C., and Falkow, S. (1975). Translocation of a plasmid DNA sequence which mediates ampicillin resistance: Molecular nature and specificity of insertion. *Proc. Natl. Acad. Sci. U.S.A.* **72**, 3623–3627.

Heidrich, H., and Olsen, W. (1975). Deoxyribonucleic acid–envelope complexes from *Escherichia coli:* A complex-specific protein and its possible function for the stability of the complex. *J. Cell Biol.* **67**, 444–460.

Hendrix, R. W., and Casjens, S. R. (1975). Assembly of bacteriophage lambda heads: Protein processing and its genetic control in petit λ assembly. *J. Mol. Biol.* **91**, 187–200.

Hershey, A. D., and Burgi, E. (1965). Complementary structure of interacting sites and the ends of λ DNA molecules. *Proc. Natl. Acad. Sci. U.S.A.* **53**, 325–328.

Herskowitz, I. (1973). Control of gene expression in bacteriophage lambda. *Annu. Rev. Genet.* **7**, 289–324.

Hilliker, S., and Botstein, D. (1976). Specificity of genetic elements controlling regulation of early function in temperate bacteriophages. *J. Mol. Biol.* **106**, 537–566.

Hirsch, H. J., Starlinger, P., and Brachet, P. (1972). Two kinds of insertions in bacterial genes. *Mol. Gen. Genet.* **119**, 191–206.

Hohn, B. (1975). DNA as substrate for packaging into bacteriophage lambda *in vitro. J. Mol. Biol.* **98**, 93–106.

Hohn, B., and Hohn, T. (1974). Activity of empty, headlike particles for packaging of DNA of bacteriophage λ *in vitro. Proc. Natl. Acad. Sci. U.S.A.* **71**, 2372–2376.

Hohn, B., and Korn, D. (1969). Cosegration of a sex factor with the *Escherichia coli* chromosome during curing by acridine orange. *J. Mol. Biol.* **45**, 385–395.

Hohn, T., Flick, H., and Hohn, B. (1975). Petit λ, a family of particles from coliphage lambda-infected cells. *J. Mol. Biol.* **98**, 107–120.

Holloman, W. K., and Radding, C. M. (1976). Recombination promoted by superhelical DNA and the *rec*A gene of *Escherichia coli. Proc. Natl. Acad. Sci. U.S.A.* **73**, 3910–3914.

Howe, M. M., and Bade, E. G. (1975). Molecular biology of bacteriophage mu. *Science* **190**, 624–632.

Hu, S., Ohtsubo, E., and Davidson, N. (1975a). Electron microscope heteroduplex studies of sequence relations among plasmids of *E. coli:* Structure of F13 and related F-primes. *J. Bacteriol.* **122**, 749–763.

Hu, S., Ohtsubo, E., Davidson, N., and Saedler, H. (1975b). Electron microscope heteroduplex studies of sequence relations among bacterial plasmids: Identification and mapping of the insertion sequences IS1 and IS2 in F and R plasmids. *J. Bacteriol.* **122**, 764–775.

Hu, S., Ptashne, K., Cohen, S., and Davidson, N. (1975c). $\alpha\beta$ Sequence of F is IS3. *J. Bacteriol.* **123,** 687–692.

Huberman, J. (1968). Visualization of replicatious mammalian and T4 bacteriophage DNA. *Cold Spring Harbor Symp. Quant. Biol.* **33,** 509–524.

Hyman, R. W., and Summers, W. C. (1972). Isolation and physical mapping of T7 gene 1 messenger RNA. *J. Mol. Biol.* **71,** 573–582.

Inman, R. B. (1966). A denaturation map of the λ phage DNA molecule determined by electron microscopy, *J. Mol. Biol.* **18,** 464–476.

Inman, R. B. (1967). Denaturation maps of the left and right sides of the lambda DNA molecules determined by electron microscopy. *J. Mol. Biol.* **28,** 103–116.

Jacob, F., Ryter, A., and Cuzin, F. (1966). On the association between DNA and membrane in bacteria. *Proc. R. Soc. London, Ser. B* **164,** 267–278.

Jaskunas, S. R., Burgess, R. R., and Nomura, M. (1975). Identification of a gene for the α-subunit of RNA polymerase at the *str–spc* region of the *Escherichia coli* chromosome. *Proc. Natl. Acad. Sci. U.S.A.* **72,** 5036–5040.

Kaiser, D., and Masuda, T. (1973). *In vitro* assembly of bacteriophage lambda heads. *Proc. Natl. Acad. Sci. U.S.A.* **70,** 260–264.

Kaiser, D., Syvanen, M., and Matsuda, T. (1975). DNA packaging steps in bacteriophage λ head assembly. *J. Mol. Biol.* **91,** 175–186.

Kavenoff, R., and Bowen, B. (1976). Electron microscopy of membrane-free folded chromosomes from *Escherichia coli. Chromosoma* **59,** 89–101.

Kavenoff, R., and Ryder, O. (1976). Electron microscopy of membrane-associated folded chromosomes of *Escherichia coli. Chromosoma* **55,** 13–25.

Kellenberger, E., Ryter, A., and Séchaud, J. (1958). Electron microscopic study of DNA-containing plasms. II. Vegetative and phage DNA as compared with normal bacterial nucleoids in different physiological states. *J. Biophys. Biochem. Cytol.* **4,** 671–676.

Kim, J.-S., and Davidson, N. (1974). Electron microscope heteroduplex study of sequence relations of T2, T4, and T6 bacteriophage DNA's. *Virology* **57,** 93–111.

Kleckner, N., Chan, R., Tye, B.-K., and Botstein, D. (1975). Mutagenesis by insertion of a drug-resistance element carrying an inverted repetition. *J. Mol. Biol.* **97,** 561–575.

Kleinschmidt, A. K. (1968). Monolayer techniques in electron microscopy of nucleic acid molecules. *In* "Methods in Enzymology" (L. Grossman and K. Moldave, eds.), Vol. 12, Part B, pp. 361–377. Academic Press, New York.

Kline, B., and Miller, J. (1975). Detection of nonintegrated plasmid deoxyribonucleic acid in the folded chromosome of *Escherichia coli:* Physicochemical approach to studying the unit of segregation. *J. Bacteriol.* **121,** 165–172.

Kline, B., Miller, J., Cress, D., Wlodarczyk, J., Manis, J., and Otten, M. (1976). Nonintegrated plasmid–chromosome complexes in *Escherichia coli. J. Bacteriol.* **127,** 881–889.

Kopecko, D., and Cohen, S. (1975). Site specific *rec*A-independent recombination between bacterial plasmids: Involvement of palindromes at the recombinational loci. *Proc. Natl. Acad. Sci. U.S.A.* **72,** 1373–1377.

Kopecko, D. J., Brevet, J., and Cohen, S. N. (1976). Involvement of multiple translocating DNA segments and recombinational hotspots in the structural evolution of bacterial plasmids. *J. Mol. Biol.* **108,** 333–360.

Kornberg, T., Lockwood, A., and Worcel, A. (1974). Replication of the *Escherichia coli* chromosome with a soluble enzyme system. *Proc. Natl. Acad. Sci. U.S.A.* **71,** 3189.

Kubitschek, H., and Freedman, M. (1971). Chromosome replication and the division cycle of *E. coli* B/r. *J. Bacteriol.* **107,** 95–99.

Laemmli, U. K. (1975). Characterization of DNA condensates induced by poly (ethylene oxide) and polylysine. *Proc. Natl. Acad. Sci. U.S.A.* **72,** 4288–4292.

Laemmli, U. K., and Favre, M. (1973). Maturation of the head of bacteriophage T4. I. DNA Packaging events. *J. Mol. Biol.* **80,** 575–599.

Laemmli, U. K., Teaf, N., and Ambrosia, D. (1974). Maturation of the head of bacteriophage T4. III. DNA packaging into preformed heads. *J. Mol. Biol.* **88,** 749–765.

Lee, H. J., Ohtsubo, E., Deonier, R. C., and Davidson, N. (1974). Electron microscope heteroduplex studies of sequence relations among plasmids of *Escherichia coli*. V. *ilv*+ Deletion mutants of F14. *J. Mol. Biol.* **89,** 585–597.

Leibowitz, P., and Schaechter, M. (1975). The attachment of the bacterial chromosome to the cell membranc. *Int. Rev. Cytol.* **41,** 1–28.

Lerman, L. (1973). Chromosomal analogues: Long-range order in ψ-condensed DNA. *Cold Spring Harbor Symp. Quant. Biol.* **38,** 59–73.

Lindahl, L., Jaskunas, S. R., Dennis, P. P., and Nomura, M. (1975). Cluster of genes in *Escherichia coli* for ribosomal proteins, ribosomal RNA and RNA polymerase subunits. *Proc. Natl. Acad. Sci. U.S.A.* **72,** 2743–2747.

Lindahl, L., Zengel, J., and Nomura, M. (1976). Organization of ribosomal protein genes in *Escherichia coli*. II. Mapping of ribosomal protein genes by *in vitro* synthesis of ribosomal proteins using DNA fragments of a transducing phage as templates. *J. Mol. Biol.* **106,** 837–856.

Lindahl, L., Yamamoto, M., Nomura, M., Kirschbaum, J. B., Allet, B., and Rochaix, J. D. (1977). Mapping of a cluster of genes for components of the transcription and translational machineries of *Escherichia coli*. *J. Mol. Biol.* **109,** 23–47.

Low, K. B. (1972). *Escherichia coli* K-12 F-prime factors, old and new. *Bacteriol. Rev.* **36,** 587–607.

Luria, S., and Human, M. (1950). Chromation staining of bacteria during bacteriophage infection. *J. Bacteriol.* **59,** 551–560.

Lydersen, B., and Pettijohn, D. (1977). Interactions stabilizing DNA tertiary structure in the *E. coli* chromosome investigated with ionizing radiation. *Chromosoma* **62,** 199–215.

Malamy, M., Fiandt, M., and Szybalski, W. (1972). Electron microscopy of polar insertions in the *lac* operon of *E. coli*. *Mol. Gen. Genet.* **119,** 207–222.

Maniatis, T., Ptashne, M., Backman, K., Kleid, D., Flashman, S., Jeffrey, A., and Maurer, R. (1975). Recognition sequences of repressor and polymerase in the operators of bacteriophage lambda. *Cell* **5,** 109–113.

Maxam, A. M., and Gilbert, W. (1977). A new method for sequencing DNA. *Proc. Natl. Acad. Sci. U.S.A.* **74,** 560–564.

Meyer, M., DeJong, M., Woldringh, C., and Nanninga, N. (1976). Factors affecting the release of folded chromosomes from *Escherichia coli*. *Eur. J. Biochem.* **63,** 469–475.

Miller, O., Hamkalo, B., and Thomas, C. (1970). Visualization of bacterial genes in action. *Science* **169,** 392–394.

Musso, R., DiLauro, R., Rosenberg, M., and DeCrombrugghe, B. (1977). Nucleotide sequence of the operator–promoter region of the galactose operon of *Escherichia coli*. *Proc. Natl. Acad. Sci. U.S.A.* **74,** 106–110.

Ohtsubo, E., Deonier, R. C., Lee, H. J., and Davidson, N. (1974a). Electron microscope heteroduplex studies of sequence relations among plasmids of *Escherichia coli*. IV. The F sequences in F14. *J. Mol. Biol.* **89,** 565–584.

Ohtsubo, E., Lee, H. J., Deonier, R. C., and Davidson, N. (1974b). Electron microscope heteroduplex studies of sequence relations among plasmids of *Escherichia coli*. VI. Mapping of F14 sequences homologous to ϕ80 dmet BJF and ϕ80 darg ECBH bacteriophages. *J. Mol. Biol.* **89,** 599–618.

Ohtsubo, E., Soll, L., Deonier, R. C., Lee, H. J., and Davidson, N. (1974c). Electron microscope heteroduplex studies of sequence relations among plasmids of *Escherichia coli.* VIII. The structure of bacteriophage $\phi 80$ d_3 ilv^+ su^+ 7, including the mapping of the ribosomal RNA genes. *J. Mol. Biol.* **89,** 631–646.

Padmanabhan, R., and Wu, R. (1972). Nucleotide sequence analysis of DNA. IV. Complete nucleotide sequence analysis of the left-hand cohesive end of coliphage 186 DNA. *J. Mol. Biol.* **65,** 447–467.

Padmanabhan, R., Wu, R., and Bode, V. C. (1972). Arrangement of DNA in lambda bacteriophage heads. III. Location and number of nucleotides cleaved from λ DNA by micrococcal nuclease attack on heads. *J. Mol. Biol.* **69,** 201–207.

Pettijohn, D. E. (1976). Prokaryotic DNA in nucleoid structure. *CRC Crit. Rev. Biochem.* **4,** 175–202.

Pettijohn, D., and Hecht, R. (1973). RNA molecules bound to the folded bacterial genome stabilize DNA folds and segregate domains of supercoiling. *Cold Spring Harbor Symp. Quant. Biol.* **38,** 31–41.

Pettijohn, D., Stonington, O., and Kossman, C. (1970a). Chain termination of ribosomal RNA synthesis *in vitro. Nature* (*London*) **228,** 235–239.

Pettijohn, D., Clarkson, K., Kossman, C., and Stonington, O. (1970b). Synthesis of ribosomal RNA on a protein–DNA complex isolated from bacteria: A comparison of ribosomal RNA synthesis *in vitro* and *in vivo. J. Mol. Biol.* **52,** 281–300.

Pettijohn, D., Hecht, R., Stonington, O., and Stamato, T. (1973). Factors stabilizing DNA folding in bacterial chromosomes. *In* "DNA Synthesis in Vitro" (R. Wells and R. Inman, eds.), pp. 145–162. Univ. Park Press, Baltimore, Maryland.

Pribnow, D. (1975a). Nucleotide sequence of an RNA polymerase binding site at an early T7 promoter. *Proc. Natl. Acad. Sci. U.S.A.* **72,** 784–788.

Pribnow, D. (1975b). Bacteriophage T7 early promotors: Nucleotide sequences of two RNA polymerase binding sites. *J. Mol. Biol.* **99,** 419–443.

Ptashne, K., and Cohen, S. (1975). Occurrence of insertion sequence (IS) regions on plasmid deoxyribonucleic acid as direct and inverted nucleotide sequence duplications. *J. Bacteriol.* **122,** 776–781.

Ptashne, M., Backman, K., Humagun, M. Z., Jeffry, A., Maurer, R., Meyer, B., and Sauer, R. T. (1976). Autoregulation and function of a repressor in bacteriophage lambda. *Science* **194,** 156–161.

Reanney, D. (1976). Extrachromosomal elements as possible agents of adaptation and development. *Bacteriol. Rev.* **40,** 552–590.

Reichardt, L. F. (1975a). Control of bacteriophage lambda repressor synthesis after phage infection: The role of the *N, cII, cIII* and *cro* products. *J. Mol. Biol.* **93,** 267–288.

Reichardt, L. F. (1975b). Control of bacteriophage lambda repressor synthesis: Regulation of the maintenance pathway by the *cro* and *cI* products. *J. Mol. Biol.* **93,** 289–309.

Reif, H. J., and Saedler, H. (1975). ISI is involved in deletion formation in the *gal* region of *E. coli* K-12. *Mol. Gen. Genet.* **137,** 17–28.

Roberts, R. J. (1976). Restriction endonucleases. *CRC Crit. Rev. Biochem.* **4,** 123–164.

Rouvière-Yaniv, J., and Gros, F. (1975). Characterization of a novel, low molecular weight DNA-binding protein from *Escherichia coli. Proc. Natl. Acad. Sci. U.S.A.* **72,** 3428–3432.

Rownd, R. H., Perlman, D., and Goto, N. (1975). Structure and replication of R-factor deoxyribonucleic acid in *Proteus mirabilis. In* "Microbiology-1974" (D. Schlessinger, ed.), pp. 76–94. Am. Soc. Microbiol., Washington, D.C.

Ryder, O., and Smith, D. (1974). Isolation of membrane-associated folded chromosomes from *Escherichia coli:* Effect of protein synthesis inhibition. *J. Bacteriol.* **120,** 1356–1363.

Ryter, A. (1968). Association of the nucleus and the membrane of bacteria: A morphological study. *Bacteriol. Rev.* **32,** 39–54.

Ryter, A., and Chang, A. (1975). Localization of transcribing genes in the bacterial cell by means of high resolution autoradiography. *J. Mol. Biol.* **98,** 797–810.

Sadler, J. R., Tecklenburg, M., Betz, J. L., Goeddel, D. V., Yansura, D. G., and Caruthers, M. H. (1977). Cloning of chemically synthesized lactose operators. *Gene* **1,** 305–321.

Saedler, H., and Heiss, B. (1973). Multiple copies of the insertion DNA sequences IS1 and IS2 in the chromosome of *E. coli* K-12. *Mol. Gen. Genet.* **122,** 267–277.

Saedler, H., Reif, H. J., Hu, S., and Davidson, N. (1974). IS2, a genetic element for turn-off and turn-on of gene activity in *E. coli. Mol. Gen. Genet.* **132,** 265–289.

Sanger, F., and Coulson, A. R. (1975). A rapid method for determining sequences in DNA by primed synthesis with DNA polymerase. *J. Mol. Biol.* **94,** 441–448.

Sanger, F., Air, G. M., Barrell, G. B., Brown, N. L., Coulson, A. R., Fiddes, J. C., Hutchison, C. A., III, Slocombe, P. M., and Smith, M. (1977). Nucleotide sequence of bacteriophage ϕX174 DNA. *Nature (London)* **265,** 687–695.

Saucier, J.-M., and Wang, J. C. (1972). Angular alteration of the DNA helix by *E. coli* DNA polymerase. *Nature (London), New Biol.* **239,** 167–170.

Schaller, H., Gray, C., and Herrmann, K. (1975). Nucleotide sequences of an RNA polymerase binding site from the DNA of bacteriophage fd. *Proc. Natl. Acad. Sci. U.S.A.* **72,** 737–741.

Scherer, G., Hobom, G., and Kössel, H. (1977). DNA base sequence of the p_0 promotor region of phage λ. *Nature (London)* **265,** 117–121.

Schnös, M., and Inman, R. B. (1970). Position of branch points in replicating λ DNA. *J. Mol. Biol.* **51,** 61–73.

Sekiya, T., and Khorana, H. G. (1974). Nucleotide sequence in the promotor region of the *Escherichia coli* tyrosine tRNA gene. *Proc. Natl. Acad. Sci. U.S.A.* **71,** 2978–2982.

Serwer, P. (1974). Fast sedimenting bacteriophage T7 DNA from T7-infected *Escherichia coli. Virology* **59,** 70–88.

Sharp, P. A., Hsu, M. T., Ohtsubo, E., and Davidson, N. (1972). Electron microscope heteroduplex studies of sequence relations among plasmids of *Escherichia coli.* I. Structure of F prime factors. *J. Mol Biol.* **71,** 471–497.

Sharp, P. A., Cohen, S. N., and Davidson, N. (1973). Electron microscope heteroduplex studies of sequence relations among plasmids of *Escherichia coli.* II. Structure of drug resistance (R) factors and F factors. *J. Mol. Biol.* **75,** 235–255.

Simon, M. N., Davis, R. W., and Davidson, N. (1971). Heteroduplexes of DNA molecules of lambdoid phages: Physical mapping of their base sequence relations by electron microscopy. *In* "The Bacteriophage Lambda" (A. D. Hershey, ed.), pp. 313–328. Cold Spring Harbor Lab., Cold Spring Harbor, New York.

Sirotkin, K., Wei, J., and Snyder, L. (1977). A T4 bacteriophage-coded RNA polymerase subunit blocks host transcription and unfolds the host chromosome. *Nature (London)* **265,** 28–32.

Skalka, A., Poonian, M., and Barte, P. (1972). Concatemers in DNA replication: Electron microscopic studies of partially denatured intracellular lambda DNA. *J. Mol. Biol.* **64,** 541–550.

Smith, M., Brown, N. L., Air, G. M., Barrell, B. G., Coulson, A. R., Hutchison, C. A., III, and Sanger, F. (1977). DNA sequence at the C termini of the overlapping genes A and B in bacteriophage ϕX174. *Nature (London)* **265,** 702–705.

Snustad, D., and Conroy, L. (1974). Mutants of bacteriophage T4 deficient in the ability to induce nuclear disruption. I. Isolation and genetic characterization. *J. Mol. Biol.* **89,** 663–673.

Snustad, D., Parson, K., Warner, H., Tutas, D., Wehner, J., and Koerner, J. (1974). Mutants of bacteriophage T4 deficient in the ability to induce nuclear disruption. II. Physiological state of the host nucleoid in infected cells. *J. Mol. Biol.* **89,** 675–687.

Snustad, D., Tigges, M., Parson, K., Bursch, C., Caron, F., Koerner, J., and Tutas, D. (1976). Identification and preliminary characterization of a mutant defective in the bacteriophage T4-induced unfolding of the *Escherichia coli* nucleoid. *J. Virol.* **17,** 622–642.

Stahl, F. W., McMilin, K. D., Stahl, M. M., Malone, R. E., Nozur, Y., and Russo, V. E. A. (1972). A role for recombination in the production of "free loader" lambda bacteriophage particles. *J. Mol. Biol.* **68,** 57–67.

Stonington, O., and Pettijohn, D. (1971). The folded genome of *Escherichia coli* isolated in a protein–DNA–RNA complex. *Proc. Natl. Acad. Sci. U.S.A.* **68,** 6–9.

Syvanen, M. (1975). The processing of bacteriophage lambda DNA during its assembly into heads. *J. Mol. Biol.* **91,** 165–174.

Szybalski, W. (1974). Bacteriophage lambda. *Handb. Genet.* **1,** 309–325.

Takanami, M., Sugimoto, K., Sugisaki, H., and Okamoto, T. (1976). Sequence of promoter for coat protein gene of bacteriophage fd. *Nature* (*London*) **260,** 297–302.

Taylor, A. L. (1963). Bacteriophage-induced mutation in *Escherichia coli*. *Proc. Natl. Acad. Sci. U.S.A.* **50,** 1043–1050.

Thomas, M., White, R. L., and Davis, R. W. (1976). Hybridization of RNA to double-strand DNA: Formation of R-loops. *Proc. Natl. Acad. Sci. U.S.A.* **73,** 2294–2298.

Tikchonenko, T., Budowsky, E., Sklyadneva, V., and Khromov, I. (1971). The secondary structure of bacteriophage DNA in situ. III. Reaction with O-methyl hydroxylamine. *J. Mol. Biol.* **55,** 535–547.

Tye, B. K., Chan, R. K., and Botstein, D. (1974). Packaging of an oversize transducing genome by *Salmonella* phage P22. *J. Mol. Biol.* **85,** 485–500.

Vinograd, J., Lebowitz, J., Radloff, R., Watson, R., and Lapis, P. (1965). The twisted circular form of *Polyowa* DNA. *Proc. Natl. Acad. Sci. U.S.A.* **53,** 1104–1111.

Walz, A., and Pirrotta, V. (1975). Sequence of the P_R promoter of phage λ. *Nature* (*London*) **254,** 118–121.

Walz, A., Pirrotto, V., and Ineichen, K. (1976). λ repressor regulates the switch between P_R and P_{RM} promotors. *Nature* (*London*) **262,** 665–669.

Wang, J. C. (1971). Interaction between DNA and an *Escherichia coli* protein ω. *J. Mol. Biol.* **55,** 523–533.

Wang, J. C. (1973). Protein ω: A DNA swivelase from *Escherichia coli*. *In* "DNA Synthesis in Vitro" (R. D. Wells and R. B. Inman, eds.), pp. 163–174. Univ. Park Press, Baltimore, Maryland.

Wang, J. C. (1974). Interactions between twisted DNA's and enzymes: The effects of superhelical turns. *J. Mol. Biol.* **87,** 797–816.

Wang, J. C., and Davidson, N. (1966). Thermodynamic and kinetic studies on the interconversion between the linear and circular forms of phage lambda DNA. *J. Mol. Biol.* **15,** 111–123.

Wang, J. C., and Davidson, N. (1968). Cyclization of phage DNA's. *Cold Spring Harbor Symp. Quant. Biol.* **33,** 409–415.

Wang, J. C., Barkley, M. D., and Bourgeois, S. (1974). Measurements of unwinding of *lac* operator by repressor. *Nature* (*London*) **251,** 247–249.

Weigel, P. H., Englund, P. T., Murray, K., and Old, R. W. (1973). The 3′-terminal nucleotide sequences of bacteriophage λ DNA. *Proc. Natl. Acad. Sci. U.S.A.* **70,** 1151–1155.

Westmoreland, B. C., Szybalski, W., and Ris, H. (1969). Mapping of deletions and substitutions in heteroduplex DNA molecules of bacteriophage lambda by electron microscopy. *Science* **163,** 1343–1348.

Worcel, A., and Burgi, E. (1972). On the structure of the folded chromosome of *Escherichia coli*. *J. Mol. Biol.* **71,** 127–147.

Worcel, A., and Burgi, E. (1974). Properties of a membrane-attached form of the folded chromosome of *Escherichia coli*. *J. Mol. Biol.* **82,** 91–105.

Worcel, A., and Portalier, R. (1976). Association of the folded chromosome with the cell envelope of *Escherichia coli:* Characterization of the proteins at the DNA membrane attachment site. *Cell* **8,** 245.

Wu, R., and Taylor, E. (1971). Nucleotide sequence analysis of DNA. II. Complete nucleotide sequence of the cohesive ends of bacteriophage λ DNA. *J. Mol. Biol.* **57,** 491–511.

Younghusband, H. B., and Inman, R. B. (1974). The electron microscopy of DNA. *Annu. Rev. Biochem.* **43,** 605–619.

Zipkas, D., and Riley, M. (1975). Proposal concerning mechanism of the genome of *Escherichia coli*. *Proc. Natl. Acad. Sci. U.S.A.* **72,** 1354–1358.

2

The Molecular Principles and the Enzymatic Machinery of DNA Replication

Leland H. Johnston, Friedrich Bonhoeffer, and Peter Symmons

ISBN 0-12-289502-9

I. MOLECULAR PRINCIPLES

A. Introduction

The enzymatic machinery responsible for the replication of the genetic material of the cell is capable of producing copies of the parental DNA with extremely high fidelity and at a very high rate. Considered at a superficial level, the structure of DNA suggests an appealing and simple mechanism for DNA replication (Watson and Crick, 1953a); the strands of the double helix are separated, followed by polymerization of deoxyribonucleotides on the parental strands to form two identical copies. However, purely from a detailed consideration of the structure of DNA, as discussed below, it quickly becomes apparent that the replication process requires a complex enzymatic apparatus that is considerably more involved than a simple template-directed polymerization of nucleotides.

Figure 1 shows the structure of DNA in three magnifications. From the highest magnification (Figure 1c), it can be seen that each DNA strand has a polarity that is caused by the phosphodiester bond between C-3′ and C-5′ atoms of two neighboring nucleosides. The polarity is usually indicated by arrows (Figure 1a, 1b). We can see that the two strands of the

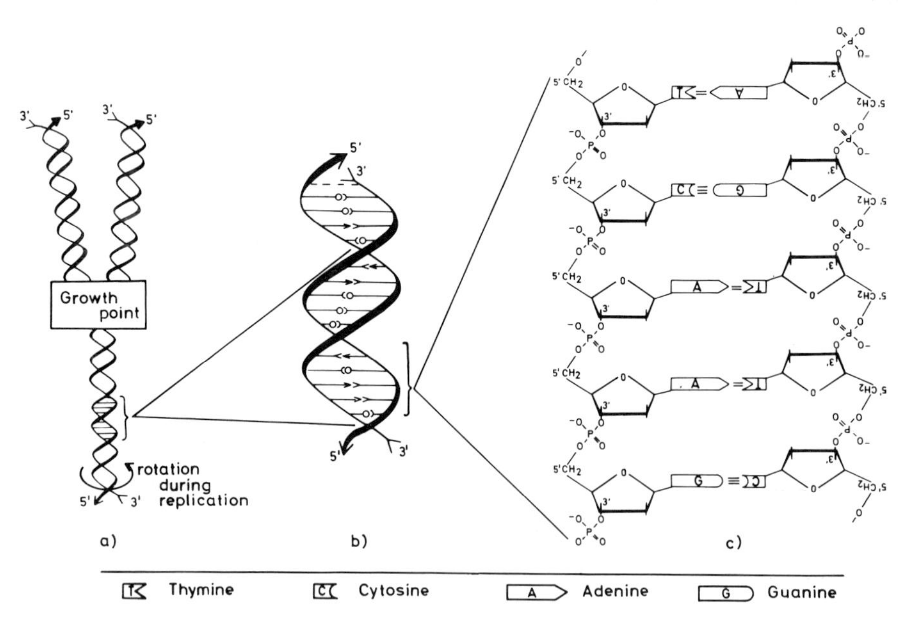

Fig. 1. Structure of DNA at three magnifications.

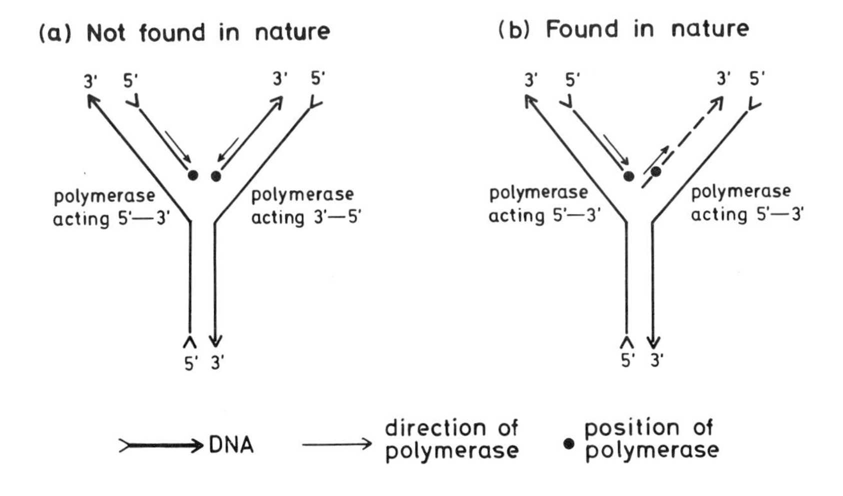

Fig. 2. Logical solutions to the polarity problem.

(a) Not found in nature	(b) Found in nature
i. Requires two different polymerase activities	i. Requires only one type of polymerase activity
ii. Direction of synthesis of both polymerases the same as growth point movement	ii. Discontinuous synthesis on at least one strand
iii. Chemically, strands synthesized in opposite directions	iii. Chemically, both strands synthesized in the same direction
iv. Physically, both strands synthesized in same direction	iv. Physically, strands synthesized in opposite directions

double helix have opposing polarities. Thus, during replication, two daughter strands of opposite polarity must be produced (Figure 1a). Logically, this requires either two different polymerizing activities (Figure 2, left) (clearly not found in nature) or some more complicated replication process. If both parental strands are copied in the same region, at the same time, by the same type of polymerizing activity, the synthesis of the second strand must be discontinuous (Figure 2, right). This mechanism is found in nature. One new strand is synthesized in the same physical direction as the movement of the growth point along the DNA molecule, whereas the other (second strand) is synthesized in short pieces (discontinuously) in the opposite direction. This mechanism requires, in addition to the polymerizing activity, an enzymatic activity that can join the pieces formed during discontinuous synthesis and, furthermore, a mechanism (priming) for reiterant starts of DNA synthesis on the strand that is synthesized discontinuously.

From Figure 1, two other problems become apparent, both of which involve the separation of the parental strands. The first problem is: How are the forces that hold the two parental strands together broken during replication? In this chapter, we shall call this process "strand separation." The strand separation process may or may not be driven by polymerization. It almost certainly requires some special strand-separation proteins. The second problem is a topological one. The parental strands are wound around each other such that they cannot be separated by a simple translational movement (see Figure 5). Separation of the parental strands can occur only by rotation of the parental molecule (Figure 1a). Whether the total or only fractions of the parental DNA molecule must rotate depends on whether the parental molecule itself contains sites where parts of the parental DNA can rotate relative to one another. Sites of free rotation are called swivels. Such sites of free rotation could be scissions or breaks within one of the parental DNA strands. Whereas for linear DNA molecules a swivel is logically not required for replication, it *must* exist for the replication of covalently closed, circular, double-stranded DNA ring molecules for topological reasons.

Thus far, we have referred to problems that concern the mechanism of DNA synthesis during replication. Still another type of problem is involved in DNA replication that has little to do with DNA synthesis per se and can be treated separately from it, namely, the question of how replication is initiated. The process of initiation is certainly a very specific enzymatic process. It occurs only at specific sites on the genome, is subject to regulation by the cell, and is intimately coupled with cell growth.

In agreement with the theoretical considerations, biochemical and genetic evidence clearly indicates that DNA synthesis and its initiation are multicomponent processes in which the enzymes involved might possibly act as complexes. These processes have been investigated in a variety of replication systems: in prokaryotic and eukaryotic organisms, in viruses with double- or single-stranded DNA, and in systems in which reverse transcription of viral RNA into DNA occurs. All of these have been investigated both *in vivo* and as cell-free systems.

Both the principles and theoretical problems of DNA replication appear to be ubiquitous; therefore, in the first part of this chapter, the various experimental systems will be dealt with jointly, under headings relating to the major mechanistic problems of replication. In the second part of the chapter, we will describe in detail the enzymes and proteins which are or may be involved in replication.

B. Polymerization

The polymerization process during DNA replication comprises the formation of a phosphodiester bond between phosphorylated deoxyribonucleoside precursors and the end of a growing strand of DNA (the primer) directed by a single strand of DNA (the template) to which the primer is hydrogen bonded. This reaction is catalyzed by enzymes, template-directed DNA polymerases, that have been characterized in a wide variety of organisms; however, the clear identification of any of these with the enzyme responsible for the polymerization of deoxyribonucleotides in DNA replication has only been achieved for some prokaryotic polymerases. The difficulty in demonstrating this relationship can be attributed to two major causes. First, both prokaryotic and eukaryotic organisms may contain as many as three distinctly different polymerases which must be clearly differentiated from one another before any analysis of their function can begin. Second, many DNA polymerases have been shown to possess a number of enzymatic activities apart from their polymerization activities (see Section II). Thus, the isolation of a conditional lethal mutation in the structural gene of a polymerase, while indicating that the enzyme is essential to the cell, is no guarantee that the enzyme is primarily responsible for the polymerization involved in DNA replication.

However, three important generalizations may be made about the polymerases that have been isolated from both eukaryotes and prokaryotes: *in vitro*, they have an absolute requirement for a primer; they utilize 5′-deoxyribonucleoside triphosphates (dNTP's) as precursors; and they synthesize only in a $5' \rightarrow 3'$ direction (see Figure 2). Two sets of observations suggest that these properties are possessed by the polymerase responsible for replication *in vivo*. First, 5′- and not 3′-phosphorylated nucleosides are used as precursors for replication *in vivo* (Price *et al.*, 1967). Second, analysis of newly synthesized DNA strands of either polarity reveals a $5' \rightarrow 3'$ direction of synthesis (Okazaki and Okazaki, 1969; Sugino and Okazaki, 1972). The fact that all polymerases appear to synthesize in the same direction imposes limitations on the mechanisms by which replication can occur. As explained in the introduction, this means that strand growth of one daughter strand, at least, must occur discontinuously.

Escherichia coli possesses three known DNA polymerases: polymerases I, II, and III (pol I, II, III, respectively) [each of which exhibits at least one exonuclease activity (see below) in addition to being able to catalyze polymerization] that correspond to the structural genes *pol A, B,* and *C*. These may be used to provide more specific examples of the different functional roles of polymerases.

The polymerizing activity known as pol I, first observed in crude extracts (Kornberg *et al.*, 1956) and subsequently highly purified (see Section II), was widely assumed to be the main enzyme responsible for DNA polymerization in replication until a mutant (*pol A1*) defective in this enzyme was isolated (De Lucia and Cairns, 1969) and the mutation was found not to be lethal. Subsequently, another polymerase, pol III, was isolated (Kornberg and Gefter, 1971, 1972) and identified (Gefter *et al.*, 1971; Nüsslein *et al.*, 1971) with the product of a gene *pol C* (formerly gene *dna E*), known to be essential for DNA replication (Wechsler and Gross, 1971). This result strongly suggests that pol III is responsible for polymerization *in vivo*. One feature of the purified enzyme is also consistent with this role. *In vitro*, a pol III molecule can polymerize 1.5×10^4 nucleotides per minute at 37°C (Kornberg and Gefter, 1972), a rate which approaches the expected *in vivo* polymerization rate per polymerase of 5×10^4 nucleotides per minute.[1] The maximum rate of synthesis, *in vitro*,

[1] Rates of growth point movement, polymerization, and unwinding.

The minimum rate of growth point movement of *E. coli* can be calculated as follows: the molecular weight of the *E. coli* chromosome = 2.6×10^9; average molecular weight of one nucleotide = 330; number of nucleotides per chromosome (N) = 8×10^6 nucleotides; replication time of chromosome (T) = 40 min (Helmstetter and Cooper, 1968). Replication in *E. coli* is bidirectional, therefore, there are two growth points (values in nucleotides/min/growth point)

$$\therefore \text{rate of growth point movement} = \frac{N}{2T} = \frac{8 \times 10^6}{80} = 10^5$$

If it is assumed that each growth point contains two polymerases, one polymerizing on each parental strand, then the minimum rate of polymerization is half the growth point movement rate = 5×10^4 nucleotides/min/polymerase. The above calculation, in fact, gives the overall rate of polymerization per round of replication. If, however, another step in the process of replication is rate limiting, e.g., joining of Okazaki pieces, then the actual rate of polymerization could be higher.

The rate of unwinding of the chromosome at each growth point can also be calculated: one turn of the double helix occupies ten base pairs (i.e., 20 nucleotides); therefore one turn will be unwound for every 20 nucleotides polymerized.

$$\therefore \text{rate of unwinding} = \frac{10^5}{20} = 5 \times 10^3 \text{ rpm}$$

by the other two *E. coli* polymerases is at least an order of magnitude lower.

Pol III has other interesting properties. *In vitro,* the enzyme can act only on single-stranded DNA, a form which the template could feasibly assume *in vivo* (see Section I,F). In contrast, pol I can utilize both single-stranded and (nicked) double-stranded DNA. Although pol III is necessary for *in vivo* DNA synthesis, the purified enzyme synthesizes poorly *in vitro*; provided with a primed single-stranded template, it makes only short DNA chains about 50 nucleotides in length (Kornberg and Gefter, 1972; Tamblyn and Wells, 1975; Livingston *et al.,* 1975). However, in the presence of protein cofactors, single-stranded (SS) bacteriophage DNA molecules some 6000 nucleotides in length, each primed with only one short oligonucleotide, can be almost completely copied *in vitro* by pol III (W. Wickner *et al.,* 1973; Hurwitz and Wickner, 1974).[2] The protein cofactors are of two types, namely, those that probably interact with the polymerase (e.g., copolymerase III) and those that may modify the template, such as *E. coli* strand-separation protein (SSP) (see Section III for details of both).

It may also be relevant to an understanding of the *in vivo* function of pol III that, when assayed *in vitro* under the salt conditions thought to prevail in the cell ("physiological" conditions), very high concentrations of the enzyme are necessary for maximum rates of synthesis to be achieved (Nüsslein *et al.,* 1971; Otto *et al.,* 1973). In dilute solution, the enzyme requires a tenfold lower salt concentration for optimal activity (Kornberg and Gefter, 1972).

What roles do the other two polymerases play? After the isolation of the *pol A1* mutant, there was some speculation that the *pol A* gene product was dispensable (Editorial, *Nature,* 1971) although *pol A1* cells do contain very low levels of residual pol I polymerase activity and wild-type levels of its normal 5′ → 3′ exonuclease activity (Section II,A,1,a) (Lehman and Chien, 1973). However, *pol A1* mutants show poor joining of the products of discontinuous synthesis, "Okazaki pieces" (see footnote 3) (Kuempel and Veomett, 1970; Okazaki *et al.,* 1971), and are unable to replicate

[2] The *in vitro* single-stranded phage replication systems consist of purified soluble enzymes and cofactors incubated with the covalently closed single-stranded DNA genome of the coliphages φX174, M13, or G4 (*not* to be confused with phage T4). The single-stranded (SS) DNA is converted to double-stranded [replicative form (RF)] DNA, mimicking the *in vivo* conversion of SS-DNA to RF-DNA (SS → RF conversion). The process, *in vitro,* requires only host enzymes. Since only one strand of DNA is copied, *in vitro* SS → RF conversions do not represent replication in the strictest sense; they do, however, require a priming event, extensive polymerization, primer excision, and DNA joining. All these features are also relevant to the replication of duplex DNA. For a review of these systems, see Scheckman *et al.* (1974).

certain plasmids (Kingsbury and Helinski, 1970). Since the 5′ → 3′ exonuclease activity of pol I is apparently normal, this leaves open the question of whether the polymerizing activity of pol I takes part in normal replication (see Section I,D). Recently, conditional lethal mutants defective in the 5′ → 3′ exonuclease have been isolated (Konrad and Lehman, 1974; Olivera and Bonhoeffer, 1974) that are also defective in the joining of "Okazaki fragments," implying that it is this activity of pol I that is essential for replication (see Section I,D).

Mutants defective in pol II have been isolated (*pol B*), but they are phenotypically indistinguishable from their wild-type counterparts (Campbell *et al.*, 1972, 1974; Molineux *et al.*, 1974). Pol II thus would not seem to be necessary for replication. However, final judgment on this should be reserved; low levels of activity remain in *pol B* mutants, and these may be sufficient for an essential role of pol II *in vivo* (Campbell *et al.*, 1972; Hirota *et al.*, 1972). A possible role for pol II is also suggested by comparison with the DNA polymerase of bacteriophage T4; this enzyme, which is known to be responsible for the polymerization in T4 DNA replication, is specifically stimulated by the T4 gene 32 SSP (see Section I,F), as is pol II by the analogous *E. coli* SSP (Molineux *et al.*, 1974; Sigal *et al.*, 1972).

Of the nonpolymerizing activities of polymerases, the 5′ → 3′ exonuclease of pol I may be involved in the joining of discontinuously synthesized DNA (see Section I,D). Another nonpolymerizing activity of some polymerases, a 3′ → 5′ exonuclease, plays a role in ensuring the accuracy of the copying mechanism (see Section I,E).

The T4 DNA polymerase and others that are also clearly necessary for polymerization during replication (avian virus reverse transcriptases, *B. subtilis* pol III) are described in Section II together with a brief summary of some of the known properties of eukaryotic polymerases.

C. Priming

The start of synthesis of new chains (priming) of any linear polymer involves either an initial reaction between two free monomers, or alternatively, a preexisting structure (primer) to which a monomer can be added. In the latter case, the primer may be removed after some polymerization has occurred. The chemical composition of a primer need not, *a priori*, be similar to that of the polymer.

An example of the first mechanism is provided by RNA synthesis catalyzed by the *E. coli* enzyme RNA polymerase. This enzyme both catalyzes an initial reaction between two monomers and then polymerizes further from the initial dinucleotide product (see Section II). DNA synthesis, on the other hand, probably requires the second mechanism. As we noted in the previous section, all DNA polymerases appear to have an absolute requirement for a primer, specifically a polynucleotide chain with a free 3′-OH end (Kier, 1962; Atkinson *et al.*, 1968).

Reiterated priming, as we outlined in Section I,A, is essential for DNA replication. It is needed in discontinuous synthesis, which is itself the only means of avoiding the apparent paradox that one of the two newly synthesized strands grows in a direction in which no known DNA polymerase can synthesize (see Figure 2b). The size of the immediate products (Okazaki pieces[3]) of discontinuous synthesis makes it clear that priming is a frequent event in replication. If DNA polymerases cannot start synthesis by themselves, then how can this requirement for priming be satisfied?

DNA polymerases, with two known exceptions (Spadari and Weissbach, 1975), appear to be capable of using either DNA or RNA as primer *in vitro* (Keller, 1972; Roychoudhury, 1973). Since, as mentioned above, RNA polymerases are capable of starting synthesis without a primer, one attractive possibility is that an RNA polymerase first synthesizes a short RNA primer which is then extended by a DNA polymerase. Support for the general idea of RNA-primed DNA synthesis is given by a number of observations, which fall into two classes. First, nascent small pieces of DNA (of the same size as Okazaki pieces) with covalently attached ribonucleotides have been isolated from both prokaryotes (Sugino *et al.*, 1972; Kurosawa *et al.*, 1975) and a variety of eukaryotes (Waqar and Huberman, 1975; Sato *et al.*, 1973; Fox *et al.*, 1973; Pigiet *et al.*, 1974; Hunter and Francke, 1974). These RNA–DNA molecules are probably intermediates in DNA replication. Second, there is definitive evidence for the role of an RNA primer in the initiation of DNA synthesis by widely differing systems. Avian virus RNA-directed DNA

[3] The Okazaki pieces of discontinuous synthesis have been isolated from a wide variety of organisms after short pulse labeling (Sakabe and Okazaki, 1966; Okazaki *et al.*, 1968; Fareed and Salzman, 1972; Huberman and Horowitz, 1973; Sakamaki *et al.*, 1975). In bacteria, they are normally about 1000–2000 nucleotides long—about 7–11 S by sedimentation analysis (Okazaki *et al.*, 1968). In eukaryotes, they are of variable size (cf. Kidwell and Miller, 1969; Magnusson *et al.*, 1973), but can be as much as tenfold smaller—between 3 and 5 S (Huberman and Horowitz, 1973; Magnusson *et al.*, 1973). The reason for size differences between the various systems is obscure. Under certain conditions, "4 S" pieces can also be found in *E. coli*, but it is not yet clear whether these bacterial "4 S" pieces are normal intermediates in replication (Lark and Wechsler, 1975; Konrad and Lehman, 1975; Diaz *et al.*, 1975; Geider and Hoffmann-Berling, 1971). Two *E. coli* gene products, *dna B* and *dna S*, have been implicated in the control of the size of *E. coli* Okazaki pieces, but it is not yet known whether this control is exerted directly (see Section II). There is also evidence that under certain circumstances discontinuous synthesis occurs on both strands (Kurosawa and Okazaki, 1975; Olivera and Bonhoeffer, 1972).

polymerases (reverse transcriptases) require a specific RNA primer when synthesizing on their natural templates (Dahlberg *et al.*, 1974; Waters *et al.*, 1975), and RNA synthesis is probably also required for the initiation of replication of *B. subtilis, E. coli* chromosomes, and a number of *E. coli* plasmids and phages (see Section I,I).

The question now arises as to which enzyme is responsible for the formation of primer RNA. The eukaryotic enzyme is unknown: however, in *E. coli* there is evidence for the operation of two different activities acting independently of one another and at different stages of replication. Primer synthesis for the initiation of replication is normally, but not always (cf. phages ϕX174 and G4; Silverstein and Billen, 1971; Zechel *et al.*, 1975), carried out by a cellular RNA polymerase similar or identical to the one responsible for transcription (see Section I,I). However, another activity may be required for discontinuous synthesis. DNA replication in progress is relatively insensitive to rifampicin (an inhibitor of the RNA polymerase responsible for transcription), as is the synthesis of the RNA found attached to DNA (Sugino *et al.*, 1972). This raises the question of whether the usual transcriptional RNA polymerase could be responsible for synthesis of this RNA. There are a number of possibilities: the normal RNA polymerase might become rifampicin-insensitive if it were modified by an additional factor, or if it were to act as part of a multienzyme replication complex; alternatively, there could be an entirely different RNA-polymerizing activity present.

The *dna G* gene product has been implicated in the formation of Okazaki pieces (Lark, 1972a; Klein *et al.*, 1973; Louarn, 1974), possibly at the level of the start of synthesis. *In vitro,* the *dna G* gene product alone is required for the initiation of DNA synthesis in the conversion of single-stranded bacteriophage G4 DNA to replicative form (see footnote 2) (Scheckman *et al.*, 1974; Zechel *et al.*, 1975). Thus, the *dna G* gene product may be involved, possibly as a rifampicin-insensitive RNA polymerase, in the priming event necessary for synthesis of Okazaki pieces (Bouché *et al.*, 1975; also see Section II,E,3).

D. The Joining of Discontinuously Synthesized Pieces

In the previous section (Section I,C) we discussed the evidence which showed that covalently linked RNA–DNA molecules are probably intermediates at two stages in DNA replication. If this is the case, then the ribonucleotides must be excised and replaced with deoxyribonucleotides before the pieces can be joined to each other to form a final DNA product, since most of the genomes we have been discussing contain exclusively

DNA.[4] The processes of excision, DNA substitution, and joining are all discussed in this section. The necessary enzymes for such a scheme have been found in nature: 5′ → 3′ exonucleases which can remove RNA primers, at least one polymerase which is implicated specifically in the replacement of ribonucleotides, and DNA ligases, which in a duplex DNA structure catalyze the formation of a phosphodiester bond between the 3′-OH group of a deoxynucleotide and the 5′-phosphoryl group of a directly adjacent one.

Candidates for the RNA excision activity must be capable of degrading the RNA moiety of a double-stranded RNA–DNA hybrid, since, in the cases of both initiation of replication and discontinuous synthesis, it would seem reasonable that the RNA portion of RNA–DNA molecules would hybridize with the template strand to form a duplex (i.e., double-helical) structure. Such nucleases, collectively known as RNase H (H for hybrid), have been found in both prokaryotes (Mölling *et al.,* 1971; Grandgenett and Green, 1974; Miller *et al.,* 1973; Henry *et al.,* 1973) and eukaryotes (Stein and Hausen, 1969; Büsen and Hausen, 1975; Doenecke *et al.,* 1972; Banks, 1974). Some of the isolated enzymes are exonucleases (Grandgenett and Green, 1974; Leis *et al.,* 1973) and others endonucleases (Henry *et al.,* 1973; Miller *et al.,* 1973; Banks, 1974; Keller and Crouch, 1972). Both types of nuclease presumably could remove the RNA portion from a covalently linked RNA–DNA molecule. There is, however, no direct evidence that they are necessary for DNA replication *in vivo.*

In *E. coli,* a more likely candidate for an excision activity is provided by the 5′ → 3′ exonuclease activity of pol I (see Section I,B and Section II,A,1,a), an activity which also can degrade RNA in a duplex DNA–RNA hybrid (Keller and Crouch, 1972; Leis *et al.,* 1973; Roychoudhury,

[4] A clear example of the exclusively deoxyribonucleotide composition of certain genomes is provided by the resistance of some small covalently closed double-stranded DNA molecules of viruses and plasmids to alkaline digestion or RNase treatment (cf., e.g., Vinograd *et al.,* 1968; Blair *et al.,* 1972). Alkali causes the hydrolysis of phosphodiester bonds between ribonucleotides or between ribonucleotides and deoxyribonucleotides, but polydeoxyribonucleotides are not attacked. The lack of alkali or RNase sensitivity of exclusively deoxyribonucleotide polymers is contrasted by the sensitivity of certain "DNA" genomes which do contain RNA such as those of some mitochondria (Grossman *et al.,* 1973) and of some plasmids grown under certain experimental conditions (Blair *et al.,* 1972).

A minimum estimate of the length of exclusively DNA-containing genetic material in mammalian chromosomes is 2×10^4 nucleotides long [measured as single-stranded nucleic acid, Hozier and Taylor (1975)], which is considerably longer than the DNA in an Okazaki piece. The synthesis of such long stretches would clearly require the excision of RNA primers. Longer stretches of the mammalian genome may however contain nonDNA components, whose nature is still uncertain (Hozier and Taylor, 1975).

1973). This would not seem to be the ideal enzyme for excision, however. In contrast to the RNase H of *E. coli* (Henry *et al.*, 1973; Miller *et al.*, 1973), this exonuclease is not specific for RNA, and acts on DNA as well. Nevertheless, because of its associated polymerase, the enzyme, if it has an appropriate primer, can immediately replace excised ribonucleotides with deoxyribonucleotides. The simultaneous occurrence of degradation and synthesis is known as "nick translation" (Kelly *et al.*, 1970). The fact that excision and sealing can be accomplished only by the two enzymes pol I and ligase has been shown in a model system. In this system, a single-stranded circular phage DNA was first primed with RNA and then converted by *E. coli* DNA pol III into a double-stranded molecule except for a short gap between the 3′-OH terminal end of the DNA and the 5′-phosphoryl end of the RNA. When pol III was removed and pol I and ligase were added, a completely covalently closed duplex DNA was formed without any extensive degradation of the DNA (Westergaard *et al.*, 1973).

Direct support for an *in vivo* role of the *E. coli* pol I 5′ → 3′ exonuclease in replication has been provided by the isolation of mutants conditionally lethal due to a lesion in the *pol A* gene (Konrad and Lehman, 1974; Olivera and Bonhoeffer, 1974). The mutants, which are grossly defective only in the 5′ → 3′ exonuclease activity of pol I, accumulate small pieces of DNA at the nonpermissive temperature (Konrad and Lehman, 1974; Olivera and Bonhoeffer, 1974), and, in addition, they contain larger amounts of RNA–DNA pieces than wild-type cells (Kurosawa *et al.*, 1975). There is also *in vivo* evidence that the polymerizing activity of pol I is required for efficient joining of Okazaki pieces (see Section I,B).

The use of a multifunctional enzyme similar to pol I for the removal and replacement of RNA primers may not, however, be universal. The combination of a 5′ → 3′ exonuclease and a polymerase thus far has been found only in prokaryotic organisms, and most eukaryotic polymerases appear to lack any associated exonuclease activity (see Section II,A).

In contrast, DNA ligases appear to be widely distributed in nature (see Section II). However, firm evidence that the enzyme is required in replication has been obtained only from *E. coli*. Conditional lethal mutants of *E. coli* defective in this enzyme have been isolated (Pauling and Hamm, 1968; Horiuchi *et al.*, 1975) and have the property, under nonpermissive conditions, of accumulating newly synthesized DNA in the form of Okazaki pieces (Pauling and Hamm, 1969; Gottesman *et al.*, 1973; Konrad *et al.*, 1973). DNA ligase is, therefore, an indispensable enzyme for replication in *E. coli*.

E. Fidelity of Replication

A consideration of the enzymology of DNA replication must account for the observed fidelity of replication, i.e., the accuracy with which incorrectly base-paired nucleotides are excluded from newly synthesized DNA. The occurrence of tautomeric forms of the bases gives rise to the possibility of mismatching, e.g., a tautomeric change in adenine could result in its pairing with cytosine instead of with thymine (Figure 3; Watson and Crick, 1954). The tautomeric equilibrium constants of the bases have been estimated to be approximately 10^4 (Koch and Miller, 1965), so the theoretical mutation rate per base pair (pbp) is far higher than that observed *in vivo*, 2×10^{-10} pbp (Drake, 1969). Hydrogen bonding alone accounting for the observed fidelity of replication therefore seems unlikely. It has been proposed that DNA polymerase may participate in the selection of a correct base pair during replication (Koch and Miller, 1965).

Support for this idea has been provided by the isolation of phage T4 mutants defective in their DNA polymerase, mutants which have an increased mutation frequency throughout the T4 genome—a mutator

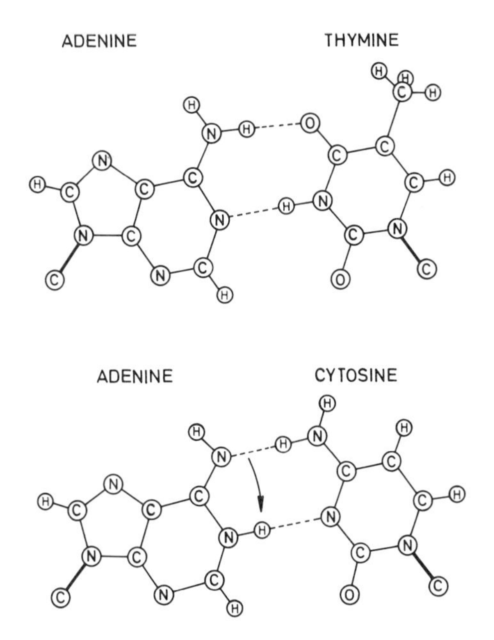

Fig. 3. The way in which a tautomeric shift in adenine can allow the molecule to base pair with cytosine. (Reproduced with permission from J. D. Watson and F. H. C. Crick, 1954, *Cold Spring Harbor Symp. Quant. Biol.* **18,** 123.)

phenotype (Speyer, 1965). *In vitro* studies have shown that the DNA polymerase itself is responsible for these altered mutation rates. Highly purified polymerase from a mutator strain of T4 incorporates incorrect bases into a template with a fourfold higher frequency than does a wild-type polymerase (Hall and Lehman, 1968).

There are two known activities associated with polymerases that participate in the maintenance of fidelity during replication: (1) the polymerizing activity itself, and (2) a $3' \rightarrow 5'$ exonuclease activity which is only found associated with some prokaryotic polymerases.

1. The Polymerizing Activity

A polymerase from a T4 mutant which is both a mutator and is temperature-sensitive for DNA replication has been purified to homogeneity (Hershfield, 1973). The polymerizing activity of the mutant enzyme is unstable at high temperature. By a factor of six, the mutant enzyme is also more likely than wild-type to incorporate incorrect nucleotides, and there is an increase in the utilization of incorrect nucleotides with increasing temperature. This suggests that both the temperature-sensitive replication and the mutator phenotypes reside in the polymerizing activity of the DNA polymerase.

Studies on eukaryotic polymerases provide further support for the participation of the polymerizing activity in the selection of base pairs during DNA synthesis. The 3.4 S calf thymus polymerase has no associated $3' \rightarrow 5'$ exonuclease (Chang and Bollum, 1973), yet fidelity of copying *in vitro* is very high; in the presence of a poly(dA) template, the incorporation of cytidine compared to thymidine is 6×10^{-6}, and that of guanosine to thymidine is 10^{-5}. This accuracy in copying must lie purely in the ability of the enzyme to recognize correct base-pairing. Also, Springgate and Loeb (1973) have found that a polymerase from acute lymphoblastic leukemic cells has mutator activity; it produces tenfold more polymerization errors than does the polymerizing activity from normal lymphocytes. Since this polymerase also does not have an associated $3' \rightarrow 5'$ exonuclease activity, the mutator defect must be in the polymerizing activity of the molecule.

Thus, the polymerizing activity participates in the selection of correct bases during replication. However, the manner in which the polymerase senses the presence of a correct base pair is not known.

2. The $3' \rightarrow 5'$ Exonuclease Activity

This exonuclease is an integral component of both *E. coli* and T4 DNA polymerases (see Section II,A,1,a, and 3) and has a "proofreading" function in DNA copying (Goulian *et al.*, 1968; Brutlag and Kornberg, 1972; Hershfield and Nossal, 1972). For polymerization by *E. coli* and T4 DNA polymerases *in vitro,* the terminal nucleotide of the primer must be correctly paired with the opposite nucleotide on the template. If the terminal nucleotide on the primer is not base-paired to the template, no addition of further nucleotides occurs (Brutlag and Kornberg, 1972) (Figure 4). Addition of nucleotides to such unpaired termini occurs after the $3' \rightarrow 5'$ exonuclease has removed the mismatched nucleotides and a correctly base-paired terminus is reached. It has also been shown that the $3' \rightarrow 5'$ exonuclease acts during synthesis (Hershfield and Nossal, 1972). Since the polymerases do not extend a mismatched terminus, the $3' \rightarrow 5'$ exonuclease removes an incorrect nucleotide inserted by the polymerase itself; in this manner, the enzyme "proofreads" its own activity. Enzymes with an altered $3' \rightarrow 5'$ exonuclease activity have been isolated from T4 mutator strains (Bessman *et al.*, 1974).

Whether these two activities are the only mechanisms for maintaining fidelity in DNA replication is not certain. While the fidelity of replication of purified polymerases *in vitro* may be as high as 10^{-6} pbp (e.g., Hall and Lehman, 1968), this is still less than *in vivo* fidelity, which is estimated to be from 1.7×10^{-8} pbp (for phage T4) to 7×10^{-12} (for *Neurospora*) (Drake, 1969). Thus, other functions may also be involved; e.g., in *E. coli,*

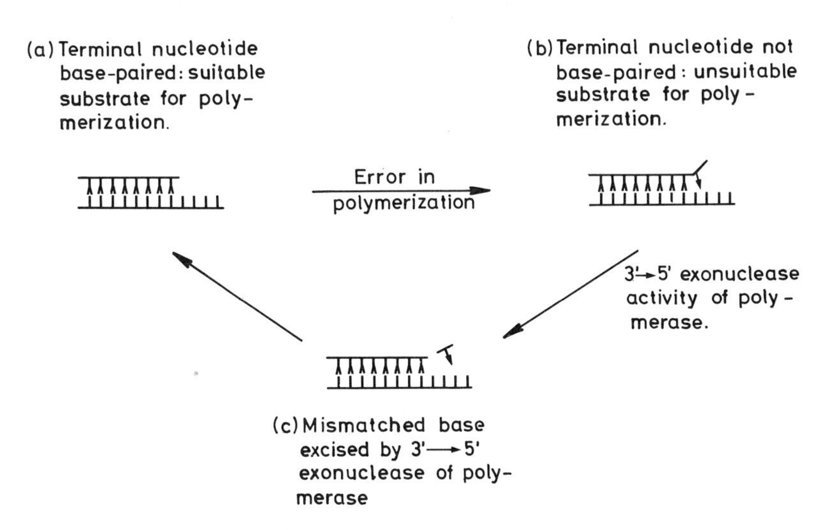

Fig. 4. The "proofreading" function of *E. coli* pol I and phage T4 polymerase.

mutator genes are known which, when defective, cause an increased mutation rate (Siegel and Bryson, 1967; Helling, 1968). These genes are distinct from those known to be involved in DNA replication. If the products of these genes do not act directly at the growing point during replication, there must also exist a mechanism whereby they can determine which strand of the duplex is the newly synthesized strand. Without such a mechanism, a repair activity could not determine which base of a mismatched pair is incorrect, and the mutation frequency would be approximately equal to the mismatch frequency. In whatever manner fidelity is maintained, it must be finely balanced (together with the efficiency of mechanisms for repair of DNA damage); too great an efficiency would prevent the mutations necessary for evolution (i.e., adaptation to environmental change), while inefficiency would result in too many mutations and populations that would be at a selective disadvantage because of a high incidence of mutationally based defects.

F. Strand Separation

Replication requires strand separation, thereby posing the energetic problem of how the forces holding the two parental strands together can be overcome. *In vitro,* at physiological temperatures and under physiological ionic conditions, DNA exists almost entirely as a double helix (Marmur and Doty, 1962). Under the same *in vitro* ionic conditions, the helix–coil transition from the duplex structure to single-stranded DNA occurs only at high temperature (>70°C) (Marmur and Doty, 1962). The transition is marked by a large positive standard enthalpy (approximately 8 kcal per mole base pair) and is a highly cooperative process (Bloomfield *et al.,* 1974). However, local short regions of single-stranded DNA do occur transiently, even under conditions in which the double helix predominates (Printz and von Hippel, 1965; Utiyama and Doty, 1971; von Hippel and McGhee, 1972b).

The Watson–Crick model of DNA structure suggests an attractive mechanism for DNA replication (Watson and Crick, 1953a,b) whereby strand separation of the parental duplex exposes two template strands to which incoming complementary mononucleotides can be base-paired before being subsequently polymerized. However, since the bases on each strand can be recognized even when they are in duplex form (von Hippel and McGhee, 1972a), other mechanisms of replication involving strand separation *after* the synthesis of new strands cannot *a priori* be ruled out, although such mechanisms have not been supported experimentally. In contrast, there is experimental support for the idea that strand separation occurs before polymerization, in that two polymerases known to be responsible for replication, T4 polymerase and *E. coli* pol III, *in vitro* can use only single-stranded DNA as a template.

The isolation of a class of proteins, strand separation proteins (SSP),[5] from both prokaryotes and eukaryotes (see Section II,C) suggests a solution to the strand separation problem. *In vitro*, under physiological ionic conditions, these proteins bind tightly (K_b = 10^6–10^9 M^{-1}) and cooperatively to single-stranded DNA. This binding facilitates the formation of single-stranded regions in duplex DNA, even at physiological temperatures, since the negative free energy change involved in the binding of strand separation proteins is greater than that required for the maintenance of the double helix (von Hippel and McGhee, 1972b; Alberts and Frey, 1970). When catalyzing strand separation, these proteins are postulated to bind initially to single-stranded regions in duplex DNA and then to extend these regions by sequential cooperative binding (von Hippel and McGhee, 1972b; Alberts, 1970). Whether the initial single-stranded regions that are bound are the spontaneous, transiently occurring, single-stranded regions referred to above, or whether these proteins cause an additional initial melting of double-stranded DNA is not yet clear. During replication, strand separation proteins could operate by separating the parental DNA ahead of DNA polymerase to provide fresh template (Alberts, 1973). This function in replication would require that the cooperative binding found *in vitro* also takes place *in vivo*, and that the DNA–SSP complex could be used as a template by the DNA polymerase responsible for replication.

The gene 32 protein of bacteriophage T4 is a SSP that provides a clear example of the essential role of such a protein in the replication process. Studies with temperature-sensitive gene 32 mutants show that it is necessary for a functioning growth point (Riva *et al.*, 1970; Alberts, 1971). Large amounts of gene 32 protein per cell are produced, and it has been calculated that up to 100–200 molecules of it could be present for each T4 growth point (Alberts and Frey, 1970; Alberts, 1973). Each molecule of gene 32 protein binds ten nucleotides (Alberts and Frey, 1970); therefore, the protein could bind quantitatively to provide adequate amounts of template for replication. The single-stranded DNA gene 32 protein complex is known to be an efficient template for the T4 DNA polymerase *in vitro* (Huberman *et al.*, 1971; Nossal, 1974; also Section II,A,3).

A protein analogous to the T4 gene 32 product has been isolated from *E. coli* (Sigal *et al.*, 1972). However, no *E. coli* mutants defective in this

[5] These proteins are referred to either as melting (von Hippel and McGhee, 1972b) or unwinding (Sigal *et al.*, 1972) proteins in the literature.

protein have yet been isolated, so whether it is necessary for *E. coli* replication is not known. Nevertheless, it is required specifically by *E. coli* DNA pol III to give optimal polymerization in the *in vitro* single-stranded phage systems (see footnote 2) (Geider and Kornberg, 1974; S. Wickner and Hurwitz, 1974; Weiner *et al.*, 1975), a positive indication that it may be similarly required *in vivo*.

Whether strand separation proteins alone can account for the rapid separation of strands that must occur during replication is not yet clear. Some other mechanism could actually initiate strand separation, and the strand separation protein function could be merely to stabilize single-strand DNA and prevent reannealing.

G. Swiveling

There are two types of double helices that differ in their topology. In one (plectonemic), the two helices are wound around each other as if two parallel strands had been coiled around a central axis. This can be illustrated by the simultaneous winding of two parallel strands around a rod (Figure 5). The two helices can only be separated from each other through rotational movement of one about the other. The other kind of double helix (paranemic) can be produced by pushing two *existing* helices into each other. (This can be visualized by pushing two wire springs into one another.) Here, the helices can of course be separated from each other simply by translational movement.

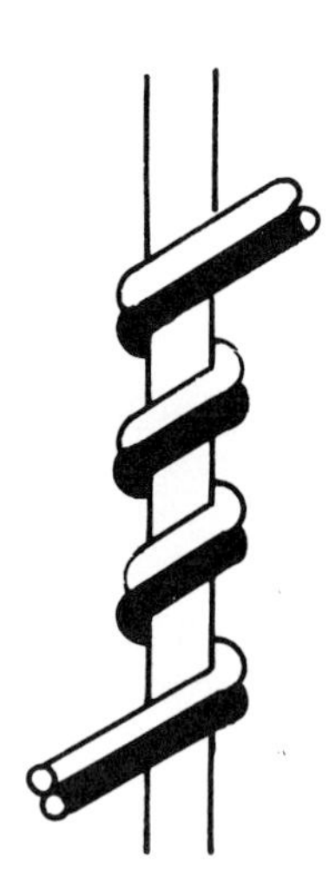

Fig. 5. A plectonemic double helix. (The central rod has been drawn in merely for purposes of illustration.)

X-ray diffraction data have shown (Watson and Crick, 1953b) that DNA is a double helix of the first type (plectonemic). Thus, semiconservative replication, which involves separation of the two parental strands, requires, in addition to strand separation, a rotational movement of the parental DNA around its axis relative to the two daughter double-stranded molecules (Figure 1a).

From the rate of DNA synthesis (10^5 nucleotides incorporated/min/growth point), it follows that the parental DNA must rotate around its own axis with about 5×10^3 rpm in *E. coli* (see footnote 1). Since it seems extremely unlikely that the total parental double helix could rotate [especially if one considers that to pack inside the cell, which has a long axis of 2 μm, the chromosome, of length 1.4 mm, must be highly folded (cf. Stonington and Pettijohn, 1971; Worcel and Burgi, 1972)], one must assume that there exist special sites (swivels) within the DNA that allow free rotation of one part of the molecule relative to the adjacent part. Thus, only the DNA between the growth point and the nearest swivel would have to rotate during replication.

It has long been recognized that the simplest swivel is a break in one of the strands of the DNA duplex (Delbrück and Stent, 1957). The existence of such a site would leave the helix free to rotate around the single phosphodiester bond in the strand opposite the break. However, there is no experimental evidence for the existence of such breaks in isolated DNA (Tomizawa and Ogawa, 1968; Sebring *et al.*, 1971; Jaenisch *et al.*, 1971). Yet they must exist, at least for circular DNA molecules, since a covalently closed circular double-stranded DNA molecule, such as the chromosomes of small viruses, could never replicate without a swivel. Therefore, it is assumed that the swivel has not been detected because it is of a very transient nature.

Enzymes that produce transient swivels are known (Wang, 1971b; Champoux and Dulbecco, 1972). They act endonucleolytically on one of the strands, thus allowing free rotation within the DNA molecule. The endonucleolytic action is followed immediately by a sealing action which erases the swivel. Thus, the presence of such a "swivelase" during replication leaves the parental DNA seemingly intact and yet prevents the development of a constraint on parental strand separation which would occur if swiveling were not allowed.

In vitro, the activity of a "swivelase" is detected by its ability to relieve the constraint in circular covalently closed DNA without causing an observable break (Champoux and Dulbecco, 1972; Wang, 1973). In principle, there are two types of constraints, one due to an increased, the other one to a decreased, number of helical turns per DNA molecule compared to the normal number of one helical turn per ten base pairs (Watson and

Crick, 1953b). Such constraints lead to changes in the configuration of the molecules resulting in the formation of positive superhelices or negative superhelices, respectively (Vinograd *et al.*, 1968; also see Figure 7). Under the appropriate conditions, closed circular molecules under constraint can be distinguished from unconstrained molecules by their different sedimentation coefficients (Crawford and Waring, 1967).

The putative swivelase acting during replication should be an enzyme which is capable of relaxing positive superhelices, because the pulling apart of the two parental strands during replication produces a constraint which would cause positive superhelices (See Figure 7).

The first swivelase activity found, the ω protein in *E. coli* (Wang, 1971b), as currently isolated and characterized, acts only on negative superhelices and is, therefore, not a likely candidate for an enzyme involved in DNA replication. Another enzyme isolated from mammalian cells that can relax both positive and negative superhelices (Champoux and Dulbecco, 1972) is a better candidate.

H. A DNA Replication Complex

Processes at the growth point are complex—on the order of ten kinds of proteins[6] are directly involved (apart from enzymes necessary for production of precursors)—and polymerization is a highly repetitive and rapid reaction (with a rate that can be as high as 1000 nucleotides polymerized per DNA strand per second in *E. coli*) (see footnote 1). This suggests at least a high degree of organization. The most efficient manner in which so many proteins can be organized in order to carry out a fast repetitive process may well be as a complex consisting of at least some of these proteins. However, present evidence of functional associations among the enzymes of DNA replication is limited. Some examples of such associations are listed below.

1. A complex between pol III and copol III in *E. coli* (see Section II).
2. A complex between the *dna B* and *C* gene products of *E. coli* (S. Wickner and Hurwitz, 1975).
3. A specific physical interaction between some of the phage T4 gene products which are essential for DNA synthesis. Products of genes 44 and 62, whose functions in DNA replication are unknown, purify together as a

[6] In *E. coli*, for instance, known participants are the *dna B, G,* and *Z* gene products, pol I and III, and DNA ligase. In addition, the *dna C* gene product, *E. coli* strand separation protein, and the "other factors" X, Y, Z (or i and n) isolated in the single-stranded phage systems (see footnote 2) are probably involved. All these factors are described in more detail in Section II.

complex (Barry *et al.*, 1973). In addition, sucrose gradient sedimentation data indicate a weak but specific interaction between T4 DNA polymerase and the gene 32 product (the strand separation protein) (Huberman *et al.*, 1971).

4. In phage T4, genes 1 and 42 are a deoxyribonucleotide kinase and a deoxycytidine hydroxymethylase, respectively, which function in the production of precursors for DNA synthesis. However, they also may play a more direct role in DNA synthesis (Wovcha *et al.*, 1973). This points to a possible close involvement of enzymes responsible for precursor production with events at the growth point.

In bacteria, it seems probable that the origin of replication is attached to the membrane (best evidence for this is available in *Bacillus subtilis;* see, for example, Yamaguchi and Yoshikawa, 1973).

In spite of an intensive search, there as yet is no compelling evidence to implicate the bacterial cytoplasmic membrane in other steps in replication (for a review, see Leibowitz and Schaechter, 1975). Current indications are that replication in eukaryotes does not involve the nuclear membrane (see, for example, Wise and Prescott, 1973; Comings and Okada, 1973).

I. Initiation

The process by which a new growth point for replication is created is known as initiation. In all organisms investigated, initiation occurs only at a limited number of sites on the genome, and frequency of initiations at those sites is strictly regulated so that a complete genome normally is duplicated only once per cell cycle (see Chapters 3 and 7).

Some operationally defined terms are used in discussing initiation. These include *replicon,* the unit of DNA that is replicated after a single initiation; *replicator,* a site on a DNA molecule at which initiation takes place; and *initiator,* enzyme or enzymes that recognize a replicator.[7] In prokaryotes, the genome itself usually constitutes a replicon, and the single replicator is probably a unique nucleotide sequence (see replicators, Section II,G,1).

An initiator and a replicator constitute the formal requirements that differentiate initiation from normal events at the growth point. Dependent upon the mode of action of the initiator, other proteins may also be needed before a growth point can be established (see examples below). In

[7] These terms were originally more stringently defined (Jacob *et al.*, 1963), but these definitions reflect current usage. The term origin (of replication) is often used instead of the term replicator.

addition, some means for regulating initiator action must exist, but this is a separate enzymatic problem (see below).

It is likely that in some prokaryotes RNA polymerase acts as an initiator. This idea is attractive, because RNA polymerase is known to be capable of recognizing specific nucleotide sequences (see Section II,G,2). The evidence for the involvement of RNA polymerase comes mainly from experiments in which specific inhibitors of the enzyme (e.g., rifampicin) block the initiation of DNA replication, but not ongoing replication (Silverstein and Billen, 1971; Lark, 1972b; Laurent, 1973; Clewell *et al.*, 1972; Brutlag *et al.*, 1971). Following recognition of the replicator, the RNA polymerase probably fulfills its initiator function by providing an initial RNA primer for DNA synthesis. Such an "initiation primer" may be differentiated from the RNA primers involved in ongoing DNA synthesis (cf. Section I,C) by the specific localization of the site on which it is produced and the rifampicin sensitivity of the RNA polymerase which produces it. Evidence for an RNA primer in initiation comes from both *in vivo* and *in vitro* work with the *E. coli* plasmid colicin E1 (Clewell *et al.*, 1972; Sakakibara and Tomizawa, 1974), the SS → RF conversion (see footnote 2 for an explanation of this terminology) of *E. coli* phage M13 (Brutlag *et al.*, 1971; Tabak *et al.*, 1974), and with *E. coli* bacteriophage λ (Dove *et al.*, 1969; Klein and Powling, 1972). Observations on the latter two systems also suggest that other proteins may be necessary for the normal transcriptional RNA polymerase to function as an initiator.

The initiation of the *in vitro* SS → RF conversion of phage M13 requires a form of RNA polymerase (RNA pol III) that is still rifampicin-sensitive but that contains a novel subunit not found in the RNA polymerase that is normally isolated (W. Wickner and Kornberg, 1974a; Section II,G). This novel subunit appears to cause RNA pol III to transcribe more selectively than normal RNA polymerase. Normal RNA polymerase can transcribe from both M13 and ϕX174 single-stranded DNA *in vitro*, whereas RNA pol III cannot transcribe at all on ϕX174 DNA but does transcribe a single specific region of the M13 genome; this specific RNA product from M13 primes DNA synthesis *in vitro*. This discrimination between the two phage genomes mimics that found *in vivo*, i.e., M13 SS → RF conversion is rifampicin-sensitive and therefore very probably requires a transcriptional RNA polymerase activity (Brutlag *et al.*, 1971), whereas the ϕX174 SS → RF conversion is rifampicin-insensitive and may depend on another enzyme(s) (Silverstein and Billen, 1971). In addition to the requirement for the novel subunit of RNA pol III to ensure template specificity, the *in vitro* transcription of only a specific region of the M13 genome to form an "initiation primer" also requires the *E. coli* SSP (see Section II,C,2).

In the case of bacteriophage λ, a putative "initiation primer," an 80 nucleotide-long piece of RNA known as "oop RNA," has been isolated (Dahlberg and Blattner, 1973). *In vivo,* "oop" synthesis is dependent not only upon RNA polymerase (as determined by rifampicin sensitivity) and a replicator (Section II,G,1), but also upon the active products of *E. coli* genes *dna B* and *dna G* (but not *dna A* or *pol C*) and upon the phage λ *O* and *P* gene products (Section II,F,1,3 and G,5) (Hayes and Szybalski, 1973a). However, it is not known how a putative RNA initiation primer might be distinguished from the numerous other transcripts in the cell such that it alone is used specifically for initiation.

In *E. coli,* as many as five different gene products may be involved in initiation (the products of genes *dna A, C, H, "I,"* and *P;* see Section II,G,3). A similar number are found for *B. subtilis* (Section II,G,4). While it is not known how these products are involved, there are indications that the process of initiation is more complex than simply the provision of an initiation primer by an RNA polymerase. For instance, under nonpermissive temperatures, *dna* P^{ts} and *dna* C^{ts} mutants of *E. coli* do not initiate DNA replication, even though an initiation primer has probably already been synthesized (Wada and Yura, 1974; Hiraga and Saitoh, 1974). This suggests that initiation includes a step following the synthesis of primer. In contrast, certain proteins may be required prior to initiation primer synthesis, as experiments in *E. coli* with protein synthesis inhibitors have suggested (Lark and Renger, 1969).

Apart from RNA polymerase, only one other protein has been isolated that is known to be necessary for the initiation of DNA replication. It is the gene *A* product of the bacteriophage φX174, a protein which exhibits a nuclease activity that is specific for only one of the two strands of the covalently closed circular (RF) form of this phage. However, its function in initiation is still far from clear (see Section II,G,5).

In addition to the formal requirements of a replicator and an initiator, the following steps also have been suggested as possibly necessary for initiation: (1) an initial strand separation (Alberts, 1970, 1973) or (2) if a replication complex exists, an enzyme involved in its initial formation (Wechsler, 1975). The involvement of the *E. coli dna B* and *dna G* gene products in bacteriophage λ "oop" RNA synthesis might result from the participation of growth point enzymes in initiation.

Since initiation is a strictly regulated process, some of the genetically defined functions may also play a purely regulatory role. However, current ignorance of the details of initiation prevents further analysis. Positive and/or negative regulation of replicator recognition, or, if initiation is a multistep process, of some step prior or subsequent to replicator recognition are all possibilities.

The dearth of information about initiation is even more marked for eukaryotes. As of this writing, there is no evidence for RNA primers in initiation in eukaryotes, although there is some evidence that certain proteins, which act at particular stages in the cell cycle, are required (Muldoon *et al.,* 1971; Williamson, 1973; Hereford and Hartwell, 1973, 1974). In addition, some yeast mutants defective in initiation have been isolated (see Section II,G,6). The regulation of initiation in eukaryotes is, however, likely to be more complex than that in prokaryotes. Eukaryotic genomes may comprise many thousands of replicons (compared to the single one of prokaryotes), with replicators spaced between 3×10^3 and 3×10^5 nucleotides apart, depending on the organism (Huberman and Riggs, 1968; Blumenthal *et al.,* 1973; Callan, 1973). Nonadjacent replicons appear to be replicated in a controlled temporal sequence (Braun and Wili, 1969; Muldoon *et al.,* 1971; Dawes and Carter, 1974), and the size of the replicon may vary with different stages of development of the organism (Blumenthal *et al.,* 1973; Callan, 1973). At least these two features of additional complexity will have to be explained by an eventual molecular description of eukaryotic initiation—apart from any steps that may be common to both prokaryotes and eukaryotes.

II. MOLECULAR MACHINERY

A. DNA Polymerases

1. E. coli DNA Polymerases

a. Polymerase I (Pol I). *i. General properties.* Single polypeptide chain, molecular weight 109,000, amino acid composition known (Jovin *et al.,* 1969). Contains a free sulfhydryl group that can be substituted without loss of activity (Englund *et al.,* 1968). Optimal ionic strength for polymerization: 0.1 *M* KCl. Has a requirement for Zn^{2+} (Springgate *et al.,* 1973b). An approximately spherical molecule with a diameter of 65 Å (Griffith *et al.,* 1971). Pol I can utilize as a template-primer, nicked duplex DNA, duplex DNA containing large gaps (>100 nucleotides), or a primed single strand of DNA. Polymerization on nicked duplex DNA is normally accompanied by simultaneous hydrolysis of the strand ahead of the nick ("nick translation") (Kelly *et al.,* 1970; Masamune *et al.,* 1971); however, polymerization from nicks can occur independently of $5' \rightarrow 3'$ exonuclease activity (Setlow *et al.,* 1972; Masamune and Richardson, 1971). The products polymerized on nicked duplex DNA, when nick translation does not take place, contain sequences copied from both template strands covalently linked to one another (Shildkraut *et al.,* 1964; Masamune and

Richardson, 1971). Polymerization and 3′ → 5′ exonuclease activity by pol I are inhibited by the addition of *E. coli* "strand separation protein (SSP)" (Molineux *et al.*, 1974). Pol I has no affinity for duplex DNA but binds at nicks, gaps, the ends of chains, and along single-strand regions (Englund *et al.*, 1969b). Approximately 400 molecules per cell (Richardson *et al.*, 1964). Turnover number ca. 1000 (number of nucleotides polymerized at 37°C/min/molecule enzyme; Englund *et al.*, 1968).

ii. Structural gene. Pol A, 75 min on *E. coli* genetic map (Taylor and Trotter, 1972).

iii. Catalytic properties. The enzyme has five binding sites (Kornberg, 1969): (1) for part of the DNA template; (2) for the growing DNA chain; (3) for the primer terminus, essential for the recognition of 3′-OH terminal nucleotide of the primer. Also in this region are the sites for hydrolytic cleavage of the 3′-OH terminated primer chain (in a 3′ → 5′ direction) and for the pyrophosphorolytic cleavage [reaction (c) below] of the terminal nucleotide from the chain; (4) for the triphosphate moiety of a deoxyribonucleoside triphosphate; requires Mg^{2+} (Englund *et al.*, 1969a); (5) for the 5′ end of a DNA chain, base-paired to the template where hydrolytic cleavage of a DNA chain occurs in the 5′ → 3′ direction (Setlow and Kornberg, 1972).

These sites participate in the following enzymatic reactions:

(a) Polymerization of nucleotides

$$\text{DNA}_n + \text{dNTP} \rightleftharpoons \text{DNA}_{n+1} + \text{PP}_i$$

This is the template-directed addition of nucleotides, activated as 5′-triphosphates, to a growing chain of DNA. The reaction involves the condensation of substrate molecules bound at sites (3) and (4) and may occur as follows: the triphosphates must be bound to the enzyme so that correct hydrogen bonding can occur with the complementary base on the template. Precisely how the enzyme recognizes that a correct base pair has been formed is not known. No specific hydrogen bonding requirements must be satisfied between the nucleotide at the primer terminus and the incoming nucleotide. However, correct base-pairing is essential between this terminal nucleotide and its partner in the template. When base-pairing conditions are satisfied, a bridge is formed between the α-phosphate (the 5′-phosphate) of the nucleoside triphosphate and the 3′-OH on the nucleotide at the primer terminus. Inorganic pyrophosphate is released during the reaction. At diester bond formation, the primer terminus loses its 3′hydroxyl group and so is no longer held in the primer terminus site on the enzyme. Through movement of the DNA chain relative to the enzyme, the 3′-OH of the newly added nucleotide comes to occupy the primer terminus site, and the reaction can be repeated. The enzyme can dissociate from the template after the addition of each nucleotide, i.e., it is nonprocessive[8] (McClure and Jovin, 1975).

(b) 3′ → 5′ exonuclease (Lehman and Richardson, 1964)

$$\mathrm{DNA}_n - 3'\text{-OH} + \mathrm{H_2O} \rightleftharpoons \mathrm{DNA}_{n-1} + \mathrm{dNMP}$$

Removes only a single nucleotide at a time by hydrolytic cleavage of a diester bond, and specifically requires a 3′-OH terminus. The function specifically removes incorrectly paired bases ("proofreading") from the primer terminus (Brutlag and Kornberg, 1972). It is unable to digest beyond a thymine dimer (Kelly *et al.*, 1969), so presumably cannot remove thymine dimers.

(c) Pyrophosphorolysis

$$\mathrm{DNA}_n + \mathrm{PP_i} \rightleftharpoons \mathrm{DNA}_{n-1} + \mathrm{dNTP}$$

This is the reversal of polymerization (Brutlag and Kornberg, 1972; Deutscher and Kornberg, 1969a).

(d) Pyrophosphate exchange

$$\mathrm{DNA}_n + \mathrm{dNTP} + \mathrm{PP_i^*} \rightleftharpoons \mathrm{DNA}_n + \mathrm{dNTP^*} + \mathrm{PP_i}$$

This is the exchange of pyrophosphate with the β and γ groups of a triphosphate. The significance of the reaction is unknown (Deutscher and Kornberg, 1969a).

(e) 5′ → 3′ exonuclease (Klett *et al.*, 1968; Deutscher and Kornberg, 1969b)

$$\mathrm{DNA}_n - 5'\text{-P} + \mathrm{H_2O} \rightleftharpoons \mathrm{DNA}_{n-x} - 5'\text{-P} + x\mathrm{dNMP}$$

It excises mononucleotides or oligonucleotides, as well as thymine dimers (Kelly *et al.*, 1969). It requires double-helical DNA for activity (Cozzarelli *et al.*, 1969) and is stimulated by simultaneous occurrence of the polymerization reaction (Setlow and Kornberg, 1972). It can also remove RNA from RNA–DNA hybrids (Keller and Crouch, 1972; Tamblyn and Wells, 1975).

Pol I can be cleaved by a protease to give a 36,000-dalton fragment containing the 5′ → 3′ exonuclease and a 76,000-dalton fragment contain-

[8] Processive enzymes interact with linear polymers and may be either synthetic or degradative. Once they are bound to the end of a polymer chain, individual enzyme molecules will continue sequentially to degrade or synthesize only that chain.

ing the polymerase and 3′ → 5′ exonuclease activities (Setlow *et al.*, 1972; Setlow and Kornberg, 1972). Thus, the site for hydrolytic cleavage of DNA by the 5′ → 3′ exonuclease activity is separable from the other four sites on the enzyme.

iv. Physiological role. The role of pol I in replication has been discussed in Sections I,B and D. In addition, pol I probably has a function in the repair of certain kinds of DNA damage (De Lucia and Cairns, 1969; Gross and Gross, 1969; Town *et al.*, 1971).

b. Polymerase II (Pol II). *i. General properties.* Molecular weight 120,000; inhibited by reagents which react with sulfhydryl groups (R. B. Wickner *et al.*, 1972a). Optimal ionic strength for polymerization: 80 m*M* KCl. It is unable to initiate synthesis without a primer or to utilize a nicked duplex as a template-primer for *in vitro* synthesis. The most suitable template-primer is double-stranded DNA containing short gaps (100 nucleotides or less); larger gaps are filled inefficiently, if at all (R. B. Wickner *et al.*, 1972b). The rate of synthesis on long gaps can be stimulated by the addition of the *E. coli* SSP, an effect not found for pols I or III (Sigal *et al.*, 1972; Molineux *et al.*, 1974). It can also form a complex with *E. coli* DNA SSP (Molineux and Gefter, 1974). Antiserum against pol I has no effect on pol II (Kornberg and Gefter, 1971; R. B. Wickner *et al.*, 1972a). Inhibited by low concentrations of arabinosylcytosine 5′-triphosphate—in contrast to pol I, which is unaffected, and to pol III, which is affected only at much higher concentrations (Reddy *et al.*, 1971). Twenty–100 molecules/cell (Kornberg and Gefter, 1971; R. B. Wickner *et al.*, 1972a). Turnover number approx. 250–1000 (in the presence of *E. coli* DNA SSP).

ii. Structural gene. Pol B, 2.5 min on *E. coli* genetic map (Taylor and Trotter, 1972).

iii. Catalytic properties. Similar to those described for pol I as far as is known (Gefter *et al.*, 1972; R. B. Wickner *et al.*, 1972a), but no 5′ → 3′ exonuclease has been found.

iv. Physiological role. Mutants with low levels of pol II activity *in vitro* have been isolated (Campbell *et al.*, 1972; Hirota *et al.*, 1972). As yet, no phenotypic differences between these mutants and their *pol* B^{+} parents have been found. However, in mutants deficient in pol I and III, pol II can serve in the repair of ultraviolet light-induced DNA lesions (Masker *et al.*, 1973).

c. Polymerase III (Pol III). Since this enzyme was first isolated only in 1970, and because of its structural complexity, low concentrations per cell, and instability, its exact structure is still a matter of some debate. We expect the details of its molecular structure to be clarified over the next few years.

At present, on the basis of subunit structure and the ability to utilize long single-stranded DNA as a template *in vitro,* three forms of the enzyme are recognized, namely, DNA pol II, DNA pol III*, and holoenzyme.

DNA Pol III

i. General properties. Molecular weight estimations vary between 140,000 and 180,000 (Kornberg and Gefter, 1972; Otto *et al.,* 1973; Livingston *et al.,* 1975). The subunit structure of pol III is not yet clear (Kornberg and Gefter, 1972; Livingston *et al.,* 1975). This form of the enzyme is active only on duplex DNA with short gaps, the efficiency falls off with large gaps (Kornberg and Gefter, 1972; Livingston and Richardson, 1975), and it is inactive in the *in vitro* conversion of ϕX174 or M13 primed single-stranded DNA to the double-stranded replicative form (see footnote 2) (Livingston and Richardson, 1975; W. Wickner *et al.,* 1973). When assayed under standard conditions similar to those employed with pol I and pol II, the optimal ionic strength is 20 m*M* salt (Kornberg and Gefter, 1972; Otto *et al.,* 1973; Livingston *et al.,* 1975). Sensitive to reagents reacting with sulfhydryl groups (Kornberg and Gefter, 1972; Otto *et al.,* 1973; Livingston *et al.,* 1975). Polymerizing activity can be stimulated by a variety of organic solvents (Kornberg and Gefter, 1972; Heinze and Carl, 1975). Approximately ten molecules per cell (Kornberg and Gefter, 1972; Otto *et al.,* 1973). Turnover number 15,000 (Kornberg and Gefter, 1972). Antiserum against pol I has no effect on pol III (Kornberg and Gefter, 1972).

ii. Structural gene. Pol C (formerly *dna E*), 2.5 min on the *E. coli* genetic map (Hirota *et al.,* 1972; Taylor and Trotter, 1972).

iii. Catalytic properties. Has all the catalytic activities of pol I (Gefter and Kornberg, 1972; Otto *et al.,* 1973; Livingston *et al.,* 1975; Livingston and Richardson, 1975; Gefter, 1974), except that it can only synthesize oligonucleotide (10–50 nucleotides) products (Livingston *et al.,* 1975; Tamblyn and Wells, 1975).

DNA Pol III* (Pol III*: Pol Three Star)

i. General properties. Molecular weight ca. 360,000, an oligomer of pol III (W. Wickner *et al.,* 1973; W. Wickner and Kornberg, 1974b). This form of the enzyme is very similar to DNA pol III and may be converted to pol III by heating, dilution, or aging. Pol III* alone can only synthesize on duplex DNA with short gaps. However, in association with copolymerase III*, a single polypeptide of 77,000 daltons, pol III* is able to synthesize on long, primed, single-stranded templates (W. Wickner *et al.*, 1973). Copolymerase III* has no known independent enzymatic activity, and it has no effect on the activity of pols I, II, or III.

Factor I (Hurwitz *et al.*, 1973; Hurwitz and Wickner, 1974) is probably identical to copolymerase III* (Hurwitz and Wickner, 1974; S. Wickner and Hurwitz, 1974).

Factor II (Hurwitz and Wickner, 1974) (MW 150,000) is probably distinct from factor I, and is not thermolabile when isolated from a *pol* C^{ts} mutant (Hurwitz and Wickner, 1974). Essential for the *in vitro* conversion of the primed SS DNA of phages fd or ϕX174 to duplex DNA (see footnote 2) (Hurwitz and Wickner, 1974; S. Wickner and Hurwitz, 1974; Livingston and Richardson, 1975). Factor II + pol III appear to be equivalent to pol III* (W. Wickner and Kornberg, 1974b; S. Wickner and Hurwitz, 1974).

Pol III* can be separated from pol III by size and by chromatography on phosphocellulose (W. Wickner *et al.*, 1973).

ii. Structural gene. Pol C (W. Wickner *et al.*, 1973; Taylor and Trotter, 1972).

iii. Catalytic properties. Unlike pol III, pol III* is able to synthesize on a primed, single-stranded molecule; this reaction requires ATP and spermidine in addition to the normal cofactors (W. Wickner and Kornberg, 1973).

HOLOENZYME (W. Wickner and Kornberg, 1974b)

i. General properties. Molecular weight ca. 330,000; holoenzyme can be dissociated into pol III* and copolymerase III* subunits by chromatography on phosphocellulose. Holoenzyme is able to convert primed single-stranded coliphage DNA templates to replicative forms without the addition of further copolymerase III*.

Holoenzyme has optimal activity at low salt concentrations and in the presence of phospholipid. Requires ATP for synthesis on long single-stranded DNA, and this activity is also sensitive to antibody directed against copolymerase III*. Synthesis on duplex DNA containing short gaps does not require ATP and is not sensitive to copolymerase III* antibody.

ii. Structural gene. The pol III subunit is a product of the *pol C* gene (Taylor and Trotter, 1972). No structural gene is known for the 77,000-dalton copolymerase III* subunit.

iii. Physiological role. Pol C mutants (Wechsler and Gross, 1971) cease DNA replication at elevated temperatures and are conditional lethal mutants. The *pol C* gene product has been shown to be responsible for pol III activity (Gefter *et al.*, 1971; Nüsslein *et al.*, 1971; Livingston *et al.*, 1975). Thus, pol III (in one of the three forms) is essential for replication of the bacterial chromosome.

In addition, pol III can participate in the excision repair of thymine dimers in DNA (Youngs and Smith, 1973).

2. Other Bacterial Polymerases

Three DNA polymerases have been found in *Bacillus subtilis* (Okazaki and Kornberg, 1964; Ganesan *et al.*, 1973a,b; Gass and Cozzarelli, 1973). They are designated pol I, pol II, and pol III. They have not been as well characterized as the *E. coli* polymerases; however, pol I has similarities with pol I of *E. coli*, although no 5′ → 3′ exonuclease activity has yet been detected (Gass *et al.*, 1971; Laipis and Ganesan, 1972), and pol III plays an essential role in DNA replication (Gass *et al.*, 1973; Cozzarelli and Low, 1973).

A polymerase with similarities to *E. coli* pol I, including a 5′ → 3′ exonuclease, has been isolated from *Micrococcus luteus* (Zimmerman, 1966; Miller and Wells, 1972).

3. T4 DNA Polymerase

i. General properties. A single polypeptide chain, molecular weight 102,000–115,000 (Nossal and Hershfield, 1971; Goulian *et al.*, 1968; Huang and Lehman, 1972). Contains 13–15 sulfhydryl groups and is inhibited by reagents reacting with them (Goulian *et al.*, 1968; Huang and Lehman, 1972). Strongly inhibited by arabinoside cytosine triphosphate (araCTP) (Reddy *et al.*, 1971). Optimum ionic strength 50 m*M* NaCl, strongly inhibited by 200 m*M* NaCl, needs Mg^{2+} for activity (Goulian *et al.*, 1968).

Requires primed single-stranded DNA as template-primer and efficiently fills short gaps (Masamune and Richardson, 1971) or copies long single-stranded stretches (longer than 10^4 nucleotides) (Goulian *et al.*, 1968; Englund, 1971). T4 polymerase binds to single-stranded DNA (Masamune *et al.*, 1971). Can only polymerize on nicked DNA in the presence of the T4 gene 32 (strand separation) protein (Nossal, 1974) and even then not on all templates (Huberman *et al.*, 1971). Gene 32 protein also stimulates the rate and extent of DNA synthesis on gapped DNA, especially under ionic conditions that would otherwise favor the stabilization of duplex DNA (Huberman *et al.*, 1971). Artifactual products (copies of both parental strands covalently joined) are formed on nicked duplex DNA in the presence of gene 32 protein, probably by the same mechanism as similar artifacts of *E. coli* pol I synthesis (see above) (Masamune *et al.*, 1971; Nossal, 1974). At T4 gene 32 protein concentrations that saturate the template, the 3′ → 5′ exonuclease of T4 polymerase is completely inhibited (Huang and Lehman, 1972). T4 DNA polymerase forms a complex with gene 32 protein under certain conditions (Huberman *et al.*, 1971).

ii. Structural gene. Gene 43 (DeWaard *et al.*, 1965) on T4 genetic map (Wood, 1974).

iii. Catalytic properties. Similar to *E. coli* pol I, so far as is known, except that it has no 5′ → 3′ exonuclease. The 3′ → 5′ exonuclease can also "proofread" (Cozzarelli *et al.*, 1969; Hershfield and Nossal, 1972; Nossal and Hershfield, 1973; Huang and Lehman, 1972; Brutlag and Kornberg, 1972).

iv. Physiological role. Essential for the replication of T4 DNA (Epstein *et al.*, 1963; Swart *et al.*, 1972). The polymerase activity and probably the 3′ → 5′ exonuclease activity play a role in maintaining the fidelity of T4 DNA replication (see Section I,E). In addition to its catalytic role, T4 DNA polymerase is a direct (negative) regulator of its own synthesis (Russel, 1973).

4. Eukaryotic Polymerases

i. General properties. DNA polymerases have been isolated from a wide variety of eukaryotic cells, e.g., calf thymus (Bollum, 1960), Ehrlich ascites tumor cells (Smellie *et al.*, 1960), HeLa cells (Bach, 1972), regenerating rat liver (Mantsaninos, 1964), *Tetrahymena pyriformis* (Crerar and Pearlman, 1971), *Saccharomyces cerevisiae* (Wintersberger and Wintersberger, 1970a), from mitochondria in rat liver (Meyer and Simpson, 1968) and *Saccharomyces cerevisiae* (Wintersberger and Wintersberger, 1970b); for a review, see Fansler (1974). Broadly speaking, these polymerases are of two types, namely, large (5–8 S) and small (3–4 S)[9], both of which can be found within the same cell. Major differences between them are as follows:

1. Size—the small polymerase from calf thymus is a single

[9] A standard scheme of nomenclature for vertebrate polymerases has been formulated by a group of workers in this field. According to this scheme, with a few exceptions, most large polymerases = DNA polymerase α, most small polymerases = DNA polymerase β, and mitochondrial polymerases = DNA polymerase mt (Weissbach *et al.*, 1975).

polypeptide of 45,000 daltons (3.4 S) (Chang, 1973a), while that from human blood lymphocytes is 30,000 daltons (Smith and Gallo, 1972). Two large polymerases from sea urchin embryos are 150,000 daltons each (Loeb, 1969), and one from human blood lymphocytes is also 150,000 daltons (Smith and Gallo, 1972).

2. Many eukaryotic polymerases are inactivated by reagents that react with sulfhydryl groups. Large polymerases are more sensitive to these reagents than small polymerases (e.g., Weissbach *et al.*, 1971; Smith and Gallo, 1972; Sedwick *et al.*, 1972).

3. Small polymerases have isoelectric points of 9.0, while those of large polymerases are 5.0 (e.g., Sedwick *et al.*, 1972; Smith and Gallo, 1972).

4. Cellular concentrations—the small polymerases remain relatively constant during the cell cycle, while the large polymerases increase parallel with DNA synthesis (e.g., Chang and Bollum, 1972a; Chang *et al.*, 1973).

5. There are two reports that oligoribonucleotides can serve as primers for DNA synthesis only with the large polymerase (Chang and Bollum, 1972b; Spadari and Weissbach, 1975).

The precise relationship between the large and small polymerases is not clear. The mammalian DNA polymerases appear to be related, since antiserum against the large polymerase of calf thymus inhibits both large and small DNA polymerases isolated from a number of other mammalian tissues (but not *E. coli* pol I) (Chang and Bollum, 1972c). This indicates that large and small polymerases may have polypeptide sequences in common. The isolation of enzymatically active 3.3 S subunits from a 6.8 S DNA polymerase has been reported (Tanabe and Takahashi, 1973; Lazarus and Kitron, 1973; Hecht and Davidson, 1973; Hecht, 1973); however, it is not yet clear whether this is a general phenomenon.

The large polymerases themselves, even within a particular cell type, do not appear to be homogeneous (Chang *et al.*, 1973; Persico *et al.*, 1973). This heterogeneity is apparent in affinity chromatography on denatured DNA, ion exchange chromatography, gel chromatography, and sedimentation behavior. The large polymerases thus may represent several distinct types of polymerase.

All eukaryotic polymerases require Mg^{2+} or Mn^{2+}. Many, perhaps all, also have a requirement for Zn^{2+} (Stavrianopoulos *et al.*, 1972; Slater *et al.*, 1971).

For both large and small polymerases, duplex DNA containing gaps is the best template (Loeb, 1969; Weissbach *et al.*, 1971; Smith and Gallo, 1972).

ii. Catalytic properties. Polymerization of nucleotides—the polymerization activity of the large and small polymerases is essentially similar to that of *E. coli* pol I.

$3' \rightarrow 5'$ Exonuclease activity has not yet been unambiguously established in association with any eukaryotic polymerases. Of three polymerases isolated from human KB cells, two incompletely purified polymerases exhibited $3' \rightarrow 5'$ exonuclease activities at 1–2% the level of the polymerizing activity, while the third, a small polymerase purified to homogeneity, was free of exonuclease activity (Sedwick *et al.*, 1972). Similarly, the 5–8 S and 3.4 S enzymes from calf thymus possess no significant $3' \rightarrow 5'$ exonuclease activity (Chang and Bollum, 1973), even when the 3.4 S polymerase is tested in the presence of a template with a mismatched terminal nucleotide (Chang, 1973b).

Pyrophosphate exchange, pyrophosphorolysis—the 5–8 S polymerase from calf thymus has both these activities, while the 3.4 S calf thymus polymerase has neither (Chang and Bollum, 1973).

iii. Physiological role. There is an increase in the amount of the large polymerases in parallel with DNA synthesis (Loeb and Agarwal, 1971; Chang *et al.*, 1973). The end of the DNA synthesis period is accompanied by a decline in the polymerase activity (O'Neil and Strohman, 1969; Friedman, 1970). In addition, a large polymerase is found in the cytoplasm of sea urchin eggs that translocates to the nucleus during DNA synthesis (Fansler and Loeb, 1972). The above observations suggest that the large polymerases, at least, may be involved in DNA replication.

5. *Reverse Transcriptases (RNA- and DNA-Directed DNA Polymerases)*

Enzymes which produce DNA from an RNA template have been characterized from various sources, including *E. coli* (Beljanski and Beljanski, 1974), a number of vertebrate cell types (reviewed in Temin and Mizutani, 1974), and in the virions of a large number of RNA viruses (reviewed in Temin and Mizutani, 1974; Green and Gerard, 1974). All of these activities have been referred to as reverse transcriptase, but only for certain viral enzymes, described below, is there evidence for a role in replication.

Avian Myeloblastosis Virus (AMV) and Rous Sarcoma Virus (RSV) Reverse Transcriptases

Both enzymes have very similar properties and therefore are considered together.

i. General properties. Molecular weight 160,000; consists of two subunits; α (60,000) and β (100,000) (Faras *et al.*, 1973a; Macian *et al.*, 1971; Grandgenett *et al.*, 1973; Verma *et al.*, 1974; Verma, 1975). The α subunit carries both polymerase and RNase H (see below) activities (Grandgenett *et al.*, 1973; Grandgenett and Green, 1974; Verma, 1975). The enzymes are serologically distinct from host cellular proteins (Nowinski *et al.*, 1972, Scolnick *et al.*, 1972) and form part of the mature virus particle (Bolognesi, 1974). Mg^{2+} and sulfhydryl-protecting reagents are required for activity (Leis and Hurwitz, 1972).

Homopolymer RNA and DNA can be used as template (Wells *et al.*, 1972; Kacian *et al.*, 1971; Goodman and Spiegelman, 1971; Scolnick *et al.*, 1972; Verma, 1975). Heteropolymer RNA and DNA can also be used as template (Goodman and Spiegelman, 1971; Verma *et al.*, 1972; Kacian *et al.*, 1971; Faras *et al.*, 1972). A primer with a free 3′-OH end that can be either RNA or DNA is always required for synthesis, and the newly synthesized DNA is covalently linked to the primer (Wells *et al.*, 1972; Faras *et al.*, 1972; Bishop *et al.*, 1973). Short, but not long, single-stranded regions of DNA can be copied (Keller, 1972; Hurwitz and Leis, 1972). Nicked duplex DNA cannot be used as template (Hurwitz and Leis, 1972). The presumed natural template of these enzymes is the viral genome. In the virion, this is found associated with low molecular weight RNA's, one of which, a 4 S tRNA of cellular origin, serves specifically as primer *in vitro* (Erikson and Erikson, 1971; Waters *et al.*, 1975; Faras *et al.*, 1973b; Faras and Dibble, 1975). *In vitro,* using viral RNA, a very large portion (greater than 85%) of the final product is a double-stranded DNA which represents only a small (5%) portion of the genome; other parts of the genome are also copied, but only to a very minor extent (Taylor *et al.*, 1972; Garapin *et al.*, 1973; Bishop *et al.*, 1973). In the presence of actinomycin D, the entire viral genome is copied with greater frequency (Bishop *et al.*, 1973; Garapin *et al.*, 1973). Incomplete copying is also observed on other templates, suggesting that it is a property of the enzyme (Bishop *et al.*, 1973; Taylor *et al.*, 1972). The fidelity of copying by the AMV enzyme on homopolymer templates is poor, and the error frequency can exceed 10^{-3} (Springgate *et al.*, 1973a; Battula and Loeb, 1974). This is also a feature of other reverse transcriptases (Sirover and Loeb, 1974). Double-stranded DNA is synthesized from RNA via intermediates that include RNA–DNA hybrids, some with single-stranded regions of DNA (Leis and Hurwitz, 1972; Taylor *et al.*, 1972). Duplex DNA synthesis is prevented by actinomycin D (Ruprecht *et al.*, 1973). The enzyme is inhibited by arabinosyl UTP and CTP (Taylor *et al.*, 1974) and also by several rifamycin derivatives (Green and Gerard, 1974).

The enzymes also possess an RNase H activity, which removes RNA specifically from RNA–DNA hybrids (Mölling *et al.*, 1971).

ii. Catalytic properties. The mechanism of polymerization is assumed to be similar to that described for pol I, except for the ability of the enzyme to utilize an RNA template. The RNase H is an exonuclease which acts in either a 5′ → 3′ or a 3′ → 5′ direction to produce, initially at least, oligonucleotides (Keller and Crouch, 1972; Baltimore and Smoler, 1972; Leis *et al.*, 1973; Grandgenett *et al.*, 1972; Grandgenett and Green, 1974). The α subunit acts randomly, but the $\alpha\beta$ holoenzyme acts processively (see footnote 8) (Grandgenett and Green, 1974; Leis *et al.*, 1973). The RNase H does not require concomitant polymerase activity (Grandgenett and Green, 1974; Leis *et al.*, 1973).

iii. Physiological role. ts mutants of RSV, in which the *ts* mutation has been shown to lie in the α subunit and to affect the RNA → DNA, DNA → DNA polymerase, and RNase H activities, do not replicate in their host (Linial and Mason, 1974; Verma *et al.*, 1974; Verma, 1975). RSV either may multiply in the host cell or integrate its genome into that of the host (Tooze, 1973). Both pathways are blocked in the reverse transcriptase mutants (Linial and Mason, 1974). There is evidence that the RNA genome is replicated via both RNA–DNA hybrids and duplex DNA intermediates (Green *et al.*, 1974; Varmus *et al.*, 1974). It is not yet clear whether reverse transcriptase is required for one or both of these steps or how full-length DNA copies of the genome could be produced *in vivo* by the enzyme alone (Green *et al.*, 1974; Leis *et al.*, 1975).

B. Polynucleotide Ligases

Ligases catalyze the synthesis of a phosphodiester bond between adjacent chains in duplex DNA. The reaction is coupled to the cleavage of the pyrophosphate bond of NAD or ATP. They have been found in *E. coli* (Gellert, 1967; Olivera and Lehman, 1967), T4-infected *E. coli* (Weiss and Richardson, 1967), and in a number of eukaryotic cell types (Lindahl and Edelman, 1968; Sambrook and Shatkin, 1969; Tsukada and Ichimura, 1971). The *E. coli* ligase requires NAD as cofactor (Olivera and Lehman, 1967; Zimmerman *et al.*, 1967), while the T4 and eukaryotic ligases require ATP (Weiss and Richardson, 1967; Lindahl and Edelman, 1968; Sambrook and Shatkin, 1969; Tsukuda and Ichimura, 1971). The *E. coli* and T4 enzymes have been studied the most extensively, and will be described as examples of an NAD-requiring ligase and an ATP-requiring ligase.

1. *E. coli* Ligase

i. General properties. Single polypeptide chain, molecular weight 77,000; amino acid composition known (Modrich *et al.*, 1973). Markedly stimulated by monovalent cations (NH_4^+ best) (Modrich and Lehman, 1973). Turnover number 25 (number of phosphodiester bonds formed at 30°C/min/molecule enzyme) (Modrich and Lehman, 1973). Two hundred–400 molecules/cell (Modrich *et al.*, 1973).

ii. Structural gene. Lig/dna L, 46 min on *E. coli* genetic map (Gottesman *et al.*, 1973; Nagata and Horiuchi, 1974; Taylor and Trotter, 1972).

iii. Catalytic properties. Catalyzes the formation of a phosphodiester bond between the 3′-OH group of a deoxyribonucleotide and the 5′-phosphoryl group of a directly adjacent deoxyribonucleotide, with NAD as cofactor.

The reaction takes place in a sequence of three steps (see Figure 6) involving two separate intermediates (from Lehman, 1974):

1. Nucleophilic attack of the ε-amino group of a lysine residue of the enzyme on the adenylyl phosphorus of NAD to form ligase–adenylate, with the release of NMN [upper pathway in stage (i), Figure 6].

2. The adenylyl group is transferred from the ligase–adenylate to the DNA to form a new pyrophosphate linkage between the adenosine

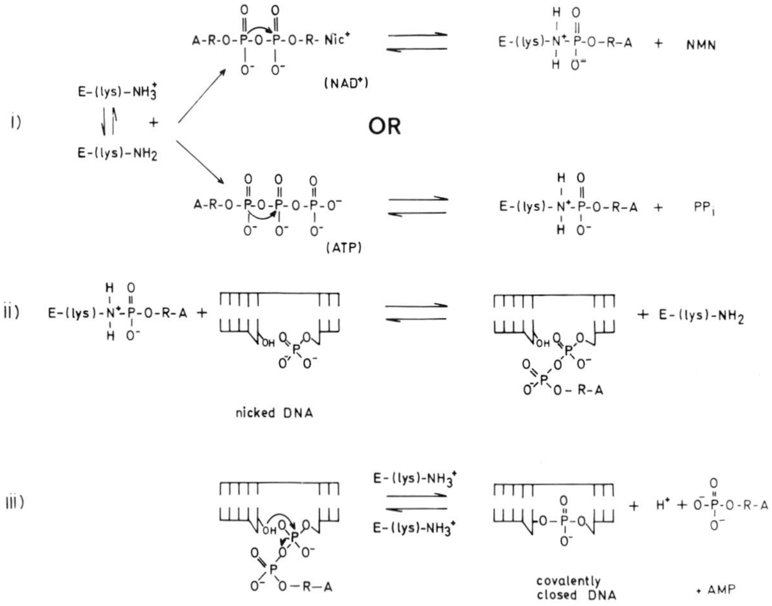

Fig. 6. Mechanism of DNA ligase reaction (for explanation see text). Key to abbreviations: A, adenosine; R, ribose; Nic, nicotinamide; E-(lys)-NH_2, enzyme; NMN, nicotinamine mononucleotide; PP_i, pyrophosphate. (Adapted from Lehman, 1974. Copyright 1974 by the American Association for the Advancement of Science.)

monophosphate and the 5′-phosphoryl group on one of the nucleotides at the single-strand scission in the DNA.

3. The 5′-phosphoryl group is attacked by the 3′-OH group on the other nucleotide at the nick to form a phosphodiester bond and so to close the nick in the DNA. AMP is released. For a discussion of the evidence supporting this mechanism, see Lehman (1974).

iv. Physiological role. Two conditional lethal mutants with a defective DNA ligase have been isolated—*lig ts* 7 (Pauling and Hamm, 1968, 1969; Modrich and Lehman, 1971) and *lig* 321(amber) (Nagata and Horiuchi, 1974). Under nonpermissive conditions, both are defective in DNA synthesis and accumulate newly synthesized DNA as Okazaki pieces (Pauling and Hamm, 1969; Gottesman *et al.*, 1973; Konrad *et al.*, 1973; Horiuchi *et al.*, 1975). Thus, ligase is probably essential for joining intermediates in discontinuous DNA synthesis. In addition, ligase is probably involved in the repair of certain kinds of DNA damage (Nagata and Horiuchi, 1974; Horiuchi *et al.*, 1975; Konrad *et al.*, 1973).

2. Phage T4 Ligase

i. General properties. Single polypeptide chain, molecular weight 68,000 (Panct *et al.*, 1973). Has an obligatory requirement for Mg^{2+} (Weiss *et al.*, 1968).

ii. Structural gene. Gene 30 on the T4 genetic map (Fareed and Richardson, 1967; Wood, 1974).

iii. Catalytic properties. Essentially the same as the *E. coli* ligase, except that the adenylate moiety of the enzyme–adenylate intermediate is derived from ATP [lower pathway in stage (i) of Figure 6] (Lehman, 1974).

iv. Physiological role. Apparently T4 ligase has the same function as that of *E. coli* ligase (Okazaki *et al.*, 1968). However, physiological complications make it unclear whether it is essential for T4 growth (Berger and Kozinski, 1969).

C. Strand Separation Proteins

These proteins have been isolated from a wide variety of sources, e.g., bacteriophages T7 (Reuben and Gefter, 1974) and fd—the fd SSP is also referred to as the fd gene 5 protein (Henry and Pratt, 1972)—adenovirus (Van der Vliet *et al.*, 1975), and calf thymus (Alberts *et al.*, 1971), and are known as strand-separation proteins or single-stranded DNA binding proteins. They are also referred to as DNA melting proteins or DNA unwinding proteins (see footnote 5), but these latter terms are somewhat misleading. Two of the most studied examples are described.

1. Bacteriophage T4 Gene 32 Protein

i. General properties. A single polypeptide, molecular weight 34,000–38,000 (Alberts *et al.*, 1968; Alberts, 1970; Carroll *et al.*, 1972), ellipsoid, longest dimension approximately 120 Å (Alberts *et al.*, 1968). Aggregates to form either relatively stable dimers or more unstable higher aggregates (Carroll *et al.*, 1972). Stimulates the DNA polymerizing activity (Huberman *et al.*, 1971; Nossal, 1974) and inhibits the $3' \rightarrow 5'$ exonuclease activity (Huang and Lehman, 1972) of T4 DNA polymerase. It does not affect the polymerizing activity of *E. coli* pol I (Huberman *et al.*, 1971). In the absence of DNA, forms a complex with T4 DNA polymerase, but not with *E. coli* pol I (Huberman *et al.*, 1971). Up to 10,000 molecules per cell may accumulate (Alberts and Frey, 1970).

ii. DNA binding properties. Binds cooperatively to single-stranded DNA; i.e., unbound molecules of gene 32 protein bind preferentially to single-stranded DNA contiguous to a previously bound molecule (Alberts *et al.*, 1968; Alberts and Frey, 1970). The predicted clustering has been observed electron microscopically (Delius *et al.*, 1972). Binds to homopolymers poly(dT), poly(dA), and poly(dI) (Alberts, 1971) and weakly to single-stranded RNA (Delius *et al.*, 1973). The cooperative binding of gene 32 protein is tight ($K_d < 10^{-9}$ *M;* Alberts and Frey, 1970) as compared to noncooperative binding, which is the only form of binding to double-stranded DNA (Alberts *et al.*, 1968). Noncooperative binding also occurs with single-stranded DNA at low protein concentrations (Alberts *et al.*, 1968; Alberts and Frey, 1970). Cooperatively bound protein can only be removed from DNA at salt concentrations greater than 0.6 *M* (Alberts *et al.*, 1968). At saturation binding, one molecule of gene 32 protein is bound per 10 nucleotides (Alberts and Frey, 1970; Alberts, 1971).

Single-stranded DNA normally adopts a coiled conformation (Studier, 1969), but when gene 32 protein is cooperatively bound, the conformation is rigid and extended (ten nucleotides extend to approximately 70 Å, Alberts and Frey, 1970; Delius *et al.*, 1972—compare this with the dimensions of the DNA duplex shown in Figure 1). The selective binding of gene 32 protein destabilizes the double helical form of DNA (see Section I,F). Destabilization leads to the denaturation (i.e., strand separation) of the duplex alternating copolymer poly (dA–dT) (Alberts and Frey, 1970) and the dAT-rich regions of the phage λ and T4 genomes (Delius *et al.*, 1972). Denaturation may be partially counteracted by divalent cations or polycations which stabilize double helices (Alberts and Frey, 1970; Delius *et al.*, 1972; Mahler and Mehrota, 1963).

At low temperatures, gene 32 protein increases the rate of renaturation of complementary DNA single strands, probably because the extended conformation makes all the bases available for base pairing (Alberts and Frey, 1970; Alberts, 1971).

iii. Structural gene. Gene 32 on the T4 genetic map (Alberts *et al.*, 1968; Wood, 1974).

iv. Physiological role. The gene 32 protein is essential for T4 DNA replication (Epstein *et al.*, 1963; Riva *et al.*, 1970; Alberts, 1971). Certain conditional lethal mutants of T4, defective in the gene 32 protein, stop DNA synthesis immediately (less than one minute) after transfer to nonpermissive conditions (Alberts, 1971). In addition, gene 32 protein is required for genetic recombination (Tomizawa *et al.*, 1966; Kozinski and Felgenhauer, 1967) and may protect T4 DNA from some form of nucleolytic breakdown (Kozinski and Felgenhauer, 1967; Littler, 1973). There is some evidence that gene 32 protein may regulate (negatively) the rate of its own synthesis (Krisch *et al.*, 1974).

2. *E. coli* Strand Separation Protein

i. General properties. Native molecular weight 78,000–90,000 (Molineux *et al.*, 1974; Weiner *et al.*, 1975), a tetramer of 18,500–22,000 dalton subunits (Molineux *et al.*, 1974; Sigal *et al.*, 1972; Weiner *et al.*, 1975). Heat-resistant, amino acid composition known, contains no cysteine, and is slightly acidic (pI = 6.0); hydrodynamic properties indicate an asymmetric molecule (Weiner *et al.*, 1975). Antibody to the *E. coli* protein cross-reacts slightly with the phage fd gene 5 protein but not with T4 gene 32 protein (Weiner *et al.*, 1975). Lowers the polymerizing activity of normal *E. coli* RNA polymerase on single-stranded, but not on double-stranded, DNA (W. Wickner and Kornberg, 1974a; Molineux *et al.*, 1974; Weiner *et al.*, 1975). Saturation binding to phage M13 single-stranded circular DNA limits RNA synthesis (by *E. coli* RNA polymerase or RNA polymerase III) (cf. Sections I,I and II,G,2) to a unique site (Tabak *et al.*, 1974; Geider and Kornberg, 1974). At saturation binding, it prevents synthesis by RNA pol III on ϕX174 single-stranded DNA (Wickner and Kornberg, 1974a). Stimulates polymerization by *E. coli* DNA pol III* or DNA pol III holoenzyme in the single-stranded phage systems (see footnote 2) (S. Wickner and Hurwitz, 1974; Geider and Kornberg, 1974; Scheckman *et al.*, 1974; Weiner *et al.*, 1975). Less stimulation (on G4 DNA) or none (on M13 DNA) is observed on substituting the T4 gene 32 protein or the fd gene 5 protein (Geider and Kornberg, 1974; Weiner *et al.*, 1975). Stimulates polymerization by *E. coli* DNA pol II on a gapped DNA template, but inhibits that by *E. coli* pol I or T4 DNA polymerase (Sigal *et al.*, 1972; Molineux *et al.*, 1974). Stimulates the 3′ → 5′ exonuclease of *E. coli* DNA pol II, but inhibits those of *E. coli* DNA pols I and III (Molineux and Gefter, 1974). In the absence of DNA, forms a specific complex with *E. coli* DNA pol II (Molineux and Gefter, 1974). Two hundred–300 molecules per cell, enough to bind to 10^4 nucleotides (Sigal *et al.*, 1972; Weiner *et al.*, 1975).

ii. DNA binding properties. Binds tightly and cooperatively to single-stranded DNA (cf. gene 32 protein), one molecule covering 32 nucleotides; the binding is inhibited by high salt or spermidine (Sigal *et al.*, 1972; Weiner *et al.*, 1975). Binds to homopolymers poly(dT) and poly(dI), but in contrast to T4 gene 32 protein not to poly(dA) (Weiner *et al.*, 1975). Catalyzes the denaturation of dAT-rich regions of duplex DNA at 37°C (Sigal *et al.*, 1972).

iii. Physiological role. There is no direct evidence to show that this protein is required for DNA replication in *E. coli*.

D. DNA "Swivelases"

"Swivelase" activity is detected *in vitro* by removal of superhelical twists from covalently closed circular DNA. A superhelical twist is a twist in the entire helical DNA molecule (see Figure 7) and occurs in a covalently closed circular DNA molecule (or a linear DNA molecule if any restrictions to rotation exist) when the number of helical turns per base pair is greater or smaller than that found in a relaxed circle (see Figure 7). By convention, the sense of the twists found when the helical turns per base pair are greater than in the relaxed form is termed positive, and when it is smaller, negative (see Figure 7). All supercoiled circles isolated from nature have been found to be negatively supercoiled (Vinograd *et al.*, 1968; Bauer and Vinograd, 1968; Wang, 1969b), but positively supercoiled molecules can be produced in the laboratory (Wang, 1969a, 1971a). Removal of supercoils can be measured sensitively by alterations in the sedimentation behavior of closed circles in the presence of intercalating dyes (Crawford and Waring, 1967). "Swivelase" activity is relevant to DNA replication, since strand separation tends to cause the formation of positive superhelices (see Section I,G). Two activities capable of removing superhelices have been described. One is from *E. coli* (ω protein) (Wang, 1971b), and one is from secondary mouse embryo cells (Champoux and Dulbecco, 1972).

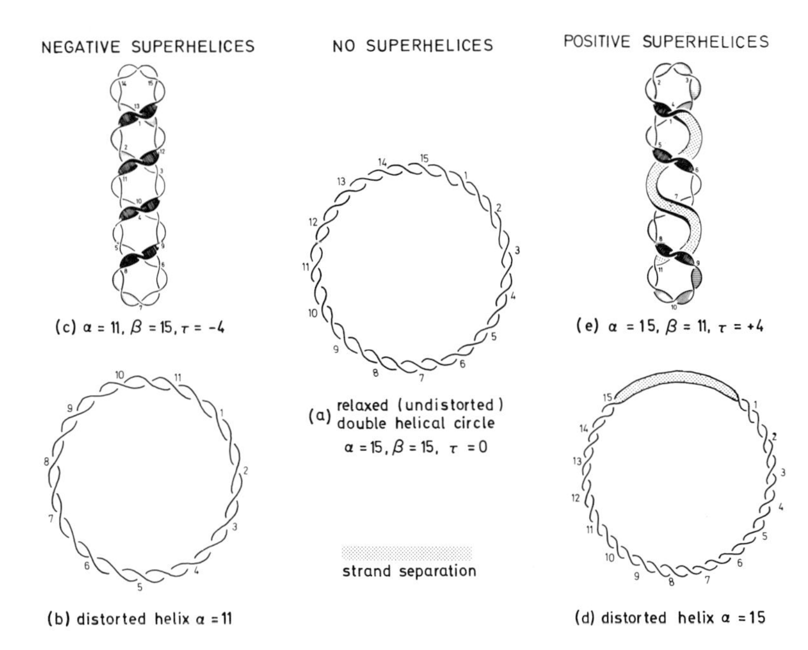

Fig. 7. Positive and negative superhelices. The sign and number of superhelices in a covalently closed circular DNA molecule can be derived from the following relationship (Vinograd *et al.*, 1968):

$$\tau = \alpha - \beta$$

where τ = the number of superhelices; α = the number of 360° revolutions of one strand about the duplex axis in a closed circular molecule when the duplex axis is constrained to lie in one plane; β = the number of 360° revolutions of one stand about the duplex axis when the molecule is nicked or not constrained to lie in one plane.

By convention, when τ is negative the sense of twist of the superhelix is taken to be right-handed and when τ is positive, left-handed. The consequences of this relationship on the physical structure of a hypothetical covalently closed double helical molecule with 15 double helical turns is illustrated. No superhelices: (a) A relaxed, covalently closed circle containing 15 turns. Negative superhelices: (b) The same circle as in (a) but with four duplex turns (360°) removed (this transformation would require the operations of unwinding, cutting, and sealing, not shown here for the sake of clarity). When the circle is constrained to lie in a plane as shown here, the double helix is distorted and contains only 11 duplex turns. (c) The same circle as in (b) allowed to adopt a three-dimensional configuration. Fifteen duplex turns can now occur and are compensated for by four negative superhelical turns. Note the right-handed direction of twist. Positive superhelices: (d) The strands are separated over a length equivalent to four duplex turns in the relaxed circle shown in (a). When constrained to lie in one plane, as shown here, the double helix is distorted, but still contains 15 turns. (e) The same circle as in (d) is allowed to adopt a three-dimensional configuration. Eleven duplex turns of the same dimensions as in the relaxed circle can now occur and the remaining number of 360° turns is made up by four positive superhelical turns. Note the left-handed sense of the superhelix.

1. ω Protein

i. General properties. Molecular weight approximately 100,000 (Carlson and Wang, 1974). Has no cofactor requirements and no DNA polymerase or DNA ligase activity (Wang, 1971b). Specific for DNA with negative superhelical turns (Wang, 1971b, 1973). The reaction with DNA is very sensitive to temperature and is inhibited by single-stranded DNA, native DNA being much less inhibitory and tRNA being without effect (Wang, 1971b).

ii. Structural gene. Not known.

iii. Catalytic properties. Removal of negative superhelical twists from covalently closed DNA. The DNA remains covalently closed and no physical change can be detected other than a reduction in negative superhelices (Wang, 1971b, 1973). Does not bind irreversibly to DNA (Wang, 1973). The loss of superhelices follows a gradual course rather than a one-hit mechanism (Wang, 1971b, 1973). This is consonant with a transient and movable swivel (Wang, 1973). A mechanism for the action of a "swivelase" has been proposed (Wang, 1971b; Champoux and Dulbecco, 1972): The enzyme generates a nick and remains bound at the site of the nick. While this structural intermediate lasts, the helix is free to rotate around the single phosphodiester bond in the strand opposite the nick. Thus, any constraint in the DNA relative to the nicked, circular form will be removed. Reversal of the nicking action regenerates the original closed circular DNA molecule.

iv. Physiological role. Unknown, but see Section I,G.

2. The Activity from Mouse Embryo Cells

This activity is very similar to ω protein; however, it removes both positive and negative superhelices from DNA (Champoux and Dulbecco, 1972). Therefore, it could act *in vivo* to remove from parental DNA the constraints which, as a consequence of the topology of the DNA helix, arise as the two parental strands separate.

E. Genetically Defined Factors Involved at the Growth Point in *E. coli* DNA Replication

1. dna B Gene Product

i. General properties. Molecular weight: 250,000 (Wright *et al.*, 1973), or 48,000 in the presence of SDS (S. Wickner *et al.*, 1974). The native molecule is probably an oligomer (S. Wickner *et al.*, 1974; Lark and Wechsler, 1975). Not affected by reagents which react with sulfhydryl groups [i.e., NEM (*N*-ethylmaleimide) Wright *et al.*, 1973]. Required for the *in vitro* SS → RF conversion of φX174 (see footnote 2) (R. B. Wickner *et al.*, 1972b; Wright *et al.*, 1973; Scheckman *et al.*, 1974, 1975). Forms a complex with the *E. coli dna C* gene product (S. Wickner and Hurwitz, 1975). About ten molecules per cell (S. Wickner *et al.*, 1974).

ii. Structural gene. dna B, 79 min on the *E. coli* genetic map (Wechsler and Gross, 1971; Taylor and Trotter, 1972). The *dna B* locus appears to be a single cistron (Wechsler, 1973).

iii. Catalytic properties. Contains a DNA-dependent ribonucleoside triphosphatase (rNTP) which is specific for rNTP's and is unaffected by NEM. It cleaves rNTP to rNDP + P_i and is stimulated more effectively by SS- than duplex DNA (Wright *et al.*, 1973; S. Wickner *et al.*, 1974). There is also a DNA-independent rNTPase activity which has 10% of the activity of the DNA-dependent rNTPase (S. Wickner *et al.*, 1974). The DNA-independent rNTPase is inhibited in the complex with the *dna C* gene product (S. Wickner and Hurwitz, 1975).

iv. Physiological role. Essential for DNA replication in *E. coli* (Gross, 1972). Required for the *in vivo* SS → RF conversion of phage φX174 (see footnote 2) (Steinberg and Denhardt, 1968; Dumas and Miller, 1974). "4 S" DNA pieces (see footnote 3) accumulate in mutants severely restricted by temperature (Lark and Wechsler, 1975). No *in vivo* role is known for the *dna B*-associated rNTPase.

2. dna C *Gene Product*

Previously thought to be involved only in initiation, but probably also involved at the growth point (see Section II,G,3).

3. dna G *Gene Product*

i. General properties. Molecular weight 60,000 (W. Wickner *et al.*, 1973; Bouché *et al.*, 1975). Resistant to NEM (S. Wickner *et al.*, 1973a). Binds weakly to single- and double-stranded DNA (S. Wickner *et al.*, 1973a). Required for *in vitro* DNA synthesis in the SS → RF conversion of phage φX174 and G4 DNA (see footnote 2) (R. B. Wickner *et al.*, 1972b; S. Wickner *et al.*, 1973a; S. Wickner and Hurwitz, 1974; Scheckman *et al.*, 1974, 1975; Zechel *et al.*, 1975). For φX174 DNA synthesis during the SS → RF conversion, a number of other proteins are required (S. Wickner and Hurwitz, 1974; Scheckman *et al.*, 1975). With phage G4 DNA (see footnote 2), apart from pol III holoenzyme and the *E. coli* SSP, only the *dna G* gene product is required and appears to be necessary for initiation of DNA synthesis (Zechel *et al.*, 1975; Bouché *et al.*, 1975). Approximately ten molecules/cell (S. Wickner *et al.*, 1974).

ii. Structural gene. dna G, 60–65 min on the *E. coli* genetic map (Wechsler and Gross, 1971; Taylor and Trotter, 1972).

iii. Catalytic properties. May be a rifampicin-insensitive RNA polymerase, which *in vitro* is dependent upon strand-separation protein for activity (Bouché *et al.*, 1975).

iv. Physiological role. Essential for DNA replication in *E. coli* (Gross, 1972). Required for the *in vivo* SS → RF conversion of phage ϕX174 (see footnote 2) (McFadden and Denhardt, 1974). At the nonpermissive temperature, the synthesis of Okazaki pieces is inhibited (Lark, 1972a; Klein *et al.*, 1973; Louarn, 1974). This inhibition may be at the level of the start of synthesis of Okazaki pieces.

4. *dna S* Gene Product

i. General properties. Defined by mutation only, not essential for DNA synthesis, but under nonpermissive conditions very low molecular weight DNA pieces are produced instead of Okazaki pieces (see footnote 3) (Konrad and Lehman, 1975).

ii. Structural gene. dna S, 72 min on the *E. coli* genetic map (Konrad and Lehman, 1975; Taylor and Trotter, 1972).

5. *dna Z* Gene Product

i. General properties. Defined by mutation only (Filip *et al.*, 1974; Truitt and Walker, 1974). Conditional lethal; at the nonpermissive temperature, the rate of DNA synthesis decreases immediately (Filip *et al.*, 1974).

ii. Structural gene. Maps away from *dna B, G,* or *E* (Filip *et al.*, 1974).

F. Other Factors

The *in vitro* SS → RF conversion of phage ϕX174 DNA (see footnote 2) requires, other factors in addition to the known genetically defined factors and the *E. coli* SSP. These have been isolated and physically characterized by two groups of workers. They are called X, Y, and Z (S. Wickner and

Hurwitz, 1974, 1975) or i and n (Scheckman *et al.*, 1974, 1975); X is probably the same as i (Scheckman *et al.*, 1975).

G. Factors Involved in the Initiation of DNA Replication

1. *Replicators*

The origin of replication (replicator) of a number of prokaryotes has been localized, by a variety of methods, to a single, small, physically or genetically defined region of the genome. Organisms whose origins of replication are known include the bacteria *E. coli* (Bird *et al.*, 1972), *B. subtilis* (Yoshikawa and Sueoka, 1963; O'Sullivan *et al.*, 1975), the *E. coli* plasmid Col E1 (Tomizawa *et al.*, 1974; Inselberg, 1974; Lovett *et al.*, 1974), the *E. coli* bacteriophages λ (Schnös and Inman, 1970; Dove *et al.*, 1971; Hayes and Szybalski, 1973b), and T7 (Dressler *et al.*, 1972), as well as the animal viruses SV40 (Fareed *et al.*, 1972; Nathans and Danna, 1972) and polyoma (Crawford *et al.*, 1973; Griffin *et al.*, 1974). In the cases of Col E1 plasmid and polyoma, the accuracy of the mapping indicates that initiation takes place within a region less than 200 base pairs long. In the case of bacteriophage λ, mutants designated either *ori* (Dove *et al.*, 1971; Stevens *et al.*, 1971) or *rep* (Rambach, 1973) have been isolated that are replication defective as the result of a *cis*-acting mutation mapping at, or near, the origin of replication. These mutants probably have defective replicators (Dove *et al.*, 1971; Stevens *et al.*, 1971; Rambach, 1973).

2. *E. coli RNA Polymerase*

i. General properties. The enzyme usually isolated is a multimeric molecule known as holoenzyme (Burgess, 1969), with the structure $\alpha_2\beta\beta'\sigma$, where the Greek characters refer to characterizable subunits. Catalyzes the formation of polyribonucleotides on a single- or double-stranded DNA template using ribonucleoside triphosphates in a reaction absolutely dependent on Mg^{2+} or Mn^{2+} and on Zn^{2+}; the latter is tightly bound to the enzyme (Scrutton *et al.*, 1971). Molecular weight of subunits: α, 39,000–41,000; β, 145,000–150,000; β', 150,000–165,000, and σ, 86,000–95,000, and that of the complete enzyme between 460,000 and 500,000 (Chamberlin, 1974a,b). The holoenzyme initiates RNA synthesis on intact duplex DNA at specific sites (promotors). Promotors recognized *in vivo* are also recognized *in vitro* (see, e.g., Blattner and Dahlberg, 1972), but some additional promotors are apparently recognized only *in vitro* (Chamberlin and Ring, 1972; Minkley and Pribnow, 1973). The σ subunit is necessary for promotor recognition (Burgess, 1969; Berg *et al.*, 1971); in its absence, the $\alpha_2\beta\beta'$ apoenzyme, known as core enzyme (Burgess,

1969), still synthesizes RNA, but without any specificity and extensively only on SS–DNA or nicked duplex DNA (Hinkle *et al.*, 1972). Approximately 2000–7000 molecules per cell (Matzura *et al.*, 1973; Iwakura *et al.*, 1974).

Apart from the normal holoenzyme, other forms of RNA polymerase have been found, both in normal *E. coli* cells (Chao and Speyer, 1973; W. Wickner and Kornberg, 1974a) and in phage-infected cells (Goff, 1974; Brown and Cohen, 1974). These forms have altered subunit composition and/or altered template specificity. One such modified form, known as RNA pol III, discriminates *in vitro* between the SS–DNA of phages M13 and ϕX174 in the presence of the *E. coli* strand-separation protein. In contrast to normal holoenzyme, the modified form does not synthesize on ϕX174 SS–DNA (W. Wickner and Kornberg, 1974a).

ii. Catalytic properties. Synthesis by RNA polymerase holoenzyme may be conveniently considered in four stages: (a) site recognition and template binding; (b) RNA chain initiation; (c) RNA chain elongation; (d) RNA chain termination.

a. Site recognition and template binding. This noncovalent interaction depends both upon the enzyme and the sequence of DNA in the promotor. Promotors may be functionally defined as regions containing an RNA polymerase recognition site and a polymerase-binding site, at which RNA chains are initiated (Reznikoff, 1972; Dickson *et al.*, 1975). These two sites may be separate (Zillig *et al.*, 1970; Blattner *et al.*, 1972; Dickson *et al.*, 1975).

The RNA polymerase-binding sites are about 40 nucleotides long, and RNA synthesis initiates about half-way along their length (Schaller *et al.*, 1975; Dickson *et al.*, 1975). Under standard conditions, holoenzyme binds to promotors very tightly ($t_{\frac{1}{2}}$ dissociation ca. 30–60 hr) (Hinkle and Chamberlin, 1972). Tight binding is favored by agents which destabilize the double helix and probably involves strand separation over a region of 6–8 base pairs (Saucier and Wang, 1970; Mangel and Chamberlin, 1974). Under the same standard conditions, the core enzyme binds tightly ($t_{\frac{1}{2}}$ dissociation ca. 20–60 min), but the holoenzyme weakly ($t_{\frac{1}{2}}$ dissociation less than one second) to nonpromotor regions (Hinkle and Chamberlin, 1972). Thus, σ subunit confers specificity by allowing the enzyme to dissociate and rebind rapidly until it encounters a proper promotor region.

b. Chain initiation. After binding to the template, the enzyme catalyzes the formation of an initial dinucleotide whose 5′ terminal member is either adenosine or guanosine (Maitra *et al.*, 1967; Terao *et al.*, 1972). The 5′ terminal nucleotide is normally a triphosphate *in vivo*, but *in vitro* it can be provided by 5′-mono- or diphosphate (Krakow and Fronk, 1969). The initiation reaction can be represented as follows:

$$(p)(p)p\ A/G + pppN \rightarrow (p)(p)pA/G\ pN + PP_i$$

It has been proposed that the enzyme contains a specific site for the binding of a 5′-terminal adenosine or guanosine triphosphate (Wu and Goldthwait, 1969). Binding of purine rNTP at this site does not require Mg^{2+} and is relatively weak (K_d 1–2 × 10^{-4} M); the site has an even lower affinity for di- and monophosphates, and pyrimidine nucleotides are not detectably bound. Binding at this site is blocked by rifampicin (Wu and Goldthwait, 1969); the binding site is probably located on the β subunit (Zillig *et al.*, 1970).

c. Chain elongation. After initiation (the formation of the first phosphodiester bond), a ternary complex of DNA-growing RNA chain-protein exists, which is stable under conditions in which holoenzyme alone cannot bind strongly to DNA (Richardson, 1966; Chamberlin, 1974a). In this complex, the enzyme catalyzes the addition of rNMPs (with bases complementary to one strand of the DNA template) to the 3′-OH end of the growing RNA chain. The mechanism of the reaction is known in detail (Chamberlin, 1974b; Rhodes and Chamberlin, 1974). The enzyme does not dissociate from this complex and therefore continues to add nucleotides processively (see footnote 8) until chain termination occurs. Elongation is insensitive to rifampicin but is inhibited by the drug streptolidigin (Chamberlin, 1974a). The observed rate of elongation, both *in vitro* and *in vivo,* is in the range of 20–50 nucleotides/sec at 37°C (Richardson, 1970; Bremer and Yuan, 1968).

d. Chain termination. Termination involves the cessation of polymerization, and normally involves the release of newly synthesized RNA and the enzyme from a duplex DNA template (Chamberlin and Berg, 1963). It depends on specific sequences in the DNA template (see, e.g., Dahlberg and Blattner, 1973), but is also influenced by ionic conditions and some DNA-binding proteins (Chamberlin, 1974a; Roberts, 1969). The product RNA is usually released from duplex DNA, but on a supercoiled template or on SS–DNA it may remain noncovalently bound (Chamberlin and Berg, 1963; Champoux and McConaughty, 1975).

iii. Structural gene. The β and β' subunits map at 79 min on the *E. coli* genetic map (Khesin *et al.*, 1969; Austin, 1974; Taylor and Trotter, 1972). The location of the other subunits is not known.

iv. Physiological role. Escherichia coli RNA polymerase has been implicated in the initiation of DNA synthesis, both of *E. coli* itself and of some of its phages and plasmids (Section I,I). In addition, the enzyme is necessary for gene expression within the cell.

3. *Factors Involved in the Initiation of DNA Replication of E. coli*

All of these factors, except *dna C*, are defined by mutation only. The mutant phenotype is a temperature-sensitive conditional lethal. Four criteria by which a defect can be identified as participating in initiation of DNA replication are as follows:

1. When mutants are shifted from permissive to nonpermissive conditions: (a) only DNA synthesis should be inhibited (i.e., not RNA or protein synthesis); (b) DNA synthesis in an asynchronously growing population should stop in approximately the time it takes to replicate a replicon (i.e., they are "slow-stopping" DNA synthesis mutants in comparison to those with defective growth points);
2. When DNA synthesis in such mutants has ceased as outlined in (a) and (b) above and they are returned to permissive conditions: (c) reinitiation of DNA synthesis should occur at a replicator. Satisfaction of this criterion is usually inferred if reinitiation occurs only at one site on each genome; (d) integration of a replicon defective in initiation within one that is competent heterologous, should, under certain conditions, lead to phenotypic suppression of the initiation defect. This phenomenon, known as integrative suppression (Nishimura *et al.*, 1971), will occur if the competent replicon can provide a replicator and a suitably regulated initiator (Pritchard *et al.*, 1969; Timmis *et al.*, 1974). In *E. coli*, an integrated F sex factor often provides integrative suppression.

The stage of initiation at which the defective gene product is active can be further defined by the sensitivity of reinitiation [see point (c) above] to inhibitors of RNA or protein synthesis. If reinitiation occurs in the presence of an RNA synthesis inhibitor, then the step affected by the mutation does not require RNA synthesis. The alternative result, that reinitiation is prevented by RNA or protein synthesis inhibitors, does not lead to simple conclusions; either the mutant gene product could be irreversibly denatured under nonpermissive conditions and requires resynthesis, or the mutant gene product could be responsible for the synthesis of RNA or protein essential for initiation. General methods for determining the order of action of gene products required for initiation are discussed by Hereford and Hartwell (1974).

Gene products, which have been characterized according to some or all of the above criteria, are discussed below.

a. *dna A* Gene Product. *i. Structural gene. dna A*, 73 min on the *E. coli* genetic map (Hirota *et al.*, 1970; Beyersmann *et al.*, 1974; Taylor and Trotter, 1972).

ii. Physiological role. Satisfaction of criteria (a)–(c) listed above has been observed by Hirota *et al.* (1970) and Abe and Tomizawa (1971). Integrative suppression has also been found (Nishimura *et al.*, 1971; Beyersmann *et al.*, 1974). At the nonpermissive temperature, phage M13 replicates abnormally (Bouvier and Zinder, 1974), and some membrane properties are altered in *dna A* mutants (Hirota *et al.*, 1970; Bell *et al.*, 1972). Reinitiation is blocked by protein synthesis inhibitors (Kuempel, 1969; Hirota *et al.*, (1970) (see above for the significance of this last observation).

b. *dna C* Gene Product. *i. General properties.* The enzyme is required for the *in vitro* SS → RF conversion of phage ϕX174 (see footnote 2) and has been partially purified; it is sensitive to reagents reacting with sulfhydryl groups (S. Wickner *et al.*, 1973b).

ii. Catalytic properties. Forms a complex with the *E. coli dna B* gene product and inhibits its DNA-independent ATPase activity (S. Wickner and Hurwitz, 1975).

iii. Structural gene. dna C, 39 min on the *E. coli* genetic map (Wechsler and Gross, 1971, Taylor and Trotter, 1972). All mutants previously designated as "*dna D*" have also been shown to be *dna C* mutants (Wechsler, 1975).

iv. Physiological role. Some mutants with defective *dna C* products fulfill the criteria (a)–(c) outlined above (Carl, 1970; Wolf, 1972; Schubach *et al.*, 1973). Mutants with other *dna C* alleles, however, stop DNA synthesis immediately on transfer to nonpermissive conditions (Wechsler, 1975). The *dna C* gene product of the fast-stopping mutants is more defective than that of the slow-stopping mutants when assayed by *in vitro* SS → RF ϕX174 conversion (S. Wickner *et al.*, 1973b). Reinitiation is resistant to rifampicin (Hiraga and Saitoh, 1974). Both fast- and slow-stopping mutants are subject to integrative suppression (Beyersmann *et al.*, 1974). The above observations may be explained if the *dna C* product is involved both in initiation and in growth point function.

c. *dna P* Gene Product. *i. Structural gene. dna P* approximately 75 min on the *E. coli* genetic map (Wada and Yura, 1974; Taylor and Trotter, 1972).

ii. Physiological role. Criteria (a)–(c) listed above are satisfied; "integrative suppression" has not yet been tested; some altered membrane properties; reinitiation resistant to rifampicin (Wada and Yura, 1974).

d. *dna H* Gene Product. *i. Structural gene. dna H* approximately 54 min on the *E. coli* genetic map (Sakai *et al.,* 1974, Taylor and Trotter, 1972).

ii. Physiological role. Criteria (a)–(c) listed above are satisfied; reinitiation blocked by protein synthesis inhibitors; integrative suppression not yet tested (Sakai *et al.,* 1974).

e. *dna "I"* Gene Product. *i. Structural gene. dna "I"* between 27 and 38 min on the *E. coli* genetic map (Beyersmann *et al.,* 1974, Tayler and Trotter, 1972).

ii. Physiological role. Not yet fully characterized, but stops DNA synthesis slowly (Beyersmann *et al.,* 1974).

4. Factors Involved in the Initiation of DNA Replication in Other Bacteria

A number of *B. subtilis* gene products, defined by mutation, have been implicated in initiation by fulfillment of some of the criteria outlined in Section II,G,3 above (see, e.g., Mendelson and Gross, 1967; Riva *et al.,* 1975).

5. Bacteriophage DNA Replication Initiation Factors

Conditional lethal mutants defective in these factors can efficiently infect cells under nonpermissive conditions, but no detectable replication of duplex DNA follows.

a. Bacteriophage λ *O* and *P* Gene Products. *i. Structural genes. O* and *P,* close to the origin of λ replication (Campbell, 1961; Szybalski, 1974).

ii. Physiological role. Required for initiation of the first round of replication after infection (Ogawa and Tomizawa, 1968; Dove *et al.,* 1971) and also for subsequent replication (Takahashi, 1975). The *O* and *P* gene products may interact with one another (Tomizawa, 1971), and the *P* gene product may also interact with the host (*E. coli*) *dna B* gene product (Georgopoulos and Herskowitz, 1971). An unstable nuclease activity, to which λ genomes containing mutations in the putative replicator are partially refractory, is controlled by genes *O* and *P* (Freifelder and Kirschner, 1971; Inokuchi *et al.*, 1973), but it is not known how direct this control is. The *O* and *P* gene products are essential for the synthesis of "oop" RNA (Hayes and Szybalski, 1973b) that has been implicated in the initiation of λ DNA replication (see Section I,I).

b. Bacteriophage φX174 Gene A Product. *i. General properties.* Molecular weight (in SDS) 55,000–65,000 (Henry and Knippers, 1974). A 35,000-dalton product (not a degradation product) whose sequence overlaps with that of the 55,000–65,000-dalton product is also synthesized from the gene *A* cistron (Linney and Hayashi, 1973). Both products bind strongly to SS-DNA (Linney *et al.*, 1972), but catalytic activity has only been found in the large product (Henry and Knippers, 1974). The protein is hydrophobic and may be located in the membrane (Van der Mei *et al.*, 1972).

ii. Catalytic properties. The high molecular weight polypeptide has an endonuclease activity highly specific for the viral strand (the single strand of DNA found in the virus particle) of φX174 DNA, either as a single strand or when it occurs in the covalently closed supercoiled duplex form of φX174 DNA; other covalently closed double stranded DNA circles (e.g., SV40, Col E1) are not attacked; the protein acts once and is then inactive (Henry and Knippers, 1974).

iii. Structural gene. Gene A on the φX174 genetic map (Sinsheimer, 1974).

iv. Physiological role. The gene product acts in the "cis" position *in vivo* and is required for initiation when the phage replicates as duplex DNA (Lindqvist and Sinsheimer, 1967). When DNA replication is blocked by a defect in the host replication enzymes, the φX174 genome accumulates as duplex DNA circles. In *E. coli* cells infected with gene A^+, but not with gene A^- phage, these circles contain a single endonucleolytic cut in the viral strand (Francke and Ray, 1971). Whether this exposed nick is required for initiation is not, however, clear.

6. *Animal Virus DNA Replication Initiation Factors*

Possible initiation mutants of SV40–complementation group A (Tegtmeyer and Ozer, 1971; Chou *et al.*, 1974) and polyoma–tsa complementation group (Fried, 1965; Francke and Eckhart, 1973) have been isolated. "Nicking" accompanying initiation does not appear to occur in these viruses (Sebring *et al.*, 1971).

7. *Factors Involved in the Initiation of DNA Replication in Eukaryotes*

Three temperature-sensitive replication mutants *cdc 28, cdc 4,* and *cdc 7* have been isolated from the yeast *Saccharomyces cerevisiae* (Hartwell, 1971, 1973). If these mutants are transferred to the nonpermissive temperature after having initiated DNA synthesis, replication of the genome is completed before the block in DNA synthesis can be observed. A comparison with the "slow-stopping" replication mutants of *E. coli* (see Section II,G,3 above) suggests that these mutants are, therefore, initiation defective (Hartwell, 1973). Sequential gene function in the order *cdc 28, cdc 4, cdc 7* is required for initiation (Hereford and Hartwell, 1974).

ACKNOWLEDGMENTS

We would like to thank Drs. J. H. Weiner and J. A. Wechsler, and Professor C. C. Richardson for sending us copies of their work prior to publication. Thanks are also due to Mr. G. Berger for drawing the figures.

RECOMMENDED READING

In addition to the references cited in the text, for more detailed and extensive treatments of the phenomenology and enzymology of DNA replication than is possible in this short chapter, the reader is referred to the following sources:

General

Gefter, M. L. (1975). DNA replication. *Annu. Rev. Biochem.* **44,** 45.
Kornberg, A. (1974). "DNA Synthesis." Freeman, San Francisco, California.

Enzymology

Detailed descriptions of the DNA polymerases of bacteria, eukaryotes and RNA tumor viruses, of DNA ligases, and of RNA polymerases are in "The Enzymes" (P. D. Boyer, ed.), 3rd ed., Vol. 10. Academic Press, New York, 1974.

REFERENCES

Abe, M., and Tomizawa, J.-I. (1971). Chromosome replication in *Escherichia coli* K12 mutant affected in the process of DNA initiation. *Genetics* **69,** 1.
Alberts, B. M. (1970). Function of gene 32 protein, a new protein essential for the genetic

recombination and replication of T4 bacteriophage DNA. *Fed. Proc., Fed. Am. Soc. Exp. Biol.* **29,** 1154.

Alberts, B. M. (1971). On the structure of the replication apparatus. *Nucleic Acid–Protein Interact. Nucleic Acid Synth. Viral Infect. Proc. Miami Winter Symp., 1971* Vol. 2, p. 128.

Alberts, B. M. (1973). Studies on the replication of DNA. *In* "Molecular Cytogenetics" (B. A. Hamkalo and J. Papconstantinou, eds.), p. 233. Plenum, New York.

Alberts, B. M., and Frey, L. (1970). T4 bacteriophage gene 32: A structural protein in the replication and recombination of DNA. *Nature (London)* **227,** 1313.

Alberts, B. M., Amodio, F. J., Jenkins, M., Gutman, E. D., and Ferris, F. L. (1968). Studies with DNA–cellulose chromatography. I. DNA-binding proteins from *Escherichia coli. Cold Spring Harbor Symp. Quant. Biol.* **33,** 289.

Alberts, B. M., Herrick, G., Sigal, N., and Frey, L. (1971). Proteins that unwind DNA and their role in genetic processes. *Fed. Proc., Fed. Am. Soc. Exp. Biol.* **30,** 1036.

Atkinson, M. R., Deutscher, M. P., Kornberg, A., Russell, A. F., and Moffatt, J. G. (1968). Enzymatic synthesis of deoxyribonucleic acid. XXXIV. Termination of chain growth by a 2′, 3′-dideoxyribonucleotide. *Biochemistry* **8,** 4897.

Austin, S. (1974). Coordinate and differential *in vitro* syntheses of two RNA polymerase subunits. *Nature (London)* **252,** 596.

Bach, M. K. (1962). The incorporation of tritiated thymidine into a microsomal fraction from HeLa cells during a short exposure time. *Proc. Natl. Acad. Sci. U.S.A.* **48,** 1031.

Baltimore, D., and Smoler, D. F. (1972). Association of an endoribonuclease with the avian myeloblastis virus deoxyribonucleic acid polymerase. *J. Biol. Chem.* **247,** 7282.

Banks, G. R. (1974). A ribonuclease H from *Ustilago maydis. Eur. J. Biochem.* **47,** 499.

Barry, J., Hama-Inaba, H., Moran, L., Alberts, B., and Wilberg, J. (1973). Proteins of the T4 bacteriophage replication apparatus. *In* "DNA Synthesis in Vitro" (R. D. Wells and R. B. Inman, eds.), p. 195. Univ. Park Press, Baltimore, Maryland.

Battula, N., and Loeb, L. A. (1974). The infidelity of avian myeloblastosis virus deoxyribonucleic acid polymerase in polynucleotide replication. *J. Biol. Chem.* **249,** 4086.

Bauer, W., and Vinograd, J. (1968). The intercalation of closed circular DNA with intercalative dyes. I. The superhelix density of SV40 DNA in the presence and absence of dye. *J. Mol. Biol.* **33,** 141.

Beljanski, M., and Beljanski, M. (1974). RNA-bound reverse transcriptase in *Escherichia coli* and *in vitro* synthesis of a complementary DNA. *Biochem. Genet.* **12,** 163.

Bell, R. M., Davis, R. D., and Vagelos, P. R. (1972). Altered phospholipid metabolism in a temperature-sensitive mutant of *E. coli* CR34 T46. *Biochim. Biophys. Acta* **270,** 504.

Berg, D., Barrett, K., and Chamberlin, M. (1971). Purification of two forms of *Escherichia coli* RNA polymerase and of sigma component. *In* "Methods in Enzymology" (L. Grossman and K. Moldave, eds.), Vol. 21, p. 506. Academic Press, New York.

Berger, H., and Kozinski, A. W. (1969). Suppression of T4D ligase mutations by rIIA and rIIB mutations. *Proc. Natl. Acad. Sci. U.S.A.* **64,** 897.

Bessman, M. J., Muzyczka, N., Goodman, M. F., and Schnaar, R. L. (1974). Studies on the biochemical basis of spontaneous mutation. II. The incorporation of a base and its analogue into DNA by wild-type, mutator and antimutator DNA polymerases. *J. Mol. Biol.* **88,** 409.

Beyersmann, D., Messer, W., and Schlict, M. (1974). Mutants of *Escherichia coli* B/r defective in deoxyribonucleic acid initiation: DNA I, a new gene for replication. *J. Bacteriol.* **118,** 783.

Bird, R. E., Louarn, J., Martuscelli, J., and Caro, L. (1972). Origin and sequence of chromosome replication in *Escherichia coli. J. Mol. Biol.* **70,** 549.

Bishop, J. M., Faras, A. J., Garapin, A. C., Goodman, H. M., Levinson, W. E., Stavnezer,

J., Taylor, J. M., and Varmus, H. E. (1973). Characteristics of the transcription of RNA by the DNA polymerase of Rous sarcoma virus. *In* "DNA Synthesis in Vitro" (R. D. Wells and R. B. Inman, eds.), p. 341. Univ. Park Press, Baltimore, Maryland.

Blair, D. G., Sherratt, D. J., Clewell, D. B., and Helinski, D. R. (1972). Isolation of supercoiled colicinogenic factor E1 DNA sensitive to ribonuclease and alkali. *Proc. Natl. Acad. Sci. U.S.A.* **69,** 2518.

Blattner, F. R., and Dahlberg, J. E. (1972). RNA synthesis startpoints in bacteriophage λ: Are the promotor and operator transcribed? *Nature (London), New Biol.* **237,** 227.

Blattner, F. R., Dahlberg, J. E., Boettiger, J. K., Fiandt, M., and Szybalski, W. (1972). Distance from a promotor mutation to an RNA synthesis startpoint on bacteriophage λ DNA. *Nature (London), New Biol.* **237,** 232.

Bloomfield, V. A., Crothers, D. M., and Tinoco, I. (1974). "The Physical Chemistry of Nucleic Acids," Chapter 6, p. 293. Harper, New York.

Blumenthal, A. B., Kriegstein, H, J., and Hogness, D. S. (1973). The units of DNA replication in *Drosophila melanogaster* chromosomes. *Cold Spring Harbor Symp. Quant. Biol.* **38,** 205.

Bollum, F. J. (1960). Calf thymus polymerase. *J. Biol. Chem.* **235,** 2399.

Bolognesi, D. P. (1974). Structural components of RNA tumor viruses. *Adv. Virus Res.* **19,** 315.

Bouché, J. P., Zechel, K., and Kornberg, A. (1975). *Dna G* gene product, a rifampicin-resistant RNA polymerase, initiates the conversion of a single-stranded coliphage DNA to its duplex replicative form. *J. Biol. Chem.* **250,** 5995.

Bouvier, F., and Zinder, N. D. (1974). Effects of the *dna A* thermosensitive mutation of *Escherichia coli* on bacteriophage f1 growth and DNA synthesis. *Virology* **60,** 139.

Braun, R., and Wili, H. (1969). The time sequence of DNA replication in *Physarum*. *Biochim. Biophys. Acta* **174,** 246.

Bremer, H., and Yuan, D. (1968). Chain growth rate of messenger RNA in *Escherichia coli* infected with bacteriophage T4. *J. Mol. Biol.* **34,** 527.

Brown, A., and Cohen, S. N. (1974). Effect of λ development on template specificity of *Escherichia coli* RNA polymerase. *Biochim. Biophys. Acta* **335,** 123.

Brutlag, D., and Kornberg, A. (1972). Enzymatic synthesis of deoxyribonucleic acid. XXXVI. A proofreading function for the 3′ → 5′ exonuclease activity in deoxyribonucleic acid polymerases. *J. Biol. Chem.* **247,** 241.

Brutlag, D., Schekman, R., and Kornberg, A. (1971). A possible role for RNA polymerase in the initiation of M13 DNA synthesis. *Proc. Natl. Acad. Sci. U.S.A.* **68,** 2826.

Burgess, R. R. (1969). Separation and characterization of the subunits of ribonucleic acid polymerase. *J. Biol. Chem.* **244,** 6168.

Büsen, W., and Hausen, P. (1975). Distinct ribonuclease H activities in calf thymus. *Eur. J. Biochem.* **52,** 179.

Callan, H. G. (1973). DNA replication in the chromosomes of eukaryotes. *Cold Spring Harbor Symp. Quant. Biol.* **38,** 195.

Campbell, A. (1961). Sensitive mutants of bacteriophage λ. *Virology* **14,** 22.

Campbell, J. L., Soll, L., and Richardson, C. C. (1972). Isolation and partial characterization of a mutant of *Escherichia coli* deficient in DNA polymerase II. *Proc. Natl. Acad. Sci. U.S.A.* **69,** 2090.

Campbell, J. L., Shizuya, H., and Richardson, C. C. (1974). Mapping of a mutation, *pol B 100,* affecting deoxyribonucleic acid polymerase II in *Escherichia coli* K-12. *J. Bacteriol.* **119,** 494.

Carl, P. C. (1970). *Escherichia coli* mutants with temperature-sensitive synthesis of DNA. *Mol. Gen. Genet.* **109,** 107.

Carlson, J. O., and Wang, J. C. (1974). A DNA–cellulose procedure for the purification of the *Escherichia coli* protein ω. *Methods Mol. Biol.* **7,** 231.

Carrol, R. B., Neet, K. E., and Goldthwait, D. A. (1972). Self-association of gene 32 protein of bacteriophage T4. *Proc. Natl. Acad. Sci. U.S.A.* **69,** 2741.

Chamberlin, M. J. (1974a). Bacterial DNA-dependent RNA polymerase. *In* "The Enzymes" (P. D. Boyer, ed.), 3rd ed., Vol. 10, p. 333. Academic Press, New York.

Chamberlin, M. J. (1974b). The selectivity of transcription. *Annu. Rev. Biochem.* **43,** 721.

Chamberlin, M. J., and Berg, P. (1963). Studies of DNA-directed RNA polymerase; formation of DNA–RNA complexes with single-stranded φX174 DNA as template. *Cold Spring Harbor Symp. Quant. Biol.* **28,** 43.

Chamberlin, M. J., and Ring, J. (1972). Studies of the binding of *Escherichia coli* RNA polymerase to DNA. V. T7 RNA chain initiation by enzyme–DNA complexes. *J. Mol. Biol.* **70,** 221.

Champoux, J. J., and Dulbecco, R. (1972). An activity from mammalian cells that untwists superhelical DNA—a possible swivel for DNA replication. *Proc. Natl. Acad. Sci. U.S.A.* **69,** 143.

Champoux, J. J., and McConaughty, B. L. (1975). Priming of superhelical SV40 DNA by *Escherichia coli* RNA polymerase for *in vitro* DNA synthesis. *Biochemistry* **14,** 307.

Chang, L. M. S. (1973a). Low molecular weight deoxyribonucleic acid polymerase from calf thymus chromatin. I. Preparation of homogeneous enzyme. *J. Biol. Chem.* **248,** 3789.

Chang, L. M. S. (1973b). Low molecular weight deoxyribonucleic acid polymerase from calf thymus chromatin. II. Initiation and fidelity of homopolymer replication. *J. Biol. Chem.* **248,** 6983.

Chang, L. M. S., and Bollum, F. J. (1972a). Variation of deoxyribonucleic acid polymerase activities during rat liver regeneration. *J. Biol. Chem.* **247,** 7948.

Chang, L. M. S., and Bollum, F. J. (1972b). A chemical model for transcriptional initiation of DNA replication. *Biochem. Biophys. Res. Commun.* **46,** 1354.

Chang, L. M. S., and Bollum, F. J. (1972c). Antigenic relationships in mammalian DNA polymerase. *Science* **175,** 1116.

Chang, L. M. S., and Bollum, F. J. (1973). A comparison of associated enzyme activities in various deoxyribonucleic acid polymerases. *J. Biol. Chem.* **248,** 3398.

Chang, L. M. S., Brown, M., and Bollum, F. J. (1973). Induction of DNA polymerase in mouse L cells. *J. Mol. Biol.* **74,** 1.

Chao, L., and Speyer, J. F. (1973). A new form of RNA polymerase isolated from *Escherichia coli*. *Biochem. Biophys. Res. Commun.* **51,** 399.

Chou, J. Y., Avila, J., and Martin, R. G. (1974). Viral DNA synthesis in cells infected by temperature-sensitive mutants of simian virus 40. *J. Virol.* **14,** 116.

Clewell, D. B., Evenchik, B. G., and Cranston, J. W. (1972). Direct inhibition of Col E1 plasmid DNA replication in *Escherichia coli* by rifampicin. *Nature (London), New Biol.* **237,** 29.

Comings, D. E., and Okada, T. A. (1973). DNA replication and the nuclear membrane. *J. Mol. Biol.* **75,** 609.

Cozzarelli, N. R., and Low, R. L. (1973). Mutational alteration of *Bacillus subtilis* DNA polymerase III to hydroxyphenylazo-pyrimidine resistance: Polymerase III is necessary for DNA replication. *Biochem. Biophys. Res. Commun.* **51,** 151.

Cozzarelli, N. R., Kelly, R. B., and Kornberg, A. (1969). Enzymic synthesis of DNA. XXXIII. Hydrolysis of a 5′-triphosphate terminated polynucleotide in the active center of DNA polymerase. *J. Mol. Biol.* **45,** 513.

Crawford, L. V., and Waring, M. J. (1967). Supercoiling of polyoma virus DNA measured by its interaction with ethidium bromide. *J. Mol. Biol.* **25,** 23.

Crawford, L. V., Syrett, C., and Wilde, A. (1973). The replication of polyoma DNA. *J. Gen. Virol.* **21**, 515.

Crerar, M., and Pearlman, R. E. (1971). DNA polymerase activity from *Tetrahymena pyriformis*. *FEBS Lett.* **18**, 231.

Dahlberg, J. E., and Blattner, F. R. (1973). *In vitro* transcription products of λ DNA, nucleotide sequences and regulatory sites. *In* "Virus Research" (C. F. Fox and W. S. Robinson, eds.), p. 533. Academic Press, New York.

Dahlberg, J. E., Sawyer, R. C., Taylor, J. M., Faras, A. J., Levinson, W. E., Goodman, H. M., and Bishop, J. M. (1974). Transcription of DNA from the 70 S RNA of RSV. I. Identification of a specific 4 S RNA which serves as primer. *J. Virol.* **13**, 1126.

Dawes, I. W., and Carter, B. L. A. (1974). Nitrosoguanidine mutagenesis during nuclear and mitochondrial gene replication. *Nature (London)* **250**, 709.

Delbruck, M., and Stent, G. S. (1975). On the mechanism of DNA replication. *In* "The Chemical Basis of Heredity" (W. D. McElroy and B. Glass, eds.), p. 699. Johns Hopkins Press, Baltimore, Maryland.

Delius, H., Mantell, J., and Alberts, B. (1972). Characterization by electron microscopy of the complex formed between T4 bacteriophage gene 32 protein and DNA. *J. Mol. Biol.* **67**, 341.

Delius, H., Westphal, H., and Axelrod, N. (1973). Length measurements of RNA synthesized *in vitro* by *Escherichia coli* RNA polymerase. *J. Mol. Biol.* **74**, 677.

De Lucia, P., and Cairns, J. (1969). Isolation of an *E. coli* strain with a mutation affecting DNA polymerase. *Nature (London)* **224**, 1164.

Deutscher, M. P., and Kornberg, A. (1969a). Enzymatic synthesis of deoxyribonucleic acid. XXVIII. The pyrophosphate exchange and pyrophosphorylysis reactions of deoxyribonucleic acid polymerase. *J. Biol. Chem.* **244**, 3019.

Deutscher, M. P., and Kornberg, A. (1969b). Enzymatic synthesis of deoxyribonucleic acid. XXIX. Hydrolysis of deoxyribonucleic acid from the 5′ terminus by an exonuclease function of deoxyribonucleic acid polymerase. *J. Biol. Chem.* **244**, 3029.

DeWaard, A., Paul, A. V., and Lehman, I. R. (1965). The structural gene for deoxyribonucleic acid polymerase in bacteriophages T4 and T5. *Proc. Natl. Acad. Sci. U.S.A.* **54**, 1241.

Diaz, A. T., Wiener, D., and Werner, R. (1975). Synthesis of small polynucleotide chains in thymine-depleted bacteria. *J. Mol. Biol.* **95**, 45.

Dickson, R. C., Abelson, J., Barnes, W. M., and Reznikoff, W. S. (1975). Genetic regulation: The *lac* control region. *Science* **187**, 27.

Doenecke, D., Marmaras, V. J., and Sekeris, C. E. (1972). Increased RNase H (hybridase) activity in the integument of blowfly larvae during development and under the influence of β-ecdysone. *FEBS Lett.* **22**, 261.

Dove, W. F., Hargrove, E., Ohashi, M., Haugli, F., and Guha, A. (1969). Replicator activation in Lambda. *Jpn. J. Genet.* **44**, Suppl. 1, 11.

Dove, W. F., Inokuchi, H., and Stevens, W. F. (1971). Replication control in phage Lambda. *In* "The Bacteriophage Lambda" (A. D. Hershey, ed.), p. 747. Cold Spring Harbor Lab., Cold Spring Harbor, New York.

Drake, J. W. (1969). Comparative rates of spontaneous mutation. *Nature (London)* **221**, 1132.

Dressler, D., Wolfson, J., and Magazin, M. (1972). Initiation and reinitiation of DNA synthesis during replication of bacteriophage T7. *Proc. Natl. Acad. Sci. U.S.A.* **69**, 998.

Dumas, L. B., and Miller, C. A. (1974). Inhibition of bacteriophage φX174 DNA replication in *dna B* mutants of *Escherichia coli* C. *J. Virol.* **14**, 1369.

Editorial (1971). Lifting replication out of the rut. *Nature (London), New Biol.* **233**, 97.

Englund, P. T. (1971). The initial step of *in vitro* synthesis of deoxyribonucleic acid by the T4 deoxyribonucleic acid polymerase. *J. Biol. Chem.* **246,** 5684.

Englund, P. T., Deutscher, M. P., Jovin, T. M., Kelly, R. B., Cozzarelli, N. R., and Kornberg, A. (1968). Structural and functional properties of *Escherichia coli* DNA polymerase. *Cold Spring Harbor Symp. Quant. Biol.* **33,** 1.

Englund, P. T., Huberman, J. A., Jovin, T. M., and Kornberg, A. (1969a). Enzymatic synthesis of deoxyribonucleic acid. XXX. Binding of triphosphates to deoxyribonucleic acid polymerase. *J. Biol. Chem.* **244,** 3038.

Englund, P. T., Kelly, R. B., and Kornberg, A. (1969b). Enzymatic synthesis of deoxyribonucleic acid. XXXI. Binding of deoxyribonucleic acid to deoxyribonucleic acid polymerase. *J. Biol. Chem.* **244,** 3045.

Epstein, R. H., Bolle, A., Steinberg, C. M., Kellenberger, E., Boy de la Tour, E., Chevalley, R., Edgar, R. S., Susman, M., Denhardt, G. H., and Lielausis, A. (1963). Physiological studies of conditional lethal mutants of bacteriophage T4D. *Cold Spring Harbor Symp. Quant. Biol.* **28,** 375.

Erikson, E., and Erikson, R. L. (1971). Association of 4 S ribonucleic acids with oncornavirus ribonucleic acids. *J. Virol.* **8,** 254.

Fansler, B. S. (1974). Eukaryotic DNA polymerases: Their association with the nucleus and relationship to DNA replication. *Int. Rev. Cytol., Suppl.* **4,** 363.

Fansler, B. S., and Loeb, L. A. (1972). Sea urchin nuclear DNA polymerase. IV. Reversible association of DNA polymerase with nuclei during the cell cycle. *Exp. Cell Res.* **75,** 433.

Faras, A. J., and Dibble, N. A. (1975). RNA-directed DNA synthesis by the DNA polymerase of Rous sarcoma virus: Structural and functional identification of 4 S primer RNA in uninfected cells. *Proc. Natl. Acad. Sci. U.S.A.* **72,** 859.

Faras, A. J., Taylor, J. M., McDonnell, J. P., Levinson, W. E., and Bishop, J. M. (1973a). Purification and characterization of the deoxyribonucleic acid polymerase associated with Rous sarcoma virus. *Biochemistry* **11,** 2334.

Faras, A. J., Taylor, J. M., Levinson, W. E., Goodman, H. M., and Bishop, J. M. (1973b). RNA-directed DNA polymerase of Rous sarcoma virus: Initiation of synthesis with 70 S viral RNA as template. *J. Mol. Biol.* **79,** 163.

Fareed, G. C., and Richardson, C. C. (1967). Enzymatic breakage and joining of deoxyribonucleic acid. II. The structural gene for polynucleotide ligase in bacteriophage T4. *Proc. Natl. Acad. Sci. U.S.A.* **58,** 665.

Fareed, G. C., and Salzman, N. P. (1972). Intermediate in SV40 DNA chain growth. *Nature (London), New Biol.* **238,** 274.

Fareed, G. C., Argon, C. F., and Salzman, N. P. (1972). Origin and direction of simian virus 40 deoxyribonucleic acid replication. *J. Virol.* **10,** 484.

Filip, C. C., Allen, J. S., Gustafson, R. A., Allen, R. G., and Walker, J. R. (1974). Bacterial division regulation: Characterization of the *dna H* locus of *Escherichia coli. J. Bacteriol.* **119,** 443.

Fox, R. M., Mendelsohn, J., Barbosa, E., and Goulian, M. (1973). RNA in nascent DNA from cultured human lymphocytes. *Nature (London), New Biol.* **245,** 234.

Francke, B., and Eckhart, W. (1973). Polyoma gene function required for viral DNA synthesis. *Virology* **55,** 127.

Francke, B., and Ray, D. S. (1971). Formation of the parental replicative form DNA of bacteriophage ϕX174 and initial events in its replication. *Proc. Natl. Acad. Sci. U.S.A.* **69,** 475.

Freifelder, D., and Kirschner, I. R. (1971). A phage λ endonuclease controlled by genes *O* and *P*. *Virology* **44,** 223.

Fried, M. (1965). Cell-transforming ability of a temperature-sensitive mutant of polyoma virus. *Proc. Natl. Acad. Sci. U.S.A.* **53,** 486.

Friedman, D. L. (1970). DNA polymerase from HeLa cell nuclei: Levels of activity during a synchronized cell cycle. *Biochem. Biophys. Res. Commun.* **39,** 100.

Ganesan, A. T., Laipis, P. J., and Yehle, C. O. (1973a). *In vitro* DNA synthesis and function of DNA polymerases in *Bacillus subtilis. In* "DNA Synthesis in Vitro" (R. D. Wells and R. B. Inman, eds.), p. 405. Univ. Park Press, Baltimore, Maryland.

Ganesan, A. T., Yehle, C. O., and Yu, C. C. (1973b). DNA replication in a polymerase I deficient mutant and the identification of DNA polymerases II and III in *Bacillus subtilis. Biochem. Biophys. Res. Commun.* **50,** 155.

Garapin, A. C., Varmus, H. E., Faras, A. J., Levinson, W. E., and Bishop, J. M. (1973). RNA-directed DNA synthesis by virions of Rous sarcoma virus: Further characterization of the templates and the extent of their transcription. *Virology* **52,** 264.

Gass, K. B., and Cozzarelli, N. R. (1973). Further genetic and enzymological characterization of the three *Bacillus subtilis* deoxyribonucleic acid polymerases. *J. Biol. Chem.* **248,** 7688.

Gass, K. B., Hill, T. C., Goulian, M., Strauss, B. S., and Cozzarelli, N. R. (1971). Altered deoxyribonucleic acid polymerase activity in a methyl methanesulfonate-sensitive mutant of *Bacillus subtilis. J. Bacteriol.* **108,** 364.

Gass, K. B., Low, R. L., and Cozzarelli, N. R. (1973). Inhibition of a DNA polymerase from *Bacillus subtilis* by hydroxyphenyl-azopyrimidine. *Proc. Natl. Acad. Sci. U.S.A.* **70,** 103.

Gefter, M. L. (1974). DNA polymerases II and III of *Escherichia coli. Prog. Nucleic Acid Res. Mol. Biol.* **14,** 101.

Gefter, M. L., Hirota, Y., Kornberg, T., Wechsler, J. A., and Barnoux, C. (1971). Analysis of DNA polymerases II and III in mutants of *Escherichia coli* thermosensitive for DNA synthesis. *Proc. Natl. Acad. Sci. U.S.A.* **68,** 3150.

Gefter, M. L., Molineux, I. J., Kornberg, T., and Khorana, H. G. (1972). Deoxyribonucleic acid synthesis in cell-free extracts. III. Catalytic properties of deoxyribonucleic acid polymerase II. *J. Biol. Chem.* **247,** 3321.

Geider, K., and Hoffman-Berling, H. (1971). DNA synthesis in nucleotide permeable *Escherichia coli* cells. Chain elongation in specific regions of the bacterial chromosome. *Eur. J. Biochem.* **21,** 374.

Geider, K., and Kornberg, A. (1974). Conversion of the M13 viral single strand to the double-stranded replicative forms by purified proteins. *J. Biol. Chem.* **249,** 3999.

Gellert, M. (1967). Formation of covalent circles of Lambda DNA by *E. coli* extracts. *Proc. Natl. Acad. Sci. U.S.A.* **57,** 148.

Georgopoulos, C. P., and Herskowitz, I. (1971). *Escherichia coli* mutants blocked in λ DNA synthesis. *In* "The Bacteriophage Lambda" (A. D. Hershey, ed.), p. 553. Cold Spring Harbor Lab., Cold Spring Harbor, New York.

Goff, C. G. (1974). Chemical structure of a modification of the *Escherichia coli* ribonucleic acid polymerase α polypeptides induced by bacteriophage T4 infection. *J. Biol. Chem.* **249,** 6181.

Goodman, N. C., and Spiegelman, S. (1971). Distinguishing reverse transcriptase of an RNA tumor virus from other known DNA polymerases. *Proc. Natl. Acad. Sci. U.S.A.* **68,** 2203.

Gottesman, M. M., Hicks, M. L., and Gellert, M. (1973). Genetics and function of DNA ligase in *E. coli. J. Mol. Biol.* **77,** 531.

Goulian, M., Lucas, Z. J., and Kornberg, A. (1968). Enzymatic synthesis of deoxyribonucleic acid. XXV. Purification and properties of deoxyribonucleic acid polymerase induced by infection with phage T4. *J. Biol. Chem.* **243,** 627.

Grandgenett, D. P., and Green. M. (1974). Different mode of action of ribonuclease H in purified α and $\alpha\beta$ ribonucleic acid-directed deoxyribonucleic acid polymerase from avian myeloblastosis virus. *J. Biol. Chem.* **249,** 5148.

Grandgenett, D. P., Gerard, G. F., and Green, M. (1972). Ribonuclease H: A ubiquitous activity in virions of ribonucleic acid tumor viruses. *J. Virol.* **10,** 1136.

Grandgenett, D. P., Gerard, G. F., and Green, M. (1973). A single subunit from avian myeloblastosis virus with both RNA-directed DNA polymerase and ribonuclease H activity. *Proc. Natl. Acad. Sci. U.S.A.* **70,** 230.

Green, M., and Gerard, G. F. (1974). RNA-directed DNA polymerase properties and functions in oncogenic RNA viruses and cells. *Prog. Nucleic Acid Res. Mol. Biol.* **14,** 187.

Green, M., Grandgenett, D., Gerard, G., Rho, H. M., Loui, M. C., Salzberg, S., Shanmugam, G., Bhadhuri, S., and Vecchio, G. (1974). Properties of oncornavirus RNA-directed DNA polymerase, the RNA template, and the intracellular products formed early during infection and cell transformation. *Cold Spring Harbor Symp. Quant. Biol.* **34,** 975.

Griffin, B. E., Fried, M., and Cowie, A. (1974). Polyoma DNA: A physical map. *Proc. Natl. Acad. Sci. U.S.A.* **71,** 2077.

Griffith, J., Huberman, J. A., and Kornberg, A. (1971). Electron microscopy of DNA polymerase bound to DNA. *J. Mol. Biol.* **55,** 209.

Gross, J. D. (1972). DNA replication in bacteria. *Curr. Top. Microbiol. Immunol.* **57,** 40.

Gross, J. D., and Gross, M. (1969). Genetic analysis of an *E. coli* strain with a mutation affecting DNA polymerase. *Nature (London)* **224,** 1166.

Grossman, L. I., Watson, R., and Vinograd, J. (1973). The presence of ribonucleotides in mature closed circular mitochondrial DNA. *Proc. Natl. Acad. Sci. U.S.A.* **70,** 3339.

Hall, Z. W., and Lehman, I. R. (1968). An *in vitro* transversion by a mutationally altered T4-induced DNA polymerase. *J. Mol. Biol.* **36,** 321.

Hartwell, L. H. (1971). Genetic control of the cell division cycle in yeast. II. Genes controlling DNA replication and its initiation. *J. Mol. Biol.* **59,** 183.

Hartwell, L. H. (1973). Three additional genes required for DNA synthesis in *S. cerevisiae. J. Bacteriol.* **115,** 966.

Hayes, S., and Szybalski, W. (1973a). Synthesis of RNA primer for Lambda DNA replication is controlled by phage and host. *In* "Molecular Cytogenetics" (B. A. Hamkalo and J. Papaconstantinou, eds.), p. 277. Plenum, New York.

Hayes, S., and Szybalski, W. (1973b). Control of short leftward transcripts from the immunity and *ori* regions in induced coliphage Lambda. *Mol. Gen. Genet.* **126,** 275.

Hecht, N. B. (1973). Enzymatically active intermediate in the conversion between the low and high molecular weight DNA polymerases. *Nature (London), New Biol.* **245,** 199.

Hecht, N. B., and Davidson, D. (1973). The presence of a common active subunit in low and high molecular weight murine DNA polymerases. *Biochem. Biophys. Res. Commun.* **51,** 299.

Heinze, J. E., and Carl, P. L. (1975). The effects of organic solvents in *Escherichia coli* DNA polymerase III. *Biochim. Biophys. Acta* **402,** 35.

Helling, R. B. (1968). Selection of a mutant of *Escherichia coli* which has high mutation rates. *J. Bacteriol.* **96,** 975.

Helmstetter, C. E., and Cooper, S. (1968). DNA synthesis during the division cycle of rapidly growing *Escherichia coli* B/r. *J. Mol. Biol.* **31,** 507.

Henry, C. M., Ferdinand, F. J., and Knippers, R. (1973). A hybridase from *Escherichia coli. Biochem. Biophys. Res. Commun.* **50,** 603.

Henry, T. J., and Knippers, R. (1974). The isolation and function of the *gene A* initiator of bacteriophage ϕX174, a highly specific DNA endonuclease. *Proc. Natl. Acad. Sci. U.S.A.* **71,** 1549.

Henry, T. J.,and Pratt, D. (1972). The proteins of bacteriophage M13. *Proc. Natl. Acad. Sci. U.S.A.* **62,** 800.

Hereford, L. M., and Hartwell, L. H. (1973). Role of protein synthesis in the replication of yeast DNA. *Nature (London), New Biol.* **244,** 129.

Hereford, L. M., and Hartwell, L. H. (1974). Sequential gene function in the initiation of *Saccharomyces cerevisiae* DNA synthesis. *J. Mol. Biol.* **84,** 445.

Hershfield, M. S. (1973). On the role of deoxyribonucleic acid polymerase in determining mutation rates. Characterization of the defect in the T4 deoxyribonucleic acid polymerase caused by the *ts L88* mutation. *J. Biol. Chem.* **248,** 1417.

Hershfield, M. S., and Nossal, N. G. (1972). Hydrolysis of template and newly synthesized deoxyribonucleic acid by the 3′ to 5′ exonuclease activity of the T4 deoxyribonucleic acid polymerase. *J. Biol. Chem.* **247,** 3393.

Hinkle, D. C., and Chamberlin, M. J. (1972). Studies of the binding of *Escherichia coli* RNA polymerase to DNA. I. The role of sigma subunit in site selection. *J. Mol. Biol.* **70,** 157.

Hiraga, S., and Saitoh, T. (1974). Initiation of DNA replication in *Escherichia coli.* I. Characteristics of the initiation process in *dna* mutants. *Mol. Gen. Genet.* **132,** 49.

Hirota, Y., Mordoh, J., and Jacob, F. (1970). On the process of cellular division in *Escherichia coli.* III. Thermosensitive mutants of *Escherichia coli* altered in the process of DNA initiation. *J. Mol. Biol.* **53,** 369.

Hirota, Y., Gefter, M., and Mindlich, L. (1972). A mutant of *Escherichia coli* defective in DNA polymerase II activity. *Proc. Natl. Acad. Sci. U.S.A.* **69,** 3238.

Horiuchi, T., Sato, T., and Nagata, T. (1975). DNA degradation in an amber mutant of *Escherichia coli* K12 affecting DNA ligase and viability. *J. Mol. Biol.* **95,** 271.

Hozier, J. C., and Taylor, J. H. (1975). Length distributions of single-stranded DNA in Chinese hamster ovary cells. *J. Mol. Biol.* **93,** 181.

Huang, W. M., and Lehman, I. R. (1972). On the exonuclease activity of phage T4 DNA polymerase. *J. Biol. Chem.* **247,** 3139–3146.

Huberman, J. A., and Horowitz, H. (1973). Discontinuous DNA synthesis in mammalian cells. *Cold Spring Harbor Symp. Quant. Biol.* **38,** 233.

Huberman, J. A., and Riggs, A. D. (1968). On the mechanism of DNA replication in mammalian chromosomes. *J. Mol. Biol.* **32,** 327.

Huberman, J. A., Kornberg, A., and Alberts, B. M. (1971). Stimulation of T4 bacteriophage DNA polymerase by the protein product of T4 gene 32. *J. Mol. Biol.* **62,** 39.

Hunter, T., and Francke, B. (1974). *In vitro* polyoma DNA synthesis: Involvement of RNA in discontinuous chain growth. *J. Mol. Biol.* **83,** 123.

Hurwitz, J., and Leis, J. P. (1972). RNA-dependent DNA polymerase activity of RNA tumor viruses. I. Directing influence of DNA in the reaction. *J. Virol.* **9,** 116.

Hurwitz, J., and Wickner, S. (1974). Involvement of two protein factors and ATP in *in vitro* DNA synthesis catalyzed by DNA polymerase III in *E. coli. Proc. Natl. Acad. Sci. U.S.A.* **71,** 6.

Hurwitz, J., Wickner, S., and Wright, M. (1973). Studies on *in vitro* DNA synthesis. II. Isolation of a protein which stimulates deoxyribonucleotide incorporation catalyzed by DNA polymerases of *E. coli. Biochem. Biophys. Res. Commun.* **51,** 257.

Inokuchi, H., Dove, W. F., and Freifelder, D. (1973). Physical studies of RNA involvement in bacteriophage λ DNA replication and prophage excision. *J. Mol. Biol.* **74,** 721.

Inselburg, J. (1974). Replication of colicin E1 plasmid DNA in minicells from a unique replication initiation site. *Proc. Natl. Acad. Sci. U.S.A.* **71,** 2256.

Iwakura, Y., Ito, K., and Ishihama, A. (1974). Biosynthesis of RNA polymerase in *Escherichia coli.* I. Control of RNA polymerase content at various growth rates. *Mol. Gen. Genet.* **133,** 1.

Jacob, F., Brenner, S., and Cuzin, F. (1963). On the regulation of DNA replication in bacteria. *Cold Spring Harbor Symp. Quant. Biol.* **28,** 329.

Jaenisch, R., Mayer, A., and Levine, A. (1971). Replicating SV40 molecules containing closed circular template DNA strands. *Nature* (*London*), *New Biol.* **233,** 72.

Jovin, T. M., Englund, P. T., and Bertsch, L. L. (1969). Enzymatic synthesis of deoxyribonucleic acid. XXVI. Physical and chemical studies of a homogeneous deoxyribonucleic acid polymerase. *J. Biol. Chem.* **244,** 2996.

Kacian, D. L., Watson, K. F., Burny, A., and Spiegelman, S. (1971). Purification of the DNA polymerase of avian myeloblastis virus. *Biochim. Biophys. Acta* **246,** 365.

Keller, W. (1972). RNA-primed DNA synthesis *in vitro*. *Proc. Natl. Acad. Sci. U.S.A.* **69,** 1560.

Keller, W., and Crouch, R. (1972). Degradation of DNA hybrids by ribonuclease H and DNA polymerases of cellular and viral origin. *Proc. Natl. Acad. Sci. U.S.A.* **69,** 3360.

Kelly, R. B., Atkinson, M. R., Huberman, J. A., and Kornberg, A. (1969). Excision of thymine dimers and other mismatched sequences by DNA polymerase of *Escherichia coli*. *Nature* (*London*) **224,** 495.

Kelly, R. B., Cozzarelli, N. R., Deutscher, M. P., Lehman, I. R., and Kornberg, A. (1970). Enzymatic synthesis of deoxyribonucleic acid. XXXII. Replication of duplex deoxyribonucleic acid by polymerase at a single strand break. *J. Biol. Chem.* **245,** 39.

Khesin, R. B., Gorlenko, Zh. M., Shemyakin, M. F., Strolinsky, S. L., Mindlin, S. Z., and Ilyina, T. S. (1969). Studies of the functions of RNA polymerase components by means of mutations. *Mol. Gen. Genet.* **105,** 243.

Kidwell, W. R., and Miller, G. C. (1969). The synthesis and assembly of DNA subunits in isolated HeLa cell nuclei. *Biochem. Biophys. Res. Commun.* **36,** 756.

Kier, H. M. (1962). Stimulation and inhibition of deoxyribonucleic acid nucleotidyltransferase by oligodeoxyribonucleotides. *Biochem. J.* **85,** 265.

Kingsbury, D., and Helinski, D. R. (1970). DNA polymerase as a requirement for the maintenance of the bacterial plasmid colicogenic factor E1. *Biochem. Biophys. Res. Commun.* **41,** 1538.

Klein, A., and Powling, A. (1972). Initiation of λ DNA replication *in vitro*. *Nature* (*London*), *New Biol.* **239,** 71.

Klein, A., Nüsslein, V., Otto, B., and Powling, A. (1973). *In vitro* studies on *Escherichia coli* DNA replication factors and on the initiation of phage λ DNA replication. *In* "DNA Synthesis in Vitro" (R. D. Wells and R. B. Inman, eds.), p. 185. Univ. Park Press, Baltimore, Maryland.

Klett, R. P., Cerami, A., and Reich, E. (1968). Exonuclease. VI. A new nuclease activity associated with *E. coli* DNA polymerase. *Proc. Natl. Acad. Sci. U.S.A.* **60,** 943.

Koch, A. L., and Miller, C. (1965). A mechanism for keeping mutations in check. *J. Theor. Biol.* **8,** 71.

Konrad, E. B., and Lehman, I. R. (1974). A conditional lethal mutant of *Escherichia coli* K12 defective in the 5′ → 3′ exonuclease associated with DNA polymerase I. *Proc. Natl. Acad. Sci. U.S.A.* **71,** 2048.

Konrad, E. B., and Lehman, I. R. (1975). Novel mutants of *Escherichia coli* that accumulate very small DNA replicative intermediates. *Proc. Natl. Acad. Sci. U.S.A.* **72,** 2150.

Konrad, E. B., Modrich, P., and Lehman, I. R. (1973). Genetic and enzymatic characterisation of a conditional lethal mutant of *E. coli* K12 with a temperature-sensitive DNA–ligase. *J. Mol. Biol.* **77,** 519.

Kornberg, A. (1969). Active center of DNA polymerase. *Science* **163,** 1410.

Kornberg, A., Lehman, I. R., and Sims, E. S. (1956). Polydesoxyribonucleotide synthesis by enzymes from *Escherichia coli*. *Fed. Proc., Fed. Am. Soc. Exp. Biol.* **15,** 291.

Kornberg, T., and Gefter, M. L. (1971). Purification and DNA synthesis in cell-free extracts: Properties of DNA polymerase II. *Proc. Natl. Acad. Sci. U.S.A.* **68,** 761.

Kornberg, T., and Gefter, M. L. (1972). Deoxyribonucleic acid synthesis in cell-free extracts. IV. Purification and catalytic properties of deoxyribonucleic acid polymerase III. *J. Biol. Chem.* **247,** 5369.

Kozinski, A. W., and Felgenhauer, Z. Z. (1967). Molecular recombination in T4 bacteriophage deoxyribonucleic acid. II. Single-strand breaks and exposure of uncomplemented areas as a prerequisite for recombination. *J. Virol.* **1,** 1193.

Krakow, J., and Fronk, E. (1969). *Azotobacter vinelandii* ribonucleic acid polymerase. VIII. Pyrophosphate exchange. *J. Biol. Chem.* **244,** 5988.

Krisch, H. M., Bolle, A., and Epstein, R. H. (1974). Regulation of the synthesis of bacteriophage T4 gene 32 protein. *J. Mol. Biol.* **88,** 89.

Kuempel, P. L. (1969). Temperature-sensitive initiation of chromosome replication in a mutant of *Escherichia coli. J. Bacteriol.* **100,** 1302.

Kuempel, P. L., and Veomett, G. E. (1970). A possible function of DNA polymerase in chromosome replication. *Biochem. Biophys. Res. Commun.* **41,** 973.

Kurosawa, Y., and Okazaki, R. (1975). Mechanism of DNA chain growth. XIII. Evidence for discontinuous replication of both strands of P2 phage DNA. *J. Mol. Biol.* **94,** 229.

Kurosawa, Y., Ogawa, T., Hirose, S., Okazaki, T., and Okazaki, R. (1975). Mechanism of DNA chain growth. XV. RNA-linked nascent DNA pieces in *Escherichia coli* strains assayed with spleen exonuclease. *J. Mol. Biol.* **96,** 653.

Laipis, P. J., and Ganesan, A. T. (1972). A deoxyribonucleic acid polymerase 1-deficient mutant of *Bacillus subtilis. J. Biol. Chem.* **247,** 5867.

Lark, K. G. (1972a). Genetic control over the initiation of the synthesis of the short deoxynucleotide chains in *E. coli. Nature (London), New Biol.* **240,** 237.

Lark, K. G. (1972b). Evidence for the direct involvement of RNA in the initiation of DNA replication in *Escherichia coli* 15T⁻. *J. Mol. Biol.* **64,** 47.

Lark, K. G., and Renger, H. (1969). Initiation of DNA replication in *E. coli* 15T⁻: Chronological dissection of the three physiological processes required for initiation. *J. Mol. Biol.* **42,** 221.

Lark, K. G., and Wechsler, J. A. (1975). DNA replication in *dna B* mutants of *Escherichia coli:* Gene product interaction and synthesis of 4 S pieces. *J. Mol. Biol.* **92,** 145.

Laurent, S. J. (1973). Initiation of deoxyribonucleic acid replication in a temperature-sensitive mutant of *B. subtilis:* Evidence for a transcriptional step. *J. Bacteriol.* **116,** 141.

Lazarus, L. H., and Kitron, N. (1973). Cytoplasmic DNA polymerase: Polymeric forms and their conversion into an active monomer resembling nuclear DNA polymerase. *J. Mol. Biol.* **81,** 529.

Lehman, I. R. (1974). DNA ligase: Structure, mechanism and function. *Science* **186,** 790.

Lehman, I. R., and Chien, J. R. (1973). Persistence of deoxyribonucleic acid polymerase I and its 5′ → 3′ exonuclease activity in *pol A* mutants of *E. coli* K12. *J. Biol. Chem.* **248,** 7717.

Lehman, I. R., and Richardson, C. C. (1964). The deoxyribonucleases of *Escherichia coli.* IV. An exonuclease activity present in purified preparations of deoxyribonucleic acid polymerase. *J. Biol. Chem.* **239,** 233.

Leibowitz, P. J., and Schaechter, M. (1975). The attachment of the bacterial chromosome to the cell membrane. *Int. Rev. Cytol.* **41,** 1.

Leis, J. P., and Hurwitz, J. (1972). RNA-dependent DNA polymerase activity of RNA tumor viruses. II. Directing influence of RNA in the reaction. *J. Virol.* **9,** 130.

Leis, J. P, Berkhower, I., and Hurwitz, J. (1973). Mechanism of action of ribonuclease H isolated from avian myeloblastis virus and *Escherichia coli*. *Proc. Natl. Acad. Sci. U.S.A.* **70,** 466.

Leis, J. P., Shincariol, A., Ishizaki, R., and Hurwitz, J. (1975). RNA-dependent DNA polymerase activity of RNA tumor viruses. V. Rous sarcoma virus single-stranded RNA–DNA covalent hybrids in infected chicken embryo fibroblast cells. *Virology* **15,** 484.

Lindahl, T., and Edelman, T. (1968). Polynucleotide ligases from myeloid and lymphoid tissues. *Proc. Natl. Acad. Sci. U.S.A.* **61,** 680.

Lindqvist, B. H., and Sinsheimer, R. L. (1967). The process of infection with bacteriophage ϕX174. XV. Bacteriophage DNA synthesis in abortive infections with a set of conditional lethal mutants. *J. Mol. Biol.* **30,** 69.

Linial, M., and Mason, W. S. (1974). Characterization of two conditional early mutants of Rous sarcoma virus. *Virology* **53,** 258.

Linney, E., and Hayashi, M. (1973). Two proteins of gene A of ϕX174. *Nature (London), New Biol.* **245,** 6.

Linney, E., Hayashi, M. N., and Hayashi, M. (1972). Gene A of ϕX174. I. Isolation and identification of its products. *Virology* **50,** 381.

Little, J. W. (1973). Mutants of bacteriophage T4 which allow amber mutants of gene 32 to grow in ochre-suppressing hosts. *Virology* **53,** 47.

Livingston, D. M., and Richardson, C. C. (1975). Deoxyribonucleic acid polymerase III of *Escherichia coli*. Characterization of associated exonuclease activities. *J. Biol. Chem.* **250,** 470.

Livingston, D. M., Hinkle, D. C., and Richardson, C. C. (1975). Deoxyribonucleic acid polymerase III of *Escherichia coli*. Purification and properties. *J. Biol. Chem.* **250,** 461.

Loeb, L. A. (1969). Purification and properties of deoxyribonucleic acid polymerase from nuclei of sea urchin embryos. *J. Biol. Chem.* **244,** 1672.

Loeb, L. A., and Agarwal, S. S. (1971). DNA polymerase. Correlation with DNA replication during transformation of human lymphocytes. *Exp. Cell Res.* **66,** 299.

Louarn, J.-M. (1974). Size distribution and molecular polarity of nascent DNA in a temperature-sensitive *dna G* mutant of *Escherichia coli*. *Mol. Gen. Genet.* **133,** 193.

Lovett, M. A., Katz, L., and Helinski, D. R. (1974). Unidirectional replication of plasmid Col E1 DNA. *Nature (London)* **251,** 337.

McClure, W. R., and Jovin, T. M. (1975). The steady state kinetic parameters and non-processivity of *Escherichia coli* deoxyribonucleic acid polymerase I. *J. Biol. Chem.* **250,** 4073.

McFadden, G., and Denhardt, D. T. (1974). Mechanism of replication of ϕX174 single-stranded DNA. IX. Requirements for the *Escherichia coli dna G* protein. *J. Virol.* **14,** 1070.

Magnusson, G., Pigiet, V., Winnacker, E. L., Abrams, R., and Reichard, P. (1973). RNA-linked short DNA fragments during polyoma replication. *Proc. Natl. Acad. Sci. U.S.A.* **70,** 412.

Mahler, H. R., and Mehrota, B. D. (1963). The interaction of nucleic acids with diamines. *Biochim. Biophys. Acta* **68,** 211.

Maitra, U., Nakata, Y., and Hurwitz, J. (1967). The role of deoxyribonucleic acid in ribonucleic acid synthesis. XIV. A study of the initiation of ribonucleic acid synthesis. *J. Biol. Chem.* **242,** 4908.

Mangel, W. F., and Chamberlin, M. J. (1974). Studies of ribonucleic acid chain initiation by *Escherichia coli* ribonucleic acid polymerase bound to T7 deoxyribonucleic acid. III. The effect of temperature on ribonucleic acid chain initiation and on the conformation of binary complexes. *J. Biol. Chem.* **249,** 3007

Mantsaninos, R. (1964). Studies on the synthesis of deoxyribonucleic acid by mammalian enzymes. I. Incorporation of deoxyribonucleotide 5′-triphosphates into deoxyribonucleic acid by a partially purified enzyme from regenerating rat liver. *J. Biol. Chem.* **239,** 3431.
Marmur, J., and Doty, P. (1962). Determination of the base composition of deoxyribonucleic acid from its thermal denaturation temperature. *J. Mol. Biol.* **5,** 109.
Masamune, Y., and Richardson, C. C. (1971). Strand displacement during deoxyribonucleic acid synthesis at single strand breaks. *J. Biol. Chem.* **246,** 2692.
Masamune, Y., Fleischman, R. A., and Richardson, C. C. (1971). Enzymatic removal and replacement of nucleotides at single strand breaks in deoxyribonucleic acid. *J. Biol. Chem.* **246,** 2680.
Masker, W., Hanawalt, P., and Shizuya, H. (1973). Role of DNA polymerase II in repair replication in *Escherichia coli*. *Nature* (*London*), *New Biol.* **244,** 242.
Matzura, H., Hausen, B., and Zeuthen, J. (1973). Biosynthesis of the β and β' subunits of RNA polymerase in *Escherichia coli*. *J. Mol. Biol.* **74,** 9.
Mendelson, N. H., and Gross, J. D. (1967). Characterization of a temperature-sensitive mutant of *Bacillus subtilis* defective in deoxyribonucleic acid replication. *J. Bacteriol.* **94,** 1603.
Meyer, R. R., and Simpson, M. V. (1968). DNA biosynthesis in mitochondria: Partial purification of a distinct DNA polymerase from isolated rat liver mitochondria. *Proc. Natl. Acad. Sci. U.S.A.* **61,** 130.
Miller, H. I., Riggs, A. D., and Gill, G. N. (1973). Ribonuclease H (hybrid) in *Escherichia coli*. Identification and characterization. *J. Biol. Chem.* **248,** 2621.
Miller, L. K., and Wells, R. D. (1972). Properties of the exonucleolytic activities of the *Micrococcus luteus* deoxyribonucleic acid polymerase. *J. Biol. Chem.* **247,** 2667.
Minkley, E., and Pribnow, D. (1973). Transcription of the early region of bacteriophage T7: Selective initiation with dinucleotides. *J. Mol. Biol.* **77,** 255.
Modrich, P., and Lehman, I. R. (1971). Enzymatic characterization of a mutant of *Escherichia coli* with an altered DNA ligase. *Proc. Natl. Acad. Sci. U.S.A.* **68,** 1002.
Modrich, P., and Lehman, I. R. (1973). Deoxyribonucleic acid ligase. A steady state kinetic analysis of the reaction catalyzed by the enzyme from *Escherichia coli*. *J. Biol. Chem.* **248,** 7502.
Modrich, P., Anraku, Y., and Lehman, I. R. (1973). Deoxyribonucleic acid ligase. Isolation and physical characterization of the homogeneous enzyme from *Escherichia coli*. *J. Biol. Chem.* **248,** 7495.
Molineux, I. J., and Gefter, M. L. (1974). Properties of the *Escherichia coli* DNA binding (unwinding) protein: Interaction with DNA polymerase and DNA. *Proc. Natl. Acad. Sci. U.S.A.* **71,** 3858.
Molineux, I. J., Friedman, S., and Gefter, M. L. (1974). Purification and properties of the *Escherichia coli* deoxyribonucleic acid-unwinding protein. Effects on deoxyribonucleic acid synthesis *in vitro*. *J. Biol. Chem.* **249,** 6090.
Mölling, K., Bolognesi, D. P., Bauer, H., Büsen, W., Plassman, H. W., and Hausen, P. (1971). Association of viral reverse transcriptase with an enzyme degrading the RNA moiety of RNA–DNA hybrids. *Nature* (*London*), *New Biol.* **234,** 240.
Muldoon, J. J., Evans, T. E., Nygaard, O. F., and Evans, H. H. (1971). Control of DNA replication by protein synthesis at defined times during the S period in *Physarum polycephalum*. *Biochim. Biophys. Acta* **247,** 310.
Nagata, T., and Horiuchi, T. (1974). An amber *dna* mutant of *Escherichia coli* K12 affecting DNA ligase. *J. Mol. Biol.* **87,** 369.

Nathans,D., and Danna, K. J. (1972). Specific origin of SV40 DNA replication. *Nature (London), New Biol.* **236,** 200.

Nishimura, Y., Caro, L., Berg, C. M., and Hirota, Z. (1971). Chromosome replication in *Escherichia coli.* IV. Control of chromosome replication and cell division by an integrated episome. *J. Mol. Biol.* **55,** 441.

Nossal, N. G. (1974). DNA synthesis on a double-stranded DNA template by the T4 bacteriophage DNA polymerase and the T4 gene 32 unwinding protein. *J. Biol. Chem.* **249,** 5668.

Nossal, N. G., and Hershfield, M. S. (1971). Nuclease activity in a fragment of bacteriophage T4 deoxyribonucleic acid polymerase induced by the amber mutant *am B22. J. Biol. Chem.* **246,** 5414.

Nossal, N. G., and Hershfield, M. S. (1973). Exonuclease activity of wild-type and mutant T4 DNA polymerases: Hydrolysis during DNA synthesis *in vitro. In* "DNA Synthesis in Vitro" (R. D. Wells and R. B. Inman, eds.), p. 47. Univ. Park Press, Baltimore, Maryland.

Nowinski, R. C., Watson, K. F., Yaniv, A., and Spiegelman, S. (1972). Serological analysis of the deoxyribonucleic acid polymerase of avian oncornaviruses. II. Comparison of avian deoxyribonucleic acid polymerases. *J. Virol.* **10,** 959.

Nüsslein, V., Otto, B., Bonhoeffer, F., and Schaller, H. (1971). Function of DNA polymerase III in DNA replication. *Nature (London)* **234,** 285.

Ogawa, T., and Tomizawa, J.-I. (1968). Replication of bacteriophage DNA. I. Replication of DNA of Lambda phage defective in early functions. *J. Mol. Biol.* **38,** 217.

Okazaki, R., Okazaki, T., Sakabe, K., Sugimoto, K., Kainuma, R., Sugino, A., and Iwatsuki, N. (1968). *In vivo* mechanism of DNA chain growth. *Cold Spring Harbor Symp. Quant. Biol.* **33,** 129.

Okazaki, R., Anisawa, M., and Sugino, A. (1971). Slow joining of newly replicated DNA chains in DNA polymerase I-deficient *Escherichia coli* mutants. *Proc. Natl. Acad. Sci. U.S.A.* **68,** 2954.

Okazaki, T., and Kornberg, A. (1964). Enzymatic synthesis of deoxyribonucleic acid. XV. Purification and properties of a polymerase from *Bacillus subtilis. J. Biol. Chem.* **239,** 259.

Okazaki, T., and Okazaki, R. (1969). Mechanism of DNA chain growth. IV. Direction of synthesis of T4 short DNA chains as revealed by exonucleolytic degradation. *Proc. Natl. Acad. Sci. U.S.A.* **64,** 1242.

Olivera, B. M., and Bonhoeffer, F. (1972). Discontinuous DNA replication *in vitro.* I. Two distinct classes of intermediates. *Nature (London), New Biol.* **240,** 233.

Olivera, B. M., and Bonhoeffer, F. (1974). Replication of *Escherichia coli* requires DNA polymerase I. *Nature (London)* **250,** 513.

Olivera, B. M., and Lehman, I. R. (1967). Linkage of polynucleotides through phosphodiester bonds by an enzyme from *E. coli. Proc. Natl. Acad. Sci. U.S.A.* **57,** 1426.

O'Neil, M., and Strohman, T. (1969). Changes in DNA polymerase activity associated with cell fusion in cultures of embryonic muscle. *J. Cell. Physiol.* **73,** 61.

O'Sullivan, A., Howard, K., and Sueoka, N. (1975). Location of a unique replication terminus and genetic evidence for partial bidirectional replication in the *Bacillus subtilis* chromosome. *J. Mol. Biol.* **91,** 15.

Otto, B., Bonhoeffer, F., and Schaller, H. (1973). Purification and properties of DNA polymerase III. *Eur. J. Biochem.* **34,** 440.

Panet, A., van der Sande, J. H., Loewen, P. C., Khorana, H. G., Raae, A. J., Lillehaug,

J. R., and Kleppe, K. (1973). Physical characterization and simultaneous purification of bacteriophage T4 induced polynucleotide kinase, polynucleotide ligase and deoxyribonucleic acid polymerase. *Biochemistry* **12,** 5045.

Pauling, C., and Hamm, L. (1968). Properties of a temperature-sensitive radiation-sensitive mutant of *Escherichia coli. Proc. Natl. Acad. Sci. U.S.A.* **60,** 1495.

Pauling, C., and Hamm, L. (1969). Properties of a temperature-sensitive, radiation-sensitive mutant of *Escherichia coli.* II. DNA replication. *Proc. Natl. Acad. Sci. U.S.A.* **64,** 1195.

Persico, F. J., Nicholson, D. E., and Gottlieb, A. A. (1973). Isolation and partial characterization of multiple DNA polymerases of the murine myeloma, MOPC21. *Cancer Res.* **33,** 1210.

Pigiet, V., Eliasson, R., and Reichard, P. (1974). Replication of polyoma DNA in isolated nuclei. III. The nucleotide sequence at the RNA–DNA junction of nascent strands. *J. Mol. Biol.* **84,** 217.

Price, T. D., Darmstadt, R. A., Hinds, H. A., and Zamenhof, S. (1967). Mechanism of synthesis of deoxyribonucleic acid *in vivo.* The heterogeneity of incorporation of ^{32}P into the deoxyribonucleotidyl units in *Escherichia coli. J. Biol. Chem.* **242,** 140.

Printz, M. P., and von Hippel, P. H. (1965). Hydrogen exchange studies of DNA structure. *Proc. Natl. Acad. Sci. U.S.A.* **53,** 363.

Pritchard, R. H., Barth, P. T., and Collins, J. (1969). Control of DNA synthesis in bacteria. *Symp. Soc. Gen. Microbiol.* **19,** 263.

Rambach, A. (1973). Replication mutants of bacteriophage λ; characterization of two subclasses. *Virology* **54,** 270.

Reddy, G. V. R., Goulian, M., and Hendler, S. S. (1971). Inhibition of *E. coli* DNA polymerase II by Ara-CTP. *Nature (London), New Biol.* **234,** 286.

Reuben, R. C., and Gefter, M. L. (1974). A deoxyribonucleic acid-binding protein induced by bacteriophage T7. Purification and properties. *J. Biol. Chem.* **249,** 3843.

Reznikoff, W. (1972). The operon revisited. *Annu. Rev. Genet.* **6,** 133.

Rhodes, G., and Chamberlin, M. J. (1974). Ribonucleic acid chain elongation by *Escherichia coli* RNA polymerase. I. Isolation of ternary complexes and the kinetics of elongation. *J. Biol. Chem.* **249,** 6675.

Richardson, C. C., Shildkraut, C. L., Aposhian, H. V., and Kornberg, A. (1964). Enzymatic synthesis of deoxyribonucleic acid. XIV. Further purification and properties of deoxyribonucleic acid polymerase of *Escherichia coli. J. Biol. Chem.* **239,** 222.

Richardson, J. P. (1966). Enzymic synthesis of RNA from T7 DNA. *J. Mol. Biol.* **21,** 115.

Richardson, J. P. (1970). Rates of bacteriophage T4 RNA chain growth *in vitro. J. Mol. Biol.* **49,** 235.

Riva, S., Cascino, A., and Geiduschek, E. P. (1970). Coupling of late transcription to viral replication in bacteriophage T4 development. *J. Mol. Biol.* **54,** 85.

Riva, S., van Sluis, C., Mastromei, G., Attolini, C., Mazza, G., Polsinelli, M., and Falaschi, A. (1975). A new mutant of *Bacillus subtilis* altered in the initiation of chromosome replication. *Mol. Gen. Genet.* **137,** 185.

Roberts, J. W. (1969). Termination factor for RNA synthesis. *Nature (London)* **224,** 1168.

Roychoudhury, R. (1973). Transcriptional role in deoxyribonucleic acid replication. Nature of the primer function of newly synthesized ribonucleic acid *in vitro. J. Biol. Chem.* **248,** 8465.

Ruprecht, R. M., Goodman, N. C., and Spiegelman, S. (1973). Conditions for selective synthesis of DNA complementary to template RNA. *Biochim. Biophys. Acta* **294,** 192.

Russel, M. (1973). Control of bacteriophage T4 DNA polymerase synthesis. *J. Mol. Biol.* **79,** 83.

Sakabe, K., and Okazaki, R. (1966). A unique property of the replicating region of chromosomal DNA. *Biochim. Biophys. Acta* **129,** 651.

Sakai, H., Hashimoto, S., and Komano, T. (1974). Replication of deoxyribonucleic acid in *Escherichia coli* C mutants temperature sensitive in the initiation of chromosome replication. *J. Bacteriol.* **119,** 811.

Sakakibara, Y., and Tomizawa, J.-I. (1974). Replication of colicin E1 plasmid DNA in cell extracts. *Proc. Natl. Acad. Sci. U.S.A.* **71,** 802.

Sakamaki, T., Fukinei, K., Takahashi, N., and Tanifugi, S. (1975). Rapidly labeled intermediates in DNA replication in higher plants. *Biochim. Biophys. Acta* **395,** 314.

Sambrook, J., and Shatkin, A. J. (1969). Polynucleotide ligase activity in cells infected with simian virus 40, polyoma virus, or vaccina virus. *J. Virol.* **4,** 719.

Sato, S., Aniake, S., Sato, M., and Sugimara, T. (1973). RNA bound to nascent DNA in Ehrlich ascites tumor cells. *Biochem. Biophys. Res. Commun.* **49,** 827.

Saucier, J. M., and Wang, J. C. (1970). Angular alteration of the DNA helix by *E. coli* RNA polymerase. *Nature (London), New Biol.* **239,** 167.

Schaller, H., Grey, C., and Herrman, K. (1975). Nucleotide sequence of an RNA polymerase binding site from the DNA of bacteriophage fd. *Proc. Natl. Acad. Sci. U.S.A.* **72,** 737.

Scheckman, R., Weiner, A., and Kornberg, A. (1974). Multienzyme systems of DNA replication. *Science* **186,** 987.

Scheckman, R., Weiner, J. H., Weiner, A., and Kornberg, A. (1975). Ten proteins required for conversion of ϕX174 single-stranded DNA to duplex form *in vitro:* Resolution and reconstitution. *J. Biol. Chem.* **250,** 5859.

Schnös, M., and Inman, R. B. (1970). Position of branch points in replicating DNA. *J. Mol. Biol.* **51,** 61.

Schubach, W., Whitmer, J., and Davern, C. (1973). Genetic control of DNA initiation in *Escherichia coli. J. Mol. Biol.* **74,** 205.

Scolnick, E. M., Parks, W. P., Todaro, G. J., and Aaronson, S. A. (1972). Immunological characterisation of primate C-type virus reverse transcriptases. *Nature (London), New Biol.* **235,** 35.

Scrutton, M., Wu, C., and Goldthwait, D. (1971). The presence and possible role of zinc in RNA polymerase obtained from *Escherichia coli. Proc. Natl. Acad. Sci. U.S.A.* **68,** 2497.

Sebring, E. D., Kelly, T. J., Jr., Thoren, M. M., and Salzman, N. P. (1971). Structure of replicating simian virus 40 deoxyribonucleic acid molecules. *J. Virol.* **8,** 478.

Sedwick, W. D., Wang, T. S.-F., and Korn, D. (1972). Purification and properties of nuclear and cytoplasmic deoxyribonucleic acid polymerases from human KB cells. *J. Biol. Chem.* **247,** 5026.

Setlow, P., and Kornberg, A. (1972). Deoxyribonucleic acid polymerase: Two distinct enzymes in one polypeptide. II. A proteolytic fragment containing the $5' \rightarrow 3'$ exonuclease function. Restoration of intact enzyme functions from the two proteolytic fragments. *J. Biol. Chem.* **247,** 232.

Setlow, P., Brutlag, D., and Kornberg, A. (1972). Deoxyribonucleic acid polymerase: two distinct enzymes in one polypeptide. I. A proteolytic fragment containing the polymerase and the $3' \rightarrow 5'$ exonuclease. *J. Biol. Chem.* **247,** 224.

Shildkraut, C. L., Richardson, C. C., and Kornberg, A. (1964). Enzymic synthesis of

deoxyribonucleic acid. XVII. Some unusual physical properties of the product primed by native DNA templates. *J. Mol. Biol.* **9,** 24.
Siegel, E. C., and Bryson, V. (1967). Mutator gene of *Escherichia coli* B. *J. Bacteriol.* **94,** 38.
Sigal, N., Delius, H., Kornberg, T., Gefter, M., and Alberts, B. (1972). A DNA-unwinding protein isolated from *Escherichia coli:* Its interaction with DNA and DNA polymerases. *Proc. Natl. Acad. Sci. U.S.A.* **69,** 3537.
Silverstein, S., and Billen, D. (1971). Transcription: Role in the initiation and replication of DNA synthesis in *Escherichia coli* and ϕX174. *Biochim. Biophys. Acta* **247,** 383.
Sinsheimer, R. L. (1974). Bacteriophage ϕX174. *Handb. Genet.* **1,** 323.
Sirover, M. A., and Loeb, L. A. (1974). Infidelity of DNA synthesis: A general property of RNA tumor viruses. *Biochem. Biophys. Res. Commun.* **52,** 410.
Slater, J. P., Mildvan, A. S., and Loeb, L. A. (1971). Zinc in DNA polymerases. *Biochem. Biophys. Res. Commun.* **44,** 37.
Smellie, R. M. S., Gray, E. D., Keir, H. M., Richards, J., Bell, D., and Davidson, J. N. (1960). Studies on the biosynthesis of deoxyribonucleic acid by extracts of mammalian cells. III. Net synthesis of polynucleotides. *Biochim. Biophys. Acta* **37,** 243.
Smith, R. G., and Gallo, R. L. (1972). DNA-dependent DNA polymerases I and II from normal human blood lymphocytes. *Proc. Natl. Acad. Sci. U.S.A.* **69,** 2879.
Spadari, S., and Weissbach, A. (1975). RNA-primed DNA synthesis: Specific catalysis by HeLa cell DNA polymerase α. *Proc. Natl. Acad. Sci. U.S.A.* **72,** 503.
Speyer, J. F. (1965). Mutagenic DNA polymerase. *Biochem. Biophys. Res. Commun.* **21,** 6.
Springgate, C. F., and Loeb, L. A. (1973). Mutagenic DNA polymerase in human leukemic cells. *Proc. Natl. Acad. Sci. U.S.A.* **70,** 245.
Springgate, C. F., Battula, N., and Loeb, L. A. (1973a). Infidelity of DNA synthesis by reverse transcriptase. *Biochem. Biophys. Res. Commun.* **52,** 401.
Springgate, C. F., Mildvan, A. S., Abramson, R., Engle, J. L., and Loeb, L. A. (1973b). *Escherichia coli* DNA polymerase I, a zinc metalloenzyme. *J. Biol. Chem.* **248,** 5987.
Stavrianopoulos, J. G., Karkas, J. D., and Chargaff, E. (1972). DNA polymerase of chicken embryo: Purification and properties. *Proc. Natl. Acad. Sci. U.S.A.* **69,** 1781.
Stein, H., and Hausen, P. (1969). Enzyme from calf thymus degrading the RNA moiety of DNA–RNA hybrids: Effect on DNA-dependent RNA polymerase. *Science* **166,** 393.
Steinberg, R. A., and Denhardt, D. T. (1968). Inhibition of synthesis of ϕX174 DNA in a mutant host thermosensitive for DNA synthesis. *J. Mol. Biol.* **37,** 525.
Stevens, W. F., Adhya, S., and Szybalski, W. (1971). Origin and bidirectional orientation of DNA replication in coliphage Lambda. *In* "The Bacteriophage Lambda" (A. D. Hershey, ed.), p. 515. Cold Spring Harbor Lab., Cold Spring Harbor, New York.
Stonington, O. G., and Pettijohn, D. E. (1971). The folded genome of *Escherichia coli* isolated in a protein–DNA–RNA complex. *Proc. Natl. Acad. Sci. U.S.A.* **68,** 6.
Studier, F. W. (1969). Conformational changes of single-stranded DNA. *J. Mol. Biol.* **41,** 189.
Sugino, A., and Okazaki, R. (1972). Mechanism of DNA chain growth. VII. Direction and rate of growth of T4 nascent short DNA chains. *J. Mol. Biol.* **64,** 61.
Sugino, A., Hirose, S., and Okazaki, R. (1972). RNA-linked nascent DNA fragments in *E. coli. Proc. Natl. Acad. Sci. U.S.A.* **69,** 1863.
Swart, M. N., Nakamura, H., and Lehman, I. R. (1972). Activation of a defective DNA polymerase in a temperature-sensitive mutant of bacteriophage T4. *Virology* **47,** 338.
Szybalski, W. (1974). Bacteriophage Lambda. *Handb. Genet.* **1,** 309.
Tabak, H. F., Griffith, J., Geider, K., Schaller, H., and Kornberg, A. (1974). Initiation of deoxyribonucleic acid synthesis. VII. A unique location of the gap in the M13 replicative duplex synthesized *in vitro. J. Biol. Chem.* **249,** 3049.

Takahashi, S. (1975). Role of genes *O* and *P* in the replication of bacteriophage λ DNA. *J. Mol. Biol.* **94,** 385.

Tamblyn, T. M., and Wells, R. D. (1975). Comparative ability of RNA and DNA to prime DNA synthesis *in vitro:* Role of sequence, sugar and structure of template primer. *Biochemistry* **14,** 1412.

Tanabe, K., and Takahashi, T. (1973). Conversion of DNA polymerase extracted from rat ascites hepatoma cells. *Biochem. Biophys. Res. Commun.* **53,** 295.

Taylor, A. L., and Trotter, C. D. (1972). Linkage map of *Escherichia coli* strain K12. *Bacteriol. Rev.* **36,** 504.

Taylor, J. M., Faras, A. J., Varmus, H. E., Levinson, W. E., and Bishop, J. M. (1972). Ribonucleic acid directed deoxyribonucleic acid synthesis by the purified deoxyribonucleic acid polymerase of Rous sarcoma virus. Characterisation of the enzymatic product. *Biochemistry* **11,** 2343.

Taylor, J. M., Garfin, D. E., Levinson, W. E., Bishop, J. M., and Goodman, H. M. (1974). Tumor virus ribonucleic acid directed deoxyribonucleic acid synthesis: Nucleotide sequence at the 5′ terminus of nascent deoxyribonucleic acid. *Biochemistry* **13,** 3159.

Tegtmeyer, P., and Ozer, H. L. (1971). Temperature-sensitive mutants of simian virus 40 infection of permissive cells. *J. Virol.* **8,** 516.

Temin, H. C., and Mizutani, S. (1974). RNA tumor virus DNA polymerases. *In* "The Enzymes" (P. D. Boyer, ed.), 3rd ed., Vol. 10, p 211. Academic Press, New York.

Terao, T., Dahlberg, J., and Khorana, H. G. (1972). Studies on polynucleotides. CXX. On the transcription of a synthetic 29-unit long deoxyribonucleotide. *J. Biol. Chem.* **247,** 6157.

Timmis, K., Cabello, F., and Cohen, S. N. (1974). Utilization of two distinct modes of replication by a hybrid plasmid constructed *in vitro* from separate replicons. *Proc. Natl. Acad. Sci. U.S.A.* **71,** 4556.

Tomizawa, J.-I. (1971). Functional cooperation of genes *O* and *P*. *In* "The Bacteriophage Lambda" (A. D. Hershey, ed.), p. 549. Cold Spring Harbor Lab., Cold Spring Harbor, New York.

Tomizawa, J.-I., and Ogawa, T. (1968). Replication of phage Lambda DNA. *Cold Spring Harbor Symp. Quant. Biol.* **33,** 533.

Tomizawa, J.-I., Anraku, N., and Iwama, Y. (1966). Molecular mechanisms of genetic recombination in bacteriophage. VI. A mutant defective in the joining of DNA molecules. *J. Mol. Biol.* **21,** 247.

Tomizawa, J.-I., Sakakibara, Y., and Kakefuda, T. (1974). Replication of colicin E1 plasmid DNA in cell extracts. Origin and direction of replication. *Proc. Natl. Acad. Sci. U.S.A.* **71,** 2260.

Tooze, J., ed. (1973). "The Molecular Biology of Tumor Viruses," p. 585. Cold Spring Harbor Lab., Cold Spring Harbor, New York.

Town, C. D., Smith, K. C., and Kaplan, H. S. (1971). DNA polymerase required for rapid repair of X-ray induced DNA strand breaks *in vivo*. *Science* **172,** 851.

Truitt, C. L., and Walker, J. R. (1974). Growth of phages φX174 and M13 requires the *dna Z* (previously *dna H*) gene product of *Escherichia coli*. *Biochem. Biophys. Res. Commun.* **61,** 1036.

Tsukada, K., and Ichimura, M. (1971). Polynucleotide ligase from rat liver after partial hepatectomy. *Biochem. Biophys. Res. Commun.* **42,** 1156.

Utiyama, H., and Doty, P. (1971). Kinetic studies of denaturation and reaction with formaldehyde on polydeoxyribonucleotides. *Biochemistry* **10,** 1254.

Van der Mei, D., Zandberg, J., and Jansz, H. S. (1972). The effect of chloramphenicol on synthesis of ϕX174-specific proteins and detection of the cistron A protein. *Biochim. Biophys. Acta* **287,** 312.

Van der Vliet, P. C., Levine, A. J., Ensinger, M. J., and Ginsburg, H. S. (1975). Thermolabile DNA binding proteins from cells infected with a temperature-sensitive mutant of adenovirus defective in viral DNA synthesis. *J. Virol.* **15,** 348.

Varmus, H. E., Guntaka, R. V., Deng, C. T., and Bishop, J. M. (1974). Synthesis, structure and function of avian sarcoma virus-specific DNA in permissive and nonpermissive cells. *Cold Spring Harbor Symp. Quant. Biol.* **39,** 987.

Verma, I. M. (1975). Studies on reverse transcriptase of RNA tumor viruses. I. Localisation of thermolabile DNA polymerase and RNase H activities on one polypeptide. *J. Virol.* **15,** 121.

Verma, I. M., Temple, G. F., Fan, H., and Baltimore, D. (1972). *In vitro* synthesis of DNA complementary to rabbit reticulocyte 10 S RNA. *Nature (London), New Biol.* **235,** 163.

Verma, I. M., Mason, W. S., Drost, S. D., and Baltimore, D. (1974). DNA polymerase activity from two temperature-sensitive mutants of Rous sarcoma virus is thermolabile. *Nature (London)* **251,** 27.

Vinograd, J., Lebowitz, J., and Watson, R. (1968). Early and late helix-coil transitions in closed circular DNA. The number of superhelical turns in polyoma DNA. *J. Mol. Biol.* **33,** 173.

von Hippel, P. H., and McGhee, J. D. (1972a). DNA–protein interactions. *Annu. Rev. Biochem.* **41,** 237.

von Hippel, P. H., and McGhee, J. D. (1972b). DNA–protein interactions. *Annu. Rev Biochem.* **41,** 253.

Wada, C., and Yura, T. (1974). Phenethyl alcohol resistance in *Escherichia coli.* III. A temperature-sensitive mutation (*dna P*) affecting DNA replication. *Genetics* **77,** 199.

Wang, J. C. (1969a). Variation of the average rotation angle of the DNA helix and the superhelical turns of covalently closed cyclic λ DNA. *J. Mol. Biol.* **43,** 25.

Wang, J. C. (1969b). Degree of superhelicity of covalently closed cyclic DNA's from *Escherichia coli. J. Mol. Biol.* **43,** 263.

Wang, J. C. (1971a). Use of intercalating dyes in the study of superhelical DNA's. *Procedures Nucleic Acid Res.* **2,** 407.

Wang, J. C. (1971b). Interaction between DNA and an *Escherichia coli* protein ω. *J. Mol. Biol.* **55,** 523.

Wang, J. C. (1973). Protein ω: A DNA swivelase from *Escherichia coli? In* "DNA Synthesis in Vitro" (R. D. Wells and R. B. Inman, eds.), p. 163. Univ. Park Press, Baltimore, Maryland.

Waqar, M. A., and Huberman, J. A. (1975). Covalent linkage between RNA and nascent DNA in the slime mold *Physarum polycephalum. Biochim. Biophys. Acta* **383,** 410.

Waters, L. C., Mullin, B. C., Ho, T., and Yang, W.-K. (1975). Ability of tryptophan tRNA to hybridize with 35 S RNA of avian myeloblastosis virus and to prime reverse transcription *in vitro. Proc. Natl. Acad. Sci. U.S.A.* **72,** 2155.

Watson, J. D., and Crick, F. H. C. (1953a). Genetical implications of the structure of deoxyribonucleic acid. *Nature (London)* **171,** 964.

Watson, J. D., and Crick, F. H. C. (1954). The structure of DNA. *Cold Spring Harbor Symp. Quant. Biol.* **18,** 123.

Wechsler, J. A. (1973). Complementation analysis of mutations at the *dna B, dna C,* and *dna*

D loci. *In* "DNA Synthesis in Vitro" (R. D. Wells and R. B. Inman, eds.), p. 375. Univ. Park Press, Baltimore, Maryland.

Wechsler, J. A. (1975). Genetic and phenotypic characterization of *dna C* mutations. *J. Bacteriol.* **121,** 594.

Wechsler, J. A., and Gross, J. D. (1971). *Escherichia coli* mutants temperature-sensitive for DNA synthesis. *Mol. Gen. Genet.* **113,** 273.

Weiner, J. H., Bertsch, L. L., and Kornberg, A. (1975). The deoxyribonucleic acid unwinding protein of *Escherichia coli*. *J. Biol. Chem.* **250,** 1972.

Weiss, B., and Richardson, C. C. (1967). Enzymatic breakage and joining of deoxyribonucleic acid. I. Repair of single strand breaks in DNA by an enzyme system from *Escherichia coli* infected with T4 bacteriophage. *Proc. Natl. Acad. Sci. U.S.A.* **57,** 1021.

Weiss, B., Sablon, A. J., Live, T. R., Fareed, G. C., and Richardson, C. C. (1968). Enzymatic breakage and joining of deoxyribonucleic acid. VI. Further purification and properties of polynucleotide ligase from *Escherichia coli* infected with bacteriophage T4. *J. Biol. Chem.* **243,** 4543.

Weissbach, A., Schlabach, A., Friedlander, B., and Bolden, A. (1971). DNA polymerases from human cells. *Nature (London), New Biol.* **231,** 167.

Weissbach, A., Baltimore, D., Bollum, F., Gallo, R., and Korn, D. (1975). Nomenclature of eukaryotic DNA polymerases. *Science* **190,** 401.

Wells, R. D., Flugel, R. M., Larson, J. E., Schendel, P. F., and Sweet, R. W. (1972). Comparison of some reactions catalyzed by deoxyribonucleic acid polymerase from avian myeloblastis virus, *Escherichia coli,* and *Micrococcus luteus*. *Biochemistry* **11,** 621.

Westergaard, O., Brutlag, D., and Kornberg, A. (1973). Initiation of deoxyribonucleic acid synthesis. IV. Incorporation of the ribonucleic acid primer into the phage replicative form. *J. Biol. Chem.* **248,** 1361.

Wickner, R. B., Ginsberg, B., Berkower, I., and Hurwitz, J. (1972a). Deoxyribonucleic acid polymerase II of *Escherichia coli*. I. Purification and characterization of the enzyme. *J. Biol. Chem.* **247,** 489.

Wickner, R. B., Ginsberg, B., and Hurwitz, J. (1972b). Deoxyribonucleic acid polymerase II of *Escherichia coli*. II. Studies of the template requirements and the structure of the deoxyribonucleic acid product. *J. Biol. Chem.* **247,** 498.

Wickner, S., and Hurwitz, J. (1974). Conversion of ϕX174 viral DNA to double-stranded form by purified *Escherichia coli* proteins. *Proc. Natl. Acad. Sci. U.S.A.* **71,** 4120.

Wickner, S., and Hurwitz, J. (1975). Interaction of *Escherichia coli dna B* and *dna C (D)* gene products *in vitro*. *Proc. Natl. Acad. Sci. U.S.A.* **72,** 921.

Wickner, S., Wright, M., and Hurwitz, J. (1973a). Studies on *in vitro* DNA synthesis. Purification of the *dna G* gene product from *Escherichia coli*. *Proc. Natl. Acad. Sci. U.S.A.* **70,** 1613.

Wickner, S., Berkower, I., Wright, M., and Hurwitz, J. (1973b). Studies on *in vitro* DNA synthesis: Purification of *dna C* gene product containing *dna D* activity from *Escherichia coli*. *Proc. Natl. Acad. Sci. U.S.A.* **70,** 2369.

Wickner, S., Wright, M., and Hurwitz, J. (1974). Association of DNA-dependent and independent ribonucleoside triphosphatase activities with *dna B* gene product of *Escherichia coli*. *Proc. Natl. Acad. Sci. U.S.A.* **71,** 783.

Wickner, W., and Kornberg, A. (1973). DNA polymerase III star requires ATP to start synthesis on a primed DNA. *Proc. Natl. Acad. Sci. U.S.A.* **70,** 3679.

Wickner, W., and Kornberg, A. (1974a). A novel form of RNA polymerase from *Escherichia coli*. *Proc. Natl. Acad. Sci. U.S.A.* **71,** 4425.

Wickner, W., and Kornberg, A. (1974b). A holoenzyme form of deoxyribonucleic acid polymerase III. Isolation and properties. *J. Biol. Chem.* **249,** 6244.

Wickner, W., Schekman, R., Geider, K., and Kornberg, A. (1973). A new form of DNA polymerase III and a copolymerase replicate a long single-stranded primer template. *Proc. Natl. Acad. Sci. U.S.A.* **70,** 1764.

Williamson, D. H. (1973). Replication of the nuclear genome in yeast does not require concomitant protein synthesis. *Biochem. Biophys. Res. Commun.* **52,** 731.

Wintersberger, U., and Wintersberger, E. (1970a). Studies on deoxyribonucleic acid polymerases from yeast. I. Partial purification and properties of two DNA polymerases from mitochondria-free cell extracts. *Eur. J. Biochem.* **13,** 11.

Wintersberger, U., and Wintersberger, E. (1970b). Studies on deoxyribonucleic acid polymerases from yeast. II. Partial purification and characterization of mitochondrial DNA polymerase from wild-type and respiration-deficient yeast cells. *Eur. J. Biochem.* **13,** 20.

Wise, G. E., and Prescott, D. M. (1973). Initiation and continuation of DNA replication are not associated with the nuclear envelope in mammalian cells. *Proc. Natl. Acad. Sci. U.S.A.* **70,** 714.

Wolf, B. (1972). The characteristics and genetic map location of a temperature-sensitive DNA mutant of *E. coli* K12. *Genetics* **72,** 569.

Wood, W. B. (1974). Genetic map of bacteriophage T4. *Handb. Genet.* **1,** 327.

Worcel, A., and Burgi, E. (1972). On the structure of the folded chromosome of *Escherichia coli*. *J. Mol. Biol.* **71,** 127.

Wovcha, M. G., Tomich, P. K., Chiu, C.-S. C., and Greenberg, R. (1973). Direct participation of the dCMP hydroxymethylase in synthesis of bacteriophage T4 DNA. *Proc. Natl. Acad. Sci. U.S.A.* **70,** 2196.

Wright, M., Wickner, S., and Hurwitz, J. (1973). Studies on *in vitro* DNA synthesis. Isolation of the *dna B* gene product from *Escherichia coli*. *Proc. Natl. Acad. Sci. U.S.A.* **70,** 3120.

Wu, C., and Goldthwaite, D. (1969). Studies of nucleotide binding to the ribonucleic acid polymerase by a fluorescent technique. *Biochemistry* **8,** 4450.

Yamaguchi, F., and Yoshikawa, H. (1973). Topography of chromosome–membrane junction in *Bacillus subtilis*. *Nature (London), New Biol.* **244,** 204.

Yoshikawa, H., and Sueoka, N. (1963). Sequential replication of *Bacillus subtilis* chromosome. I. Comparison of marker frequencies in exponential and stationary growth phases. *Proc. Natl. Acad. Sci. U.S.A.* **49,** 559.

Youngs, D. A., and Smith, K. C. (1973). Involvement of DNA polymerase III in excision repair after ultraviolet irradiation. *Nature (London), New Biol.* **244,** 240.

Zechel, K., Bouché, J.-P., and Kornberg, A. (1975). Replication of phage G4. A novel and simple system for the initiation of deoxyribonucleic acid synthesis. *J. Biol. Chem.* **250,** 4684.

Zillig, W., Zechel, K., Rabussay, D., Schachner, M., Sethi, V., Palm, P., Heil, A., and Seifert, W. (1970). On the role of different subunits of DNA-dependent RNA polymease from *E. coli* in the transcription process. *Cold Spring Harbor Symp. Quant. Biol.* **35,** 47.

Zimmerman, B. K. (1966). Purification and properties of deoxyribonucleic acid polymerase from *Micrococcus lysodeikticus*. *J. Biol. Chem.* **241,** 2035.

Zimmerman, S. B., Little, J. W., Oshinsky, C. K., and Gellert, M. (1967). Enzymatic joining of DNA strands: A novel reaction of diphosphophyridine nucleotide. *Proc. Natl. Acad. Sci. U.S.A.* **57,** 1841.

3

Replication of the Prokaryotic Chromosome with Emphasis on the Bacterial Chromosome Replication in Relation to the Cell Cycle

Cedric I. Davern

ISBN 0-12-289502-9

I. INTRODUCTION—AN OVERVIEW OF BACTERIAL CHROMOSOME REPLICATION AND ITS REGULATION IN THE CONTEXT OF THE CELL CYCLE

Bacteria growing under a wide range of nutritional conditions and temperatures respond to these variables by assuming an environment-specific steady state with respect to cell volume, macromolecular constituents, and doubling or generation time, in which the average cell mass and the DNA content/cell increase exponentially relative to growth rate (Schaecter *et al.,* 1958). This environment-specific steady state reflects both a stable age distribution for the exponentially reproducing population (Powell, 1956) and a stable length of cell cycle. While the length of the cell cycle (i.e., generation time) varies from cell to cell and from generation to generation within a cell clone, the average generation time in the population remains the same (Koch, 1966).

In this steady state, cell mass is doubled and divided into two daughter cells every generation. Preeminent in this doubling and partitioning process is the doubling of the genome by DNA synthesis and the partitioning of the daughter products by segregation and cell division. Of concern to us here is the nature of the mechanisms, as well as their regulation, underlying this coordination of the cell growth and division cycle with the cycle of DNA replication and segregation.

Initiation of a round of DNA replication was early recognized as a cardinal event in the DNA replication cycle and in the cell cycle as well. Maaløe and Kjeldgaard (1966) first suggested that regulation of the frequency of initiation, and not modulation of the rate of replication, coordinated DNA synthesis rate with the cell division rate. The basis for this insight was the finding that the genome of *Escherichia coli* is encompassed in a single circular DNA molecule of about 1 mm (Cairns, 1963) that replicates from a particular origin (K. G. Lark *et al.,* 1963) in a sequential manner by means of one or two replication forks (Cairns, 1963; Bonhoeffer and Gierer, 1963). Initiation occurs when sufficient initiator protein, made anew for each replication cycle, has accumulated during the course of the cell cycle (Maaløe and Hanawalt, 1961). These findings, coupled with evidence that DNA molecules other than the bacterial chromosome sometimes found in the bacterial cell are replicated under their own distinct and specific control systems, led to the proposal of the replicon theory of replication control (Jacob *et al.*, 1963). This theory postulates that the initiation of DNA replication is positively regulated by the binding of an initiation protein to a specific site on the DNA molecule. Replication then proceeds by passage of the DNA molecule through a fixed membrane site containing the proteins that catalyze replication. Thus, the event of initiation has been singled out as a control point for the process of replication and for governing the rate of DNA synthesis.

The cardinal position of initiation of DNA replication in models of the cell cycle was confirmed further when Helmstetter and Cooper (1968) examined the rate of DNA synthesis as a function of cell age in synchronized cell populations. Just as Maaløe had originally suggested, they found that the doubling of DNA per cell per generation, as generation time is varied over a range of 22–40 minutes at 37°C, took place such that the frequency of DNA initiation matches precisely the frequency of cell division, while the period of the DNA replication cycle and the interval between termination (or initiation) and cell division are conserved at 41 and 20 min, respectively (Helmstetter *et al.*, 1968; Cooper and Helmstetter, 1968). With a doubling time greater than 60 minutes, the almost cosmic constant for the length of DNA cycle transit time (*C*) starts to increase. It was observed empirically that *C* tended to be two-thirds of the generation time for generation times >60 minutes (Cooper and Helmstetter, 1968; Helmstetter *et al.*, 1968).

When DNA synthesis was inhibited at various times throughout the cycle in a synchronized population of *E. coli*, cell division was inhibited for all cells treated up to an age in the cycle corresponding to the termination of a round of replication. Cells that had terminated a round of replication when the inhibitor was added were capable of dividing once (Clark, 1968). Similarly, inhibition of DNA synthesis in a nonsynchronized cell population allowed cell division to proceed at the normal rate for 20 minutes. Thus, it was concluded that the termination of a round of replication provided a signal for triggering a necessary step in the process of cell division.

Because the frequency of initiation of a round of replication regulates the rate of DNA synthesis, and because the transit time of the replication is constant at 41 minutes over a range of doubling times from 22 to 60 minutes, it is clear that the circular chromosomes of rapidly growing cells must be occupied by several replication forks at any one time, the number being a function of the ratio of the replication transit time (41 minutes) to the cell generation time. Assuming that the signal for initiation is received and acted upon by all replication origin sites, regardless of the number per cell, then such fast growing cells should contain chromosomes with multiple replication forks in a symmetric dichotomous array. For example, one would expect a circular chromosome that was host to three ongoing rounds of bidirectional replication (Figure 1) to have 1 + 2 + 4 = 7 pairs of replicating forks. Data compatible with such patterns of replication have been obtained both in *Bacillus subtilis* (Yoshikawa *et al.*, 1964; Oishi *et al.*, 1964) and *E. coli* (Bird and Lark, 1968).

In slow-growing cells in which the cell generation time exceeds the DNA cycle transit time, the intervals between successive initiation events must be longer than the DNA cycle transit time. Thus, at a particular instant, only a fraction of the cells should be engaged in DNA synthesis, namely, that fraction equal to the ratio of the DNA cycle transit time to the cell doubling time. For that fraction of cells engaged in DNA synthesis, the chromosome will contain only those replication forks engendered by the initiation of a single round of replication (i.e., one fork if replication were unidirectional, two if bidirectional). Again, evidence for these expectations is abundant (C. Lark, 1966; Helmstetter *et al.*, 1968; Kubitschek and Freedman, 1971; Gudas and Pardee, 1974; Chandler *et al.*, 1975).

Finally, the postulation by Jacob and co-workers (1963) in their replicon model of a cell membrane–chromosome attachment provides the physical basis for the coordination of the DNA replication cycle with the cell division cycle, particularly for the orderly segregation of the daughter chromosomes into their daughter cells. Strong evidence to indicate the

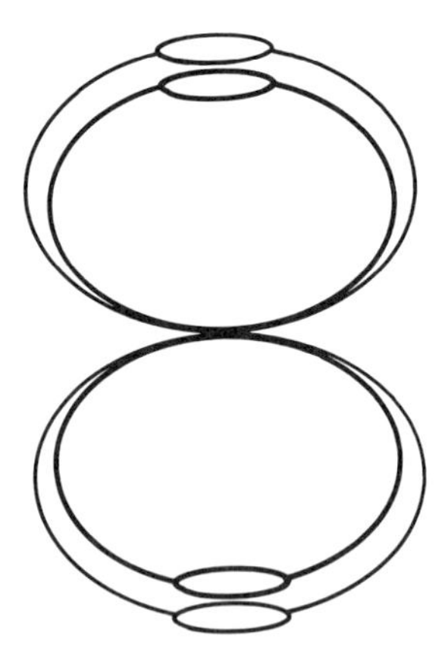

Fig. 1. A replicating circular chromosome that is supporting three successive rounds of bidirectional replication, all of which initiate at the same origin.

existence of a membrane–chromosome attachment has been obtained in the form of direct electron microscope observations (see Ryter, 1968; Ryter *et al.*, 1968); less direct biochemical evidence supports the attachment of the replication fork (Goldstein and Brown, 1961; Ganesan and Lederberg, 1965; Smith and Hanawalt, 1967; Tremblay *et al.*, 1969). Additional findings indicate attachment of the chromosome at other sites such as the origin of replication (Sueoka and Quinn, 1968; Fielding and Fox, 1970) or at both origin and terminus (Sueoka and Quinn, 1968; Yamaguchi and Yoshikawa, 1973). Membrane attachment at the origin is also suggested by a role for lipid in the process of initiation (Fralick and Lark, 1973). The particular nature of this attachment in *E. coli* with respect to the three layers (the outer lipoprotein layer, a middle murein layer, and the cell membrane) which make up its cell envelope (Costerton *et al.*, 1974) is unknown; it is known, however, that the assay system is prone to artifacts of attachment during the preparation of the membrane fraction (Parker and Glaser, 1974).

Better evidence for cell envelope involvement in segregation should lie in the cosegregation of DNA and cell envelope components of a given age, provided that both the attachment of a DNA strand to the cell envelope and the structure of the cell envelope are conserved. Unfortunately, evidence on this point has been conflicting, and only recently has there emerged some resolution of this conflict (Pierucci and Zuchowski, 1973). The findings of Pierucci and Zuchowski provide strong support for the earlier demonstrations of nonrandom segregation of DNA strands obtained by Lark and his colleagues (K. G. Lark and Bird, 1965; K. G. Lark, 1966a,b; Eberle and Lark, 1966; Chai and Lark, 1967; K. G. Lark *et al.*, 1967). These results contrast with the evidence for random segregation obtained by Ryter and associates (1968). Even so, the latter group has obtained evidence for cosegregation of an episome with the chromosome (Cuzin and Jacob, 1965), indicating that both are probably attached to some common structure, possibly membrane.

In a reductionist framework, one may summarize the foregoing lines of evidence and insight by the following two sets of statements about the cell cycle of *E. coli* as it varies in tempo with nutritional conditions. The first set of statements, which should apply to bacteria in general, attempts to minimally explain why each daughter cell is guaranteed a complete genetic heritage from its parent. These statements are listed below.

1. Initiation of DNA replication occurs when the cell has synthesized initiation components and assembled them into initiation complexes sufficient for all of the chromosome origins in the cell (Donachie, 1968).
2. The initiation complex is assembled at the origin site for replica-

tion on the bacterial chromosome and is attached to the cell membrane (Jacob *et al.*, 1963).

3. The initiation and replication of the DNA are effected by the membrane-bound replication complex, which is called, by analogy with the ribosome, the "replisome" (Bleecken, 1971).

4. Termination of a round of replication both destroys the functional integrity of the replisome and provides a necessary condition for the onset of the process of cell division (Clark, 1968; Helmstetter and Pierucci, 1968).

5. Not all components of the discarded replisome are recyclable; thus, some components must be made anew for each replication cycle (Maaløe and Hanawalt, 1961).

6. Segregation of daughter chromosomes is assured by the physical attachment of one strand of the DNA to a cell membrane or envelope site when that strand is first used as a template. For economy of hypothesis, this connection is the same as the one mentioned in statement (2). For this attachment site to segregate from the previously formed attachment site, the insertion of new cell envelope material between them must take place (K. G. Lark, 1966a,b).

The second set of statements relates to the timing of events in the cell cycle and reflects the peculiarities of *E. coli* more than it does the cybernetic constraints necessary for conserving a steady state that varies with nutritional conditions. They are as follows:

1. The nutritional condition can be described in terms of one parameter, the generation time or doubling time (T).

2. The transit time (C) for a DNA replication cycle is a strain-specific, temperature-dependent value that is conserved over a wide range of T.

3. Termination triggers a necessary but not necessarily rate-limiting step in the process of cell division. However, the time from the termination of a replication cycle to cell division is also a strain-specific value (D) that is conserved over a wide range of T.

The last two statements are derived from the Helmstetter and Cooper model (1968) and include their provision for C becoming a variable when T exceeds a critical value equal to about $C + D$. As we shall see, the question of the constancy of C, let alone that of D, is a subject of considerable controversy at this time.

With the accumulation of more data on the various aspects of the cell cycle in *E. coli*, the hope for a simple model has become increasingly faint. Rather than succumb to the temptation of encompassing all the observations extant into a model of Ptolemaic complexity, for heuristic reasons I have chosen to adhere to the model described by the statements above. This model provides a simple structure whose use can bring some order into a field of inquiry whose exceptions and contradictions can be explicitly identified and classified according to whether they invalidate or merely point out its inadequacies. In addition to this, the model will provide a convenient skeleton for fleshing out the descriptive detail about some of the entities and processes involved in controlling and implementing the DNA cycle in relation to the overall cell cycle.

II. INITIATION

A. Site of Initiation

Studies have revealed that the site of initiation of replication is localized to a particular region of the DNA molecule. Thus, the initiation complex at least must recognize, if not assemble, upon a particular sequence of nucleotides. Actinomycin D, a drug which interacts with DNA by intercalating between G-C base pairs and which inhibits RNA transcription, can inhibit the conversion of the small single-stranded circular coli phage DNA's into a double-stranded replicating form (Schekman *et al.*, 1972). Thus, a sequence of nucleotides is believed to be arranged in a self-complementary sequence at the site of initiation, allowing the formation of a double-stranded fold in the otherwise single-stranded molecule. This type of fold or "bobby pin" provides a structural feature which could be the basis of recognition or assembly of a replication complex and to which actinomycin D could bind and thus block initiation. Since they are not limited to single-stranded DNA, structural features of this sort could also play a role in conferring recognition sites on an otherwise relatively featureless double-stranded DNA (Gierer, 1966; Sobell, 1972). Such "palindromic" sequences have now been identified in double-stranded DNA in the region of the genome, where one would expect control of gene expression to be exercised (Gilbert *et al.*, 1973). More will be said about the specificity and mechanistic possibilities of palindromic sequences in the section dealing with the separation of daughter molecules at the end of a round of replication (Section IV).

B. The Requirements for and Entities Involved in Initiation—Physiological Analysis of Initiation

Since it was established that protein synthesis is necessary for initiation of a round of DNA replication (Maaløe and Hanawalt, 1961), questions naturally arose as to how many different proteins were needed for initiation. Are other kinds of macromolecules involved? Accordingly, at what point during the course of the cell cycle are these entities synthesized?

Physiological analysis of the initiation process, based on the use of specific inhibitors of protein synthesis and RNA synthesis, established that at least four classes of physiological events are consummated during the course of a cell cycle. Two of these involve the synthesis of a protein or proteins (K. G. Lark and Lark, 1966; C. Lark and Lark, 1964). One is sensitive to low concentrations of chloramphenicol and is completed approximately 30 minutes before initiation (K. G. Lark and Renger, 1969); the other is sensitive to a high concentration of chloramphenicol (K. G. Lark and Lark, 1966). This second event is completed about 15 minutes before initiation (K. G. Lark and Renger, 1969). The third event takes place immediately prior to initiation and is sensitive to RNA synthesis inhibitors (K. G. Lark, 1972; Messer, 1972).

The fourth class of physiological events is concurrent with the later protein synthesis-dependent event and is characterized by sensitivity to phenethyl alcohol. Treick and Konetzka (1964) showed that treatment of a growing culture of bacteria with this drug appeared to block selectively the initiation of DNA replication. This was confirmed by the Larks (1966).

In keeping with these observations, Wada and Yura (1974) have isolated a phenethyl-resistant mutant which turns out to be a temperature-sensitive initiation mutant. The Larks (1966) agreed with Treick and Konetzka's speculation that the vulnerable target affected by phenethyl alcohol is probably a membrane component and suggested that it may correspond to the membrane-attached "replicator" featured in the "replicon" model of DNA replication in *E. coli* (Jacob *et al.*, 1963).

The hypothesis that some membrane-located component is implicated in the initiation of DNA replication received further support from the studies of Fralick and Lark (1973) on the effect of an inhibitor of unsaturated fatty acid synthesis. They found, for example, that temperature-sensitive mutants, defective in initiation (*dna A* and *dna C*) when held for a period at the nonpermissive temperature in the presence of this unsaturated fatty acid synthesis inhibitor, cannot initiate DNA replication when they are returned to the permissive temperature. Because neither RNA nor protein synthesis is affected under these conditions, Fralick and Lark conclude that unsaturated fatty acids contribute to the structure of an initiation site located in the cell membrane.

These physiological investigations are limited to identifying components that cannot be recycled for use again in initiation, and thus define a minimum set of steps in initiation.

C. Biochemical–Genetic Analysis of Initiation

A much better measure of the complexity of initiation in terms of the number of primary gene products (RNA and/or protein) required to participate in initiation can be obtained by fusing two categories of approach. Both depend on the development of suitable *in vitro* DNA synthesis systems that are readily accessible to the manipulation of the macromolecular components, as well as to the immediate precursors involved in DNA synthesis. One approach is enzymological. It depends on the fractionation and purification of the macromolecules involved from a working *in vitro* system and then on reconstruction of a working system from these separate entities. This approach initially led to the characterization of an *in vitro* DNA synthesis system dependent upon one protein, DNA polymerase I, and subsequently, with more stringent criteria for DNA replication, to a more complex system dependent on additional components such as DNA ligase. However, although this method is powerful in identifying and characterizing the components necessary for a minimal system, such a system is not necessarily a replica of that which obtains *in vivo*.

The other approach involves an extensive mutant hunt followed by complementation analysis. This allows the enumeration of the distinct functions necessary for initiation *in vivo* by virtue of defining a set of complementing and frequently separable genetic loci. Insight into the possible functional interactions among the products of these genes can be obtained by screening for suppressor mutations and inquiring how many of these fall into complementation groups previously shown by biochemical and physiological criteria to play a direct role in initiation.

When De Lucia and Cairns (1969) isolated a DNA polymerase I-defective mutant of *E. coli* [albeit now recognized to be sufficiently "leaky" to supply the pol I function (see below) critical to DNA replication], that was nevertheless competent in DNA replication, the profitability of combining these two approaches became obvious: The laboratories of Hurwitz and of Kornberg, following the example set by Schaller and colleagues (1972), set about analyzing DNA replication in *E. coli* using well-defined viral DNA as probes. Even with these simple probes, DNA replication turned out to be unexpectedly complicated, and the definition of the *in vivo* system has not yielded readily to the two forms of analysis inherent in the combined approach. The one that attempts to identify all the components involved by complementation analysis has proved inadequate because there are necessary macromolecular components for which the corresponding genes have not been identified by mutation. In the other form of analysis, in which *in vitro* systems have been assembled *de novo* with purified components, there are situations in which genes have been identified as being necessary *in vivo,* but whose products can be dispensed with in the synthetic *in vitro* reconstruction systems (for example, see Tomizawa *et al.,* 1975). Such a situation is to be expected when regulatory, in addition to catalytic, proteins are involved in the system *in vivo.*

Additional complexity has been conferred by the discovery of more than one kind of DNA replication system in *E. coli,* even though these systems have many common components. Not only does it seem that viral DNA's, plasmids, and the bacterial chromosome depend on different systems of replication, but the replication of the chromosome itself also may come under the aegis of more than one system, depending on physiological circumstances.

D. "Replisome" Concept

Chapter 2, Section II, provides a catalog of the genetic loci which affect the initiation and replication of DNA in *E. coli.* Listed in this same chapter is the evidence for the idea that some of the gene products act in concert as part of a replication complex (Bleecken, 1971; K. G. Lark, 1972; Kornberg, 1974). Indeed, whereas the distinction between the functions of initiation and elongation is conserved, the distinction between the entities responsible for these functions is rapidly vanishing—both phenomena resulting from the discovery of more and more mutations which affect one of these functions but map by the criterion of complementation to the same gene (Wechsler, 1973, 1975; Kogoma and Lark, 1975; K. G. Lark and Wechsler, 1975). If initiation and replication are executed by a complex, then just such a blurring of distinction would be expected, as one would expect to find mutations within the same cistron affecting either the subfunction performed by that component or the functional or the structural integrity of the complex of which it is a necessary part.

For example, Wickner and Hurwitz (1974, 1975b) have shown that the purified dna B and dna C products interact physically and functionally *in vitro* in the presence of ATP. Also, mutations in the *P* locus of λ phage affecting replication of λ DNA, can be suppressed by certain mutations in the *dna B* locus of the host, *E. coli* (Georgopoulos and Herskowitz, 1971). Furthermore, Takahashi (1975) has extended the finding of Tomizawa (1971) to demonstrate by the same criterion, a three-way interaction between the lambda gene products of *O* and *P* and the host *dna B* locus product. Detailed reviews of the role of multienzyme systems in DNA replication can be found in Kornberg (1974) and Schekman and associates (1974).

E. Mechanism of Initiation

Given the inability of the known DNA polymerases to initiate DNA synthesis in the absence of a 3′-OH primer (Kornberg, 1974), two classes of initiation mechanisms have been suggested as a solution to the priming problem (see Klein and Bonhoeffer, 1972, for review).

One class involved priming either with an RNA oligonucleotide laid down by a self-priming RNA polymerase or with a priming protein molecule, a mechanism suggested by Denhardt (1972) that has received recent experimental support (Johnson and Sinsheimer, 1974) for the priming of ϕX174 replication.

The other involves nicking the parental strand and using the 3′-OH so exposed to prime replication that proceeds continuously around the remaining intact circular template strand. Concurrent with this is displacement of its complement, to which the nascent strand is an extension. This "rolling circle" (Gilbert and Dressler, 1968) mode of asymmetric replication has been invoked to explain the replication of such diverse molecules as those from the small single-stranded circular DNA viruses (Knippers *et al.*, 1969; Ray, 1969; Dressler, 1970), the circular double-stranded DNA for lambda late in the infection cycle (Skalka *et al.*, 1972), for the DNA replication that accompanies the transfer of the Hfr DNA to the F^- cell during conjugation in *E. coli* (Ohki and Tomizawa, 1968; Rupp and Ihler, 1968), and for the amplification of ribosomal RNA specifying DNA of the nucleolar region in eukaryotic chromosomes (Hourcade *et al.*, 1973).

Given that a round of replication can be initiated by transient priming by a parental strand, negative evidence against parental strand priming would have to be impossibly rigorous to rule it out (Kuempel, 1972; Stein and Hanawalt, 1972). Yoshikawa (1970) has positive evidence for parental strand priming in *B. subtilis*, demonstrated by the presence of a single-stranded DNA fragment containing material presumably laid down in the previous replication cycle and density-labeled precursors laid down in the current replication cycle. However, to interpret this evidence, it is critical that one know whether germinating spores actually initiate a round of replication or merely continue a blocked round of replication.

1. D-Loop Initiation Structures

In those DNA-replicating systems most amenable to study, it appears that the double-stranded circular template is covalently closed and negatively supercoiled (Vinograd *et al.*, 1968; Levine, 1974). Electron microscope studies (Delius and Worcel, 1974) on the isolated *E. coli* chromosome "nucleoid" confirm that it is a compact, membrane-associated structure wherein the chromosome is stabilized by RNA in a rosette of many negatively supercoiled loops (Stonington and Pettijohn, 1971; Worcel and Burgi, 1972; Pettijohn and Hecht, 1973). These loops are capable of independent relaxation by treatment with DNase, suggesting that the *E. coli* chromosome is topologically equivalent to a covalently closed, negatively supercoiled DNA molecule.

Since early replication intermediates in SV40 (Mayer and Levine, 1972) and mitochondrial DNA (Kasamatsu *et al.*, 1971) form a high frequency peak in the steady-state distribution of replication forms, initiation and early replication may be driven by the negative free energy change that accompanies partial melting out of a negatively supercoiled, covalently closed parental template. Replication proceeds along one strand paripassu with this energetically favorable melting-out process to form a structure recognizable under the electron microscope as a "displacement loop" or "D-loop" (Robberson *et al.*, 1972; Levine, 1974). (See Chapter 4 on organellar chromosome replication for further discussion of this subject.)

It seems reasonable to assume that the bidirectional replication of the negatively supercoiled *E. coli* chromosome is initiated with the formation of one (or more) D-loops. It is interesting to note that circular chloroplast DNA replicates from two specifically located and closely juxtaposed D-loops that subsequently expand by asymmetric replication toward one another to achieve ultimately symmetrical bidirectional replication (Kolodner and Tewari, 1975).

If progress of replication from the D-loop stage is contingent on, and rate-limited by, nicking of one or the other parental strand to allow unwinding, then it is possible that replication could be arrested at the transition from a D-loop structure to a replicating structure. This would constitute a control point in the regulation of further progress of an initiated round of replication. Such a situation could account for the abovementioned observations of Yoshikawa, in that the physical location for the control of a round of replication may not coincide with the molecular origin. For this possibility to be ruled out, specific molecular evidence of the kind described below is needed.

2. *Initiation by RNA Priming*

Both Messer (1972) and K. G. Lark (1972) have implicated RNA synthesis unrelated to messenger RNA synthesis in initiation in *E. coli* by showing that initiation is sensitive to inhibition by RNA synthesis-inhibiting drugs in a period of the bacterial cell cycle when protein synthesis inhibition no longer blocks subsequent initiation. Similar evidence also implicates RNA synthesis in the initiation of DNA replication in *B. subtilis* (Laurent, 1973). However, as Lark pointed out, this observation does not necessitate that the RNA play a priming role for initiation. Indeed, he favored the possibility that the RNA may form part of an initiator complex in much the same way as RNA forms part of a ribosome.

Details of the initiation process have been forthcoming from the *in vitro* DNA initiation and synthesis systems using viral and plasmid DNA's as probes for identifying the host components involved and their sequence of action.

Different virus DNA probes are strikingly different in the array of *E. coli* host components that they depend on for the initiation and synthesis of their DNA. For example, the Fd phage DNA depends on host RNA polymerase (Wickner *et al.*, 1972), DNA polymerase III (a complex of dna E protein and another protein), elongation factors I and II, and DNA-binding protein (Hurwitz and Wickner, 1974; Geider and Kornberg, 1974), while ϕX174 phage DNA depends on dna B protein, dna C protein, DNA binding protein, and two other proteins for which corresponding genes have not been identified. Together, all these proteins prosecute an initiation step in the absence of DNA precursors. Once this initiation step has been consummated, the cooperative action of the *dna G* gene product, elongation factors I and II, and DNA pol III in the presence of DNA precursors leads to synthesis of the complementary DNA strand to the single-stranded input DNA template. There is an extensive controversy about whether RNA priming is implicated in the initiation of ϕX174 DNA replication. The Hurwitz group has not established a need for RNA precursors in the *in vitro* reaction, while the Kornberg group has consistently done so. This probably reflects differences in the workup procedures in these two laboratories and, no doubt, as their respective systems become more accurately defined, the basis for this difference will emerge.

Another probe, which should be more akin to the chromosome of *E. coli* in terms of its initiation and replication needs, is the plasmid. Most work has been done on the replication of the small nontransferable plasmid Col E1. The replication of this plasmid depends on the presence of dna A protein (Goebel, 1970), DNA pol I (Kingsbury and Helinski, 1970), but not DNA pol III (Goebel, 1972). It is dependent on RNA priming for initiation both *in vivo* (Blair *et al.*, 1972) and *in vitro* (Tomizawa *et al.*, 1975). The plasmid DNA can be isolated in a supercoiled state complexed with a molecule of protein; treatment with either ionic detergent, protease, or alkali somehow causes this protein to nick a particular strand of the supercoiled plasmid, leaving the protein bound to the 5′-end (Blair *et al.*, 1971; Clewell and Helinski, 1972). Since this plasmid replicates in the closed theta configuration (Sakakibara and Tomizawa, 1974a,b) from a fixed origin unidirectionally (Tomizawa *et al.*, 1974) and is initiated by RNA primer (Tomizawa *et al.*, 1975), the nicking protein probably performs the role of a "swivelase," allowing unwinding of the otherwise topologically closed molecule.

Although Col E1 is unable to replicate in a temperature-sensitive dna A host at the nonpermissive temperature, it can do so at the nonpermissive temperature in the presence of a low concentration of chloramphenicol. Therefore, the role of *dna A* gene product in initiation may be to act as an anti-repressor of a constitutive repressor of initiation (Goebel, 1974). In this connection, Schekman and associates (1975) have noted that, while *dna A* gene product is needed in crude extracts for the initiation of ϕX174 replication, it is not essential in the reconstituted replication system prepared from purified components, a system in which the initiation repressor presumably is not present. Goebel further found that the *in vivo* replication of Col E1 at the nonpermissive temperature in the presence of chloramphenicol loses its sensitivity to rifampicin and, paradoxically, actually is stimulated by this drug.

A model proposed by Messer and co-workers (1975) can explain this paradoxical behavior of Col E1 replication in *ts dna A* mutants at the nonpermissive temperature. In this model, the function of the dna A protein is to overcome the inhibition of RNA polymerase transcription after the RNA polymerase has bound to a promotor site. This inhibitor is a labile protein, which, when in combination with the dna A protein, leads to a positive stimulation of RNA polymerase-catalyzed transcription. In the absence of a dna A product and diminished levels of the inhibitor protein, conditions are more appropriate for *dna G* RNA polymerase-catalyzed RNA transcription than they are for normal RNA polymerase transcription. It is suggested that prevention of binding of the latter polymerase by the addition of rifampicin leaves the way open for a rifampicin-resistant RNA polymerase to act. Support for this suggestion comes from the identification of the dna G product in *E. coli* as a rifampicin-resistant RNA polymerase (Bouché *et al.*, 1975). This polymerase is responsible for the initiation priming of the coliphage G4 DNA replication and possibly of ϕX174 DNA replication. Thus, it may also be responsible for the RNA priming of Okazaki piece initiation during ongoing replication (Sugino *et al.*, 1972).

Direct studies on the role of RNA priming in the initiation of the *E. coli* replication cycle have led to the isolation of an RNA–DNA joint molecule, whose formation is contingent upon the presence of both active *dna C* and *dna A* gene products (Messer *et al.*, 1975). The dna A protein may have the function outlined in the model above, while the dna C protein may be required for the transition from RNA priming to DNA synthesis to be effected.

Further evidence for the involvement of RNA priming in another *E. coli* initiation system has been obtained by Hayes and Szybalski (1973) for lambda phage DNA replication. Here, the *E. coli* proteins specified by genes *dna B, dna G,* and possibly the beta subunit of RNA polymerase cooperate with the proteins specified by the lambda phage genes *O* and *P* to transcribe an RNA oligonucleotide from a specific site on the lambda DNA. This RNA can be isolated and used *in vitro* to initiate lambda DNA replication, and there is evidence that it also can act in the trans configuration as an initiating primer *in vivo* (Hayes, 1975).

In summary, it appears that the initiation of a normal round of replication in *E. coli* involves the cooperative participation of a number of proteins synthesized at different times in the cell cycle. Some of these proteins must be synthesized anew for each round of replication, whereas others may be recyclable. Initiation occurs at a specific site on the chromosome and in the cell envelope. Explicit evidence for RNA priming has been obtained which involves, after binding of RNA polymerase, a regulatory step controlling transcription in which the dna A protein is probably an anti-repressor. The transition from RNA to DNA synthesis appears to be governed by the dna C protein.

It is unlikely that nicking of the parental template is involved directly in initiation in *E. coli,* although there has been a suggestion that nicking and parental strand priming may be involved in *B. subtilis* initiation. This seems unlikely, since evidence for an RNA step in initiation has been observed by Laurent (1973). Nor does nicking seem necessary at the beginning of a round of replication, since the template is probably already thrown into negative supercoils favoring melting out and the formation of D-loops by initiation and partial replication along one template strand.

Initiation and subsequent replication probably are prosecuted by a complex of the proteins involved in both processes. The degree to which proteins are added or detached in the processes of forming a membrane–template site–replisome initiation complex that initiates, replicates, and finally terminates replication is not known.

F. Abnormal Initiation

In some instances, perturbation of the cell cycle can lead to initiation of a cycle or cycles of DNA replication that is anomalous either in its requirements for initiation or in the pattern of the ensuing replication.

The first of these to be described was premature initiation brought on in a population of *E. coli* subjected to a period of DNA synthesis inhibition, such as by depriving a thymine auxotroph of thymine. After recovering from the inhibition, these cells are observed to initiate a round of replication before the ongoing round has finished. This "premature" round of replication apparently involves only one of the two available origins on the two daughter segments of the chromosome (Pritchard and Lark, 1964).

Temperature cycling between the restrictive and permissive temperatures for temperature-sensitive *dna B* mutants also leads to the onset of premature and cumulative initiation events (Worcel, 1970; Schwartz and Worcel, 1971). The initiation events are asymmetric in that only one of the two available origins is a substrate for initiation. Worcel's evidence indicates that the source of this asymmetry is inherent in the DNA molecule, and gives rise to the "rolling circle" mode of replication. By contrast, if the premature replication induced by thymine starvation is asymmetric, then K. G. Lark's evidence (1969) suggests that the source of this asymmetry neither is inherent in the polarity of the DNA molecule, nor is based on the relative age of its component strands.

Once the anomalous round of DNA replication induced by a period of thymine starvation has terminated, the pattern of DNA replication in the culture returns to normal. However, if the culture is transferred to growth medium containing chloramphenicol for about one generation period after the end of the thymine starvation period, the culture continues to synthesize DNA for hours, even in the continued presence of chloramphenicol, acting as if the replisome were not inactivated at the end of a round of replication (Rosenberg *et al.*, 1969; Kogoma and Lark, 1970).

Kogoma and Lark (1975) have contemplated the differences between the anomalous persistent replication and normal replication, namely, in the responses to the effects of various temperature-sensitive mutants carrying mutations in functions directly concerned with DNA initiation and replication. They have proposed that, while the replication machinery responsible for persistent replication requires all the known initiation and replication proteins, the anomalous replisome is somehow structurally different from the normal replisome.

G. Regulation of Initiation

Both positive and negative models have been invoked to account for the regulation of initiation. The earliest model, postulating the accumulation of initiation protein to a triggering threshold during cell growth (Maaløe and Kjeldgaard, 1966), still survives as a strong contender, elaborated by Donachie (1968) in a particularly elegant manner.

In this positive model, Donachie assumes that cells grow exponentially over the cell cycle, and, as Schaecter and associates (1958) found for *Salmonella*, the average cell mass increases as an exponential function of the frequency of cell division (I/T). Using Helmstetter and Cooper's constants for the DNA replication cycle time (C) and the time to cell division from the end of a replication cycle (D), Donachie showed that, if initiation took place when the ratio of cell mass to the number of initiation sites reached a critical value, then the time of initiation within the cell cycle would vary with growth rate (Figure 2). The predictions of this model agree with the observations made by Cooper and Helmstetter (1968) in their studies of *E. coli* B/r.

The negative models of regulation involve the intermittent production of an inhibitor of initiation by the transient expression of a gene early in

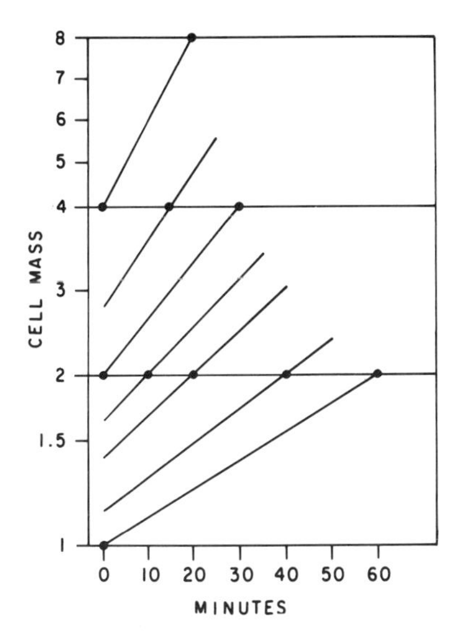

Fig. 2. Course of the increase in mass of individual cells with different rates of growth. The cells are assumed to grow exponentially over a single cell cycle which starts at the end of the previous division at 0 minutes. The next division, at the end of the cycle, takes place after the initial mass has doubled. The cell mass doubles in 20, 25, 30, 35, 40, 50, or 60 minutes. The initial mass at Time 0 is taken to be proportional to the average mass of a population of cells growing at the same growth rate. Each line therefore shows the course of mass increase of individual cells from division at 0 minutes to the next division after the mass has doubled. Because there is a constant time (60 minutes for *E. coli* B/r[1]) between the initiation of a round of DNA replication and cell division, it is possible to calculate the time when initiation occurs relative to cell division in cells growing at different rates. These times are marked as solid circles on the corresponding curves of mass increase. It therefore can be seen that the masses at which initiations take place are the same or multiples of the same cell mass for cells growing at all growth rates. (From Donachie, 1968, reprinted with permission.)

the DNA replication cycle. Initiation ensues with either the dilution of this inhibitor by further growth to an ineffective level (Pritchard *et al.*, 1969), or with the continuous synthesis of an anti-inhibitor to a concentration sufficient to neutralize the inhibitor (Rosenberg *et al.*, 1969). Pritchard and colleagues (1969) felt that a particularly attractive feature of their negative model of control was its property of self-regulation, since chance initiation at an inappropriate time in the cell cycle would be compensated for by a negatively correlated adjustment to the cell cycle in the next generation.

Dalbey (1975) points out that perturbations in the initiator accumulation model also will be damped. This occurs not by compensation, but by gradual dilution of the perturbation effect over successive generations.

Since negative correlations in the length of generation times have been observed between successive generation intervals within a cell lineage (Koch, 1966), the weight of the evidence favors some form of negative control mechanism, such as dilution of repressor or its titration by a continuously produced anti-repressor.

Some inkling of the difficulties that will be encountered in unraveling the initiation problem is provided by a recent observation (Helmstetter, 1974): that the sensitivity of initiation to inhibitors varies with the time at which initiation occurs in the cell cycle.

Thus, the nature of the definitive experiments that will be capable of ruling out one or another model of initiation, let alone the identification of the regulatory elements of the system, seem far from obvious at this time.

III. DNA REPLICATION

A. Topology

One would expect bidirectional replication to be ubiquitous among prokaryotes as well as eukaryotes, given that both initiation of a round of replication and the initiation of Okazaki small pieces during the course of replication appear to be primed by RNA (Kornberg, 1974). The exception to this general rule may reflect breakage of the parental template strand to give an open mode of unidirectional replication resembling the "rolling circle" replication (Figure 3). This seems to be the case for the replication accompanying the conjugational transfer of the *E. coli* chromosome or episomes (Ohki and Tomizawa, 1968; Rupp and Ihler, 1968; Ihler and Rupp, 1969). Alternatively, it may reflect the intervention of some control system that actively blocks the progress of replication in one of the two directions.

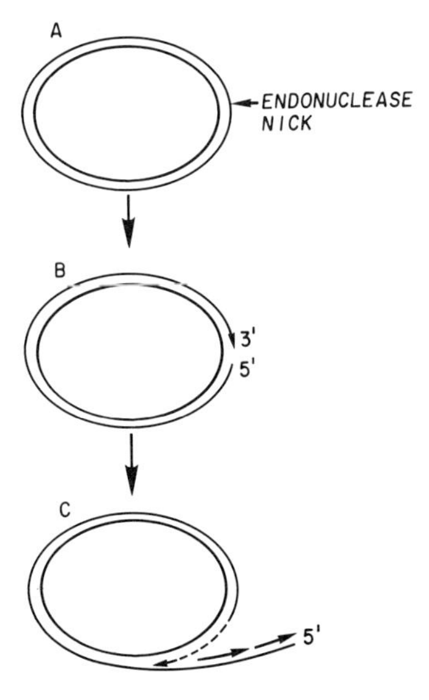

Fig. 3. A replicating circular double-stranded DNA molecule showing the rolling circle mode of replication that is initiated by an endonuclease-introduced nick into one of the strands (A), followed by DNA synthesis from the 3′-hydroxy terminus of the nicked strand with concurrent displacement of the nicked parental strand from its 5′-hydroxy terminus as replication proceeds (B). Replication of the displaced strand template presumably is initiated intermittently by RNA primer sequences laid down by an RNA polymerase. (C) The arrows denote the $5' \rightarrow 3'$ direction of synthesis of the nascent DNA strands.

In the latter alternative, this hypothetical control system may be identical to that which determines where in a circular DNA molecule a round of replication ends. (Whether such specific termination sites exist will be taken up in Section IV.) Thus, in those instances in which DNA replication appears to be grossly unidirectional, closer inspection may reveal it to be a rather extreme form of asymmetrical bidirectional replication. Such unidirectional replication could arise through the intervention of some type of control system which actively inhibits replication in one direction. Unidirectional replication on a closed circular template that may reflect such a control has been seen in the plasmid Col E1 (Lovett) *et al.*, 1974; Tomizawa *et al.*, 1974) and in the coliphage P2 (Schnös and Inman, 1971).

Just as *E. coli* employs two modes of replication, one associated with the normal cell cycle, which is closed and bidirectional, and the other associated with DNA transfer during conjugation, which is open (rolling circle) and unidirectional, so is this option available to some other organisms. For example, early in infection bacteriophage lambda DNA replicates in a closed bidirectional mode from a fixed origin (Schnös and

Inman, 1970). Then it replicates unidirectionally on the open rolling circle mode to produce long linear concatemers (Skalka *et al.*, 1972; Takahashi, 1974; Bastia *et al.*, 1975). Chloroplast DNA also replicates bidirectionally in the closed form, and also in the open rolling circle form, with the latter initiating in the region where the bidirectional mode of replication terminates (Kolodner and Tewari, 1973). The adaptiveness of this topological plasticity for DNA replication is probably related to the need to mature the λ phage DNA into the phage head, but its role in chloroplast DNA replication is quite obscure.

There is some controversy as to whether asymmetric bidirectional replication occurs. Because this asymmetry implies a specific site for termination of a replication cycle, as opposed to a casual site where the opposing replication forks meet, the question of the existence of asymmetrical bidirectional replication is important.

O'Sullivan and colleagues (1975a) have retracted (O'Sullivan *et al.*, 1975b) their findings to support earlier indications (e.g., Hara and Yoshikawa, 1973) that, while it is bidirectional (Wake, 1973) in the circular *B. subtilis* chromosome (Wake, 1972), DNA replication is only partially or asymmetrically so. Wake (1974) now has obtained direct autoradiographic evidence for symmetric bidirectional replication in the *B. subtilis* chromosome. In *E. coli,* replication is bidirectional (Prescott and Kuempel, 1972) and appears to be symmetrically so at the gross level of whole chromosome autoradiography (Rodriguez *et al.*, 1973; Kuempel *et al.*, 1973).

Helinski (1975) refers to an example of asymmetric bidirectional replication in the *E. coli* plasmid R6K, which has a terminus 20% of the plasmid length from the origin of replication, and which appears to replicate initially in one direction to the terminus, and then in the other. Thus, replication not only is asymmetric, but is asynchronous as well.

These events argue for the existence of an initiation control system that determines not only the time and direction of replication, but also a specific site of termination.

B. Replication Rate

We have seen in *E. coli* that the rate of DNA synthesis varies, accompanying changes in cell growth rate as nutritional conditions are varied. The cell accommodates these changes in DNA synthesis rate largely by altering the frequency of initiation of rounds of replication, rather than by altering the rate of progress of the replication fork (Helmstetter *et al.*, 1968), at least in cells growing with a generation time less than 60 minutes.

The data of Cooper and Helmstetter (1968) indicate that, under slow growth conditions, the time for a round of replication (C) assumes a value that approximates two-thirds of one round of generation time. That is, for generation times greater than 60 minutes, the DNA replication cycle time changes from an absolute constant to a proportional value. However, there are two published experiments that purport to demonstrate that the DNA replication cycle time is conserved as a constant over a wide range of generation times.

In the first of these experiments, the average DNA content per cell was measured as a function of generation time (Kubitschek and Freedman, 1971). It can be calculated that the average DNA content of cells taken from a culture in which the DNA replication time is two-thirds of a generation time and in which the initiation of a round of synthesis occurs at the time of cell division is 1.61 genome equivalents. This value was estimated from the data for cultures with a generation time (T) of 63 minutes (where $T = C + D$, and D is the time from the end of a round of replication to cell division). With increasing generation time, the average DNA content was observed to move toward an asymptotic value that corresponded to one genome equivalent. Earlier data appear to indicate that DNA replication takes place late in the cell division cycle (C. Lark, 1966; Kubitschek *et al.*, 1967). As generation times become larger, and the replication period occupies a smaller proportion of the cell cycle, the average DNA content per cell would asymptote to one genome equivalent.

There are two problems with this interpretation. First, the estimate of the DNA content per cell as equivalent to 1.6 genomes is based on a few scattered points, and a reliable absolute determination of one genome's worth of DNA just does not exist. Second, the evidence that the replication period is late in the cell division cycle is far from consistent. A reexamination of C. Lark's data (1966) indicates that the replication period increases as the cell generation time increases, and if anything, the replication occurs earlier rather than later in the cell cycle. Since Lark's data are based on an autoradiographic estimation of the proportion of the cell cycle devoted to DNA replication, they are inherently more reliable than the data of Kubitschek and Freedman. While the latter may have ruled out an extreme and unlikely hypothesis that the DNA replication cycle occupies a constant proportion of the cell cycle when cells are growing slowly, it is far from established that the DNA replication cycle is, as claimed, a constant 47 minutes for generation times up to 50 hours.

The second experiment that purports to demonstrate the constancy of C depends on measuring, by estimation of gene dosage, the proportion of cells in a culture that have a replication fork located between two markers on the *E. coli* chromosome (Chandler *et al.*, 1975). Again, the data rule out the unlikely hypothesis that C occupies a constant proportion of the cell cycle, since this exceeds a certain critical value of $T = C + D$. Unfortunately, the data are sparse and do not extend beyond 220 minutes for the cell cycle time and once again provide only feeble support for the unlikely hypothesis that C is a constant irrespective of generation time. In addition, the data do not discriminate among a host of empirical and uninteresting hypotheses in which C increases with generation time in some nonlinear fashion.

If, as seems likely from C. Lark's data (1966) and confirmed by Bird and Lark (1968, 1970), the length of the replication does vary with nutritional conditions, then the question is immediately raised: How does the cell regulate the rate of DNA synthesis, given that it is a discontinuous process involving the initiation and the elongation of Okazaki pieces? This problem was studied by Bird and Lark (1970), who found that the rate of fork progress was modulated by the frequency of initiation of Okazaki pieces, rather than to the rate of their elongation.

That the rate of progress of the replication fork can be altered by precursor availability has been explicitly demonstrated by Pritchard and Zaritsky (1970) and Zaritsky and Pritchard (1971). These workers showed that, when the thymine concentration in the growth medium was limiting for thymine-requiring auxotrophs, the overall growth rate and the DNA synthesis rate did not appear to be affected. However, closer examination revealed that the replication cycle took longer. This was compensated for by maintaining frequency of initiation, whereupon the average number of replication forks per chromosome increased.

C. Other Factors Affecting the Rate of DNA Replication

Urban and Lark (1971) report that when DNA replication is initiated at one temperature and the culture is transferred to another temperature, the rate of replication is influenced by the temperature at which initiation occurs. This led to the speculation that the lipid composition of the membrane site of the replisome varies with the temperature at which the replisome site is assembled, an alteration which in turn affects subsequent replisome function. Fralick and Lark (1973) have obtained evidence to support this possibility.

D. Nutritional Status and Cell Cycle Strategies

Now that the topics of initiation and replication have been covered, it is appropriate to examine the validity of one of the assumptions made in Section I, namely, that the effect of variation in nutritional conditions can be accounted for in terms of a single descriptive parameter, the generation or doubling time (T). In short, are there different nutritional circumstances which give the same T, but which give rise to different cell cycle strategies? We have already seen that the answer is yes when the availability of thymine is manipulated. An observation that casts doubt on the validity of this assumption is the puzzling behavior of the $15T^-$ strain of *E. coli* when grown on succinate as a carbon source (K. G. Lark and Lark, 1965). While T (succinate) is 70 minutes and T (glucose) is 40 minutes, the average DNA content of the cells in each of these conditions is approximately the same, indicating that T cannot be used as a single descriptor of the cell cycle.

Finally, Urban (1974) has isolated a mutant of *E. coli* $15T^-$ that grows faster than its wild-type counterpart does on nutritionally poor carbon sources. This increase in growth rate does not extend to a decrease in the DNA replication cycle time, again serving to caution against use of T as the single determinant of cell cycle strategy.

IV. TERMINATION OF A ROUND OF REPLICATION AND SEPARATION OF DAUGHTER MOLECULES

Two central problems of termination merit consideration. First, how does a bidirectionally replicating circular DNA molecule, subject to intermittent nicking in the remaining unreplicated segment, manage to allow unwinding of the parental template and to complete a round of replication while its separate daughter molecules conserve their circularity? Second, is the act of termination site specific with respect to the DNA molecule?

Corollary problems stem from each of these central questions. From the first arises the question of whether separation of the daughter molecules precedes or follows the completion of a replication cycle. From the second, there is the question whether termination is catalyzed or controlled by termination-specific proteins.

First, let us consider the topological problem involved in two replicating forks approaching each other as the replication cycle draws to a close. Unwinding of the remaining parental template is contingent upon the introduction of a nick in one or the other strand. If such a nick persists, then the daughter molecule with the nicked template strand will loose its circularity as replication terminates and will separate from the other daughter molecule. Given the frequently observed asymmetry in the degree of advance of daughter strands into the fork, it is highly likely that each daughter molecule will not be completely replicated at the moment of separation. Thus, the linear daughter should have at least one protruding parental single strand (3′-end), while the circular daughter molecules will have a gap in its nascent strand.

Formally, then, separation precedes termination and one of the two daughter molecules is linear. How does it circularize? Or is there a way in which replication can be terminated and the products separated with conservation of circularity in both daughter molecules?

Circularization could be achieved if the nick occurred within one of the two symmetrically equivalent sites of palindromic termination sequence. Accordingly, the linear daughter molecules could then separate with its parental strand being single stranded at both the 5′- and 3′-ends.

Such a model implies stringent requirements, not only for a specific termination site (the palindromic sequence) and for a specific endonuclease to nick this site, but also for some means of ensuring that the linear daughter molecule always separates with its two mutually complementary ends in a single-stranded state (Figure 4). How the latter requirement could be met is not known. It should be commented here that this type of

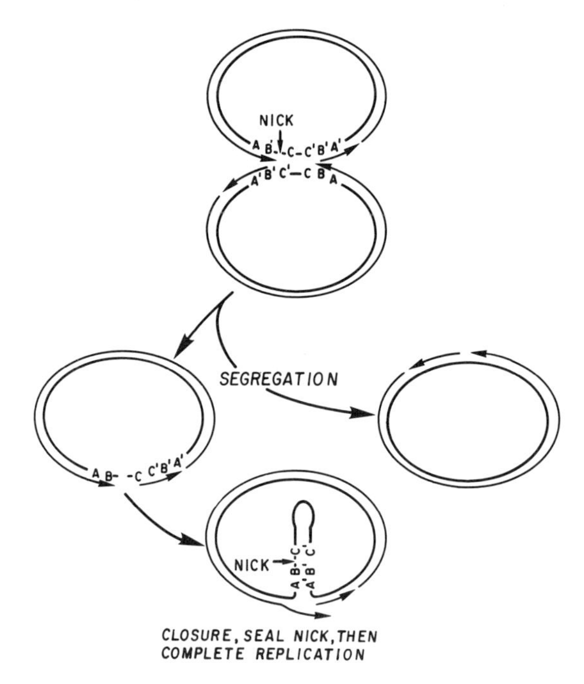

Fig. 4. Segregation of daughter molecules after nicking within a palindromic termination sequence. One daughter is circular and the other noncircular. The latter possesses single-stranded termini, both 3′-hydroxy and 5′-hydroxy encompassing the palindromic sequence, thus allowing recircularization by self-annealing of these complementary single-stranded extensions.

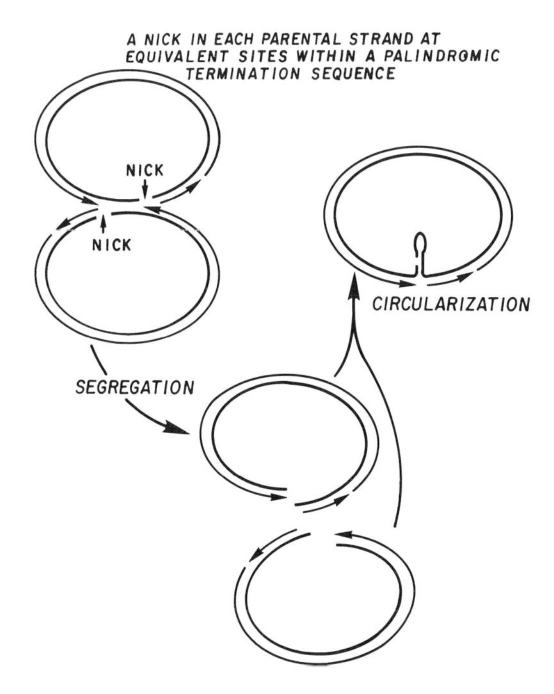

Fig. 5. Segregation of two noncircular daughter molecules from a parental molecule that has been nicked by a sequence-specific endonuclease in each of the equivalent nucleotide sequences of a palindromic termination region. Circularization occurs as in Fig. 3 by self-annealing within the palindromic sequence.

separation and termination mechanism would work equally well if the restriction endonuclease introduced two staggered nicks on opposite parental strands, with separation of both daughter molecules in linear but circularizable forms (Figure 5). This would work only if the endonuclease nicks are disposed on the parental strands such that each extending nascent strand encounters the nick on the opposite parental strand first.

If the two opposing replication forks met and passed each other without the last few turns of the helix being unwound because a nick did not occur between the opposing forks during termination, the daughter molecules would be topologically interlocked with several twists. The number of twists would correspond to the number of turns of the parental helix that were replicated in the absence of a relaxing nick. While such an interlock could be resolved by breaking one of the daughter molecules and circularity regained by the mechanism previously discussed, there is a way in which resolution can be obtained while the circularity of the daughter molecules is conserved. Again, the properties of a palindrome are employed in this model, but in a way that relates to its possible ability to exist in two alternate forms—the energetically favorable orthodox form, or the less probable cruciform state (Gierer, 1966).

In this model (Figure 6), it is assumed that the terminal region of the replicating DNA molecules enters the cruciform state before opposing replication forks reach it. The positive supercoils so produced then are dissipated by a transient nick in the vicinity. Further replication in the absence of a nick will neutralize the negative supercoil potential of this cruciform structure and will allow the separation of the resulting daughter molecules.

Such elaborate site-specific termination models can be bypassed if the "relaxing" protein (a) found in *E. coli* (Wang, 1971) performs the transient nicking function. This protein can relax supercoiled circular DNA molecules by reversible nicking while conserving the circularity.

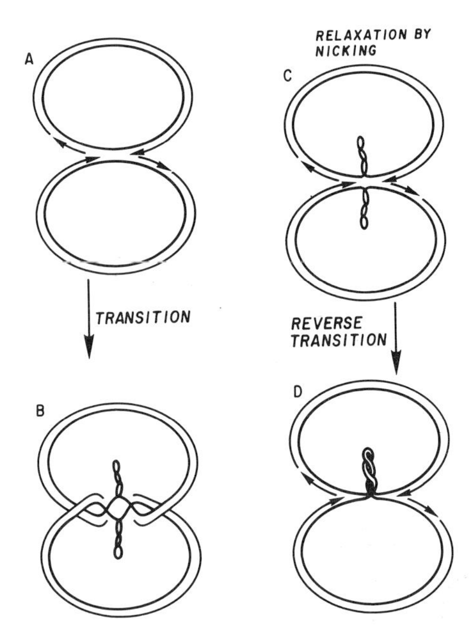

Fig. 6. Completion of a replication cycle in the absence of nicks utilizing "stored" negative hypercoiling arising out of a transition of the palindromic sequence from the orthodox Watson–Crick structure to the "cruciform" structure. Near the end of the replication cycle (A) a palindromic termination sequence undergoes a transition to the cruciform state with consequent introduction of positive supercoiling (B). This supercoiling is then relaxed by a nicking–releasing event (C), and the molecule undergoes a reverse transition to the orthodox Watson–Crick duplex structure, with the introduction of negative supercoils (D). Bidirectional replication then neutralizes these negative supercoils to allow topological separation of the daughter molecules in an intact circular state.

Since the detailed topology of termination of bacterial chromosomes has not become accessible to study, little can be said at this time about the relevance of the foregoing speculations to the termination and separation processes in bacteria. However, much more is known about the details of termination and separation for viral and plasmid DNA molecules. First, SV40 replicates with the separation of its daughter molecules before replication is completed, giving rise to daughter molecules with single-stranded gaps (Fareed *et al.,* 1973). Second, bacterial plasmid DNA molecules (Inselburg and Fuke, 1971; Kupersztoch and Helinski, 1973; Novick *et al.,* 1973; Sheehy and Novick, 1975), SV40 (Jaenisch and Levine, 1973), polyoma (Meinke and Goldstein, 1971), and bacteriophage lambda DNA (Sogo *et al.,* 1975) occur during replication as interlocked dimers, in which one molecule is frequently supercoiled and the other relaxed. Such structures would be expected to occur as intermediates in the previously described models for termination. This would occur in cases in which separation is dependent on endonucleolytic cleavage of one of the daughter molecules, or it would take place as a consequence of delayed nicking, in which replication was terminated before separation. Since inhibition of protein synthesis in SV40-infected monkey cells increases the steady-state frequency of interlocked dimers, Levine (1974) argues that specific proteins may be needed to effect separation of the daughter molecules.

The data of Marunouchi and Messer (1973) support termination in *E. coli* as a regulated process, rather than the casual consequence of the collision of two opposing replicating forks. They observed that, if bacteria were allowed to "terminalize" their rounds of DNA replication during a period of amino acid starvation and then were challenged with the DNA synthesis inhibitor, nalidixic acid, before the amino acid was replenished, the cells would fail to divide. When the nalidixic acid challenge was delayed until after the amino acids had been replenished, cell division ensued. These results suggest that completion of a round of replication is contingent upon the presence of a particular newly synthesized protein, namely, a "termination-specific" protein; the completion of a round of replication in turn provides a necessary signal for cell division.

Dix and Helmstetter (1973) have cast some doubt on this interpretation; they observed that thymine starvation would not substitute for nalidixic acid challenge. Since the amount of DNA that is supposed to be involved in this "termination" step is equivalent to 0.5% of the chromosome (Marunouchi and Messer, 1973), it is necessary for Dix and Helmstetter to demonstrate that, under their conditions, the DNA equivalent of thymine precursor pool is less than 0.5%. The same can be said for the results of Jones and Donachie (1973), who also found that thymine starvation applied concurrently with relief from a period of rifampicin inhibition did not prevent subsequent cell division. Thus, while the evidence for the existence of a particular termination site, and of proteins that regulate and catalyze termination of DNA replication is not very convincing, it would seem that the need for precision in the termination process dictates the necessity of a regulatory mechanism.

V. SEGREGATION

A. The Cell Division Cycle in Relation to the Replication Cycle

We have seen that the event which coordinates the cell division cycle and the replication cycle is the termination of a round of replication (Clark, 1968; Pierucci and Helmstetter, 1969; Hoffman *et al.*, 1972; Marunouchi and Messer, 1973). Jones and Donachie (1973) have clearly shown that two categories of protein synthesis must be completed in order for cell division to take place. In the undisturbed cell cycle, proteins necessary for cell division are synthesized during the replication cycle. These proteins can accumulate during a period of DNA synthesis arrest. The synthesis of the second protein category is contingent upon the completion of the DNA replication cycle.

Since the interval between termination of the replication cycle and subsequent cell division can be shortened by experimentally lengthening the DNA replication cycle time, Meacock and Pritchard (1975) argue that the event determining the time of cell division is not contingent upon termination of the DNA replication cycle. Rather, it presumably is dependent upon the last protein synthesized in the set constituting the first category, i.e., those synthesized before the termination of the replication cycle.

In Section I we assumed that the interval (D) between the end of a round of DNA synthesis and cell division was a constant. It now would be appropriate for this supposition to be withdrawn and replaced with one allowing D to vary not only with nutritional circumstances, but also with the strain of *E. coli* (Jones and Donachie, 1973).

B. Segregation

If cell division ensues and is contingent upon the completion of a round of replication, then no matter how rapidly a culture is growing, there will be two separable chromosomal structures, irrespective of their state of replication at the time of division. We already have seen that the complexity of these structures in terms of the number of replicating forks that they carry at any instant is clearly a function of growth rate. But having two separable chromosomal structures at the time of cell division is not sufficient to guarantee their orderly segregation into each daughter cell. A specific mechanism is needed for this to occur. The nature of this mechanism and the evidence for it have already been discussed in Section I, where a physical connection was shown to exist between the daughter strand and some part of the cell envelope. Growth of the cell leads to a separation of these attachment sites with their associated daughter molecules and to their ultimate segregation into each daughter cell.

Two kinds of evidence are necessary to support directly such a picture of segregation. The first involves a demonstration of the "growing apart" of the attachment sites. Since it is not known to which of the three layers of the gram-negative cell wall the chromosome of *E. coli* is attached, a review of our knowledge of the pattern of cell envelope growth is useful (see below) to see if any layer has a growth pattern that would fulfill the need for segregation.

The second kind of evidence relates to the nature of the cell envelope–chromosome connection. Its basis is the assumption of regularity to the rules determining this attachment, and that, once made, the attachment is permanent. Neither of these prejudices need obtain, but if a regularity in segregation pattern is observed, then these assumptions are justified.

1. Cell Envelope Growth Patterns

Because of the mobility of the components of cell membranes, it would seem unlikely that the innermost layer of the cell envelope, the cell membrane, can provide a physical reference point for chromosome segregation. Yet there is considerable evidence, especially in the gram-positive bacterium, *B. subtilis,* that the nucleoid is attached to an invagination of the cell membrane, the mesosome (see review by Ryter, 1968). The insertion of new material into the membrane occurs equatorially (Jacob *et al.*, 1966), thus providing a mechanism for separation of the mesosomes (Ryter and Jacob, 1964). The outer cell envelope of gram-positive bacteria also appears to grow by equatorial intercalation of new material (Cole and Hahn, 1962). However, Chung and co-workers (1964) have obtained evidence that the new material can be inserted equatorially or at the poles, at least in *Bacillus megatherium.*

The situation appears to be much more complex in gram-negative bacteria, in which there are two distinct layers in the cell envelope external to the cell membrane. The layer adjacent to the cell membrane, the sacculus, is virtually one gigantic netlike macromolecule. It is composed of murein (peptidoglycan) and gives the bacterial cell its shape and rigidity. Early studies seemed to indicate that the sacculus grows by intercalation of material at random sites over its entire surface (Van Tubergen and Setlow, 1961; Lin *et al.,* 1971). Later, Ryter and colleagues (1973) revealed that the sacculus grows by equatorially localized intercalation, but the newly deposited murein then is dispersed over the sacculus as it grows. Thus, both processes, i.e., sacculus extension to accommodate cell growth and sacculus partition formation to accommodate cell division, can be achieved by equatorial deposition. The significance of the subsequent randomization of this material is elusive.

Again, the sacculus growth pattern satisfies the need for providing an orderly segregation mechanism, as required by the replicon hypothesis (Jacob *et al.,* 1963). That chromosome attachment to the sacculus may provide the physical basis for segregation is supported further by the observation that DNA synthesis inhibition blocks penicillin-induced localized blebbing in the region of the sacculus where cell division occurs (Schwarz *et al.,* 1969).

Even the outermost (lipoprotein) layer of the gram-negative cell envelope may grow by equatorial intersusception (Ryter *et al.,* 1975). There is, however, conflicting evidence on this point (Begg and Donachie, 1973).

Thus, all three layers may have the appropriate growth pattern to serve as the mechanical base for segregation. Further information is required to establish which layer of the cell envelope—if it is just one layer—performs this function for the cell.

2. *Segregation Patterns*

Given the double-stranded structures of DNA and its semiconservative replication, the simplest mechanism of segregation would involve attachment of the older template strand to an envelope site. This mechanism also would require the formation of an attachment to the other side of the cell-envelope growth zone by the younger identical template strand during the replication cycle.

If, once formed, these attachments are conserved, one would expect to see a regularity in the pattern of disposition of the DNA strands present in the founding cell among the cells in the clone derived from it, provided the relative positions of the cells in the resulting clone are a faithful reflection of their lineage. Furthermore, if the regions of the cell envelope to which the chromosomes are attached are also conserved in substance, then one would expect to see cosegregation of the DNA and cell envelope material that were contemporaneous in synthesis. Both of these expectations have experimental support.

Eberle and Lark (1966) and Chai and Lark (1967) obtained clear evidence for such a regular segregation pattern in *B. subtilis* and for the cosegregation of contemporaneously synthesized cell envelope material and DNA in *Lactobacillus acidophilus*, respectively. In contrast, similar studies performed in the Pasteur laboratories indicated random segregation for both *E. coli* and *B. subtilis* (Ryter *et al.*, 1968). One reason for this conflict may lie in the cultural conditions used to grow the cells. In Lark's laboratory, cells were cultured in minimal media, whereas the Pasteur group apparently cultured theirs in rich media. In the latter instance, the regularity of the segregation patterns could be obscured, at least initially, by the contemporaneous presence of more than one set of replication forks per replicating chromosome structure.

Paying careful attention to the nutritional status of the cells and to ensuring an accurate reflection of cell lineage, Pierucci and Zuchowski (1973) established unambiguously that the segregation pattern for *E. coli* cells is orderly. This confirms the expectation that a strand of DNA is permanently attached to the cell envelope when it first is used as a template.

VI. CONCLUSIONS

From the foregoing, we have seen that, apart from a minor modification involving the substitution of bidirectional for unidirectional replication, the replicon model of Jacob and associates (1963) has served as an excellent guide to subsequent experimental design. This model has survived essentially intact, receiving some compatible embellishments.

We also have seen a tendency to confuse constancy with order whereby expectations are generated which are as easy to test as they are unreal. Perhaps a constant kernel exists in the innermost physiological recess of the bacterial cell; however, all that has been observed to date is the variation of the cell's attributes with the vicissitudes of its existence, mayhap to "canonize" that kernel which as yet remains to be revealed.

ACKNOWLEDGMENTS

The author is indebted to George S. and Dolores Dore Eccles for the support afforded him while writing this manuscript as the incumbent of their presidential endowed chair; to the National Science Foundation for their support; and to Lou Ann Thomas for her able editorial assistance.

REFERENCES

Bastia, D., Sueoka, N., and Cox, E. C. (1975). Studies on the late replication of phage lambda: Rolling-circle replication of the wild type and a partially suppressed strain, *O*am29 *P*am80. *J. Mol. Biol.* **98**, 305–320.

Begg, K. J., and Donachie, W. D. (1973). Topography of outer membrane growth in *E. coli*. *Nature (London), New Biol.* **245,** 38–39.

Bird, R., and Lark, K. G. (1968). Initiation and termination of DNA replication after amino acid starvation of *E. coli* $15T^-$. *Cold Spring Harbor Symp. Quant. Biol.* **33,** 799–808.

Bird, R. E., and Lark, K. G. (1970). Chromosome replication in *Escherichia coli* $15T^-$ at different growth rates: Rate of replication of the chromosome and the role of formation of small pieces. *J. Mol. Biol.* **49,** 343–366.

Blair, D. G., Clewell, D. B., Sheratt, D. J., and Helinski, D. R. (1971). Strand-specific supercoiled DNA–protein relaxation complexes: Comparison of the complexes of bacterial plasmids $ColE_1$ and $ColE_2$. *Proc. Natl. Acad. Sci. U.S.A.* **68,** 210–214.

Blair, D. G., Sherratt, D. J., Clewell, D. B., and Helinski, D. R. (1972). Isolation of supercoiled colicinogenic factor E DNA sensitive to ribonuclease and alkali. *Proc. Natl. Acad. Sci. U.S.A.* **69,** 2518–2522.

Bleecken, S. (1971). "Replisome"-controlled initiation of DNA replication. *J. Theor. Biol.* **32,** 81–92.

Bonhoeffer, F., and Gierer, A. (1963). On the growth mechanism of the bacterial chromosome. *J. Mol. Biol.* **7,** 534–540.

Bouché, J.-P., Zechel, K., and Kornberg, A. (1975). *dna G* Gene product, a rifampicin-resistant RNA polymerase, initiates the conversion of a single-stranded coliphage DNA to its duplex replicative form. *J. Biol. Chem.* **250,** 5995–6001.

Cairns, J. (1963). The chromosome of *Escherichia coli*. *Cold Spring Harbor Symp. Quant. Biol.* **28,** 43–46.

Chai, N. C., and Lark, K. G. (1967). Segregation of deoxyribonucleic acid in bacteria association of the segregating unit with the cell envelope. *J. Bacteriol.* **94,** 415–421.

Chandler, M., Bird, R. E., and Caro, L. (1975). The replication time of the *Escherichia coli* K12 chromosome as a function of cell doubling time. *J. Mol. Biol.* **94,** 127–132.

Chung, K. L., Hawirko, R. Z., and Isaac, P. K. (1964). Cell wall replication. I. Cell wall growth of *Bacillus cereus* and *Bacillus megatherium*. *Can. J. Microbiol.* **10,** 43–48.

Clark, D. J. (1968). The regulation of DNA replication and cell division in *E. coli* B/r. *Cold Spring Harbor Symp. Quant. Biol.* **33,** 823–838.

Clewell, D. B., and Helinski, D. R. (1972). Effect of growth conditions on the formation of the relaxation complex of supercoiled ColE1 deoxyribonucleic acid and protein in *Escherichia coli*. *J. Bacteriol.* **110,** 1135–1146.

Cole, R. M., and Hahn, J. J. (1962). Cell wall replication in *Streptococcus pyogenes*. Immunofluorescent methods applied during growth show that new wall is formed equatorially. *Science* **135,** 722–724.

Cooper, S., and Helmstetter, C. H. (1968). Chromosome replication and the division cycle of *Escherichia coli* B/r. *J. Mol. Biol.* **31,** 519–540.

Costerton, J. W., Ingram, J. M., and Cheng, K. J. (1974). Structure and function of the cell envelope of gram-negative bacteria. *Bacteriol. Rev.* **38,** 87–110.

Cuzin, F., and Jacob, F. (1965). Existence chez *Escherichia coli* d'une unité génétique de ségrégation formée de différents réplicons. *C.R. Hebd. Seances Acad. Sci.* **260,** 5411–5414.

Dalbey, M. (1975). Initiation of DNA replication—*Escherichia coli*. Biology Thesis, University of California, Santa Cruz.

Delius, H., and Worcel, A. (1974). Electron microscopic visualization of the folded chromosome of *E. coli*. *J. Mol. Biol.* **82,** 107–109.

De Lucia, P., and Cairns, J. (1969). Isolation of an *E. coli* strain with a mutation affecting DNA polymerase. *Nature (London)* **224,** 1164–1166.

Denhardt, D. T. (1972). A theory of DNA replication. *J. Theor. Biol.* **34,** 487–508.

Dix, D. E., and Helmstetter, C. E. (1973). Coupling between chromosome completion and cell division in *Escherichia coli*. *J. Bacteriol.* **115,** 786–795.
Donachie, W. D. (1968). Relationship between cell size and the time of initiation of DNA replication. *Nature* (*London*) **219,** 1077–1079.
Dressler, D. (1970). The rolling circle for ϕX DNA replication. II. Synthesis of single-stranded circles. *Proc. Natl. Acad. Sci. U.S.A.* **67,** 1934–1942.
Eberle, H., and Lark, K. G. (1966). Chromosome segregation in *B. subtilis*. *J. Mol. Biol.* **22,** 183–186.
Fareed, G. C., McKerlie, M. L., and Salzman, N. P. (1973). Characterization of simian virus 40 DNA component II during viral DNA replication. *J. Mol. Biol.* **74,** 95–111.
Fielding, P. E., and Fox, C. F. (1970). Evidence for stable attachment of DNA to membrane at the replication origin of *Escherichia coli*. *Biochem. Biophys. Res. Commun.* **41,** 157–162.
Fralick, J., and Lark, K. G. (1973). Evidence for the involvement of unsaturated fatty acids in initiating chromosome replication in *Escherichia coli*. *J. Mol. Biol.* **80,** 459–475.
Ganesan, A. T., and Lederberg, J. (1965). A cell-membrane bound fraction of bacterial DNA. *Biochem. Biophys. Res. Commun.* **18,** 824–835.
Geider, K., and Kornberg, A. (1974). Conversion of the M13 viral single strand to the double-stranded replicative forms by purified proteins. *J. Biol. Chem.* **249,** 3999–4005.
Georgopoulos, C. P., and Herskowitz, I. (1971). *Escherichia coli* mutants blocked in lambda DNA synthesis. *In* "The Bacteriophage Lambda" (A. D. Hershey, ed.), pp. 553–564. Cold Spring Harbor Lab., Cold Spring Harbor, New York.
Gierer, A. (1966). Model for DNA and protein interactions and the function of the operation. *Nature* (*London*) **212,** 1480–1481.
Gilbert, W., and Dressler, D. (1968). DNA replication: The rolling circle model. *Cold Spring Harbor Symp. Quant. Biol.* **33,** 473–484.
Gilbert, W., Maizels, N., and Maxam, A. (1973). Sequences of controlling regions of the lactose operon. *Cold Spring Harbor Symp. Quant. Biol.* **38,** 845–855.
Goebel, W. (1970). Studies on extrachromosomal DNA elements. Replication of the colicinogenic factor Col E_1 in two temperature-sensitive mutants of *Escherichia coli* defective in DNA replication. *Eur. J. Biochem.* **15,** 311–320.
Goebel, W. (1972). Replication of the DNA of the colicinogenic factor E_1 (Col E_1) at the restrictive temperature in a DNA replication mutant thermosensitive for DNA polymerase III. *Nature* (*London*), *New Biol.* **237,** 67–70.
Goebel, W. (1974). Studies on the initiation of plasmid DNA replication. *Eur. J. Biochem.* **41,** 51–62.
Goldstein, A., and Brown, B. J. (1961). Effect of sonic oscillation upon "old" and "new" nucleic acids in *Escherichia coli*. *Biochim. Biophys. Acta* **53,** 19–28.
Gudas, L. I., and Pardee, A. B. (1974). Deoxyribonucleic acid synthesis during the division cycle of *Escherichia coli*—a comparison of strains B/r, K-12, 15, and $15T^-$ under conditions of slow "growth." *J. Bacteriol.* **117,** 1216–1223.
Hara, H., and Yoshikawa, H. (1973). Asymmetric bidirectional replication of *Bacillus subtilis* chromosome. *Nature* (*London*), *New Biol.* **244,** 200–203.
Hayes, S. (1975). Role of oop RNA in initiation of λ DNA replication. *ICN–UCLA Symp. Mol. Cell. Biol.* Vol. 3, pp. 486–512.
Hayes, S., and Szybalski, W. (1973). Possible RNA primer for DNA replication in coliphage lambda. *Fed. Proc., Fed. Am. Soc. Exp. Biol.* **32,** 529 (abstr.).
Helinski, D. R., *et al.* (1975). Modes of plasmid DNA replication in *Escherichia coli*. *ICN–UCLA Symp. Mol. Cell. Biol.* Vol. 3, pp. 514–536.

Helmstetter, C. E. (1974). Initiation of chromosome replication in *Escherichia coli*. II. Analysis of the control mechanism. *J. Mol. Biol.* **84,** 21–36.

Helmstetter, C. E., and Cooper, S. (1968). DNA synthesis during the division cycle of rapidly growing *Escherichia coli* B/r. *J. Mol. Biol.* **31,** 507–518.

Helmstetter, C. E., and Pierucci, O. (1968). Cell division during inhibition of deoxyribonucleic acid synthesis in *E. coli*. *J. Bacteriol.* **95,** 1627–1633.

Helmstetter, C. E., Cooper, S., Pierucci, O., and Revelos, E. (1968). On the bacterial life sequence. *Cold Spring Harbor Symp. Quant. Biol.* **33,** 809–822.

Hoffman, B., Messer, W., and Schwarz, U. (1972). Regulation of polar cup formation in the life cycle of *E. coli*. *J. Supramol. Struct.* **1,** 29–37.

Hourcade, D., Dressler, D., and Wolfson, J. (1973). The amplification of ribosomal RNA genes involves a rolling circle intermediate. *Proc. Natl. Acad. Sci. U.S.A.* **70,** 2926–2930.

Hurwitz, J., and Wickner, S. (1974). Involvement of two protein factors and ATP in the *in vitro* DNA synthesis catalyzed by DNA polymerase III of *E. coli*. *Proc. Natl. Acad. Sci. U.S.A.* **71,** 6–10.

Ihler, G., and Rupp, W. D. (1969). Strand-specific transfer of donor DNA during conjugation in *E. coli*. *Proc. Natl. Acad. Sci. U.S.A.* **63,** 138–143.

Inselburg, J., and Fuke, M. (1971). Isolation of catenated and replicating DNA molecules of colicin factor E1 from minicells. *Proc. Natl. Acad. Sci. U.S.A.* **68,** 2839–2842.

Jacob, F., Brenner, S., and Cuzin, F. (1963). On the regulation of DNA replication in bacteria. *Cold Spring Harbor Symp. Quant. Biol.* **28,** 329–348.

Jacob, F., Ryter, A., and Cuzin, F. (1966). On the association between DNA and membrane in bacteria. *Proc. R. Soc. London, Ser. B* **164,** 267–278.

Jaenisch, R., and Levine, A. J. (1973). DNA replication in SV40-infected cell. VIII. formation. *J. Mol. Biol.* **73,** 199–212.

Johnson, P. H., and Sinsheimer, R. L. (1974). Structure of an intermediate in the replication of bacteriophage ϕX174 deoxyribonucleic acid. The initiation site for DNA replication. *J. Mol. Biol.* **83,** 47–61.

Jones, N. C., and Donachie, W. D. (1973). Chromosome replication, transcription and the control of cell division in *E. coli*. *Nature* (*London*), *New Biol.* **243,** 100–103.

Kasamatsu, H., Robberson, D. L., and Vinograd, J. (1971). A novel closed-circular mitochondrial DNA with properties of a replicating intermediate. *Proc. Natl. Acad. Sci. U.S.A.* **68,** 2252–2257.

Kingsbury, D. T., and Helinski, D. R. (1970). DNA polymerase as a requirement for the maintenance of the bacterial plasmid colicinogenic factor E_1. *Biochem. Biophys. Res. Commun.* **41,** 1538–1544.

Klein, A., and Bonhoeffer, F. (1972). DNA replication. *Annu. Rev. Biochem.* **41,** 301–332.

Knippers, R., Whalley, J. M., and Sinsheimer, R. L. (1969). The process of infection with bacteriophage ϕX174. XXX. Replication of double-stranded ϕX DNA. *Proc. Natl. Acad. Sci. U.S.A.* **64,** 275–282.

Koch, A. L. (1966). On evidence supporting a deterministic process of bacterial growth. *J. Gen. Microbiol.* **43,** 1–5.

Kogoma, T., and Lark, K. G. (1970). DNA replication in *Escherichia coli:* Replication in absence of protein synthesis after replication inhibition. *J. Mol. Biol.* **52,** 143–164.

Kogoma, T., and Lark, K. G. (1975). Characterization of the replication of *Escherichia coli* DNA in the absence of protein synthesis. Stable DNA replication. *J. Mol. Biol.* **94,** 243–256.

Kolodner, R., and Tewari, K. K. (1973). Replication of circular chloroplast DNA. *J. Cell Biol.* **59,** 174a.

Kolodner, R., and Tewari, K. K. (1975). Chloroplast DNA from higher plants replicates by both the Cairns and the rolling circle mechanism. *Nature (London)* **256,** 708–711.

Kornberg, A. (1974). "DNA Synthesis." Freeman, San Francisco, California.

Kubitschek, H. E., and Freedman, M. L. (1971). Chromosome replication and the division cycle of *E. coli* B/r. *J. Bacteriol.* **107,** 95–99.

Kubitschek, H. E., Bendigkeit, H. E., and Loken, M. R. (1967). Onset of DNA synthesis during the cell cycle in chemostat cultures. *Proc. Natl. Acad. Sci. U.S.A.* **57,** 1611–1617.

Kuempel, P. L. (1972). Molecular weight of deoxyribonucleic acid synthesized during initiation of chromosome replication in *Escherichia coli. J. Bacteriol.* **112,** 114–125.

Kuempel, P. L., Maglothin, P., and Prescott, D. M. (1973). Bidirectional termination of chromosome replication in *Escherichia coli. Mol. Gen. Genet.* **125,** 1–8.

Kupersztoch, Y. M., and Helinski, D. R. (1973). A catenated DNA molecule as an intermediate in the replication of the resistance transfer factor. *Biochem. Biophys. Res. Commun.* **54,** 1451–1459.

Lark, C. (1966). Regulation of deoxyribonucleic acid synthesis in *Escherichia coli.* Dependence on growth rates. *Biochim. Biophys. Acta* **119,** 517–525.

Lark, C., and Lark, K. G. (1964). Evidence for two distinct aspects of the mechanism regulating chromosome replication in *Escherichia coli. J. Mol. Biol.* **10,** 120–136.

Lark, K. G. (1966a). Regulation of chromosome replication and segregation in bacteria. *Bacteriol. Rev.* **30,** 3–32.

Lark, K. G. (1966b). Chromosome replication in *Escherichia coli. In* "Cell Synchrony" (I. L. Cameron, ed.), Chapter 4, pp. 54–80. Academic Press, New York.

Lark, K. G. (1969). Role of deoxynucleotide strand age on polarity in determining asymmetric chromosome replication after thymine starvation of *E. coli* 15T$^-$. *J. Mol. Biol.* **44,** 217–231.

Lark, K. G. (1972). Evidence for the direct involvement of RNA in the initiation of DNA replication in *Escherichia coli* 15T$^-$. *J. Mol. Biol.* **64,** 47–60.

Lark, K. G., and Bird, R. E. (1965). Segregation of the conserved units of DNA in *Escherichia coli. Proc. Natl. Acad. Sci. U.S.A.* **54,** 1444–1450.

Lark, K. G., and Lark, C. (1965). Regulation of chromosome replication in *Escherichia coli.* Alternate replication of two chromosomes at slow growth rates. *J. Mol. Biol.* **13,** 105–126.

Lark, K. G., and Lark, C. (1966). Regulation of chromosome replication in *E. coli.* A comparison of the effects of phenethyl alcohol treatment with those of amino acid starvation. *J. Mol. Biol.* **20,** 9–19.

Lark, K. G., and Renger, H. (1969). Initiation of DNA replication in *Escherichia coli* 15T$^-$: Chronological dissection of the physiological processes required for initiation. *J. Mol. Biol.* **42,** 221–235.

Lark, K. G., and Wechsler, J. A. (1975). DNA replication in *dna B* mutants of *Escherichia coli.* Gene product interaction and synthesis of 4 S pieces. *J. Mol. Biol.* **92,** 145–164.

Lark, K. G., Repko, T., and Hoffman, E. J. (1963). The effect of amino acid deprivation on subsequent deoxyribonucleic acid replication. *Biochim. Biophys. Acta* **76,** 9–24.

Lark, K. G., Eberle, H., Consigli, R. A., Minocha, H. C., Chai, N., and Lark, C. (1967). Chromosome segregation and the regulation of DNA replication. *In* "Organizational Biosynthesis" (H. J. Vogel, J. O. Lampen, and V. Bryson, eds.), pp. 63–89. Academic Press, New York.

Laurent, S. J. (1973). Initiation of DNA replication in a *ts* mutant of *Bacillus subtilis.* Evidence for a transcriptional step. *J. Bacteriol.* **116,** 141–145.

Levine, A. J. (1974). The replication of papovavirus DNA *Prog. Med. Virol.* **17,** 1–37.

Lin, E. C. C., Hirota, Y., and Jacob, F. (1971). On the process of cellular division in *Escherichia coli.* VI. Use of a Methocel–autoradiographic method for the study of cellular division in *E. coli. J. Bacteriol.* **108,** 375–385.

Lovett, M. A., Katz, L., and Helinski, D. R. (1974). Unidirectional replication of plasmid Col E1 DNA. *Nature (London)* **251,** 337–340.

Maaløe, O., and Hanawalt, P. C. (1961). Thymine deficiency and the normal DNA replication cycle. *J. Mol. Biol.* **3,** 144–155.

Maaløe, O., and Kjeldgaard, N. O. (1966). "Control of Macromolecular Synthesis." Benjamin, New York.

Marunouchi, T., and Messer, W. (1973). Replication of a specific terminal chromosome segment in *Escherichia coli* which is required for cell division. *J. Mol. Biol.* **78,** 211–228.

Mayer, A., and Levine, A. J. (1972). DNA replication in SV40-infected cells. VIII. The distribution of replicating molecules at different stages of replication in SV40-infected cells. *Virology* **50,** 328–338.

Meacock, P. A., and Pritchard, R. H. (1975). Relationship between chromosome replication and cell division in a thymineless mutant of *Escherichia coli* B/r. *J. Bacteriol.* **122,** 931–942.

Meinke, W., and Goldstein, D. A. (1971). Studies on the structure and formation of polyma DNA replicative intermediates. *J. Mol. Biol.* **61,** 543–563.

Messer, W. (1972). Initiation of DNA replication in *E. coli* B/r: Chronology of events and transcriptional control of initiation. *J. Bacteriol.* **112,** 7–12.

Messer, W., Dankwarth, L., Tippe-Schindler, R., Womack, J. E., and Zahn, G. (1975). Regulation of the initiation of DNA replication in *E. coli.* Isolation of I-RNA and the control of I-RNA synthesis. *ICN–UCLA Symp. Mol. Cell. Biol., 1975* Vol. 3, pp. 602–617.

Novick, R. P., Smith, K., Sheehy, R. J., and Murphy, E. (1973). A catenated intermediate in plasmid replication. *Biochem. Biophys. Res. Commun.* **54,** 1460–1469.

Ohki, M., and Tomizawa, J. (1968). Asymmetric transfer of DNA strands in bacterial conjugation. *Cold Spring Harbor Symp. Quant. Biol.* **33,** 651–658.

Oishi, M., Yoshikawa, H., and Sueoka, N. (1964). Synchronous and dichotomous replication of the *Bacillus subtilis* chromosome during spore germination. *Nature (London)* **204,** 1069–1073.

O'Sullivan, M. A., Howard, K., and Sueoka, N. (1975a). Location of a unique replication terminus and genetic evidence for partial bidirectional replication in the *Bacillus subtilis* chromosome. *J. Mol. Biol.* **91,** 15–38.

O'Sullivan, M. A., Howard, K., and Sueoka, N. (1975b). On the bidirectional replication of the *Bacillus subtilis* chromosome. *J. Mol. Biol.* **99,** 347–348.

Parker, D. L., and Glaser, D. A. (1974). Chromosomal sites of DNA–membrane attachment in *E. coli. J. Mol. Biol.* **87,** 153–168.

Pettijohn, D. E., and Hecht, R. (1973). RNA molecules bound to the folded bacterial genome stabilize DNA folds and segregate domains of supercoiling. *Cold Spring Harbor Symp. Quant. Biol.* **38,** 31–41.

Pierucci, O., and Helmstetter, C. E. (1969). Chromosome replication, protein synthesis and cell division in *E. coli. Fed. Proc., Fed. Am. Soc. Exp. Biol.* **28,** 1755–1760.

Pierucci, O., and Zuchowski, C. (1973). Nonrandom segregation of DNA strands in *E. coli* B/r. *J. Mol. Biol.* **80,** 477–503.

Powell, O. E. (1956). Growth rate and generation time of bacteria with special reference to continuous cultures. *J. Gen. Microbiol.* **15,** 492–511.

Prescott, D. M., and Kuempel, P. (1972). Bidirectional replication of the chromosome in *Escherichia coli*. *Proc. Natl. Acad. Sci. U.S.A.* **69**, 2840–2845.

Pritchard, R. H., and Lark, K. G. (1964). Induction of replication by thymine starvation at the chromosome origin in *Escherichia coli*. *J. Mol. Biol.* **9**, 288–307.

Pritchard, R. H., and Zaritsky, A. (1970). Effect of thymine concentration on the replication velocity of DNA in a thymineless mutant of *E. coli*. *Nature* (*London*) **226**, 126–131.

Pritchard, R. H., Barth, P. T., and Collins, J. (1969). Control of DNA synthesis in bacteria. *Symp. Soc. Gen. Microbiol.* **19**, 263–297.

Ray, D. S. (1969). Replication of bacteriophage M13. II. The role of replicative forms in single-strand synthesis. *J. Mol. Biol.* **43**, 631–643.

Robberson, D. L., Kasamatsu, H., and Vinograd, J. (1972). Replication of mitochondrial DNA. Circular replicative intermediates in mouse L cell. *Proc. Natl. Acad. Sci. U.S.A.* **69**, 737–741.

Rodriguez, R., Dalbey, M., and Davern, C. I. (1973). Autoradiographic evidence for bidirectional DNA replication in *Escherichia coli*. *J. Mol. Biol.* **74**, 599–604.

Rosenberg, B. H., Cavalieri, L. F., and Ungers, G. (1969). The negative control mechanism for *E. coli* DNA replication. *Proc. Natl. Acad. Sci. U.S.A.* **63**, 1410–1417.

Rupp, W. D., and Ihler, G. (1968). Strand selection during bacterial mating. *Cold Spring Harbor Symp. Quant. Biol.* **33**, 647–650.

Ryter, A. (1968). Association of the nucleus and the membrane of bacteria: A morphological study. *Bacteriol. Rev.* **32**, 39–54.

Ryter, A., and Jacob, F. (1964). Étude au microscopic électronique de la liaison entre noyau et mésosome chez *Bacillus subtilis*. *Ann. Inst. Pasteur, Paris* **107**, 384–400.

Ryter, A., Hirota, Y., and Jacob, F. (1968). DNA-membrane complex and nuclear segregation in bacteria. *Cold Spring Harbor Symp. Quant. Biol.* **33**, 669–676.

Ryter, A., Hirota, Y., and Schwarz, U. (1973). Process of cellular division in *Escherichia coli* growth pattern of *E. coli* murein. *J. Mol. Biol.* **78**, 185–195.

Ryter, A., Shuman, H., and Schwartz, M. (1975). Integration of the receptor for bacteriophage lambda in the outer membrane of *Escherichia coli:* Coupling with cell division. *J. Bacteriol.* **122**, 295–301.

Sakakibara, Y., and Tomizawa, J-I. (1974a). Replication of colicin E1 plasmid DNA in cell extracts. *Proc. Natl. Acad. Sci. U.S.A.* **71**, 802–806.

Sakakibara, Y., and Tomizawa, J.-I. (1974b). Replication of colicin E1 plasmid DNA in cell extracts. II. Selective synthesis of early replicative intermediates. *Proc. Natl. Acad. Sci. U.S.A.* **71**, 1403–1407.

Schaecter, M., Maaløe, O., and Kjeldgaard, N. O. (1958). Dependency on medium and temperature of cell size and chemical composition during balanced growth of *Salmonella typhimurium*. *J. Gen. Microbiol.* **19**, 592–606.

Schaller, H., Otto, B., Nüsslein, V., Huf, J., Herrmann, R., and Bonhoeffer, F. (1972). Deoxyribonucleic acid replication *in vitro*. *J. Mol. Biol.* **63**, 183–200.

Schekman, R., Wickner, W., Westergaard, O., Brutlag, D., Geider, K., Bertsch, L. L., and Kornberg, A. (1972). Initiation of DNA synthesis: Synthesis of ϕX174 RF requires RNA synthesis resistant to rifampicin. *Proc. Natl. Acad. Sci. U.S.A.* **69**, 2691–2695.

Schekman, R., Weiner, A., and Kornberg, A. (1974). Multienzyme systems of DNA replication. *Science* **186**, 987–993.

Schekman, R., Weiner, J. H., Weiner, A., and Kornberg, A. (1975). Ten proteins required for conversion of ϕX174 single-stranded DNA to duplex form. *J. Biol. Chem.* **250**, 5859–5865.

Schnös, M., and Inman, R. B. (1970). Position of branch points in replicating λ DNA. *J. Mol. Biol.* **51**, 61–73.

Schnös, M., and Inman, R. B. (1971). Starting point and direction of replication in P2 DNA. *J. Mol. Biol.* **55,** 31–38.

Schwartz, M., and Worcel, A. (1971). Reinitiation of chromosome replication in a *ts* DNA mutant of *E. coli*. Synchronization of chromosome replication after temperature shifts. *J. Mol. Biol.* **61,** 329–342.

Schwarz, U., Asmus, A., and Frank, H. (1969). Autolytic enzymes and cell division of *E. coli*. *J. Mol. Biol.* **41,** 419–429.

Sheehy, R. J., and Novick, R. P. (1975). Studies on plasmid replication. V. Replicative intermediates. *J. Mol. Biol.* **93,** 237–253.

Skalka, A., Poonian, M., and Bartl, P. (1972). Concatemers in DNA replication: Electron microscope studies of partially denatured intracellular lambda DNA. *J. Mol. Biol.* **64,** 541–550.

Smith, D. W., and Hanawalt, P. C. (1967). Properties of the growing point region of the bacterial chromosome. *Biochim. Biophys. Acta* **149,** 519–531.

Sobell, H. M. (1972). Molecular mechanism for genetic recombination. *Proc. Natl. Acad. Sci. U.S.A.* **69,** 2483–2487.

Sogo, J. M., Greenstein, M., and Skalka, A. (1976). The circle mode of replication of bacteriophage lambda: The role of covalently closed templates and the formation of mixed catenated dimers. *J. Mol. Biol.* **103,** 537–562.

Stein, G. H., and Hanawalt, P. C. (1972). Initiation of the DNA replication cycle in *Escherichia coli:* Linkage of origin daughter DNA to parental DNA? *J. Mol. Biol.* **64,** 393–407.

Stonington, O. G., and Pettijohn, D. E. (1971). The folded genome of *Escherichia coli* isolation in a protein–DNA–RNA complex. *Proc. Natl. Acad. Sci. U.S.A.* **68,** 6–9.

Sueoka, N., and Quinn, W. G. (1968). Membrane attachment of the chromosome replication origin in *Bacillus subtilis*. *Cold Spring Harbor Symp. Quant. Biol.* **33,** 695–705.

Sugino, A., Hirose, S., and Okazaki, R. (1972). RNA-linked nascent DNA fragments in *Escherichia coli*. *Proc. Natl. Acad. Sci. U.S.A.* **69,** 1863–1867.

Takahashi, S. (1974). The *rolling-circle* replicative structure of a bacteriophage λ DNA. *Biochem. Biophys. Res. Commun.* **61,** 657–663.

Takahashi, S. (1975). Role of genes *O* and *P* in the replication of bacteriophage λ DNA. *J. Mol. Biol.* **94,** 385–396.

Tomizawa, J.-I. (1971). Functional cooperation of genes *O* and *P*. *In* "The Bacteriophage Lambda" (A. D. Hershey, ed.), pp. 549–552. Cold Spring Harbor Lab., Cold Spring Harbor, New York.

Tomizawa, J.-I., Sakakibara, Y., and Kakefuda, T. (1974). Replication of colicin E1 plasmid DNA in cell extracts. Origin and direction of replication. *Proc. Natl. Acad. Sci. U.S.A.* **71,** 2260–2264.

Tomizawa, J.-I., Sakakibara, Y., and Kakefuda, T. (1975). Replication of colicin E1 plasmid DNA added to cell extracts. *Proc. Natl. Acad. Sci. U.S.A.* **72,** 1050–1054.

Treick, R. W., and Konetzka, W. A. (1964). Physiological state of *Escherichia coli* and the inhibition of deoxyribonucleic acid synthesis by phenethyl alcohol. *J. Bacteriol.* **88,** 1580–1584.

Tremblay, G. Y., Daniels, M. J., and Schaecter, M. (1969). Isolation of a cell membrane–DNA-nascent RNA complex from bacteria. *J. Mol. Biol.* **40,** 65–76.

Urban, J. E. (1974). Isolation of a new division altered mutant of *E. coli* $15T^-$. *Biochem. Biophys. Res. Commun.* **60,** 1475–1481.

Urban, J. E., and Lark, K. G. (1971). DNA replication in *Escherichia coli* $15T^-$ growing at 20°C. *J. Mol. Biol.* **58,** 711–724.

Van Tubergen, R. P., and Setlow, R. B. (1961). Quantitative radioautographic studies on exponentially growing cultures of *E. coli*. *Biophys. J.* **1,** 589–625.

Vinograd, J., Lebowitz, J., and Watson, R. (1968). Early and late helix coil transitions in closed circular DNA. The number of superhelical turns in polyoma DNA. *J. Mol. Biol.* **33,** 173–197.

Wada, C., and Yura, T. (1974). Phenethyl alcohol resistance in *E. coli*. III. A temperature-sensitive mutation (*dna P*) affecting DNA replication. *Genetics* **77,** 199–220.

Wake, R. G. (1972). Visualization of reinitiated chromosomes in *Bacillus subtilis*. *J. Mol. Biol.* **68,** 501–509.

Wake, R. G. (1973). Circularity of the *Bacillus subtilis* chromosome and further studies on its bidirectional replication. *J. Mol. Biol.* **77,** 569–575.

Wake, R. G. (1974). Termination of *Bacillus subtilis* chromosome replication as visualized by autoradiography. *J. Mol. Biol.* **86,** 223–231.

Wang, J. C. (1971). Interaction between DNA and an *Escherichia coli* protein. *J. Mol. Biol.* **55,** 523–533.

Wechsler, J. A. (1973). Complementation analysis of mutations at the *dna B*, *dna C* and *dna D* loci. *DNA Synth. in Vitro, Proc. Annu. Harry Steenbock Symp., 2nd, 1972* pp. 375–382.

Wechsler, J. A. (1975). Genetic and phenotypic characterization of *dna C* mutations. *J. Bacteriol.* **121,** 594–599.

Wickner, S., and Hurwitz, J. (1974). Conversion of ϕX174 viral DNA to double-stranded form by purified *Escherichia coli* proteins. *Proc. Natl. Acad. Sci. U.S.A.* **71,** 4120–4124.

Wickner, S., and Hurwitz, J. (1975a). *In vitro* synthesis of DNA. *ICN–UCLA Symp. Mol. Cell. Biol. 1975* Vol. 3, pp. 227–238.

Wickner, S., and Hurwitz, J. (1975b). Interaction of *Escherichia coli dna B* and *dna C* (*D*) gene products *in vitro*. *Proc. Natl. Acad. Sci. U.S.A.* **72,** 921–925.

Wickner, W., Brutlag, D., Schekman, R., and Kornberg, A. (1972). RNA synthesis initiates *in vitro* conversion of M13 DNA to its replicative form. *Proc. Natl. Acad. Sci. U.S.A.* **69,** 965–969.

Worcel, A. (1970). Induction of chromosome re-initiations in a thermosensitive DNA mutant of *Escherichia coli*. *J. Mol. Biol.* **52,** 371–386.

Worcel, A., and Burgi, E. (1972). On the structure of the folded chromosome of *Escherichia coli*. *J. Mol. Biol.* **71,** 127–147.

Yamaguchi, K., and Yoshikawa, H. (1973). Topography of chromosome membrane junction in *Bacillus subtilis*. *Nature* (*London*), *New Biol.* **244,** 204–206.

Yoshikawa, H. (1970). Initiation of DNA replication in *Bacillus subtilis*. II. Linkage between termini of parental strands and origins of daughter strands. *J. Mol. Biol.* **47,** 403–417.

Yoshikawa, H., O'Sullivan, A., and Sueoka, N. (1964). Sequential replication of the *Bacillus subtilis* chromosome. III. Regulation of initiation. *Proc. Natl. Acad. Sci. U.S.A.* **52,** 973–980.

Zaritsky, A., and Pritchard, R. H. (1971). Replication time of the chromosome in thymine-less mutants of *Escherichia coli*. *J. Mol. Biol.* **60,** 65–74.

4

Structure, Coding Capacity, and Replication of Mitochondrial and Chloroplast Chromosomes

Ruth M. Hall and Anthony W. Linnane

I. INTRODUCTION

Although genetic evidence for extranuclear genetic systems was reported as early as the 1920's, scarcely a decade has passed since specific species of DNA associated with chloroplasts and mitochondria first were positively identified. Initially, attention was focused on the identification of organelle DNA in different organisms and on establishing whether or-

ISBN 0-12-289502-9

ganelles had the capability to transcribe and translate information encoded in the organelle DNA and to duplicate that DNA. It is now generally accepted that organelles can perform all of these functions; synthesis of RNA, DNA, and protein has been shown to occur in isolated chloroplasts and mitochondria, and these organelles have been shown to contain DNA-dependent RNA polymerases, ribosomes, tRNA species and acylating enzymes, and DNA polymerases which are distinct from those employed by the nuclear genetic system. For details of these aspects, the reader is referred to earlier reviews, such as those by Ashwell and Work (1970), Rabinowitz and Swift (1970), Borst (1972), Linnane *et al.* (1972), and Tewari (1971). Some of these aspects are also considered elsewhere in this series.

This chapter is intended as a general survey of the well-established features of the structure and replication of mitochondrial and chloroplast DNA, as seen in early 1976. It is not our intention to present a comprehensive review of all information available and to discuss unresolved issues. Very recent results have been included only where it was considered that they were well substantiated and that they extend or clarify insights obtained from earlier well-established work. Controversial issues, such as whether copies of mitochondrial and chloroplast genomes are present in the nuclear genome, are not discussed. The interesting question of the evolutionary origin of organelle genomes also is considered as outside the scope of this chapter. The genetic behavior of chloroplast and mitochondrial DNA is dealt with elsewhere in this series.

II. IDENTIFICATION AND PURIFICATION OF ORGANELLE DNA

The identification of organelle DNA requires both the detection of a unique species of DNA and a demonstration that the DNA is specifically associated with the organelle. The existence of a specific species of DNA associated with the chloroplast was first demonstrated by Chun *et al.*, (1963) and by Sager and Ishida (1963). The following year, mitochondrial DNA was also identified (Luck and Reich, 1964).

A simple procedure can be used for the identification of organelle DNA, provided the following conditions are met: (1) The buoyant density of the organelle DNA in cesium chloride (CsCl) gradients differs significantly from that of the nuclear DNA; and (2) organelle-enriched subcellular fractions can be obtained. This procedure is illustrated in Figure 1, which shows evidence for the existence of mitochondrial DNA in *Saccharomyces cerevisiae*. Preparations of DNA extracted from whole yeast cells show two distinct DNA species distinguished on the basis of their

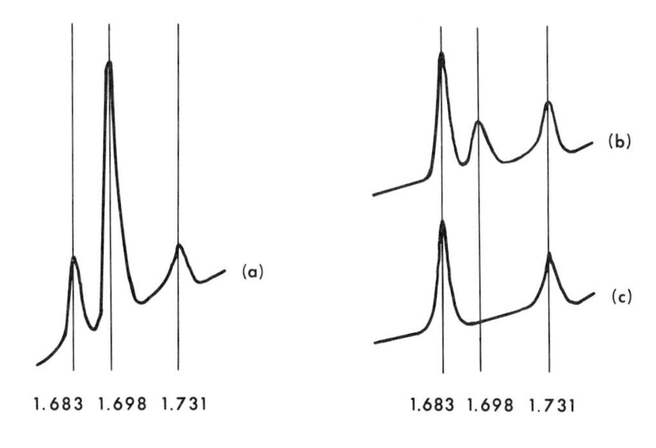

Fig. 1. Identification of mitochondrial DNA in *S. cerevisiae*. Microdensitometer tracings of DNA centrifuged to equilibrium in CsCl density gradients. DNA was extracted from (a) whoe cells, (b) a subcellular fraction enriched for mitochondria, (c) mitochondrial fraction treated with DNase I at 0°C prior to DNA extraction. The bands are: A marker, *M. lysodekticus* DNA at 1.731 gm/ml, nuclear DNA at 1.698 gm/ml, and mitochondrial DNA at 1.683 gm/ml.

density in CsCl (Figure 1a). DNA from subcellular fractions enriched for mitochondria is enriched for the DNA species of density 1.683 gm/ml (Figure 1b), and DNA from a mitochondrial preparation pretreated with DNase to remove DNA not contained within the mitochondria (see below) is enriched further for this DNA species (Figure 1c). Thus, the unique species of DNA with density 1.683 gm/ml is associated specifically with the mitochondria. In *S. cerevisiae,* mutants which are defective in mitochondrial function can be isolated since mitochondrial function is not essential for growth or viability in this organism. A complete lack of the DNA species associated with the mitochondria (Nagley and Linnane, 1970) has been demonstrated in a specific class of mutants of this type. This finding confirms the identification of mitochondrial DNA in *S. cerevisiae* and simply demonstrates the requirement of mitochondrial DNA for mitochondrial function.

The chloroplast DNA and mitochondrial DNA species of many organisms have now been identified by means of the procedure illustrated in Figure 1. However, use of this procedure is not without difficulties, and some of the pitfalls can be illustrated by considering the early attempts to identify the chloroplast DNA of higher plants. Confusion arose, first, as a result of the fact that, in many plants, chloroplast and nuclear DNA have similar buoyant densities in CsCl, and, second, because the purified chloroplast preparations used frequently were contaminated with mitochondria. As a result, mitochondrial DNA was incorrectly assigned a chloroplast origin, while the chloroplast DNA was assumed to be contaminating nuclear DNA. These problems have been discussed in detail by Kirk (1971). Confirmation of the identification of organelle DNA by the isolation of mutants which lack organelle DNA is only possible when mitochondrial or chloroplast function is not essential. Such evidence is available in the case of *Euglena gracilis;* bleached mutants which lack chloroplast function and also lack chloroplast DNA have been reported (Leff *et al.,* 1963; Ray and Hanawalt, 1965; Edelman *et al.,* 1965).

When organelle DNA clearly can be distinguished from other cellular DNA species in CsCl density gradients, highly purified organelle DNA also can be obtained by this method. However, when the density of organelle DNA is similar to that of other cellular DNA components, further characterization of putative organelle DNA is required for its positive identification, and other techniques must be used to obtain pure DNA preparations. The criterion most commonly used to distinguish organelle DNA is its ability to renature rapidly and completely after thermal or alkali denaturation. This property is a result of the low complexity of chloroplast and mitochondrial DNA. Renaturation can be monitored by density shifts in CsCl gradients. Denatured DNA has a higher density than native DNA, and after renaturation mitochondrial or chloroplast DNA again bands at a density identical to that of native DNA. Nuclear DNA normally does not renature under the same conditions and still bands at a higher density.

Another criterion used for the identification and also for the purification of organelle DNA relies on the finding that, in isolated mitochondrial or chloroplast preparations, the organelle DNA is insensitive to degradation by DNase, so long as the organelle membrane is intact. The organelle DNA is protected by the semipermeable membrane, and other DNA species contaminating the organelle preparation can be removed by DNase treatment under appropriate conditions (see, e.g., Figure 1b,c). This procedure frequently is used to purify organelle DNA when the physical properties of organelle and nuclear DNA are similar.

Other physicochemical techniques that have been used for the purification of mitochondrial and chloroplast DNA include: (1) cesium sulfate density gradient centrifugation in the presence of agents such as mercuric ions that accentuate buoyant density differences resulting from differences in base composition (Smith *et al.,* 1968; Szybalski, 1968); (2) chromatographic fractionation on hydroxyapatite (Bernardi, 1971; Bernardi *et al.,* 1972), methylated albumin kieselguhr (Mandell and Hershey, 1960), and poly-L-lysine-coated kieselguhr (Finkelstein *et al.,* 1972); and

(3) centrifugation in cesium chloride density gradients containing ethidium bromide (Radloff *et al.*, 1967), whereby covalently closed circular DNA molecules are separated from linear DNA.

III. STRUCTURE AND CODING CAPACITY OF MITOCHONDRIAL AND CHLOROPLAST CHROMOSOMES

In this section, we consider first the general features of mitochondrial and chloroplast chromosomes. The techniques used to determine the size and number of organelle chromosomes and to study important features of the structure of these chromosomes also are discussed briefly. Specific features of mitochondrial and chloroplast chromosomes from different organisms then are considered in detail. For simplicity, each subsection is subdivided further into three parts dealing with (a) the size and number of chromosomes, (b) the structure of the chromosomes, and (c) the coding capacity and the genes coded.

A. General Features of Mitochondrial and Chloroplast DNA

1. Size

Estimates of the size of mitochondrial DNA (mtDNA) and chloroplast DNA (ctDNA) have been obtained by measurement of the sizes of DNA molecules present in purified DNA preparations either by measuring the length of DNA molecules visualized by electron microscopic procedures (Younghusband and Inman, 1974) or by calculation of molecular weight of DNA from the sedimentation coefficient determined by sedimentation velocity analysis (Studier, 1965). For example, electron microscopic examination of the mtDNA from metazoa reveals a discrete population of circular molecules approximately 5 μm in length, corresponding to a molecular weight of the order of 10×10^6. The molecular weight calculated from the sedimentation coefficient of the linear form of these molecules also is approximately 10×10^6. In general, mtDNA molecules range in size from 10×10^6 daltons in metazoa up to 80×10^6 daltons in higher plants (see Table I), and ctDNA molecules are in the approximate size range $100–200 \times 10^6$ daltons (see Table II). However, these estimates alone do not define the size of the organelle genome. Estimates of the amount of organelle DNA per cell generally range from 10^9 to 10^{11} daltons, indicating that multiple copies of mtDNA and ctDNA molecules are present in the cell. Thus, the possibility that more than one class of

DNA molecule, differing in base sequence, is present in the cell must be considered.

This possibility can be investigated in two ways. First, the genetic complexity of the DNA can be estimated from the rate of renaturation of denatured DNA (Wetmur and Davidson, 1968; Britten and Kohne, 1968). The rate of renaturation under standard conditions is dependent on the number of different sequences present in the DNA, and the apparent kinetic complexity of the DNA can be calculated directly or determined by comparison with a DNA of known complexity. The value for the apparent kinetic complexity obtained from such analysis provides an estimate of the genome size. When this estimate is in reasonable agreement with the molecular weight estimates obtained by other methods, it can be assumed that the organelle genome consists of a single chromosome. In general, estimates of the sizes of DNA molecules obtained from length measurements or from sedimentation analysis are more accurate than those obtained from renaturation kinetic analysis, due to difficulties encountered in accurately measuring renaturation rates. However, estimates of the kinetic complexity of a large number of mtDNA species have been reported and in all cases are in fair agreement with estimates of the size of mtDNA molecules obtained by other methods. Results obtained with ctDNA are less conclusive (see Section III,C,2a); however, it seems probable that both mitochondrial and chloroplast genomes consist of a single chromosome.

A second procedure for determining the number of unique DNA molecules in organelle DNA preparations makes use of specific restriction nucleases. Because these nucleases cleave DNA molecules only at sites with specific base sequences, any discrete DNA molecule will yield a set of DNA fragments of characteristic size (see Nathans and Smith, 1975), and the sum of the sizes of these fragments will be equal to the size of the original DNA molecule. Owing to the high degree of specificity of restriction nucleases, it is improbable that two molecules which differ in base sequence will yield sets of fragments of identical size. Therefore, if the sum of the sizes of fragments obtained by treatment of organelle DNA with a restriction nuclease is equal to the size of individual molecules estimated by other methods, it can be assumed that the organelle genome consists of a single chromosome of this size. To date, this procedure has been used only with mtDNA from a number of metazoan species and from *S. cerevisiae* and *S. carlsbergensis,* and in these cases clear evidence for one mitochondrial chromosome has been obtained. In *S. cerevisiae,* humans, and horses, differences between the mtDNA molecules from different strains or individuals have been detected by this method (Bernardi *et al.*, 1975; Potter *et al.*, 1975).

2. *Structure*

The mtDNA and ctDNA molecules of most species studied to date have been found to be circular. Electron microscopic examination of DNA reveals both highly twisted closed circular molecules and open circular molecules. Closed circular molecules also have been detected by centrifugation in CsCl gradients containing ethidium bromide. This procedure separates closed circular molecules from other DNA forms by virtue of the fact that the topological constraints of closed circular molecules restrict the binding of dyes such as ethidium bromide, thereby restricting the size of density shifts caused by binding of the dye to DNA (Radloff *et al.*, 1967). In contrast, the mtDNA from two protozoa, *Tetrahymena pyriformis* and *Paramecium aurelia*, has been shown to consist of a linear molecule of discrete size (see Section III,B,3,b).

In many instances, mtDNA and ctDNA have been shown to form two bands in alkaline CsCl gradients, and in some cases the two bands have been shown to correspond to the complementary strands of the molecule. Strand separation yields limited information about the structure of DNA molecules, in that it indicates a bias in the base composition of the strands (Szybalski *et al.*, 1971). The importance of strand separation lies more in the fact that preparative separation of the complementary strands allows the analysis of questions such as which strand codes for rRNA, tRNA, and mRNA species (see Section III,B,1,c) and of details of the mode of replication (see Section IV). In a few instances, studies of the binding of synthetic polyribonucleotides such as poly(A), poly(U), poly(C), and poly(IG) to DNA (Szybalski *et al.*, 1971) also have been used to detect the presence of T-, A-, G-, and C-rich clusters, respectively, in mtDNA molecules.

The average base composition of mtDNA and ctDNA from different species has been found to cover a broad range, from less than 20% to greater than 50% G + C. Base composition can be estimated from the density of DNA in CsCl gradients (Mandel *et al.*, 1968), from the melting temperature (T_m) of the DNA under standard conditions (Mandel and Marmur, 1968), and by direct chemical analysis. Only values that have been obtained by direct chemical analysis can be considered as accurate estimates of the base composition of mtDNA and ctDNA, since discrepancies between the values obtained in this way and those calculated from CsCl density and T_m have been observed in a number of cases, especially when the average base composition is less than 30% G + C (see Sections III,B,2,b, and 3,b). Such discrepancies frequently arise when unusual bases such as 5-methylcytosine are present in DNA or when DNA is modified, for example, by glucosylation (see Mandel *et al.*, 1968; Mandel and Marmur, 1968). However, significant amounts of unusual

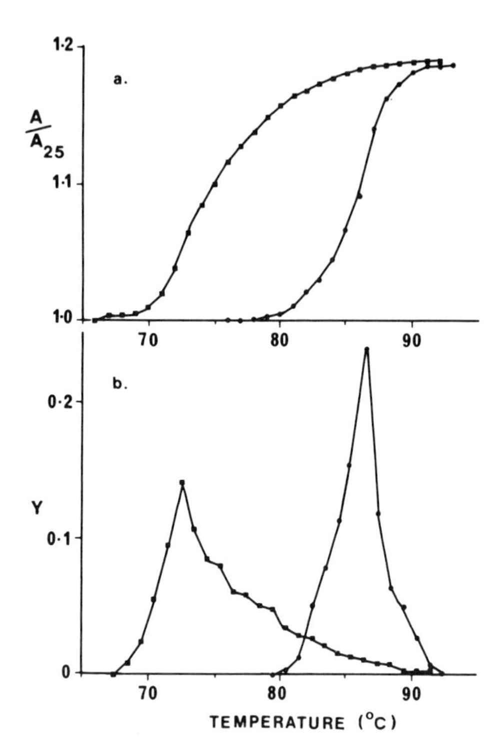

Fig. 2. Melting profiles of mtDNA and nuclear DNA from *S. cerevisiae*. (a) Melting profile: relative absorbance (A_t/A_{25}) versus temperature, where A_t, A_{25} are absorbance at 260 nm at temperatures t and 25°C. (b) First derivative of melting profile: increase in relative absorbance per degree, $Y = (A_{t_1} - A_{t_2})/(A_{100} - A_{25})(t_1 - t_2)$ versus temperature $(t_1 - t_2)/2$, where A_{t_1}, A_{t_2}, A_{100}, and A_{25} are absorbances at 260 nm at temperatures t_1, t_2, 100°, and 25°C. ■—■, mtDNA and ●—●, nuclear DNA.

bases have not been detected in mtDNA or ctDNA, and no evidence for modification of DNA has been presented. Now it is believed that the discrepancies observed with mtDNA and ctDNA result from an unusual distribution of bases that affects the physical properties of the DNA.

Analysis of the melting profiles of purified mtDNA and ctDNA preparations also can provide information concerning the distribution of bases in DNA molecules. When the base distribution is relatively homogeneous, a sharp unimodal melting profile is observed (Mandel and Marmur, 1968). In general, the melting profiles of mtDNA and ctDNA are somewhat broader than those obtained with, for example, *Escherichia coli* or T4 DNA, indicating slight heterogeneity in base composition. However, more striking heterogeneity is detected in mtDNA from *Drosophila melanogaster* and from a number of fungal and protozoan species (see

Sections III,B,1,b, 2,b, and 3,b, and in the ctDNA from algae (see Section III,C,2,b). The melting profiles of these DNA species are broad and multimodal, indicating the presence of sequences of substantially different base composition within the DNA molecules. The presence of more than one mode in a melting profile can be detected more readily in plots of the first derivative of the melting curve (change in absorbance per degree versus temperature). Unimodal melting curves yield symmetrical Gaussian first derivatives, while multimodal profiles yield first derivatives which deviate from a Gaussian distribution and which may exhibit more than one peak. Estimates of the base composition of different components may be derived from the T_m of each mode of the melting profile. This is illustrated in Figure 2, which shows the melting profiles and first derivatives of the melting profiles of mtDNA and nuclear DNA from *S. cerevisiae*. The melting profile of nuclear DNA is relatively sharp and the first derivative approximates to a Gaussian distribution; the melting profile of mtDNA is broad and the first derivative deviates markedly from a Gaussian distribution, indicating substantial heterogeneity in the base distribution within the mtDNA molecule.

An additional powerful tool for detecting base sequence heterogeneity in DNA molecules is the technique of denaturation mapping (Inman, 1974; Younghusband and Inman, 1974). This procedure relies on the fact that AT-rich sequences melt at lower temperatures than GC-rich sequences. The DNA is denatured under defined conditions, the denatured regions are fixed by reaction with formaldehyde, and the molecules are examined in the electron microscope for the presence of single-stranded regions: The less stringent the denaturation conditions, the lower the G + C content of any denatured regions detected. The presence of regions of extremely low G + C content (AT-rich sequences) in mtDNA molecules from a number of species has been detected in this way (see Sections III,B,1,b, and 2,b).

3. Coding Capacity and Genes Coded

In assessing the coding capacity of the mitochondrial and chloroplast genomes, it is essential that both the size and the average base composition of the DNA be considered. Grossman *et al.* (1971) have pointed out that a minimum base composition of about 35% G + C is required to code for an average protein, even using the codons of lowest G and C content in all cases. Thus, it cannot be assumed that all of the DNA in a genome can code for protein sequences when the average base composition is less than 30–35% G + C. It seems probable that, in the case of mtDNA and ctDNA species which have extremely low G + C contents, the genome will be found to contain AT-rich segments, as has been shown for the mtDNA's of *S. carlsbergensis, S. cerevisiae, Euglena gracilis,* and *D. melanogaster*. The presence of AT-rich regions means that the G + C content of the balance of the molecules is substantially higher than the average base composition of the DNA and therefore should be capable of coding for protein sequences.

It is well established that mitochondrial and chloroplast genomes contain genes for the rRNA species associated with the large and the small subunits of the mitochondrial or chloroplast ribosomes and also some genes for tRNA species. The base composition of rRNA genes does not appear to be as restricted as is that of genes for protein sequences. It has been found that the base composition of the rRNA species in different organisms to some extent correlates with the average base composition of the DNA. For example, in *S. cerevisiae* the base compositions of mitochondrial rRNA and mtDNA are 25% and 17% G + C, respectively, and in *N. crassa* 35–38% and 40%, respectively [see Borst and Grivell (1971) for base compositions of other mitochondrial rRNA species].

Although progress to date has been limited, the identification of the proteins coded by mitochondrial and chloroplast chromosomes is an area which is currently receiving a great deal of attention. For mitochondria, most of the information pertaining to this question has been obtained using the fungi *Neurospora crassa* and *S. cerevisiae,* and as mitochondrial mutants of *S. cerevisiae* are available, conclusive identification of proteins as gene products of the mtDNA should be possible in this organism (see Section III,B,2,c). There already is substantial evidence that a number of membrane proteins which are subunits of oligomycin-sensitive ATPase, the cytochrome *b* complex, and the cytochrome oxidase complex are mitochondrial gene products in *S. cerevisiae* and *N. crassa*. The limited information available for metazoa suggests that subunits of the same three complexes are also mitochondrial gene products in this case (see Section III,B,1,c). To date, only one protein has been identified positively as a gene product of ctDNA, the large subunit of the soluble fraction I protein. A number of membrane proteins of unidentified function may also be gene products of ctDNA (see Section III,C,1,c). Chloroplast mutants of the unicellular alga *Chlamydomonas reinhardi* are available, and evidence for the specification of some proteins of the chloroplast ribosomes also has been presented for this organism (see Section III,C,2,c).

It should be noted that mitochondria and chloroplasts contain both nuclear- and organelle-coded proteins. Indeed, all the enzyme complexes mentioned in the foregoing section (oligomycin-sensitive ATPase, complex III, cytochrome oxidase, and Fraction I protein) contain both nuclear- and organelle-coded subunits. Mitochondrial and chloroplast ribosomes are also composed of both nuclear and organelle gene products. In both cases the rRNA is of organelle origin, while evidence available to date suggests that all mitochondrial ribosomal proteins are nuclear-coded but that chloroplast ribosomes contain proteins of both nuclear and chloroplast origin.

B. Mitochondrial DNA

The mitochondrial DNA species isolated from metazoa, fungi, protozoa, and from higher plants fall into four distinct groups, on the basis of both size and other physical properties. For simplicity, these groups are considered separately.

1. Metazoa

a. Size. The mtDNA of metazoa consists of small circular molecules of relatively constant size. Molecules with contour lengths in the 4.5–6-μm range have been observed by electron microscopic examination of purified preparations of mtDNA and DNA released from mitochondria by osmotic shock procedures. To date, circles in this size range have been observed in a number of invertebrate species: Annelida (Dawid and Brown, 1970), Nematoda (Carter *et al.*, 1972; Tobler and Gut, 1974), Insecta (Van Bruggen *et al.*, 1968; Tanguay and Chaudhary, 1972; Bultmann and Laird, 1973; Peacock *et al.*, 1973; Polan *et al.*, 1973), Echinodermata (Pikó *et al.*, 1968) and Crustacea (Schmitt *et al.*, 1974), and in a variety of vertebrates, including amphibia (Wolstenhome and Dawid, 1968), birds, fish, and mammals (Sinclair *et al.*, 1967; Van Bruggen *et al.*, 1968; Kroon *et al.*, 1966; Nass, 1969a; Radloff *et al.*, 1967; Suyama and Miura, 1968). One interesting exception has been reported. Mitochondrial DNA from spermatocytes of the insect *Rhynchosciara hollaenderi* was found to consist of circles approximately 9 μm in length (Handel *et al.*, 1973). This finding is discussed further in Section III,B,1,c.

Because internal length standards were not used in most of these studies and because of the variations in the DNA spreading procedures used, comparisons of the size of mtDNA molecules from different organisms cannot be made using such data. However, mixing experiments have been employed to demonstrate real differences in the length of mtDNA molecules from different organisms (Wolstenhome and Dawid, 1968; Nass, 1969a). On the other hand, mtDNA molecules isolated from different tissues of one organism are almost certainly of identical size, as has been shown for mtDNA from mouse liver, pancreas, brain, and kidney (Sinclair *et al.*, 1967). Thus, the size of mtDNA molecules is characteristic for an individual organism.

The molecular weight of mtDNA, as calculated from the lengths 4.5–6 μm is approximately 9–12 $\times$ 10^6, and 10–11 $\times$ 10^6, as calculated from the sedimentation coefficient of the linear form of mtDNA (Borst *et al.*, 1967; Dawid and Wolstenholme, 1967). The sum of the sizes of the restriction endonuclease cleavage fragments of mtDNA from a number of mammalian species also has been found to be approximately 10 $\times$ 10^6 daltons (Robberson *et al.*, 1974; Brown and Vinograd, 1974; Potter *et al.*, 1975), and it therefore may be concluded that the mitochondrial genome of metazoa consists of a single chromosome.

b. Structure. In metazoan mtDNA preparations, three components can be distinguished by both electron microscopic and sedimentation velocity analysis. These are covalently closed circles, open circles, and linear molecules (Borst *et al.*, 1967; Dawid and Wolstenhome, 1967). In carefully isolated DNA preparations, more than 80% of molecules are found to be covalently closed circles (Borst *et al.*, 1967; Leffler *et al.*, 1970), and mtDNA is believed to exist *in vivo* in the covalently closed circular form. It is assumed that other forms arise as the result of random breaks which occur during isolation.

Isolated closed circular mtDNA molecules contain approximately 35 superhelical turns per molecule (Nass, 1969a; Van Bruggen *et al.*, 1968; Ruttenberg, *et al.*, 1968). The closed circular form of mtDNA recently has been shown to contain some ribonucleotides (Wong-Staal *et al.*, 1973; Grossman *et al.*, 1973; Miyaki *et al.*, 1973; Porcher and Koch, 1973; Lonsdale and Jones, 1974). As yet, the significance of this interesting observation is not known. Other physicochemical properties of the three DNA forms have been described in detail in reviews by Borst and Kroon (1969) and Rabinowitz and Swift (1970). Complex forms of mtDNA, such as dimers and catanes, also have been observed. For details of the properties and significance of these forms, the reader is referred to Clayton and Smith (1974).

The average base composition of mtDNA from vertebrate species is generally in the 40–50% G + C range. Values calculated from the density in CsCl and from the T_m are in good agreement, and when direct chemical analysis has been performed, values are also in agreement (see Borst and Kroon, 1969, for data). Values for the average base composition of mtDNA from most invertebrate species studied to date are in the range 30–45% G + C (Dawid and Brown, 1970; Carter *et al.*, 1972; Pikó *et al.*, 1968; Tanguay and Chaudhary, 1972; Schmitt *et al.*, 1974; Skinner and Kerr, 1971). However, the mtDNA's of two insect species are striking exceptions. *Drosophila melanogaster* mtDNA has an average base composition of 22% G + C calculated from CsCl density and T_m (Bultmann and Laird, 1973; Polan *et al.*, 1973) and determined directly (Peacock *et al.*,

1973). The mtDNA of *Rhynchosciara hollaenderi* has a density of 1.681 gm/ml in CsCl (Handel *et al.*, 1973), indicating a base composition of approximately 21% G + C. The significance of the low G + C content of these mtDNA species is discussed in detail in Section III,B,1,c. In general, no unusual bases have been detected, although the presence of some 5-methylcytosine in mtDNA from mouse, hamster, and ox has been reported (Nass, 1973; Vanyushin and Kirnos, 1974).

Only limited information on the distribution of bases in mtDNA is available. Nearest-neighbor analysis of mtDNA from mouse liver indicated that the distribution of bases was less random than in any double-stranded DNA previously studied (Antonoglou and Georgastos, 1972), and the frequency of the dinucleotide CpG was close to random, a characteristic of prokaryotic rather than eukaryotic DNA. Pyrimidine tract analysis of beef heart mtDNA also suggests a similarity to prokaryotic DNA (Vanyushin and Kirnos, 1974). The mtDNA from a number of species has been shown to form two bands in alkaline CsCl gradients (Dawid and Wolstenholme, 1967; Pikó *et al.*, 1968; Corneo *et al.*, 1968; Borst and Ruttenberg, 1969; Leffler *et al.*, 1970; Nass and Buck, 1970; Clayton *et al.*, 1970; Skinner and Kerr, 1971; Polan *et al.*, 1973), and in some cases the two bands have been shown to correspond to the complementary strands of mtDNA. For human mtDNA, the heavier (H) strand has been shown to contain 1.4 times more thymine than the lighter (L) strand (Aloni and Attardi, 1971a). Studies on the binding of the polyribonucleotides poly(U), poly(I,G), poly(C), and poly(I) to the separated strands of rat and chick mtDNA indicate the presence of G-rich and T-rich clusters in the heavy (H) strand and of C- and A-rich clusters in the light (L) strand (Borst and Ruttenberg, 1969, 1972). The separated complementary strands of HeLa cell mtDNA also have been utilized to locate the genes for rRNA, tRNA, and putative mRNA species (see Section III,B,1,c). Denaturation mapping has been employed to study the base distribution in rat mtDNA (Wolstenholme *et al.*, 1972). After heating to 49°C, approximately 25% of molecules contained one to three denatured regions. Three specific regions of denaturation confined to a segment less than half the total length of the molecule were identified. These regions are probably only slightly enriched for the bases A and T, because a relatively high denaturation temperature was required for their identification.

The mtDNA of *D. melanogaster* differs from all other metazoan mtDNA species studied to date not only in that its base composition (22% G + C) is extremely low, but also in that it exhibits a high degree of intramolecular heterogeneity in base composition. The heterogeneity is indicated by the fact that the melting curve is triphasic (Bultmann and Laird,

1973; Polan *et al.*, 1973; Peacock *et al.*, 1973); this contrasts with the relatively sharp unimodal melting profile of other metazoan mtDNA's. Peacock *et al.* (1973) have confirmed this heterogeneity by denaturation mapping procedures. In mtDNA molecules of size 6 μm, a single denatured region equal in size to 15% of the molecule was observed when the denaturation temperature was 42°C; the size of this region increased to 25% in molecules denatured at 46°C. The low denaturation temperatures required for this region to be detected indicate that its G + C content is near zero. This AT-rich region also could be detected as a separate peak in CsCl gradients when the mtDNA was sheared to fragments of size 2×10^6 daltons (Peacock *et al.*, 1973).

c. Coding Capacity and Genes Coded. The average base composition of the mtDNA of most metazoan species (30–50% G + C) is sufficiently high to allow the entire genome of $9–12 \times 10^6$ daltons double-stranded DNA potentially to code for meaningful protein and RNA sequences. In the case of *D. melanogaster* and *R. hollaenderi*, the average base composition of approximately 21–22% G + C (see Section III,B,1,b) is too low to enable the entire genome to code for meaningful protein sequences. The *Drosophila* mitochondrial genome of size 11.3×10^6 daltons includes a segment of DNA that contains almost exclusively A + T, and that accounts for 15–25% of the genome (Peacock *et al.*, 1973). The average base composition of the balance of the genome therefore should be ca. 30% G + C, a level which is sufficient to enable the majority of this portion to code for meaningful sequences. The size of this GC-rich portion is near the range of sizes for other metazoan mitochondrial genomes. The size of *Rhynchosciara* mtDNA has been reported to be 9 μm, corresponding to approximately 18×10^6 daltons of double-stranded DNA. If these molecules represent unique sequences, and not dimers, as was originally suggested (Handel *et al.*, 1973), and if substantial AT-rich segments are present, as is the case for *Drosophila* mtDNA, the potential coding capacity of the mitochondrial genome of *Rhynchosciara* should be at least as great as that of other metazoan species.

It is well established that the mitochondrial genomes of *Xenopus laevis*, HeLa cells (human), and the rat code for the small and large RNA components of the mitochondrial ribosomes and for some tRNA species. In DNA–RNA hybridization experiments, the two rRNA components and bulk 4 S RNA (assumed to be tRNA) hybridize to a total of approximately 10% of the DNA at saturation, indicating that genes for these RNA species account for about 20% of the single-stranded information content of the mitochondrial chromosomes (Aloni and Attardi, 1971b; Dawid, 1972; Avadhani *et al.*, 1974). From saturation values for hybridization of the small and the large rRNA species to mtDNA, together with the molecular weights of the rRNA molecules, it can be calculated that the mitochondrial chromosomes of *X. laevis* and HeLa cells contain one gene for each rRNA species. Saturation values for the hybridization of 4 S RNA to mtDNA are sufficient to account for approximately 10–15 tRNA species of size 25,000 daltons (Aloni and Attardi, 1971b; Dawid, 1972). In rat liver, four aminoacyl-tRNA species have been shown to hybridize specifically with mtDNA; two of these, leucyl- and phenylalanyl-tRNA, are transcribed from the H strand and two, tyrosyl- and seryl-tRNA, are transcribed from the L-strand (Nass and Buck, 1970).

Both the small and the large rRNA genes have been shown to be transcribed from the H strand of *Xenopus* and HeLa cell mtDNA, and, although some are transcribed from the L strand, the majority of 4 S RNA genes also are transcribed from the H strand (Aloni and Attardi, 1971b; Dawid, 1972). A detailed study of the relative arrangement of the two rRNA genes and the 4 S RNA genes on the chromosome has been reported in the case of HeLa cells (Robberson *et al.*, 1972a; Wu *et al.*, 1972; Attardi *et al.*, 1974). The large and small rRNA species are situated adjacent to one another, separated by approximately 160 base pairs. Of nine 4 S RNA genes transcribed from the H strand, one is situated in the space between the rRNA genes, two more are situated one on each side of the region coding for the rRNA genes, and the remaining six are spaced around the chromosome. The three 4 S RNA genes transcribed from the L strand are widely separated. A detailed map of the relative positions of the RNA genes is shown in Figure 3. As yet, it has not been established whether all twelve of the mitochondrially coded 4 S RNA species identified in HeLa cells represent tRNA species.

It can be calculated that the metazoan chromosome contains information sufficient to code for twenty proteins of size 20,000 daltons, assuming an average genome size of 10×10^6 daltons double-stranded DNA, and allowing for the fact that 20% of the single-stranded information content is accounted for by rRNA and tRNA genes. This number is a maximum, since it does not allow for the presence of spacer or regulatory sequences in the genome. Aloni and Attardi (1971a,c) have shown that the majority of both the L and H strands of HeLa mtDNA is transcribed *in vivo*, although the majority of L strand transcripts are subsequently degraded. More recently, the identification of eight discrete polyadenylic acid-containing RNA species (putative mRNA species) which hybridize specifically to mtDNA has been reported (Ojala and Attardi, 1974; Hirsch and Penman, 1973). These components, if distinct, would account for approximately 50% of the single-stranded information content of HeLa

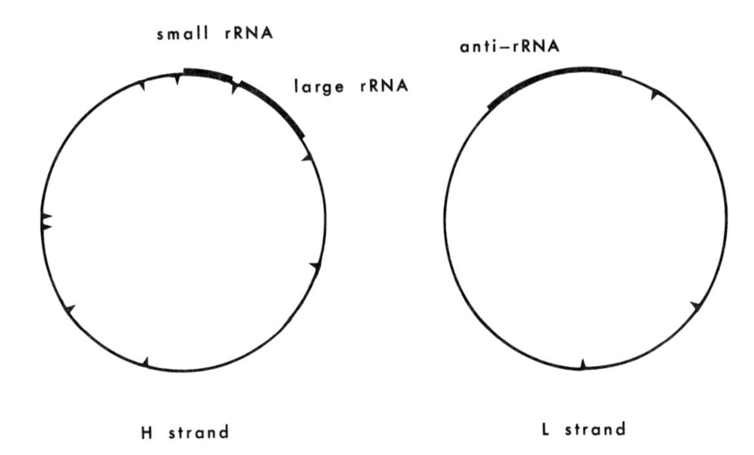

Fig. 3. Arrangement of rRNA and 4 S RNA genes on the mitochondrial genome of HeLa cells. Bars represent positions of rRNA genes, represent positions of 4 S RNA genes. Positions are as determined by Wu *et al.* (1972) and Attardi *et al.* (1974).

mtDNA, and would set a minimum of eight proteins coded. In a comparative study of the poly(A)-containing RNA species associated with the mitochondria in cell lines derived from two mammalian species (human and Chinese hamster) and two insect species (*Drosophila melanogaster* and a mosquito, *Aedes albopictus*), Hirsch *et al.* (1974) have shown that in all cases approximately eight poly(A)-containing RNA species of similar molecular weights are present. This finding supports the proposal that the coding capacity of the mitochondrial chromosomes of insect species is similar to that of other metazoan mitochondrial chromosomes (see above).

Little progress has been made toward the identification of the proteins coded by the mitochondrial genome in metazoa. In a number of species, several mitochondrial membrane proteins have been shown to be products of mitochondrial protein synthesis, by analysis of the proteins synthesized, in isolated mitochondria, or in whole cells treated with cycloheximide to inhibit cytosolic protein synthesis (Galper and Darnell, 1971; Coote and Work, 1971; Lederman and Attardi, 1973; Burke and Beattie, 1974; Kuzela *et al.*, 1975). However, these proteins have not been identified functionally. In *Locusta migratoria*, one subunit of the cytochrome oxidase complex and one subunit of the cytochrome *b* complex have been shown to be the products of mitochondrial protein synthesis (Weiss *et al.*, 1972, 1973). It generally is believed that some subunits of the ATPase, cytochrome oxidase, and cytochrome *b* complexes will prove to be encoded in metazoan mtDNA, as has been shown in the case of yeast and *Neurospora* (see Section III,B,2,c). However, it cannot be assumed that the mitochondrially coded subunits of these complexes will be the same in size and number in metazoa, although it has been calculated that the coding capacity of the metazoan mitochondrial genome is more than sufficient to encode a set of proteins identical to those of yeast and *Neurospora* (Schatz and Mason, 1974).

2. *Fungi*

a. Size. Data available from electron microscopic examination and renaturation kinetic analysis of the size of mtDNA from fungi are listed in Table I. In general, there is good agreement between the size estimates obtained by these two methods, indicating that the mitochondrial genome consists of a single chromosome. Estimates of the size of mtDNA molecules obtained by other methods also are in good agreement with the sizes listed in Table I. For *Neurospora crassa* mtDNA, the sum of the sizes of restriction nucleases fragments is approximately 40×10^6 daltons (Bernard *et al.*, 1975b); for the two closely related yeasts *S. cerevisiae* and *S. carlsbergensis*, the molecular weight of mtDNA calculated from the sedimentation coefficient is in the $45–55 \times 10^6$ range (Blamire *et al.*, 1972; Michels *et al.*, 1974) and the sum of the molecular weights of restriction nuclease cleavage fragments is $47–54 \times 10^6$ (Bernardi *et al.*, 1975; Sanders *et al.*, 1975a). The molecular weight of *Physarum polycephalum* mtDNA also has been estimated as 40×10^6 from sedimentation velocity analysis (Sonenshein and Holt, 1968).

b. Structure. The mtDNA molecules of the fungi studied to date have been shown to be circular, and highly twisted closed circular DNA molecules have been observed in mtDNA preparations. For all species except *S. cerevisiae* and *S. carlsbergensis*, a proportion of the mtDNA also has been isolated in the closed circular form in CsCl gradients containing ethidium bromide (Clayton and Brambl, 1972; O'Connor *et al.*, 1975, 1976; Sanders *et al.*, 1974). Genetic evidence for circularity has been obtained in the case of *S. cerevisiae* (Molloy *et al.*, 1975; Clark-Walker and Miklos, 1975), and the physical map of *S. carlsbergensis* mtDNA constructed by restriction fragment analysis is also circular (Sanders *et al.*, 1975a). Fungal mtDNA most probably exists *in vivo* in closed circular form. The low frequency of closed circular molecules found in DNA preparations is thought to be a result of nucleolytic cleavage during isolation. In the case of *N. crassa* and *S. carlsbergensis*, it has been shown by denaturation mapping procedures that the linear DNA molecules, which comprise the major component of mtDNA preparations, arise from a circular form by random breakage (Bernard *et al.*, 1975a; Christiansen *et al.*, 1975).

TABLE I

Size of Mitochondrial Genomes

Organism	Length (μm)	Calculated MW[a] ($\times 10^6$)	Apparent kinetic complexity[b] ($\times 10^6$)	Reference[n]
Fungi				
Neurospora crassa	25	50	—	1
	20	40	—	2–4
	—	—	40–50[g]	5
Saccharomyces cerevisiae	25	50	—	6
	21	50[c]	—	7
Saccharomyces carlsbergensis	25	50	—	6
	—	—	50[h]	8
Kluveromyces lactis	11.4	24[d]	—	9
	11.4	25	30[i]	10
Candida parapsilosis	11.1	23[d]	—	9
Hansenula wingei	8.2	17[d]	—	9
Schizosaccharomyces pombe	6.0	12.5[d]	—	9
Torulopsis glabrata	6	12.8[d]	14.8	11
Saprolegnia	14	30	—	12
Aspergillus nidulans	10–11	20	21	27
Protozoa				
Tetrahymena pyriformis	17.6	35	—	13
	14.9	30	—	14
	—	—	43[j]	15
Euglena gracilis	—	—	55[k]	16
Acanthamoeba castellanii	12.8	25[e]	26[l]	17,18
	16.7	35	28[l]	19
Paramecium aurelia	13.8	30	—	20
	—	—	45	21
Plasmodium lophurae	10.3	18	—	22
Plants				
Pea	30	66[f]	74[l]	23
Spinach	30	70[f]	70–80[l]	24
Lettuce	30	70[f]	70–80[l]	24
Potato	—	—	100[l]	25
Tobacco	—	—	60[m]	26

[a] Molecular weight/μm varies with the spreading technique used. Where an internal length standard was not included, the approximate molecular weight (to nearest 5×10^6) was calculated assuming 2×10^6 daltons/μm.

[b] Calculated from k_2, the second-order rate constant for renaturation (Wetmur and Davidson, 1968); k_2 was determined at $T_m - 25°C$ unless otherwise indicated. Apparent kinetic complexity was calculated either directly from Equation 18 of Wetmur and Davidson (1968) or relative to a marker DNA (M) of known molecular weight using the relationship: apparent kinetic complexity = $[(k_2)_M(MW)_M]/k_2$. Values for apparent kinetic complexity calculated relative to T4 (1.06×10^8 daltons) are generally much lower than those calculated relative to *E. coli* (2.5×10^9 daltons). Christiansen *et al.* (1973) have shown that this

The average base composition of mtDNA from fungal species ranges from as low as 17% up to 45% G + C. Calculated from CsCl density, the base compositions of mtDNA from fourteen fungal species cover the range 28–35% G + C (Villa and Storck, 1968), and in yeast species varies from 23% for *S. cerevisiae* (Bernardi *et al.*, 1970) and *Kluyveromyces fragilis* (Luha *et al.*, 1971) to 40% for *K. lactis* (Smith *et al.*, 1968; O'Connor *et al.*, 1975). Although the values for base composition determined by direct chemical analysis and those calculated from T_m or CsCl density have been reported to agree well in some cases, e.g., *N. crassa* (Bernard *et al.*, 1975a) and *Physarum polycephalum* (Evans and Suskind, 1971), in other instances, discrepancies between these values have been found. The base composition of *S. cerevisiae* mtDNA is 17% by direct analysis and 12 and 23% calculated from T_m and CsCl density values, respectively (Bernardi *et al.*, 1970). No unusual bases have been detected in fungal

---

results from an effect of glucosylation of T4 DNA on k_2, and not from effects of %G + C as suggested by others (Wetmur and Davidson, 1968; Seidler and Mandel, 1971). All values quoted therefore are uncorrected. Values calculated relative to T4 may be low by as much as twofold.

[c] Internal length standards λ, T4.

[d] Internal length standard λ, 30.8×10^6 daltons.

[e] Internal length standard T7, 26.4×10^6 daltons.

[f] Internal length standard ϕX174 RFII, 3.2×10^6daltons.

[g] Calculated kinetic complexity was found to be dependent on fragment size. Christiansen *et al.* (1974) have shown that fragment sizes $<2 \times 10^5$ daltons give most accurate kinetic complexity values at $T_m - 25°C$. Values obtained with fragments of $2-4 \times 10^5$ daltons therefore are quoted.

[h] Calculated from k_2 determined at $T_m - 25°C$, fragment size $<2 \times 10^5$ daltons, and from k_2 determined at $T_m - 10°C$ corrected relative to mycoplasma DNA as described by the authors.

[i] Calculated relative to T7, 25×10^6 daltons, and ϕ29, 12×10^6 daltons, from k_2 values determined as $T_m - 10°C$.

[j] Recalculated both directly and relative to *E. coli*, 2.5×10^6 daltons, from data shown by the authors.

[k] Calculated relative to PM2, 5.3×10^6 daltons, and T7, 25×10^6 daltons.

[l] Calculated relative to T4, 106×10^6 daltons.

[m] Calculated relative to T4, 180×10^6 daltons.

[n] Key to references: 1. Schäfer *et al.* (1971); 2. Clayton and Brambl (1972); 3. Agsteribbe *et al.* (1972); 4. Bernard *et al.* (1975a); 5. Wood and Luck (1969); 6. Hollenberg *et al.* (1970); 7. Petes *et al.* (1973); 8. Christiansen *et al.* (1974); 9. O'Connor *et al.* (1975); 10. Sanders *et al.* (1974); 11. O'Connor *et al.* (1976); 12. Clark-Walker and Gleason (1973); 13. Suyama and Miura (1968); 14. Arnberg *et al.* (1972); 15. Flavell and Jones (1970); 16. Talen *et al.* (1974); 17. Bohnert (1973); 18. Bohnert and Herrmann (1974); 19. Hettiarachchy and Jones (1974); 20. Flavell and Jones (1971a); 21. Goddard and Cummings (1975); 22. Kilejian (1975); 23. Kolodner and Tewari (1972a); 24. Kolodner and Tewari (1972b); 25. Vedel and Quétier (1974); 26. Wong and Wildman (1972); 27. Lopez Perez and Turner (1975).

mtDNA, except in the case of *P. polycephalum,* in which the presence of some 5-methylcytosine has been reported (Evans and Evans, 1970). Thus, the differences cannot be accounted for by the presence of unusual bases in the mtDNA, and are believed to result from an unusual distribution of bases within the mtDNA molecule. Further indication of an abnormal base distribution in fungal mtDNA comes from renaturation kinetic analysis. An unusual dependence of the apparent kinetic complexity on the fragment size of the DNA has been observed in *N. crassa* (Wood and Luck, 1969), *S. carlsbergensis* (Christiansen *et al.,* 1974), and *K. lactis* (Sanders *et al.,* 1974). Kinetic complexity is normally independent of fragment size (see Wetmur and Davidson, 1968). Christiansen *et al.* (1974) have shown that the apparent kinetic complexity of *S. carlsbergensis* mtDNA determined under standard conditions (T_m − 25°C) is independent of fragment size at sizes below 2–3 × 10^5 daltons. Complexity is also completely independent of the size of DNA if renaturation is performed at T_m − 10°C. These workers conclude that the unusual renaturation behavior is observed because at T_m − 25°C single-stranded fragments of size greater than 3 × 10^5 daltons are not fully extended random coils, possibly due to interaction between A- and T-rich clusters within the mtDNA.

In the case of *S. cerevisiae* mtDNA, intramolecular heterogeneity is further indicated from the multiphasic melting curve (Bernardi *et al.,* 1970). The first derivative of the melting curve deviates markedly from a Gaussian distribution, and a number of distinct components are distinguishable (see Figure 2). More detailed studies of *S. cerevisiae* mtDNA using a variety of procedures have demonstrated considerable intramolecular heterogeneity; the DNA was shown to consist of AT- and GC-rich segments of the order 10^5–10^6 daltons in length (Bernardi *et al.,* 1972; Piperno *et al.,* 1972; Ehrlich *et al.,* 1972; Carnevali and Leoni, 1972). Further analysis (Prunell and Bernardi, 1974) demonstrated that the mitochondrial chromosome of *S. cerevisiae* consists of equal amounts of AT-rich segments ("spacers") and GC-rich segments ("genes"). The AT-rich sequences are homogeneous in base composition and contain less than 5% G + C. The GC-rich sequences are heterogeneous, ranging in base composition from 25 to 50% G + C at a fragment size of 1.2 × 10^5 daltons. The average size of the AT- and GC-rich sequences is at least 1.5 × 10^5 daltons. Bernardi and co-workers have proposed that the mtDNA consists of interspersed "genes" and "spacers." This idea is consistent with the results of denaturation mapping studies of *S. carlsbergensis* and *S. cerevisiae* mtDNA (Christiansen *et al.,* 1975; Christiansen and Christiansen, 1976), which show a large number of readily denatured regions (AT-rich regions) interspersed with more stable (GC-rich) regions.

A similar denaturation map has been obtained with *N. crassa* mtDNA (Bernard *et al.*, 1975a), although the conditions required for denaturation in this study indicate that the more readily denatured regions are not nearly as rich in A + T as the corresponding regions in *S. carlsbergensis* and *S. cerevisiae* mtDNA.

c. Coding Capacity and Genes Coded. In general, while the mitochondrial genome size of fungi is greater than that of the metazoa, two exceptions are notable. The genome size of two yeasts, *Schizosaccharomyces pombe* and *Torulopsis glabrata,* is 12–13 × 10^6 daltons (O'Connor *et al.*, 1975, 1976), which is in the range of sizes for metazoan mitochondrial genomes. This observation raises the possibility that the capacity of yeast mitochondrial genomes to code for structural proteins and for RNA species is not substantially greater than that of metazoan genomes, and that the larger mtDNA molecules observed in other yeast species may have arisen as multimers of a basic 12–13 × 10^6-dalton unit (O'Connor *et al.*, 1976). The fact that the sizes of the yeast mitochondrial genomes studied to date are (with the exception of *Hansenula wingei*) of the order of 12 × 10^6, 24 × 10^6, and 50 × 10^6 daltons (see Table I) is consistent with this proposal. Furthermore, substantial homology between the rRNA cistrons of *S. carlsbergensis, Candida utilis,* and *K. lactis* mtDNA has been demonstrated (Groot *et al.*, 1975), a finding which provides suggestive evidence that these mitochondrial genomes arose from a common ancestor. However, even if the large yeast genomes did arise in this way, it would be premature to assume that no new genes arose in the process. Indeed, in the case of *S. cerevisiae,* in which detailed analysis of the structure of mtDNA has been reported, segments of DNA with an average base composition of less than 5% G + C have been shown to account for about 50% of the total DNA (see Prunell and Bernardi, 1974). This reduces the amount of mtDNA available to code for protein or RNA sequences from 50 × 10^6 daltons to a maximum of 25 × 10^6 daltons of double-stranded DNA, a value which is still far in excess of the 10 × 10^6 daltons DNA available for coding in metazoan mtDNA. Furthermore, in the case of *N. crassa,* the high average base composition of mtDNA (40% G + C; Bernard *et al.*, 1975a) does not indicate the presence of substantial AT-rich regions in the genome, and the majority of the genome, of size 40–50 × 10^6 daltons of double-stranded DNA, must be considered as having potential coding capacity.

To date, evidence has been presented for the presence of one copy of the gene for each of the large and small mitochondrial rRNA species in the mitochondrial genomes of *S. cerevisiae* (Morimoto *et al.*, 1971; Nagley *et al.*, 1974), *S. carlsbergensis* (Reijnders *et al.*, 1972), *Candida utilis* (Reboul and Vignais, 1974), and *N. crassa* (Schäfer and Kuntzël, 1972; Kuriyama and Luck, 1973). The presence of rRNA genes in *K. lactis* mtDNA is also indicated from the study of Groot *et al.* (1975). For *S. carlsbergensis* and *S. cerevisiae*, hybridization of bulk mitochondrial 4 S RNA (assumed to be tRNA) to mtDNA results in the saturation of 1% of mtDNA (Reijnders *et al.*, 1972; Schneller *et al.*, 1975). This value would indicate the presence of approximately 20 genes for tRNA species of average size 25,000 daltons. Hybridization to *S. cerevisiae* mtDNA of tRNA species charged with the amino acids *N*-formylmethionine, leucine, valine, alanine, phenylalanine, isoleucine, glycine, tyrosine, glutamine, aspartic acid, proline, lysine, histidine, and serine has been reported to date (Halbreich and Rabinowitz, 1971; Casey *et al.*, 1972, 1974a; Cohen and Rabinowitz, 1972; Baldacci *et al.*, 1975). Evidence for the existence of more than one species of tRNA for serine, histidine, and glutamine has been reported (Baldacci *et al.*, 1975; Martin *et al.*, 1976).

In *N. crassa*, the genes for the large and small rRNA species are located adjacent to one another on the mitochondrial genome (Kuriyama and Luck, 1973). In *S. cerevisiae* and *S. carlsbergensis*, however, these genes have been shown to be separated by the order of 30% of the genome (Sanders *et al.*, 1975b; Sriprakash *et al.*, 1976a). A possible explanation for this unusual feature of the structure of the yeast mitochondrial genome is that the genome indeed arose as a tetramer of a smaller genome, and subsequent alterations required for elimination of redundant sequences resulted in the retention of a large rRNA gene and a small rRNA gene from two different rRNA gene sets. The position of the rRNA genes on the genetic map of the mitochondrial genome of *S. cerevisiae* has been established (Sriprakash *et al.*, 1976a), and is shown in Figure 4. Preliminary studies on the location of tRNA genes (Casey *et al.*, 1974b; Martin *et al.*, 1976) do not permit their placement on this map.

In vivo labeling studies indicate that in *S. cerevisiae* and *N. crassa*, a number of mitochondrial inner membrane proteins are synthesized by the mitochondrial protein synthetic apparatus [for details, see Schatz and Mason (1974)]. The specific proteins identified to date are two to four subunits of the mitochondrial ATPase (Tzagoloff and Meagher, 1972; Tzagoloff *et al.*, 1973; Groot Obbink *et al.*, 1976; Jackl and Sebald, 1975), three of the seven subunits of the cytochrome oxidase complex (Rubin and Tzagoloff, 1973; Mason and Schatz, 1973; Weiss *et al.*, 1973; Lansman *et al.*, 1974), and one subunit of the cytochrome *b* complex (Weiss *et al.*, 1973; Weiss and Ziganke, 1974). Synthesis of the three cytochrome oxidase subunits in isolated mitochondria from *S. cerevisiae* (Poyton and Groot, 1975) has also been reported. These data are taken as evidence that the proteins are coded by the mitochondrial genome; however, such evidence, although persuasive, is not conclusive. Evidence that the mRNA

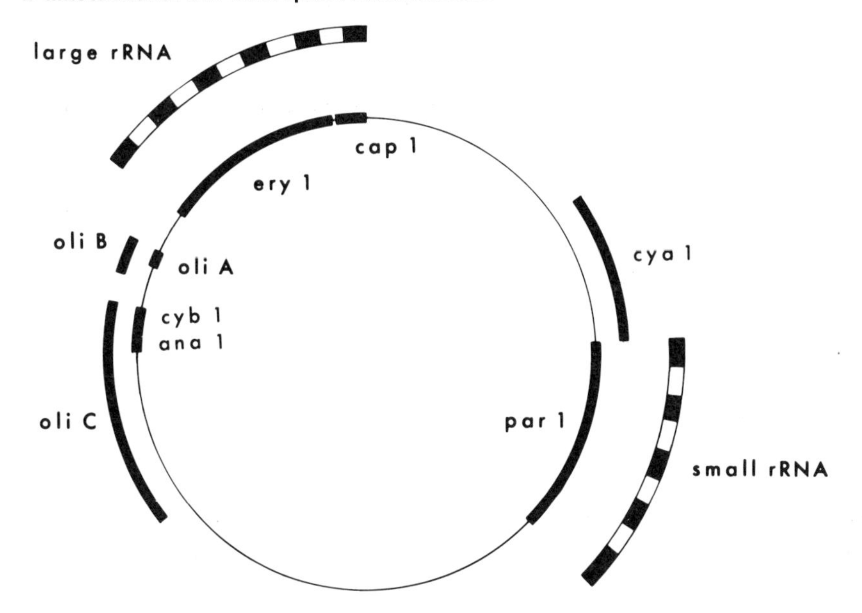

Fig. 4. Physical map of the mitochondrial genome of *S. cerevisiae*. The bars represent the region within which each mutation or rRNA cistron has been mapped. The positions of the large and small rRNA genes are as determined by Sriprakash *et al.* (1976a). Other markers are defined by resistance to the antibiotics chloramphenicol (*cap*), erythromycin (*ery*), oligomycin (*oli*), antimycin A (*ana*—formerly *mik*), and paromomycin (*par*). *cyb* and *cya* denote mutations responsible for a deficiency in cytochrome *c* reductase and cytochrome *c* oxidase, respectively. The positions of these markers are as determined by Sriprakash *et al.* (1976b) and Trembath *et al.* (1976).

used to direct mitochondrial protein synthesis is transcribed from mtDNA is also required. Such evidence has been obtained for the three mitochondrially synthesized subunits of the cytochrome oxidase complex of *S. cerevisiae*. These three polypeptides have been detected, by antibody precipitation, among the polypeptides formed in an *E. coli* cell-free system directed by a poly(A)-containing mitochondrial RNA fraction that hybridized specifically with mtDNA (Padmanaban *et al.*, 1975). Antibodies directed against whole mitochondrial membranes have also been used to demonstrate the synthesis of a number of membrane proteins by an *E. coli* cell-free system directed by *S. cerevisiae* mtDNA (Scragg and Thomas, 1975). Though these polypeptides were not identified, they were shown to be of similar size to those which are mitochondrially synthesized *in vivo*, which suggests that the majority of mitochondrially synthesized proteins are indeed gene products of mtDNA. Further support for this conclusion is now accumulating following the identification of a number of mitochondrial mutants of *S. cerevisiae* with defects in the mitochondrial

ATPase or in components of the electron transport chain (Flury *et al.*, 1974; Tzagoloff *et al.*, 1975a,b; Trembath *et al.*, 1975, 1976; Storm and Marmur, 1975). Preliminary results of studies using these mutants are already available.

Detailed genetic analysis of the known mitochondrial oligomycin resistance mutations has led to the identification of three distinct linkage groups, suggesting the existence of three mitochondrial genes whose products are associated with the oligomycin-sensitive ATPase complex (Trembath *et al.*, 1976). The location of the three linkage groups on the mitochondrial map of *S. cerevisiae* is shown in Figure 4. Biochemical analysis of mutants representative of these three groups so far has revealed specific effects on two of the mitochondrially synthesized ATPase subunits (Groot Obbink *et al.*, 1976). However, it as yet has not been established whether the mutations affect the primary structure of these proteins. Evidence indicating that a mitochondrial mutation conferring antimycin A resistance (for location, see Figure 4) affects the cytochrome *b* complex has also been reported (Trembath *et al.*, 1976; Groot Obbink *et al.*, 1976).

Tzagoloff and co-workers (1975a,b) have studied a number of mutants with defects in cytochrome oxidase and coenzyme Q–cytochrome *c* reductase. To date, analysis of the mitochondrially synthesized proteins of such mutants by gel electrophoresis has revealed a group of cytochrome oxidase mutants which lack the band corresponding to the largest cytochrome oxidase subunit and a group of mutants defective in coenzyme Q–cytochrome *c* reductase (cytochrome *b* complex) that lack another as yet unidentified band. Preliminary genetic analysis of the same mutants has classified the cytochrome oxidase mutants into three broad groups and the coenzyme Q–cytochrome *c* reductase mutants into a single group (Slonimski and Tzagoloff, 1976). This is consistent with the proposal that three cytochrome oxidase subunits and one cytochrome *b* subunit are coded mitochondrially.

3. *Protozoa and* Euglena gracilis

a. Size. Data available from electron microscopic examination and renaturation kinetic analysis of the size of the mitochondrial genome of protozoan species are listed in Table I. The agreement between the size estimates obtained by these two methods indicates a single mitochondrial chromosome. Two estimates of the molecular weight of mtDNA also have been obtained from sedimentation velocity analysis, 30–40 $\times$ 10^6 daltons for *Tetrahymena pyriformis* (Brunk and Hanawalt, 1969; Flavell and Follett, 1970) and 35 $\times$ 10^6 daltons for *Acanthamoeba castellanii* (Hettiarachchy and Jones, 1974), agreeing well with the sizes listed in Table I.

As was observed with the fungi, the mitochondrial genomes of protozoa are substantially larger than the metazoan mitochondrial genome.

b. Structure. Evidence for circular and for linear mtDNA forms has been obtained in different protozoan species. Open and closed circular DNA molecules have been observed in mtDNA preparations from *A. castellanii* (Bohnert, 1973) and *Plasmodium lophurea* (Kilejian, 1975), and closed circular mtDNA has been isolated in CsCl–ethidium bromide gradients (Bohnert and Herrmann, 1974). In contrast, a homogeneous population of linear molecules has been observed in mtDNA preparations from *T. pyriformis* (Suyama and Muira, 1968) and *Paramecium aurelia* (Goddard and Cummings, 1975). In the case of *Tetrahymena,* detailed studies have shown that the sequence of the linear molecules is unique, and that they contain a large terminal inverted-repeat sequence (Arnberg *et al.,* 1975). The form of *Euglena* mtDNA is not known.

The base composition, calculated from the CsCl density of mtDNA from the protozoa listed in Table I, is 26% G + C for *Tetrahymena* (Flavell and Jones, 1970), 40% for *Paramecium* (Flavell and Jones, 1971a), 30–35% for *A. castellanii* (Bohnert and Herrmann, 1974; Hettiarachchy and Jones, 1974), and 19% for *Plasmodium lophurae* (Kilejan, 1975). Values for the base composition obtained by direct chemical analysis are not available for these mtDNA species. The base composition of *Euglena* mtDNA is 25% G + C by direct chemical analysis and 31% calculated from CsCl density (Fonty *et al.,* 1975). No unusual bases were detected, and the discrepancy between the two estimates must indicate an unusual base distribution, as is the case for *S. cerevisiae* mtDNA (see Section III,B,2,b).

Evidence for base clustering and intramolecular heterogeneity has been reported for *Tetrahymena, Acanthamoeba,* and *Euglena* mtDNA. Strand separation in alkaline CsCl indicating a bias in the base distribution between strands has been observed with *Tetrahymena* mtDNA (Flavell and Jones, 1970). In polyribonucleotide binding studies with poly(U), poly(A), and poly(G), only poly(U) was found to bind to *Tetrahymena* mtDNA (Flavell and Jones, 1971b), indicating the presence of A-rich clusters. Separation of *Tetrahymena* mtDNA fragments of size 1–2 × 10^5 daltons by thermal elution from hydroxyapapite yielded fractions ranging in base composition from 19 to 32% G + C (Flavell and Jones, 1971b). However, while this result demonstrates some intramolecular heterogeneity in *Tetrahymena* mtDNA, the AT-rich clusters must be smaller, less common, and of higher G + C content than in *S. cerevisiae* mtDNA. Strand separation in alkaline CsCl gradients also has been observed with *Acanthamoeba* mtDNA (Bohnert and Herrmann, 1974). Furthermore, intramolecular heterogeneity is suggested from the bimodal melting curve (Bohnert and Herrmann, 1974). The first derivative of the melting curve is nonGaussian, and at least two major components are distinguishable. More detailed analysis of this mtDNA has not been reported.

Fonty *et al.* (1975) have studied the mtDNA of *Euglena gracilis* in some detail. The melting curve is bimodal and the first derivative shows two major components, the first of which has a G + C content of 10% and accounts for 40% of DNA, while the second has a G + C content of 27% and accounts for 30% of the DNA. Two minor components are also distinguishable. An analysis similar to that used by Prunell and Bernardi (1974) with *S. cerevisiae* indicates that *Euglena* mtDNA is also made up of AT-rich "spacers" and GC-rich "genes." The AT-rich segments in this case contain more G + C (10%) than the "spacers" in *S. cerevisiae* mtDNA, and the base composition of the GC-rich segments is more homogeneous. The size of these segments has not been estimated.

c. Coding Capacity and Genes Coded. The size of the mitochondrial genome of the protozoan species studied to date is two- to fivefold greater than that of the metazoan mitochondrial genome. The low average base composition of *Tetrahymena, P. lophurae,* and *Euglena* mtDNA suggests that the entire genome of these species may not have potential coding capacity. Indeed, in the case of *Euglena,* 40% of the DNA has been shown to have a base composition of 10% G + C (Fonty *et al.,* 1974), which is too low to code for protein or RNA sequences, thus reducing the potential coding capacity of *Euglena* mtDNA from an estimated 55×10^6 daltons double-stranded DNA to approximately $30–35 \times 10^6$ daltons. The fraction of AT-rich sequences in other mtDNA species has not been reported. The G + C content of *Acanthamoeba* and of *Paramecium* mtDNA is sufficiently high to allow the entire genome of $30–40 \times 10^6$-dalton double-stranded DNA potentially to code for RNA and protein species.

The mitochondrial genome of *Tetrahymena* has been shown to contain one gene for the large and small ribosomal RNA species and some genes for 4 S RNA species (Chi and Suyama, 1970). As well, the mitochondrial genome of *E. gracilis* contains one copy of the ribosomal RNA genes (Crouse *et al.,* 1974b). Similar studies for other organisms have not been reported. No protein has been identified as the product of a mitochondrial gene.

4. Plants

a. Size. Only a limited number of estimates of the size and kinetic complexity of plant mtDNA species have been reported. These are listed in Table I. To date, the size of mtDNA from all species is in the range $60–80 \times 10^6$ daltons, suggesting the possibility that the mtDNA from higher plants is of relatively constant size, as is the case for metazoan species. Although agreement between kinetic complexity and the molecular weight calculated from length measurements has been observed, estimates of kinetic complexity (with the exception of data for tobacco mtDNA) have been calculated relative to T4 DNA, assuming a complexity of 106×10^6 daltons for T4 DNA. These values must be viewed with caution, for reasons which are considered in detail in Section III,C,1,a. Nevertheless, we consider that the plant mitochondrial genome probably consists of a single chromosome.

b. Structure and Coding Capacity. A significant proportion of closed circular DNA molecules have been observed in mtDNA preparations from pea, spinach, and lettuce (Kolodner and Tewari, 1972a,b). Although it is likely that all plant mtDNA exists *in vivo* in closed circular form, further data are required before this generalization may be made.

The buoyant density of all plant mtDNA species isolated to date is 1.705–1.707 gm/ml (Suyama and Bonner, 1966; Wolstenholme and Gross, 1968; Wells and Birnstiel, 1969; Wells and Ingle, 1970; Wong and Wildman, 1972; Kolodner and Tewari, 1972a; Vedel and Quétier, 1974). This corresponds to an average base composition of 45–49% G + C. Where it is available, the base composition calculated from the T_m is in good agreement with this value. Very little additional data are available; however, no heterogeneity has been observed in melting curves of potato and pea mtDNA (Kolodner and Tewari, 1972a; Vedel and Quétier, 1974), and band splitting in alkaline CsCl gradients does not occur with potato mtDNA (Vedel and Quétier, 1974). Centrifugation of cucumber mtDNA in Cs_2SO_4 density gradients containing silver ions revealed only two very minor satellite species (Vedel *et al.*, 1972). The mtDNA of higher plants may therefore be relatively homogeneous in base composition.

The average base composition of 45–49% G + C is sufficiently high to allow the entire genome potentially to code for protein and RNA sequences. The coding capacity therefore must be considered to be equal to the genome size of $60–80 \times 10^6$ daltons double-stranded DNA, a quantity which is sufficient to code for 150–200 proteins of molecular weight 20,000. To date, no evidence for the presence of rRNA or tRNA genes in plant mtDNA has been reported. However, it seems most probable that, in common with the mtDNA of other species, the plant mitochondrial genome codes for such genes. No protein species have been identified as gene products of plant mtDNA.

C. Chloroplast DNA

As a number of differences between the structure of ctDNA from higher plants and from algae have been reported, these groups are considered separately.

1. Higher Plants

a. Size. Available estimates of the size of ctDNA from higher plants are listed in Table II. The average contour lengths of ctDNA molecules of different plants all fall in a narrow range, i.e., 40–45 μm, and although only a limited number of plant species have been studied to date, it appears that the size of ctDNA from higher plants is of relatively constant size. Estimates of the kinetic complexity of ctDNA must be considered with some caution, since all but one of the values listed in Table II have been calculated relative to T4, assuming that the kinetic complexity of T4 DNA is equal to the analytic complexity, namely, 106×10^6 daltons. The kinetic complexity of T4 DNA has been reported as $180–200 \times 10^6$ daltons by a number of workers, and Christiansen *et al.* (1973) have shown that this discrepancy results from the glucosylation of T4 DNA. True kinetic complexity values, therefore, may equal as much as twice the value calculated relative to a T4 standard. Indeed, the data of Wells and Birnstiel (1969) yield values for the complexity of lettuce ctDNA of 100×10^6 daltons relative to T4 DNA (106×10^6 daltons) and 220×10^6 daltons relative to *E. coli* DNA (2.5×10^9 daltons). These values would indicate one and two chromosomes, respectively, for ctDNA with a physical size of 40–45 μm. Thus, a firm conclusion as to the number of chloroplast chromosomes cannot be made on the basis of the available data and, although we consider it more probable that the chloroplast genome consists of a single chromosome, more careful analysis of kinetic complexity is essential. Analysis of the size of fragments produced by restriction endonucleases also should resolve this question.

b. Structure. Circular molecules have been observed in ctDNA preparations from all plant species studied to date. Closed circular molecules also have been observed (Kolodner and Tewari, 1972c; Manning *et al.*, 1972; Falk *et al.*, 1974; Herrmann *et al.*, 1975). The buoyant density in CsCl of the ctDNA of all higher plants is in the range 1.696–1.698 gm/ml, corresponding to a base composition of 37–39% G + C. Values for the base composition obtained from direct chemical analysis are in good agreement with this estimate [see Kirk (1971, 1975) for complete list of data], and 5-methylcytosine has not been detected in any ctDNA. No indication of intramolecular heterogeneity in ctDNA has been obtained from melting curves or from renaturation kinetic experiments. Melting curves are sharp and unimodal, and the first derivative of the melting curve shows a Gaussian distribution in the case of pea ctDNA (Kolodner and Tewari, 1972c). The normal relationship between fragment size and renaturation constant also has been shown to hold for lettuce and pea ctDNA (Wells and Birnstiel, 1969; Kolodner and Tewari, 1972c). The two components detected in earlier studies of renaturation kinetics (Wells and

TABLE II

Size of Chloroplast Genomes

Organism	Length (μm)	Calculated MW[a] ($\times 10^6$)	Reference	Apparent kinetic complexity[b] ($\times 10^6$)	Reference[h]
Higher plants					
Lactuca sativa	—			200[c]/100[d]	1
	40	80	2	90–100[d]	2
Pisum sativum	40	87[e]	3	95[d]	3
Spinacia oleracea	43.8	89[f]	4	100[d]	6
	45.7	90	5	—	
	40	80	2	90–100[d]	2
Beta vulgaris	44.9	90	5	100[d]	6
Antirrhinium majus	45.9	90	5	100[d]	6
Oenothera hookeri	45.2	90	5	100[d]	6
Nicotiana tabacum	—			114[c]	7
Zea mays	43	87[f]	4	—	
Narcissus pseudonarcissus	44	92[g]	8	—	
Algae					
Chlamydomonas reinhardi	—			204[c]	9
	—			194	10
Chlorella pyrenoidosa	—			235[c]	11
	—			200[c]	12
Euglena gracilis	44.5	92[f]	13	94	14
	—			180	15

[a] See footnote *a* to Table I.
[b] See footnote *b* to Table I.
[c] Calculated relative to *E. coli*, 2.5×10^9 daltons.
[d] Calculated relative to T4, 106×10^6 daltons.
[e] Internal standards λ, 29.6×10^6 daltons, and ϕX174 RFII, 3.2×10^6 daltons.
[f] Internal standard λ, 30×10^6 daltons.
[g] Internal standards T_3, 25.3×10^6 daltons, and dvl λ, 4.4×10^6 daltons.
[h] Key to references: 1. Wells and Birnstiel (1969); 2. Kolodner and Tewari (1972b); 3. Kolodner and Tewari (1972c); 4. Manning *et al.* (1972); 5. Herrmann *et al.* (1975); 6. Herrmann *et al.* (1974); 7. Tewari and Wildman (1970); 8. Falk *et al.* (1974); 9. Wells and Sager (1971); 10. Bastia *et al.* (1971); 11. Bayen and Rode (1973); 12. Dalmon and Bayen (1975); 13. Manning and Richards (1972a); 14. Slavik and Hershberger (1975); 15. Stutz (1970).

Birnstiel, 1969) have not been reported in later studies (Tewari and Wildman, 1970; Kolodner and Tewari, 1972c). Two minor satellite species have been observed in maize ctDNA centrifuged in Cs_2SO_4 gradients containing silver ions (Vedel *et al.*, 1972), indicating a small degree of intramolecular heterogeneity in this DNA. In general, however, it appears that ctDNA is relatively homogeneous in base composition.

c. Coding Capacity and Genes Coded. The G + C content of plant ctDNA is sufficiently high for the entire genome of size approximately 90×10^6 daltons double-stranded DNA to code for RNA and protein sequences.

The chloroplast genome has been shown to contain genes for the small and large RNA species of the chloroplast ribosomes, in tobacco (Tewari and Wildman, 1970), Swiss chard (Ingle *et al.*, 1971), pea (Thomas and Tewari, 1974a), and in spinach, lettuce, bean, and corn (Thomas and Tewari, 1974b). In tobacco and Swiss chard, the extent of hybridization indicates one gene for each rRNA per chromosome, while in the remaining plants two genes per chromosome are indicated. Hybridization of tRNA to ctDNA has been reported only in the case of tobacco. The extent of hybridization is sufficient to account for genes for 20–30 tRNA species of size 25,000 daltons (Tewari and Wildman, 1970).

rRNA and tRNA sequences account for 5–10% of the single-stranded information content of ctDNA. The remaining information is sufficient to code for at least 400 proteins of an average size of 20,000 daltons. That some chloroplast proteins are products of chloroplast protein synthesis has been established using differential labeling *in vivo*. These proteins include one subunit of fraction I protein (ribulose diphosphate carboxylase), which is a soluble protein and a major chloroplast constituent, as well as several membrane proteins (Ellis and Hartley, 1971; Machold, 1971; Machold and Aurich, 1972; Nielsen, 1975). Limited evidence that some proteins of the chloroplast ribosomes are also synthesized by the chloroplast has also been presented (Ellis and Harley, 1971). Synthesis of the large subunit of fraction I protein and a number of membrane proteins in isolated chloroplasts has also been reported (Blair and Ellis, 1973; Eaglesham and Ellis, 1974). The site of transcription of the mRNA used to direct the synthesis of these proteins has not been established. Only one chloroplast protein clearly has been shown to be coded by the chloroplast chromosome, the large subunit of fraction I protein. An analysis of tryptic peptides of the large subunit of fraction I protein from a number of tobacco species has revealed the presence of a peptide in one distantly related species which is not found in the other species. The difference is maternally inherited, indicating that the protein sequence is coded by the chloroplast DNA (Chan and Wildman,

1972). Synthesis of the large subunit of fraction I protein in isolated pea chloroplasts also has been reported (Blair and Ellis, 1973), indicating that the site of translation of this protein is the chloroplast ribosomes. The small subunit of fraction I protein is coded by nuclear DNA (Kawashima and Wildman, 1972). To date, no other specific proteins have been identified as products of chloroplast protein synthesis or as gene products of ctDNA.

2. *Algae and* Euglena gracilis

a. Size. Estimates of the size of ctDNA from *Chlamydomonas, Chlorella,* and *Euglena* also are listed in Table II. Circular molecules of average length 44.5 μm have been reported in the case of *Euglena gracilis* (Manning and Richards, 1972a); length measurements are not available in other cases. The two available estimates of the kinetic complexity of *Euglena* ctDNA are not in good agreement, and the number of chromosomes in the chloroplast genome therefore cannot yet be decided with certainty. Genome sizes of approximately 200×10^6 daltons have been estimated for *Chlorella* and *Chlamydomonas* ctDNA from renaturation kinetic analysis. Since considerable intramolecular heterogeneity has been demonstrated in algal ctDNA (see below), kinetic complexity may be dependent on the size of DNA fragments used, as is the case for fungal mtDNA (see Section III,B,2,b). Therefore, these data must be treated with some caution until further studies similar to those of Christiansen *et al.* (1974) are performed.

b. Structure. Closed circular molecules have been observed only in the case of *Euglena* (Manning and Richards, 1972a). The base composition of *Euglena gracilis* ctDNA is 26% G + C from CsCl density and 24% G + C by direct chemical analysis (Ray and Hanawalt, 1964). For *Chlamydomonas reinhardi* ctDNA, the base composition is 36% from the density in CsCl and 39% by direct chemical analysis (Chun *et al.,* 1963; Sager and Ishida, 1963). The base composition of *Chlorella* ctDNA calculated from the density in CsCl is 27% G + C (Bayen and Rode, 1973). Separation of the complementary strands of ctDNA in alkaline CsCl gradients has been observed with *Euglena* ctDNA (Stutz and Rawson, 1970), but does not occur with *Chlorella* ctDNA (Bayen and Rode, 1973).

Considerable evidence for intramolecular heterogeneity in the ctDNA of all three organisms has been presented. Melting curves are multiphasic, and the first derivatives show a number of distinct components. *Euglena* and *Chlorella* ctDNA show two major and three minor components with base compositions (calculated from the melting temperature) ranging from approximately 20–45% G + C (Crouse *et al.,* 1974a; Slavik and Hershberger, 1975; Bayen and Rode, 1973; Dalmon and Bayen, 1975).

Chlamydomonas ctDNA shows only two major components (Wells and Sager, 1971; Bastia *et al.,* 1971). Two major components also can be distinguished in *Chlorella* ctDNA by centrifugation in Cs_2SO_4 gradients containing silver ions (Vedel *et al.,* 1972).

A striking feature of ctDNA from *E. gracilis* is the presence of a minor satellite component detectable in neutral CsCl gradients. In general, chloroplast and mitochondrial DNA preparations form a single symmetrical band in neutral CsCl gradients. The satellite of *Euglena* ctDNA is detectable in DNA preparations of high molecular weight and has been shown to be enriched for the chloroplast ribosomal RNA sequences (Crouse *et al.,* 1974a).

c. Coding Capacity and Genes Coded. The coding capacity of algal chloroplast chromosomes cannot be accurately assessed until more detailed information is available on their size and the fraction of AT-rich sequences present. The ctDNA of *E. gracilis* has been found to contain more than one gene for both the small and large rRNA species of the chloroplast ribosomes (Scott, 1973; Crouse *et al.,* 1974b), and the rRNA genes are transcribed from the H strand of the ctDNA (Stutz, 1970). Evidence for the presence of ctRNA genes in the ctDNA of *Euglena* has also been presented (Barnett *et al.,* 1969; Schwartzbach *et al.,* 1975). In *Chlamydomonas,* only evidence for the presence of rRNA genes in the ctDNA has been presented (Surzycki and Rochaix, 1971). Several membrane proteins have been shown to be products of chloroplast protein synthesis in *Chlamydomonas* (Eytan and Ohad, 1970; Hoober, 1970), and antibiotic resistance mutations located in the chloroplast genome have been shown to affect the properties of one protein of the large ribosomal subunit (Mets and Bogorad, 1972), and one protein of the small ribosomal subunit (Ohta *et al.,* 1975). This evidence represents substantial—but not conclusive—proof for a chloroplast location of the genes for these proteins.

IV. REPLICATION OF MITOCHONDRIAL AND CHLOROPLAST DNA

In this section, we consider first the general properties of the replication of mtDNA and ctDNA and the relationships between organelle and nuclear DNA replication. The mode of replication of circular metazoan mtDNA and of the linear mtDNA from the protists *T. pyriformis* and *P. aurelia* then are considered in some detail. Details of the mode of replication of mtDNA from other species or of ctDNA have not been reported to date.

A. General Properties

That mtDNA and ctDNA are replicated within the mitochondrion or chloroplast, respectively, has been established for some time. Synthesis of DNA in isolated organelles has been reported both for mitochondria (Parsons and Simpson, 1967; Wintersberger, 1968; Neupert *et al.*, 1969; Brewer *et al.*, 1967; Nass, 1969b; Mitra and Bernstein, 1970; Koike and Kobayashi, 1973; Gause *et al.*, 1973; Ter Schegget and Borst, 1971a; Hall *et al.*, 1975), and for chloroplasts (Tewari and Wildman, 1967; Scott *et al.*, 1968; Spencer and Whitfeld, 1969). In general, DNA synthesis is dependent on the presence of four deoxyribonucleotide triphosphates and is inhibited by DNA synthesis inhibitors such as actinomycin D, acriflavin, and ethidium bromide, and the product of the reaction has the properties of organelle DNA. Synthesis of mtDNA and ctDNA in permeabilized cells also has been reported (Banks, 1973; Howell and Walker, 1972). Studies on the effects of various inhibitors on DNA synthesis *in vivo* have demonstrated differences between the nuclear and organelle replication systems. For example, ethidium bromide preferentially inhibits mtDNA synthesis in a variety of organisms (see Borst, 1972, for references) and naladixic acid preferentially inhibits ctDNA synthesis in *Euglena* (Pienkos *et al.*, 1974; Lyman *et al.*, 1975).

Differences in the control and timing of nuclear and organelle DNA synthesis also have been observed. Currently, there has been little agreement on the actual timing of organelle DNA synthesis; evidence for continuous synthesis throughout the cell cycle and for discrete bursts of DNA synthesis at specific times in the cell cycle has been presented.

Continuous synthesis of mtDNA has been reported for *Tetrahymena* (Parsons and Rustad, 1968; Charret and André, 1968), for *Physarum polycephalum* (Guttes *et al.*, 1967; Braun and Evans, 1969; Holt and Gurney, 1969), and for *S. cerevisiae* (Sena *et al.*, 1975). The evidence for continuous mtDNA synthesis is more convincing than that presented for discontinuous mtDNA synthesis, because artifically synchronized cell cultures were not used in the former case. Data presented in support of discontinuous mtDNA synthesis in *S. cerevisiae* has been critically assessed by Sena *et al.* (1975). Evidence for a discontinuous mode of mtDNA synthesis in *K. lactis* is not inconsistent with a continuous mode (Smith *et al.*, 1968). In human and mouse cells both continuous and discontinuous modes have been indicated, depending on the method of cell synchronization used (Pica-Mattoccia and Attardi, 1972; Koch and Stokstad, 1967; Bosmann, 1971; Madrieter and Mittermayer, 1972). Though we favor the conclusion that mtDNA is synthesized continuously throughout the cell cycle, further data are required to resolve this issue clearly. However, it is generally agreed that, while nuclear

DNA replication occurs only during the S phase of the cell cycle, organelle DNA synthesis is not confined to this period, indicating that the replication of organelle and of nuclear DNA are controlled independently. Furthermore, treatment of cells with cycloheximide, a specific inhibitor of cytosolic protein synthesis, stops nuclear DNA synthesis rapidly while mtDNA and ctDNA synthesis continue for some time (Grossman *et al.*, 1969; Wanka and Moors, 1970; Sussman and Rayner, 1971; Richards *et al.*, 1971; Storrie and Attardi, 1972; Werry and Wanka, 1972). These data support the conclusion that the control of organelle DNA synthesis is substantially independent of the control for nuclear DNA synthesis, and also indicate that simultaneous cytosolic protein synthesis is not required for organelle DNA replication. There is, however, no evidence to suggest that the proteins required for the continuation of organelle DNA synthesis are products of the organelle protein synthetic system, and it is generally believed that these proteins are stable nuclear-coded proteins. To date no specific protein required for organelle DNA replication has been identified conclusively.

In vivo density labeling studies have clearly demonstrated semiconservative replication of mtDNA in HeLa cells (Flory and Vinograd, 1973) and of ctDNA in *Chlamydomonas* (Chiang and Sueoka, 1967) and *Euglena* (Manning and Richards, 1972b). Similar experiments have indicated an apparent dispersive mode of mtDNA replication in *Neurospora* (Reich and Luck, 1966), *S. cerevisiae* (Williamson and Fennel, 1974), and *Euglena* (Richards and Ryan, 1974). It seems probable that replication is in fact also semiconservative in these organisms, and that the apparent dispersive replication observed *in vivo* results from either rapid turnover of mtDNA or extensive recombination of the DNA, since semiconservative mtDNA synthesis has been demonstrated *in vitro* in *S. cerevisiae* (Mattick and Hall, 1977).

Finally, a great deal of circumstantial evidence for the involvement of the membrane in organelle DNA synthesis has been presented. Strong evidence for a functional association of mtDNA synthesis with the mitochondrial membrane in *S. cerevisiae* recently has been reported (Marzuki *et al.*, 1974; Hall *et al.*, 1975). It therefore is considered likely that replication is indeed a membrane-associated process in both mitochondria and chloroplasts, as is believed to be the case in bacteria.

B. Replication of Metazoan mtDNA

The mode of replication of mtDNA in metazoa is now quite well understood; because this subject has been reviewed in detail recently (Kasamatsu and Vinograd, 1974), only the outstanding features will be considered here. In mtDNA isolated from dividing cells, a substantial proportion of the DNA molecules contain a small replication loop (see Kasamatsu and Vinograd, 1974, for references). This replicative intermediate, referred to as D-loop DNA (see Figure 5), contains a small segment (approximately 450 nucleotides) of newly synthesized H strand which is hydrogen-bonded to the parental L strand; the corresponding region of the parental H strand is single-stranded (Kasamatsu *et al.*, 1971; Ter Schegget *et al.*, 1971b). The position of the D-loop is unique, and further replication proceeds unidirectionally from this origin (Kasamatsu and Vinograd, 1973; Robberson *et al.*, 1974; Brown and Vinograd, 1974). The fact that closed circular D-loop DNA accumulates in the cell is taken to indicate that this replicative form is important in the regulation of the replication cycle. Replication proceeds as illustrated in Figure 5. Synthesis of the new H strand proceeds by displacement synthesis to form expanded D-loop intermediates. Synthesis of the new L strand proceeds by complement synthesis using the displaced parental H strand as template. Initiation of complement synthesis does not appear to be rigorously controlled. In some organisms, initiation of complement synthesis does not occur until more than half of the parental H strand has been displaced and a substantial asymmetry is observed in expanded D-loop

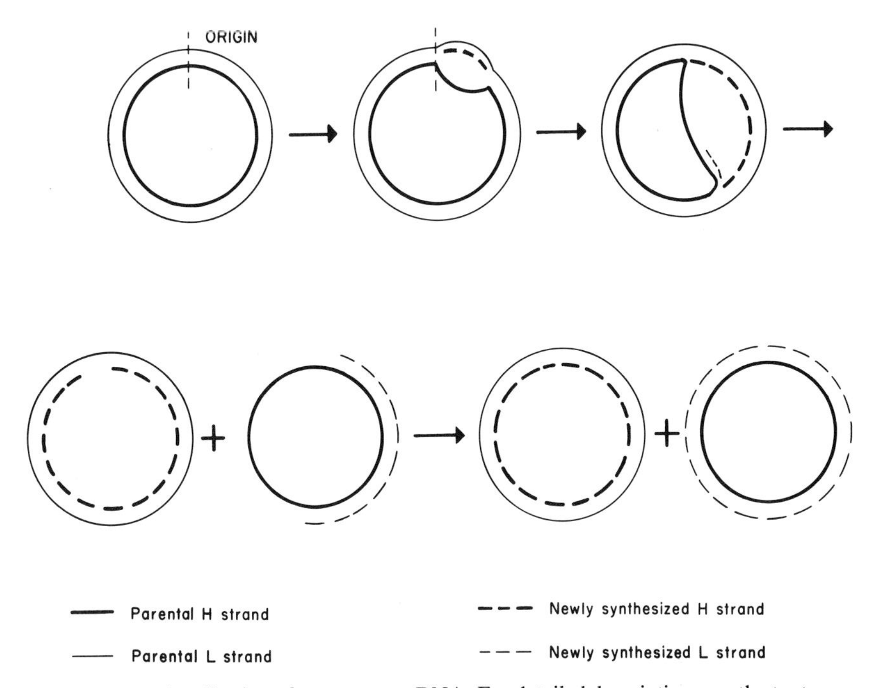

Fig. 5. Mode of replication of metazoan mtDNA. For detailed description, see the text.

forms, as illustrated in Figure 5 (Robberson *et al.*, 1972b; Berk and Clayton, 1974), while in other organisms the initiation of complement synthesis occurs earlier, and the asymmetry of expanded D-loop forms is smaller or absent (Wolstenholme *et al.*, 1973a,b). Expanded D-loop replicative forms are covalently closed circular molecules (Robberson and Clayton, 1972), and it is inferred that nicking and closing must occur continuously during D-loop expansion. The lengths of the replicative forks of expanded D-loop forms have been shown to fall into discrete size classes, which suggests that D-loop expansion occurs in a stepwise fashion, the size of each step being 4–7% of the genome (Wolstenholme *et al.*, 1973b; Koike and Wolstenholme, 1974). The size of these steps are of approximately the same size as Okazaki fragments, and it was suggested that displacement (H strand) synthesis as well as complement (L strand) synthesis may be discontinuous. Further support for this proposal has now been obtained from studies of the mode of DNA synthesis in isolated mitochondria from newborn rat liver (Koike *et al.*, 1976). In pulse and chase experiments, the newly synthesized DNA is initially found in small fragments, which are subsequently converted to high molecular weight DNA. Small fragments complementary to both the H and the L strand of the mtDNA were detected.

Berk and Clayton (1974) have shown that both daughter molecules segregate in open circular form, indicating that segregation occurs before displacement synthesis of the new H strand is complete. The daughter molecules are subsequently converted to closed circular form. Synthesis of the initiation segment to form a new D-loop intermediate occurs after closure, and it is inferred that formation of this intermediate must involve a nicking and reclosure of at least one of the parental strands (Berk and Clayton, 1974).

C. Replication of Linear mtDNA

Replicative intermediates have been identified in mtDNA isolated from exponentially growing *Tetrahymena pyriformis,* and a procedure for separating these molecules from the bulk of mtDNA has been described (Clegg *et al.*, 1974). Replicative intermediates isolated from normal cells and from cells enriched for replicative intermediates by treatment with ethidium bromide have been examined by electron microscopic procedures (Arnberg *et al.*, 1974). The major DNA forms observed were linear molecules with an internal "eye" of variable size. Analysis of the position and size of the replication "eye" in such intermediates led to the proposal that replication of *Tetrahymena* mtDNA is initiated near the center of the linear DNA molecule and proceeds bidirectionally to the ends, as illustrated on Figure 6A.

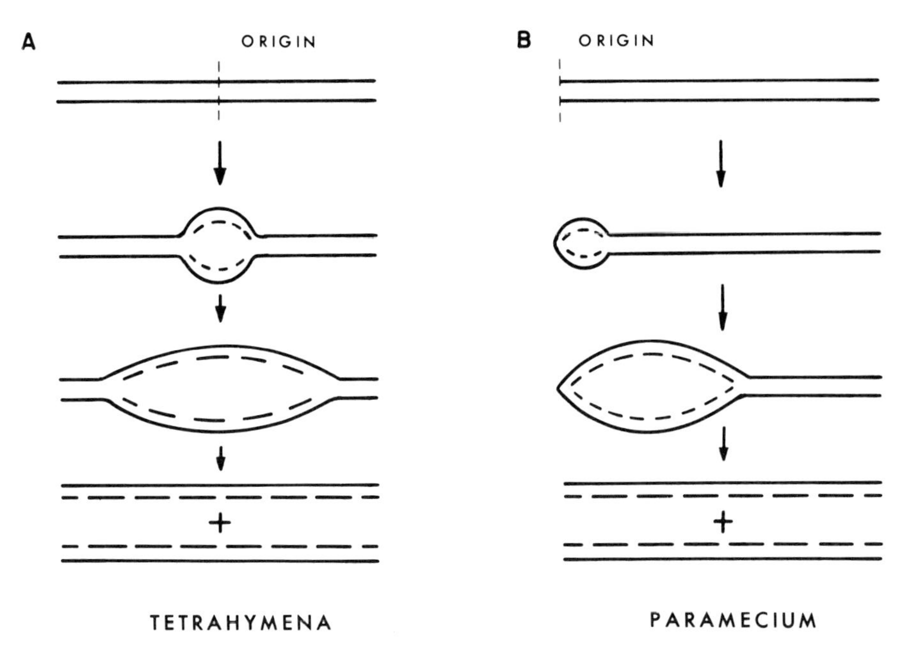

Fig. 6. Mode of replication of *Tetrahymena* (A) and *Paramecium* (B) mtDNA. Solid lines represent parental DNA strands; broken lines represent newly synthesized DNA strands. For further details, see the text.

More recently, a similar analysis of the replicative intermediates found in mtDNA isolated from *P. aurelia* has been reported (Goddard and Cummings, 1975). In this case, replication is initiated at one end of the linear mtDNA molecule and proceeds unidirectionally to the other end, as illustrated in Figure 6B.

V. CONCLUDING REMARKS

Though it is little over a decade since evidence for the existence of ctDNA and of mtDNA was first reported, many features of the structure and function of organelle DNA are now well understood, but many questions also remain to be answered. It is now accepted that in general organelle DNA codes for the small and large rRNA species of the organelle ribosomes, some tRNA species, and a limited number of proteins. For mtDNA, the proteins so far identified are all subunits of enzyme complexes associated with the inner mitochondrial membrane, while ctDNA has been shown to code for a subunit of the soluble Fraction I protein, and probably also codes for a number of membrane proteins and some ribosomal proteins.

In view of the similarity in function of mtDNA from all species, the differences observed in the size of mtDNA from different species are striking; the size of the mtDNA of most metazoa is of the order of 10×10^6 daltons, mtDNA from lower eukaryotes ranges from 12×10^6 to 50×10^6 daltons, and the mtDNA of plants is of the order of $70–80 \times 10^6$ daltons. Though in some cases (e.g., *S. cerevisiae* and *E. gracilis*) the larger mtDNA molecules have been shown to contain a substantial proportion of AT-rich sequences that cannot code for RNA or protein sequences; in other cases, where the average G + C content of the DNA is high (e.g., *Neurospora crassa* and higher plants), such sequences cannot account for a significant fraction of the size of the mtDNA molecule. The coding capacity of the smallest mtDNA molecules is close to that required to encode all the species that have so far been identified as gene products of mtDNA, and the question therefore arises of whether larger mtDNA species also encode other proteins, which remain to be identified.

The size of ctDNA molecules appears to fall within a relatively narrow range, $80–100 \times 10^6$ daltons, though available evidence suggests that algal ctDNA molecules are slightly larger than plant ctDNA. The average base composition of plant ctDNA also falls within an extremely narrow range, and this suggests that all ctDNA's code for the same products. However, the code capacity of ctDNA appears to be vastly greater than that required to encode the species so far identified as gene products of ctDNA, and further proteins probably remain to be identified.

Another interesting question for future investigation is the function of the AT-rich sequences found in organelle DNA. Though these sequences probably do not code for RNA and protein sequences, it cannot be assumed that they are without function, and it is possible that their role is in regulation of the expression of the genome.

REFERENCES

Agsteribbe, E., Kroon, A. M., and Van Bruggen, E. F. J. (1972). Circular DNA from mitochondria of *Neurospora crassa*. *Biochim. Biophys. Acta* **269,** 299–304.

Aloni, Y., and Attardi, G. (1971a). Expression of the mitochondrial genome in HeLa cells. II. Evidence for complete transcription of mitochondrial DNA. *J. Mol. Biol.* **55,** 251–270.

Aloni, Y., and Attardi, G. (1971b). Expression of the mitochondrial genome in HeLa cells. VI. Titration of mitochondrial genes for 16 S, 12 S, and 4 S RNA. *J. Mol. Biol.* **55,** 271–276.

Aloni, Y., and Attardi, G. (1971c). Symmetrical *in vivo* transcription of mitochondrial DNA in HeLa cells. *Proc. Natl. Acad. Sci. U.S.A.* **68,** 1757–1761.

Antonoglou, O., and Georgastos, J. G. (1972). Nearest-neighbor frequencies of mitochondrial deoxyribonucleic acid in mouse liver. *Biochemistry* **11,** 618–621.

Arnberg, A. C., Van Bruggen, E. F. J., Schutgens, R. B. H., Flavell, R. A., and Borst, P. (1972). Multiple D-loops in *Tetrahymena* mitochondrial DNA. *Biochim. Biophys. Acta* **272,** 487–494.

Arnberg, A. C., Van Bruggen, E. F. J., Clegg, R. A., Upholt, W. B., and Borst, P. (1974). An analysis by electron microscopy of intermediates in the replication of linear *Tetrahymena* mitochondrial DNA. *Biochim. Biophys. Acta* **361,** 266–276.

Arnberg, A. C., Van Bruggen, E. F. J., Borst, P., Clegg, R. A., Schutgens, R. B. H., Weijers, P. J., and Goldbach, R. W. (1975). Mitochondrial DNA of *Tetrahymena pyriformis* strain ST contains a long terminal duplication–inversion. *Biochim. Biophys. Acta* **383,** 359–369.

Ashwell, M., and Work, T. S. (1970). The biogenesis of mitochondria. *Annu. Rev. Biochem.* **39,** 251–290.

Attardi, G., Constantino, P., and Ojala, D. (1974). Molecular approaches to the dissection of the mitochondrial genome in HeLa cells. *In* "The Biogenesis of Mitochondria: Transcriptional, Translational and Genetic Aspects" (A. M. Kroon and C. Saccone, eds.), pp. 9–29. Academic Press, New York.

Avadhani, N. G., Lewis, F. S., and Rutman, R. J. (1974). Messenger ribonucleic acid metabolism in mammalian mitochondria. Quantitative aspects of structural information coded by the mitochondrial genome. *Biochemistry* **13,** 4638–4645.

Baldacci G., Carnevalli, F., Frontali, L., Leoni, L., Macino, G., and Palleschi, C. (1975). Heterogeneity of mitochondrial DNA from *Saccharomyces cerevisiae* and genetic information for tRNA. *Nucleic Acids Res.* **2,** 1777–1786.

Banks, G. R. (1973). Mitochondrial DNA synthesis in permeable cells. *Nature (London), New Biol.* **245,** 196–199.

Barnett, W. E., Pennington, C. J., and Fairfield, S. A. (1969). Induction of *Euglena* transfer RNA's by light. *Proc. Natl. Acad. Sci. U.S.A.* **63,** 1261–1268.

Bastia, D., Chiang, K-S., Swift, H., and Siersma, P. (1971). Heterogeneity, complexity, and repetition of the chloroplast DNA of *Chlamydomonas reinhardtii. Proc. Natl. Acad. Sci. U.S.A.* **68,** 1157–1161.

Bayen, M., and Rode, A. (1973). Heterogeneity and complexity of *Chlorella* chloroplastic DNA. *Eur. J. Biochem.* **39,** 413–420.

Berk, A. J., and Clayton, D. A. (1974). Mechanism of mitochondrial DNA replication in mouse L-cells. Asynchronous replication of strands, segregation of daughter molecules, aspects of topology and turnover of an initiation sequence. *J. Mol. Biol.* **86,** 801–824.

Bernard, U., Puhler, A., Mayer, F., and Kuntzel, H. (1975a). Denaturation map of the circular mitochondrial genome of *Neurospora crassa. Biochim. Biophys. Acta* **402,** 270–278.

Bernard, U., Bade, E., and Kuntzel, H. (1975b). Specific fragmentation of mitochondrial DNA from *Neurospora crassa* by restriction endonuclease ECO RI. *Biochem. Biophys. Res. Commun.* **64,** 783–789.

Bernardi, G. (1971). Chromatography of nucleic acids on hydroxyapatite columns. *In* "Methods in Enzymology" (L. Grossman and K. Moldave, eds.), Vol. 21, pp. 95–140. Academic Press, New York.

Bernardi, G., Faures, M., Piperno, G., and Slonimski, P. P. (1970). Mitochondrial DNA's from respiratory-sufficient and cytoplasmic respiratory-deficient mutant yeast. *J. Mol. Biol.* **48,** 23–42.

Bernardi, G., Piperno, G., and Fonty, G. (1972). The mitochondrial genome of wild-type yeast cells. I. Preparation and heterogeneity of mitochondrial DNA. *J. Mol. Biol.* **65,** 173–189.

Bernardi, G., Prunell, A., and Kopecka, H. (1975). An analysis of the mitochondrial genome of yeast with restriction enzymes. *In* "Molecular Biology of Nucleo-Cytoplasmic Relationships" (S. Puiseaux-Dao, ed.), pp. 85–90. Elsevier, Amsterdam.

Blair, G. E., and Ellis, R. J. (1973). Protein synthesis in chloroplasts. I. Light-driven synthesis of the large subunit of fraction I protein by isolated pea chloroplasts. *Biochim. Biophys. Acta* **319,** 223–234.

Blamire, J., Cryer, D. R., Finkelstein, D. B., and Marmur, J. (1972). Sedimentation properties of yeast nuclear and mitochondrial DNA. *J. Mol. Biol.* **67,** 11–24.

Bohnert, H. J. (1973). Circular mitochondrial DNA from *Acanthamoeba castelanii* (Neff-strain). *Biochim. Biophys. Acta* **324,** 199–205.

Bohnert, H. J., and Herrmann, R. G. (1974). The genomic complexity of *Acanthamoeba castellanii* mitochondrial DNA. *Eur. J. Biochem.* **50,** 83–90.

Borst, P. (1972). Mitochondrial nucleic acids. *Annu. Rev. Biochem.* **41,** 333–376.

Borst, P., and Grivell, L. A. (1971). Mitochondrial ribosomes. *FEBS Lett.* **13,** 73–88.

Borst, P., and Kroon, A. M. (1969). Mitochondrial DNA: Physicochemical properties, replication, and genetic function. *Int. Rev. Cytol.* **26,** 107–190.

Borst, P., and Ruttenberg, G. J. C. M. (1969). Mitochondrial DNA. IV. Interaction of ribopolynucleotides with the complementary strands of chick liver mitochondrial DNA. *Biochim. Biophys. Acta* **190,** 391–405.

Borst, P., and Ruttenberg, G. J. C. M. (1972). The binding of polyribonucleotides to the complementary strands of mitochondrial DNA. *Biochim. Biophys. Acta* **259,** 313–320.

Borst, P., Van Bruggen, E. F. J., Ruttenberg, G. J. C. M., and Kroon, A. M. (1967). Mitochondrial DNA. II. Sedimentation analysis and electron microscopy of mitochondrial DNA from chick liver. *Biochim. Biophys. Acta* **149,** 156–172.

Bosmann, H. B. (1971). Mitochondrial biochemical events in a synchronised mammalian cell population. *J. Biol. Chem.* **246,** 3817–3823.

Braun, R., and Evans, T. E. (1969). Replication of nuclear satellite and mitochondrial DNA in the mitotic cycle of *Physarum*. *Biochim. Biophys. Acta* **182,** 511–522.

Brewer, E. N., De Vries, A., and Rusch, H. P. (1967). DNA synthesis by isolated mitochondria of *Physarum polycephalum*. *Biochim. Biophys. Acta* **145,** 686–692.

Britten, R. J., and Kohne, D. E. (1968). Repeated sequences in DNA. *Science* **161,** 529–540.

Brown, W. M., and Vinograd, J. (1974). Restriction endonuclease cleavage maps of animal mitochondrial DNA's. *Proc. Natl. Acad. Sci. U.S.A.* **71,** 4617–4621.

Brunk, C. F., and Hanawalt, P. C. (1969). Mitochondrial DNA in *Tetrahymena pyriformis*. *Exp. Cell Res.* **54,** 143–149.

Bultmann, H., and Laird, C. D. (1973). Mitochondrial DNA from *Drosophila melanogaster*. *Biochim. Biophys. Acta* **299,** 196–209.

Burke, J. B., and Beatie, D. S. (1974). Products of rat liver mitochondrial protein synthesis: Electrophoretic analysis of the number and size of these proteins and their solubility in chloroform:methanol. *Arch. Biochem. Biophys.* **164,** 1–11.

Carnevali, F., and Leoni, L. (1972). Intramolecular heterogeneity of yeast mitochondrial DNA. *Biochem. Biophys. Res. Commun.* **47,** 1322–1331.

Carter, C. E., Wells, J. R., and MacInnis, A. J. (1972). DNA from anaerobic adult *Ascaris lambricoides* and *Hymenolepis diminuta* mitochondria isolated by zonal centrifugation. *Biochim. Biophys. Acta* **262,** 135–144.

Casey, J., Cohen, M., Rabinowitz, M., Fukahara, H., and Getz, G. S. (1972). Hybridization of mitochondrial transfer RNA's with mitochondrial and nuclear DNA of *Grande* (wild type) yeast. *J. Mol. Biol.* **63,** 431–440.

Casey, J. W., Hsu, H.-J., Getz, G. S., and Rabinowitz, M. (1974a). Transfer RNA genes in mitochondrial DNA of *Grande* (wild-type) yeast. *J. Mol. Biol.* **88,** 735–747.

Casey, J. W., Hsu, H.-J., Rabinowitz, M., and Getz, G. S. (1974b). Transfer RNA genes in mitochondrial DNA of cytoplasmic *petite* mutants of *Saccharomyces cerevisiae. J. Mol. Biol.* **88,** 717–733.

Chan, P.-H., and Wildman, S. G. (1972). Chloroplast DNA codes for the primary structure of the large subunit of fraction I protein. *Biochim. Biophys. Acta* **277,** 677–680.

Charret, R., and André, J. (1968). La synthese de l'ADN mitochondrial chez *Tetrahymena pyriformis. J. Cell Biol.* **39,** 369–381.

Chi, S. C. H., and Suyama, Y. (1970). Comparative studies on mitochondrial and cytoplasmic ribosomes of *Tetrahymena pyriformis. J. Mol. Biol.* **53,** 531–556.

Chiang, K. S., and Sueoka, N. (1967). Replication of chloroplast DNA in *Chlamydomonas reinhardi* during vegetative cell cycle: Its mode and regulation. *Proc. Natl. Acad. Sci. U.S.A.* **57,** 1506–1513.

Christiansen, C., Christiansen, G., and Bak, A. L. (1973). The influence of glucosylation on the renaturation rate of T4 phage DNA. *Biochem. Biophys. Res. Commun.* **52,** 1426–1434.

Christiansen, C., Christiansen, G., and Bak, A. L. (1974). Heterogeneity of mitochondrial DNA from *Saccharomyces carlsbergensis:* Renaturation and sedimentation studies. *J. Mol. Biol.* **84,** 65–82.

Christiansen, G., and Christiansen, C. (1976). Comparison of the fine structure of mitochondrial DNA from *Saccharomyces cerevisiae* and *S. carlsbergensis:* Electron microscopy of partially denatured molecules. *Nucleic Acids Res.* **3,** 465–476.

Christiansen, G., Christiansen, C., and Bak, A. L. (1975). Heterogeneity of mitochondrial DNA from *Saccharomyces carlsbergensis.* Denaturation mapping by electron microscopy. *Nucleic Acids Res.* **2,** 197–210.

Chun, E. H., Vaughn, L. N. H., and Rich, A. (1963). The isolation and characterization of DNA associated with chloroplast preparations. *J. Mol. Biol.* **7,** 130–141.

Clark-Walker, G. D., and Gleason, F. H. (1973). Circular DNA from the water mold *Saprolegnia. Arch. Mikrobiol.* **92,** 209–216.

Clark-Walker, G. D., and Miklos, G. (1975). Complementation in cytoplasmic *petite* mutants of yeast to form respiratory competent cells. *Proc. Natl. Acad. Sci. U.S.A.* **72,** 372–375.

Clayton, D. A., and Brambl, R. M. (1972). Detection of circular DNA from mitochondria of *Neurospora crassa. Biochem. Biophys. Res. Commun.* **46,** 1477–1482.

Clayton, D. A., and Smith, C. A. (1974). Complex mitochondrial DNA. *Int. Rev. Exp. Pathol.* **14,** 2–63.

Clayton, D. A., Davis, R. W., and Vinograd, J. (1970). Homology and structural relationships between the dimeric and monomeric circular forms of mitochondrial DNA from human leukemic leukocytes. *J. Mol. Biol.* **47,** 137–153.

Clegg, R. A., Borst, P., and Weijers, P. J. (1974). Intermediates in the replication of the mitochondrial DNA of *Tetrahymena pyriformis. Biochim. Biophys. Acta* **361,** 277–287.

Cohen, M., and Rabinowitz, M. (1972). Analysis of grande and petite yeast mitochondrial DNA by tRNA hybridization. *Biochim. Biophys. Acta* **281,** 192–201.

Coote, J., and Work, T. (1971). Proteins coded by mitochondrial DNA of mammalian cells. *Eur. J. Biochem.* **23,** 564–571.

Corneo, G., Zardi, L., and Polli, E. (1968). Human mitochondrial DNA. *J. Mol. Biol.* **36,** 419–423.

Crouse, E., Vandrey, J., and Stutz, E. (1974a). Comparative analyses of chloroplast and mitochondrial DNA's from *Euglena gracilis. In* "Proceedings of the Third International Congress on Photosynthesis'' (A. Avron, ed.), pp. 1775–1785. Elsevier, Amsterdam.

Crouse, E. J., Vandrey, J. P., and Sturz, E. (1974b). Hybridization studies with RNA and DNA isolated from *Euglena gracilis* chloroplasts and mitochondria. *FEBS Lett.* **42,** 262–266.

Dalmon, J., and Bayen, M. (1975). The chloroplastic DNA of *Chlorella pyrenoidosa* (Emerson strain): Heterogeneity and complexity. *Arch. Microbiol.* **103,** 57–61.

Dawid, I. B. (1972). Mitochondrial RNA in *Xenopus laevis*. I. The expression of mitochondrial genome. *J. Mol. Biol.* **63,** 201–216.

Dawid, I. B., and Brown, D. D. (1970). The mitochondrial and ribosomal DNA components of oocytes of *Urechis caupo*. *Dev. Biol.* **22,** 1–14.

Dawid, I. B., and Wolstenholme, D. R. (1967). Ultracentrifuge and electron microscope studies on the structure of mitochondrial DNA. *J. Mol. Biol.* **28,** 233–245.

Eaglesham, A. R. J., and Ellis, R. J. (1974). Protein synthesis in chloroplasts. II. Light-driven synthesis of membrane proteins by isolated pea chloroplasts. *Biochim. Biophys. Acta* **335,** 396–407.

Edelman, M., Schiff, J. A., and Epstein, H. T. (1965). Studies of chloroplast development in *Euglena*. XII. Two types of satellite DNA. *J. Mol. Biol.* **11,** 769–774.

Ehrlich, S. D., Thiery, J. P., and Bernardi, G. (1972). The mitochondrial genome of wild-type yeast cells. III. The pyrimidine tracts of mitochondrial DNA. *J. Mol. Biol.* **65,** 207–212.

Ellis, R. J., and Hartley, M. R. (1971). Sites of synthesis of chloroplast proteins. *Nature (London), New Biol.* **233,** 193–196.

Evans, H. E., and Evans, T. E. (1970). Methylation of the deoxyribonucleic acid of *Physarum polycephalum* at various periods during the mitotic cycle. *J. Biol. Chem.* **245,** 6436–6441.

Evans, T. E., and Suskind, D. (1971). Characterization of the mitochondrial DNA of the slime mold *Physarum polycephalum*. *Biochim. Biophys. Acta* **228,** 350–364.

Eytan, G., and Ohad, I. (1970). Biogenesis of chloroplast membranes. VI. Cooperation between cytoplasmic and chloroplast ribosomes in the synthesis of photosynthetic lamella proteins during the greening process in a mutant of *Chlamydomonas reinhardi* y-1. *J. Biol. Chem.* **245,** 4297–4307.

Falk, H., Liedvogel, B., and Sitte, P. (1974). Circular DNA in isolated chromoplasts. *Z. Naturforsch., Teil C* **29,** 541–544.

Finkelstein, D. B., Blamire, J., and Marmur, J. (1972). Isolation and fractionation of yeast nucleic acids. II. Rapid isolation of mitochondrial DNA by poly(L-lysine) kieselguhr chromatography. *Biochemistry* **11,** 4853–4858.

Flavell, R. A., and Follett, E. A. C. (1970). Size and configuration of *Tetrahymena* mitochondrial deoxyribonucleic acid. *Biochem. J.* **119,** 61P–62P.

Flavell, R. A., and Jones, I. G. (1970). Mitochondrial deoxyribonucleic acid from *Tetrahymena pyriformis* and its kinetic complexity. *Biochem. J.* **116,** 811–817.

Flavell, R. A., and Jones, I. G. (1971a). Paramecium mitochondrial DNA. Renaturation and hybridization studies. *Biochim. Biophys. Acta* **232,** 255–260.

Flavell, R. A., and Jones, I. G. (1971b). Base sequence distribution in *Tetrahymena* mitochondrial DNA. *FEBS Lett.* **14,** 354–356.

Flory, P. J., and Vinograd, J. (1973). 5-Bromodeoxyuridine labeling of monomeric and catenated circular mitochondrial DNA in HeLa cells. *J. Mol. Biol.* **74,** 81–94.

Flury, U., Mahler, H. R., and Feldman, F. (1974). A novel respiration-deficient mutant of *Saccharomyces cerevisiae*. I. Preliminary characterization of phenotype and mitochondrial inheritance. *J. Biol. Chem.* **249,** 6130–6137.

Fonty, G., Crouse, E. J., Stutz, E., and Bernardi, G. (1975). The mitochondrial genome of *Euglena gracilis*. *Eur. J. Biochem.* **54,** 367–372.

Galper, J., and Darnell, J. (1971). Mitochondrial protein synthesis. *J. Mol. Biol.* **57,** 363–367.

Gause, G. G., Dolgilevich, S. M., Fatkullina, L. G., and Mikailov, V. S. (1973). Heterogeneous rapidly labeled DNA with the properties of replicating form in isolated rat liver mitochondria. *Biochim. Biophys. Acta* **312,** 179–191.

Goddard, J. M., and Cummings, D. J. (1975). Structure and replication of mitochondrial DNA from *Paramecium aurelia*. *J. Mol. Biol.* **97,** 593–609.

Groot, G. S. P., Flavell, R. A., and Sanders, J. P. M. (1975). Sequence homology of nuclear and mitochondrial DNA's of different yeasts. *Biochim. Biophys. Acta* **378,** 186–194.

Groot Obbink, D. J., Hall, R. M., Linnane, A. W., Lukins, H. B., Monk, B. C., Spithill, T. W., and Trembath, M. K. (1976). Mitochondrial genes involved in the determination of mitochondrial membrane proteins. *In* "The Genetic Function of Mitochondrial DNA" (A. M. Kroon and C. Saccone, eds.), pp. 163–173. North-Holland Publ., Amsterdam.

Grossman, L. I., Goldring, E. S., and Marmur, J. (1969). Preferential synthesis of yeast mitochondrial DNA in the absence of protein synthesis. *J. Mol. Biol.* **46,** 367–376.

Grossman, L. I., Cryer, D. R., Goldring, E. S., and Marmur, J. (1971). The petite mutation in yeast. III. Nearest-neighbor analysis of mitochondrial DNA from normal and mutant cells. *J. Mol. Biol.* **62,** 565–575.

Grossman, L. I., Watson, R., and Vinograd, J. (1973). The presence of ribonucleotides in mature closed circular mitochondrial DNA. *Proc. Natl. Acad. Sci. U.S.A.* **70,** 3339–3343.

Guttes, E. W., Hanawalt, P. C., and Guttes, S. (1967). Mitochondrial DNA synthesis in the mitotic cycle in *Physarum polycephalum*. *Biochim. Biophys. Acta* **142,** 181–194.

Halbreich, A., and Rabinowitz, M. (1971). Isolation of *Saccharomyces cerevisiae* mitochondrial formyltetrahydrofolic acid: Methionyl-tRNA transformylase and the hybridization of mitochondrial fMet-tRNA with mitochondrial DNA. *Proc. Natl. Acad. Sci. U.S.A.* **68,** 294–298.

Hall, R. M., Mattick, J. S., Marzuki, S., and Linnane, A. W. (1975). Evidence for a functional association of DNA synthesis with the membrane in mitochondria of *Saccharomyces cerevisiae*. *Mol. Biol. Rep.* **2,** 101–106.

Handel, M. A., Papaconstantinou, J., Allison, D. P., Julku, E. M., and Chin, E. T. (1973). Synthesis of mitochondrial DNA in spermatocytes of *Rhynchosciara hollaenderi*. *Dev. Biol.* **35,** 240–249.

Herrmann, R. G., Kowallik, K. V., and Bohnert, H.-J. (1974). Structural and functional aspects of the plastome. 1. The organization of the plastome. *Port. Acta Biol. Ser. A* **14,** 91–110.

Herrmann, R. G., Bohnert, H.-J., Kowallik, K. V., and Schmitt, J. M. (1975). Size, conformation and purity of chloroplast DNA of some higher plants. *Biochim. Biophys. Acta* **378,** 305–317.

Hettiarachchy, N. S., and Jones, I. G. (1974). Isolation and characterization of mitochondrial deoxyribonucleic acid of *Acanthamoeba castellanii*. *Biochem. J.* **141,** 159–164.

Hirsch, M., and Penman, S. (1973). Mitochondrial polyadenylic acid-containing RNA: Localisation and characterisation. *J. Mol. Biol.* **80,** 379–391.

Hirsch, M., Spradling, A., and Penman, S. (1974). The messengerlike poly(A)-containing RNA species from the mitochondria of mammals and insects. *Cell* **1,** 31–35.

Hollenberg, C. P., Borst, P., and Van Bruggen, E. F. J. (1970). Mitochondrial DNA. V. A 25-μ closed circular duplex DNA molecule in wild-type yeast mitochondria. Structure and genetic complexity. *Biochim. Biophys. Acta* **209,** 1–15.

Holt, C. E., and Gurney, E. G. (1969). Minor components of the DNA of *Physarum polycephalum,* cellular location and metabolism. *J. Cell Biol.* **40,** 484–496.

Hoober, J. K. (1970). Sites of synthesis of chloroplast membrane polypeptides in *Chlamydomonas reinhardi* y-1. *J. Biol. Chem.* **245,** 4237–4334.

Howell, S. H., and Walker, L. (1972). Synthesis of DNA in toluene-treated *Chlamydomonas reinhardi. Proc. Natl. Acad. Sci. U.S.A.* **69,** 490–494.

Ingle, J., Wells, R., Possingham, J. V., and Leaver, C. J. (1971). The origins of chloroplast ribosomal RNA. *In* "Autonomy and Biogenesis of Mitochondria and Chloroplasts" (N. K. Boardman, A. W. Linnane, and R. M. Smillie, eds.), pp. 393–401. North-Holland Publ., Amsterdam.

Inman, R. B. (1974). Denaturation mapping of DNA. *In* "Methods in Enzymology" (L. Grossman and K. Moldave, eds.), Vol. 29, pp. 451–458. Academic Press, New York.

Jackl, G., and Sebald, W. (1975). Identification of two products of mitochondrial protein synthesis associated with mitochondrial adenosine triphosphatase from *Neurospora crassa. Eur. J. Biochem.* **54,** 97–106.

Kasamatsu, H., and Vinograd, J. (1973). Unidirectionality of replication in mouse mitochondrial DNA. *Nature (London), New Biol.* **241,** 103–105.

Kasamatsu, H., and Vinograd, J. (1974). Replication of circular DNA in eukaryotic cells. *Annu. Rev. Biochem.* **43,** 695–719.

Kasamatsu, H., Robberson, D. L., and Vinograd, J. (1971). A novel closed circular mitochondrial DNA with properties of a replicating intermediate. *Proc. Natl. Acad. Sci. U.S.A.* **68,** 2252–2257.

Kawashima, N., and Wildman, S. G. (1972). Studies on fraction I protein. IV. Mode of inheritance of primary structure in relation to whether chloroplast or nuclear DNA contains the code for a chloroplast protein. *Biochim. Biophys. Acta* **262,** 42–49.

Kilejian, A. (1975). Circular mitochondrial DNA from the avian malarial parasite *Plasmodium lophurae. Biochim. Biophys. Acta* **390,** 276–284.

Kirk, J. T. O. (1971). Will the real chloroplast DNA please stand up. *In* "Autonomy and Biogenesis of Mitochondria and Chloroplasts" (N. K. Boardman, A. W. Linnane, and R. M. Smillie, eds.), pp. 267–276. North-Holland Publ., Amsterdam.

Kirk, J. T. O. (1975). Chloroplast nucleic acids. *In* "Handbook of Biochemistry and Molecular Biology" (G. D. Fasman, ed.), pp. 356–374. CRC Press, Cleveland, Ohio.

Koch, J., and Stockstad, E. L. R. (1967). Incorporation of [^{3}H]-thymidine into nuclear and mitochondrial DNA in synchronised mammalian cells. *Eur. J. Biochem.* **3,** 1–16.

Koike, K., and Kobayashi, M. (1973). Synthesis of mitochondrial DNA *in vitro:* Two classes of nascent DNA's. *Biochim. Biophys. Acta* **324,** 452–460.

Koike, K., and Wolstenholme, D. R. (1974). Evidence for discontinuous replication of circular mitochondrial DNA molecules from Novikoff rat ascites hepatoma cells. *J. Cell Biol.* **61,** 14–25.

Koike, K., Kobayashi, M., and Fujisawa, T. (1976). Mode of extension of the daughter strands in replication of closed circular mitochondrial DNA *in vitro. Biochim. Biophys. Acta* **425,** 18–29.

Kolodner, R., and Tewari, K. K. (1972a). Physicochemical characterization of mitochondrial DNA from pea leaves. *Proc. Natl. Acad. Sci. U.S.A.* **69,** 1830–1834.

Kolodner, R., and Tewari, K. K. (1972b). Genome sizes of chloroplast and mitochondrial DNA's in higher plants. *Proc. Electron Microsc. Soc. Am.* **30,** 190–191.

Kolodner, R., and Tewari, K. K. (1972c). Molecular size and confirmation of chloroplast deoxyribonucleic acid from pea leaves. *J. Biol. Chem.* **247,** 6355–6364.

Kroon, A. M., Borst, P., Van Bruggen, E. F. J., and Ruttenberg, G. J. C. M. (1966). Mitochondrial DNA from sheep heart. *Proc. Natl. Acad. Sci. U.S.A.* **56,** 1836–1843.

Kuriyama, Y., and Luck, D. J. L. (1973). Ribosomal RNA synthesis in mitochondria of *Neurospora crassa*. *J. Mol. Biol.* **73,** 425–437.

Kuzela, S., Krempasky, V., Kolarov, J., and Ujhazy, V. (1975). Formation, size and solubility in chloroform/methanol of products of protein synthesis in isolated mitochondria of rat liver and Zajdela hepatoma. *Eur. J. Biochem.* **58,** 483–492.

Lansman, R., Rowe, M., and Woodward, D. (1974). Pulse-recovery studies on cycloheximide-insensitive protein synthesis in *Neurospora*. Association of products with cytochrome oxidase. *Eur. J. Biochem.* **41,** 15–23.

Lederman, M., and Attardi, G. (1973). Expression of the mitochondrial genome in HeLa cells. XVI. Electrophoretic properties of the products of *in vivo* and *in vitro* mitochondrial protein synthesis. *J. Mol. Biol.* **78,** 275–283.

Leff, J., Mandel, M., Epstein, H. T., and Schiff, J. A. (1963). DNA satellites from cells of green and aplastidic algae. *Biochem. Biophys. Res. Commun.* **13,** 126–130.

Leffler, A. T., Creskoff, E., Luborsky, S. W., McFarland, V., and Mora, P. T. (1970). Isolation and characterization of rat liver mitochondrial DNA. *J. Mol. Biol.* **48,** 455–468.

Linnane, A. W., Haslam, J. M., Lukins, H. B., and Nagley, P. (1972). The biogenesis of mitochondria in microorganisms. *Annu. Rev. Microbiol.* **26,** 163–198.

Lonsdale, D. M., and Jones, I. G. (1974). Ribonuclease-sensitivity of covalently closed rat liver mitochondrial deoxyribonucleic acid. *Biochem. J.* **141,** 155–158.

Lopez Perez, M. J., and Turner, G. (1975). Mitochondrial DNA from *Aspergillus nidulans*. *FEBS Lett.* **58,** 159–163.

Luck, D. J. L., and Reich, E. (1964). DNA in mitochondria of *Neurospora crassa*. *Proc. Natl. Acad. Sci. U.S.A.* **52,** 931–938.

Luha, A. A., Sarcoe, L. E., and Wittaker, P. A. (1971). Biosynthesis of yeast mitochondria. Drug effects on the petite negative yeast *Kluyveromyces lactis*. *Biochem. Biophys. Res. Commun.* **44,** 396–402.

Lyman, H., Jupp, A. S., and Larrinua, I. (1975). Action of nalidixic acid on chloroplast replication in *Euglena gracilis*. *Plant Physiol.* **55,** 390–392.

Machold, O. (1971). Lamellar proteins of green and chlorotic chloroplasts as affected by iron deficiency and antibiotics. *Biochim. Biophys. Acta* **238,** 324–331.

Machold, O., and Aurich, O. (1972). Sites of synthesis of chloroplast lamellar proteins in *Vicia faba*. *Biochim. Biophys. Acta* **281,** 103–112.

Madrieter, H. C., Mittermayer, C., and Rainhardt, O. (1972). ^{3}H-thymidine incorporation into mitochondria of synchronized mouse fibroblasts. *Beitr. Pathol.* **145,** 249–255.

Mandel, M., and Marmur, J. (1968). Use of ultraviolet absorbance–temperature profile for determining the guanine plus cytosine content of DNA. *In* "Methods in Enzymology" (L. Grossman and K. Moldave, eds.), Vol. 12, Part B, pp. 195–206. Academic Press, New York.

Mandel, M., Shildkraut, C. L., and Marmur, J. (1968). Use of CsCl density gradient analysis for determining the guanine plus cytosine content of DNA. *In* "Methods in Enzymology" (L. Grossman and K. Moldave, eds.), Vol. 12, Part B, pp. 184–195. Academic Press, New York.

Mandell, J. D., and Hershey, A. D. (1960). A fractionating column for analysis of nucleic acids. *Anal. Biochem.* **1,** 66–77.

Manning, J. E., and Richards, O. C. (1972a). Isolation and molecular weight of circular chloroplast DNA from *Euglena gracilis*. *Biochim. Biophys. Acta* **259,** 285–296.

Manning, J. E., and Richards, O. C. (1972b). Synthesis and turnover of *Euglena gracilis* nuclear and chloroplast deoxyribonucleic acid. *Biochemistry* **11,** 2036–2043.

Manning, J. E., Wolstenholme, D. R., and Richards, O. C. (1972). Circular DNA molecules associated with chloroplasts of spinach *Spinacia oleracea*. *J. Cell Biol.* **53,** 594–601.

Martin, N., Rabinowitz, M., and Fukuhara, H. (1976). Isoaccepting mitochondrial glutamyl-tRNA species transcribed from different regions of the mitochondrial genome of *Saccharomyces cerevisiae*. *J. Mol. Biol.* **101,** 285–296.

Marzuki, S., Hall, R. M., and Linnane, A. W. (1974). Induction of respiratory incompetent mutants by unsaturated fatty acid depletion in *Saccharomyces cerevisiae*. *Biochem. Biophys. Res. Commun.* **57,** 372–378.

Mason, T. L., and Schatz, G. (1973). Cytochrome *c* oxidase of baker's yeast. II. Site of translation of the protein components. *J. Biol. Chem.* **248,** 1355–1360.

Mattick, J. S., and Hall, R. M. (1977). Replicative deoxyribonucleic acid synthesis in isolated mitochondria from *Saccharomyces cerevisiae*. *J. Bacteriol.* **130,** 973–982.

Mets, L., and Bogorad, L. (1972). Altered chloroplast ribosomal proteins associated with erythromycin-resistant mutants in two genetic systems of *Chlamydomonas reinhardi*. *Proc. Natl. Acad. Sci. U.S.A.* **69,** 3779–3783.

Michels, C. A., Blamire, J., Goldfinger, B., and Marmur, J. (1974). A genetic and biochemical analysis of petite mutations in yeast. *J. Mol. Biol.* **90,** 431–449.

Mitra, R. S., and Bernstein, I. A. (1970). Thymidine incorporation into deoxyribonucleic acid by isolated rat liver mitochondria. *J. Biol. Chem.* **245,** 1255–1260.

Miyaki, M., Koide, K., and Ono, T. (1973). RNase and alkali sensitivity of closed circular mitochondrial DNA of rat ascites hepatoma cells. *Biochem. Biophys. Res. Commun.* **50,** 252–258.

Molloy, P. L., Linnane, A. W., and Lukins, H. B. (1975). Biogenesis of mitochondria: Analysis of deletion of mitochondrial antibiotic resistance markers in *petite* mutants of *Saccharomyces cerevisiae*. *J. Bacteriol.* **122,** 7–18.

Morimoto, H., Scragg, A. H., Nekhorocheff, J., Villa, V., and Halvorson, H. O. (1971). Comparison of the protein synthesizing systems from mitochondria and cytoplasm of yeast. *In* "Autonomy and Biogenesis of Mitochondria and Chloroplasts" (N. K. Boardman, A. W. Linnane, and R. M. Smillie, eds.), pp. 282–292. North-Holland Publ., Amsterdam.

Nagley, P., and Linnane, A. W. (1970). Mitochondrial DNA deficient *petite* mutants of yeast. *Biochem. Biophys. Res. Commun.* **39,** 989–996.

Nagley, P., Molloy, P. L., Lukins, H. B., and Linnane, A. W. (1974). Studies on mitochondrial gene purification using *petite* mutants of yeast: Characterization of mutants enriched in ribosomal RNA cistrons. *Biochem. Biophys. Res. Commun.* **57,** 232–239.

Nass, M. M. K. (1969a). Mitochondrial DNA. II. Structure and physiochemical properties of isolated DNA. *J. Mol. Biol.* **42,** 529–545.

Nass, M. M. K. (1969b). Mitochondrial DNA: Advances, problems, and goals. *Science* **165,** 25–35.

Nass, M. M. K. (1973). Differential methylation of mitochondrial and nuclear DNA in cultured mouse, hamster and virus-transformed hamster cells *in vivo* and *in vitro* methylation. *J. Mol. Biol.* **80,** 155–175.

Nass, M. M. K., and Buck, C. A. (1970). Studies on mitochondrial tRNA from animal cells. II. Hybridization of aminoacyl-tRNA from rat liver mitochondria with heavy and light complementary strands of mitochondrial DNA. *J. Mol. Biol.* **54,** 187–198.

Nathans, D., and Smith, H. O. (1975). Restriction endonucleases in the analysis and restructuring of DNA molecules. *Annu. Rev. Biochem.* **44,** 273–293.

Nielsen, N. C. (1975). Electrophoretic characterisation of membrane proteins during chloroplast development in barley. *Eur. J. Biochem.* **50,** 611–623.

O'Connor, R. M., McArthur, C. R., and Clark-Walker, G. D. (1975). Closed circular DNA from mitochondrial-enriched fractions of four *Petite*- negative yeasts. *Eur. J. Biochem.* **53,** 137–144.

O'Connor, R. M., McArthur, C. R., and Clark-Walker, G. D. (1976). Respiratory deficient mutants of *Torulopsis glabrata,* a yeast with circular mitochondrial deoxyribonucleic acid of 6 μm. *J. Bacteriol.* **126,** 959–968.

Ohta, N., Sager, R., and Inouye, M. Y. (1975). Identification of a chloroplast ribosomal protein altered by a chloroplast mutation in *Chlamydomonas. J. Biol. Chem.* **250,** 3655–3659.

Ojala, D., and Attardi, G. (1974). Identification and partial characterization of multiple discrete polyadenylic acid-containing RNA components coded for by HeLa cell mitochondrial DNA. *J. Mol. Biol.* **88,** 205–219.

Padamanaban, G., Hendler, F., Patzer, J., Ryan, R., and Rabinowitz, M. (1975). Translation of RNA that contains polyadenylate from yeast mitochondria in an *Escherichia coli* ribosomal system. *Proc. Natl. Acad. Sci. U.S.A.* **72,** 4293–4297.

Parsons, J. A., and Rustad, R. C. (1968). The distribution of DNA among dividing mitochondria of *Tetrahymena pyriformis. J. Cell Biol.* **37,** 683–693.

Parsons, P., and Simpson, M. V. (1967). Biosynthesis of DNA by isolated mitochondria: Incorporation of thymidine triphosphate-2-C^{14}. *Science* **155,** 91–93.

Peacock, W. J., Brutlag, D., Goldring, E., Appels, R., Hinton, C. W., and Lindsley, D. L. (1973). The organization of highly repeated DNA sequences in *Drosophila melanogaster* chromosomes. *Cold Spring Harbor Symp. Quant. Biol.* **38,** 405–416.

Petes, T. D., Byers, B., and Fangman, W. (1973). Size and structure of yeast chromosomal DNA. *Proc. Natl. Acad. Sci. U.S.A.* **70,** 3072–3076.

Pica-Mattoccia, L., and Attardi, G. (1972). Expression of the mitochondrial genome in HeLa cells. IX. Replication of mitochondrial DNA in relationship to the cell cycle in HeLa cells. *J. Mol. Biol.* **64,** 465–484.

Pienkos, P., Walfield, A., and Hershberger, R. C. (1974). Effect of nalidixic Acid on *Euglena gracilis:* Induced loss of chloroplast deoxyribonucleic acid. *Arch. Biochem. Biophys.* **165,** 548–553.

Pikó, L., Blair, D. G., Tyler, A., and Vinograd, J. (1968). Cytoplasmic DNA in the unfertilized sea urchin egg: Physical properties of circular mitochondrial DNA and the occurrence of catenated forms. *Proc. Natl. Acad. Sci. U.S.A.* **59,** 838–845.

Piperno, G., Fonty, G., and Bernardi, G. (1972). The mitochondrial genome of wild-type yeast cells. II. Investigation on the compositional heterogeneity of mitochondrial DNA. *J. Mol. Biol.* **65,** 191–205.

Polan, M. L., Friedman, S., Gall, J. G., and Gehring, W. (1973). Isolation and characterization of mitochondrial DNA from *Drosophila melanogaster. J. Cell Biol.* **56,** 580–589.

Porcher, H. H., and Koch, J. (1973). The anatomy of the mitochondrial DNA: The localization of the heat-induced and RNase-induced scissions in the phosphodiester backbones. *Eur. J. Biochem.* **40,** 329–336.

Potter, S. S., Newbold, J. E., Hutchison, C. A., and Edgell, M. H. (1975). Specific cleavage analysis of mammalian mitochondrial DNA. *Proc. Natl. Acad. Sci. U.S.A.* **72,** 4496–4500.

Poyton, R. O., and Groot, G. S. P. (1975). Biosynthesis of polypeptides of cytochrome *c* oxidase by isolated mitochondria. *Proc. Natl. Acad. Sci. U.S.A.* **72,** 172–176.

Prunell, A., and Bernardi, G. (1974). The mitochondrial genome of wild-type yeast cells. IV. Genes and spacers. *J. Mol. Biol.* **86,** 825–841.

Rabinowitz, M., and Swift, H. (1970). Mitochondrial nucleic acids and their relation to the biogenesis of mitochondria. *Physiol. Rev.* **50,** 376–427.

Radloff, R., Bauer, W., and Vinograd, J. (1967). A dye-buoyant-density method for the detection and isolation of closed circular duplex DNA: The closed circular DNA in HeLa cells. *Proc. Natl. Acad. Sci. U.S.A.* **57,** 1514–1521.

Ray, D. S., and Hanawalt, P. C. (1964). Properties of the satellite DNA associated with the chloroplasts of *Euglena gracilis. J. Mol. Biol.* **9,** 812–824.

Ray, D. S., and Hanawalt, P. C. (1965). Satellite DNA components in *Euglena gracilis* cells lacking chloroplasts. *J. Mol. Biol.* **11,** 760–768.

Reboul, A., and Vignais, P. (1974). Origin of mitochondrial ribosomal RNA in *Candida utilis.* Hybridization studies. *Biochimie* **56,** 269–274.

Reich, E., and Luck, D. J. L. (1966). Replication and inheritance of mitochondrial DNA. *Proc. Natl. Acad. Sci. U.S.A.* **55,** 1600–1608.

Reijnders, L., Kleisend, C. M., Grivell, L. A., and Borst, P. (1972). Hybridization studies with yeast mitochondrial RNA's. *Biochim. Biophys. Acta* **272,** 396–407.

Richards, O. C., and Ryan, R. S. (1974). Synthesis and turnover of *Euglena gracilis* mitochondrial DNA. *J. Mol. Biol.* **82,** 57–75.

Richards, O. C., Ryan, R. S., and Manning, J. E. (1971). Effects of cycloheximide and of chloramphenicol on DNA synthesis in *Euglena gracilis. Biochim. Biophys. Acta* **238,** 190–201.

Robberson, D. L., and Clayton, D. A. (1972). Replication of mitochondrial DNA in mouse L cells and their thymidine kinase$^-$ derivatives: Displacement replication on a covalently-closed circular template. *Proc. Natl. Acad. Sci. U.S.A.* **69,** 3810–3814.

Robberson, D. L., Aloni, Y., Attardi, G., and Davidson, N. (1972a). Expression of the mitochondrial genome in HeLa cells. VIII. The relative position of ribosomal RNA genes in mitochondrial DNA. *J. Mol. Biol.* **64,** 313–317.

Robberson, D. L., Kasamatsu, H., and Vinograd, J. (1972b). Replication of mitochondrial DNA. Circular replicative intermediates in mouse L cells. *Proc. Natl. Acad. Sci. U.S.A.* **69,** 737–741.

Robberson, D. L., Clayton, D. A., and Morrow, J. F. (1974). Cleavage of replicating forms of mitochondrial DNA by *Eco*RI endonuclease. *Proc. Natl. Acad. Sci. U.S.A.* **71,** 4447–4451.

Rubin, M. S., and Tzagoloff, A. (1973). Assembly of the mitochondrial membrane system. X. Mitochondrial synthesis of three of the subunit proteins of yeast and cytochrome oxidase. *J. Biol. Chem.* **248,** 4275–4279.

Ruttenberg, G. J. C. M., Smit, E. M., Borst, P., and Van Bruggen, E. F. J. (1968). The number of superhelical turns in mitochondrial DNA. *Biochim. Biophys. Acta* **157,** 429–432.

Sager, R., and Ishida, M. R. (1963). Chloroplast DNA in *Chlamydomonas. Proc. Natl. Acad. Sci. U.S.A.* **50,** 725–730.

Sanders, J. P. M., Weijers, P. J., Groot, G. S. P., and Borst, P. (1974). Properties of mitochondrial DNA from *Kluyveromyces lactis. Biochim. Biophys. Acta* **374,** 136–144.

Sanders, J. P. M., Borst, P., and Weijers, P. J. (1975a). The organization of genes in yeast mitochondrial DNA. II. The physical map of *Eco*RI and *Hin*dII + III fragments. *Mol. Gen. Genet.* **143,** 53–64.

Sanders, J. P. M., Heyting, G., and Borst, P. (1975b). The organization of genes in yeast mitochondrial DNA. I. The genes for large and small ribosomal RNA are far apart. *Biochem. Biophys. Res. Commun.* **65,** 699–707.

Schäfer, K. B., Bugge, G., Grandi, M., and Küntzel, H. (1971). Transcription of mitochondrial DNA *in vitro* from *Neurospora crassa. Eur. J. Biochem.* **21,** 478–488.

Schäfer, K. B., and Küntzel, H. (1972). Mitochondrial genes in *Neurospora:* A single cistron for ribosomal RNA. *Biochem. Biophys. Res. Commun.* **46,** 1312–1319.

Schatz, G., and Mason, T. L. (1974). The biosynthesis of mitochondrial proteins. *Annu. Rev. Biochem.* **43,** 51–87.

Schmitt, H., Beckmann, J. S., and Littauer, V. Z. (1974). Transcription of supercoiled mitochondrial DNA by bacterial RNA polymerase. *Eur. J. Biochem.* **47,** 225–234.

Schneller, J. M., Faye, G., Kujawa, C., and Stahl, A. J. C. (1975). Number of genes and base composition of mitochondrial tRNA from *Saccharomyces cerevisiae. Nucleic Acids Res.* **2,** 831–838.

Schwartzbach, S. D., Hecker, L. I., and Barnett, W. E. (1975). The transcriptional origin of *Euglena* chloroplast tRNA. *Plant Physiol.* **56,** S384.

Scott, N. S. (1973). Ribosomal RNA cistrons in *Euglena gracilis*. *J. Mol. Biol.* **81,** 327–336.

Scott, N. S., Shah, V. C., and Smillie, R. M. (1968). Synthesis of chloroplast DNA in isolated chloroplasts. *J. Cell Biol.* **38,** 151–157.

Scragg, A. H., and Thomas, D. Y. (1975). Synthesis of mitochondrial proteins in an *Escherichia coli* cell-free system directed by yeast mitochondrial DNA. *Eur. J. Biochem.* **56,** 183–192.

Seidler, R. J., and Mandel, M. (1971). Quantitative aspects of deoxyribonucleic acid renaturation: Base composition, state of chromosome replication, and polynucleotide homologies. *J. Bacteriol.* **106,** 608–614.

Sena, E. P., Welch, J. W., Halvorson, H. O., and Fogel, S. (1975). Nuclear and mitochondrial deoxyribonucleic acid replication during mitosis in *Saccharomyces cerevisiae*. *J. Bacteriol.* **123,** 497–504.

Sinclair, J. H., Stevens, B. J., Gross, N., and Rabinowitz, M. (1967). The constant size of circular mitochondrial DNA in several organisms and different organs. *Biochim. Biophys. Acta* **145,** 528–531.

Skinner, D. M., and Kerr, M. S. (1971). Characterization of mitochondrial and nuclear satellite deoxyribonucleic acids of five species of crustacea. *Biochemistry* **10,** 1864–1872.

Slavik, N. S., and Hershberger, C. L. (1975). The kinetic complexity of *Euglena gracilis* chloroplast DNA. *FEBS Lett.* **52,** 171–174.

Slonimski, P., and Tzagoloff, A. (1976). Localization in yeast mitochondrial DNA of mutations expressed in a deficiency of cytochrome oxidase and/or coenzyme QH_2–cytochrome *c* reductase. *Eur. J. Biochem.* **61,** 27–41.

Smith, D., Tauro, P., Schweizer, E., and Halvorson, H. O. (1968). The replication of mitochondrial DNA during the cell cycle in *Saccharomyces lactis*. *Proc. Natl. Acad. Sci. U.S.A.* **60,** 936–942.

Sonenshein, G. E., and Holt, C. E. (1968). Molecular weight of mitochondrial DNA in *Physarum polycephalum*. *Biochem. Biophys. Res. Commun.* **33,** 361–367.

Spencer, D., and Whitfeld, P. R. (1969). The characteristics of spinach chloroplast DNA polymerase. *Arch. Biochem. Biophys.* **132,** 477–488.

Sriprakash, K. S., Choo, K. B., Nagley, P., and Linnane, A. W. (1976a). Physical mapping of mitochondrial rRNA genes in *Saccharomyces cerevisiae*. *Biochem. Biophys. Res. Commun.* **69,** 85–91.

Sriprakash, K. S., Molloy, P. L., Nagley, P., Lukins, H. B., and Linnane, A. W. (1976b). Biogenesis of mitochondria. XII. Physical mapping of mitochondrial genetic markers in yeast. *J. Mol. Biol.* **104,** 485–503.

Storm, E. M., and Marmur, J. (1975). A temperature-sensitive mitochondrial mutation of *Saccharomyces cerevisiae*. *Biochem. Biophys. Res. Commun.* **64,** 752–759.

Storrie, B., and Attardi, G. (1972). Expression of the mitochondrial genome in HeLa cells. XIII. Effect of selective inhibition of cytoplasmic or mitochondrial protein synthesis on mitochondrial nucleic acid synthesis. *J. Mol. Biol.* **71,** 177–199.

Studier, F. W. (1965). Sedimentation studies of the size and shape of DNA. *J. Mol. Biol.* **11,** 373–390.

Stutz, E. (1970). The kinetic complexity of *Euglena gracilis* chloroplasts DNA. *FEBS Lett.* **8,** 25–28.

Stutz, E., and Rawson, J. R. (1970). Separation and characterization of *Euglena gracilis* chloroplast single-strand DNA. *Biochim. Biophys. Acta* **209,** 16–23.

Surzycki, S. J., and Rochaix, J. D. (1971). Transcriptional mapping of ribosomal RNA genes of the chloroplast and nucleus of *Chlamydomonas reinhardi*. *J. Mol. Biol.* **62,** 89–109.

Sussman, R., and Rayner, E. P. (1971). Physical characterization of deoxyribonucleic acids in *Dictyostelium discoideum*. *Arch. Biochem. Biophys.* **144,** 127–137.

Suyama, Y., and Bonner, W. D. (1966). DNA from plant mitochondria. *Plant Physiol.* **41,** 383–388.

Suyama, Y., and Miura, K. (1968). Size and structural variations of mitochondrial DNA. *Proc. Natl. Acad. Sci. U.S.A.* **60,** 235–242.

Szybalski, W. (1968). Use of cesium sulfate for equilibrium density gradient centrifugation. *In* "Methods in Enzymology" (L. Grossman and K. Moldave, eds.), Vol. 12, Part B, pp. 330–360. Academic Press, New York.

Szybalski, W., Kubinski, H., Hradecna, Z., and Summers, W. C. (1971). Analytical and preparative separation of the complementary DNA strands. *In* "Methods in Enzymology" (L. Grossman and K. Moldave, eds.), Vol. 21, pp. 383–413. Academic Press, New York.

Talen, J. L., Sanders, J. P. M., and Flavell, R. A. (1974). Genetic complexity of mitochondrial DNA from *Euglena gracilis*. *Biochim. Biophys. Acta* **374,** 129–135.

Tanguay, R., and Chaudhary, K. D. (1972). Studies on mitochondria. II. Mitochondrial DNA of thoracic muscles of *Schistocera gregaria*. *J. Cell Biol.* **54,** 295–301.

Ter Schegget, J., and Borst, P. (1971a). DNA synthesis by isolated mitochondria. I. Effect of inhibitors and characterization of the product. *Biochim. Biophys. Acta* **246,** 239–248.

Ter Schegget, J., and Borst, P. (1971b). DNA synthesis by isolated mitochondria. II. Detection of product DNA hydrogen-bonded to closed duplex circles. *Biochim. Biophys. Acta* **246,** 249–257.

Tewari, K. K. (1971). Genetic autonomy of extranuclear organelles. *Annu. Rev. Plant Physiol.* **22,** 141–168.

Tewari, K. K., and Wildman, S. G. (1967). DNA polymerase in isolated tobacco chloroplasts and nature of the polymerized product. *Proc. Natl. Acad. Sci. U.S.A.* **58,** 689–696.

Tewari, K. K., and Wildman, S. G. (1970). Information content in the chloroplast DNA. *In* "Control of Organelle Development" (P. L. Miller, ed.), pp. 147–179. Cambridge Univ. Press, London.

Thomas, J. R., and Tewari, K. K. (1974a). Ribosomal-RNA genes in the chloroplast DNA of pea leaves. *Biochim. Biophys. Acta* **361,** 73–83.

Thomas, J. R., and Tewari, K. K. (1974b). Conservation of 70 S ribosomal RNA genes in the chloroplast DNA's of higher plants. *Proc. Natl. Acad. Sci. U.S.A.* **71,** 3147–3151.

Tobler, H., and Gut, C. (1974). Mitochondrial DNA from 4-cell stages of *Ascaris lumbricoides*. *J. Cell Sci.* **16,** 593–601.

Trembath, M. K., Monk, B. C., Kellerman, G. M., and Linnane, A. W. (1975). Biogenesis of mitochondria 36: The genetic and biochemical analysis of a mitochondrially determined cold-sensitive oligomycin-resistant mutant of *Saccharomyces cerevisiae* with affected mitochondrial ATPase assembly. *Mol. Gen. Genet.* **141,** 9–22.

Trembath, M. K., Molloy, P. L., Sriprakash, K. S., Cutting, G. J., Linnane, A. W., and Lukins, H. B. (1976). Biogenesis of mitochondria. 44. Comparative studies and mapping of mitochondrial oligomycin resistance mutations in yeast based on gene recombination and petite deletion analysis. *Mol. Gen. Genet.* **145,** 43–52.

Tzagoloff, A., and Meagher, P. (1972). Assembly of the mitochondrial membrane system. VI. Mitochondrial synthesis of subunit proteins of the rutamycin-sensitive adenosine triphosphatase. *J. Biol. Chem.* **247,** 594–603.

Tzagoloff, A., Rubin, M. S., and Sierra, M. F. (1973). Biosynthesis of mitochondrial enzymes. *Biochim. Biophys. Acta* **301,** 71–104.

Tzagoloff, A., Akai, A., and Needleman, R. B. (1975a). Properties of cytoplasmic mutants of *Saccharomyces cerevisiae* with specific lesions in cytochrome oxidase. *Proc. Natl. Acad. Sci. U.S.A.* **72,** 2054–2057.

Tzagoloff, A., Akai, A., Needleman, R. B., and Zulch, G. (1975b). Assembly of the mitochondrial membrane system: Cytoplasmic mutants of *Saccharomyces cerevisiae* with lesions in enzymes of the respiratory chain and in the mitochondrial ATPase. *J. Biol. Chem.* **250,** 8236–8242.

Van Bruggen, E. F. J., Runner, C. M., Borst, P., Ruttenberg, G. J. C. M., Kroon, A. M., and Schuurmans Stekhoven, F. M. A. H. (1968). Mitochondrial DNA. III. Electron microscopy of DNA released from mitochondria by osmotic shock. *Biochim. Biophys. Acta* **161,** 402–414.

Vanyushin, B. F., and Kirnos, M. N. (1974). The nucleotide composition and pyrimidine clusters in DNA from beef heart mitochondria. *FEBS Lett.* **39,** 195–199.

Vedel, F., and Quétier, F. (1974). Physicochemical characterization of mitochondrial DNA from potato tubers. *Biochim. Biophys. Acta* **340,** 374–387.

Vedel, F., Quétier, F., Bayen, M., Rode, A., and Dalmon, J. (1972). Intramolecular heterogeneity of mitochondrial and chloroplastic DNA. *Biochem. Biophys. Res. Commun.* **46,** 972–978.

Villa, V. D., and Storck, R. (1968). Nucleotide composition of nuclear and mitochondrial deoxyribonucleic acid of fungi. *J. Bacteriol.* **96,** 184–190.

Wanka, F., and Moors, J. (1970). Selective inhibition by cycloheximide of nuclear DNA synthesis in synchronous cultures of *Chlorella*. *Biochem. Biophys. Res. Commun.* **41,** 85–90.

Weiss, H., and Ziganke, B. (1974). Cytochrome *b* in *Neurospora crassa* mitochondria. Site of translation of the heme protein. *Eur. J. Biochem.* **41,** 63–71.

Weiss, H., Lorenz, B., and Kleinow, W. (1972). Contribution of mitochondrial protein synthesis to the formation of cytochrome oxidase in *Locusta migratoria*. *FEBS Lett.* **35,** 49–51.

Weiss, H., Sebald, W., Schwab, A. J., Kleinow, W., and Lorenz, B. (1973). Contribution of mitochondrial and cytoplasmic protein synthesis to the formation of cytochrome *b* and cytochrome aa_3. *Biochimie* **55,** 815–821.

Wells, R., and Birnstiel, M. (1969). Kinetic complexity of chloroplastal deoxyribonucleic acid and mitochondrial deoxyribnoucleic acid from higher plants. *Biochem. J.* **112,** 777–786.

Wells, R., and Ingle, J. (1970). The constancy of the buoyant density of chloroplast and mitochondrial deoxyribonucleic acids in a range of higher plants. *Plant Physiol.* **46,** 178–179.

Wells, R., and Sager, R. (1971). Denaturation and the renaturation kinetics of chloroplast DNA from *Chlamydomonas reinhardi*. *J. Mol. Biol.* **58,** 611–622.

Werry, P. A. T., and Wanka, F. (1972). The effect of cycloheximide on the synthesis of major and satellite DNA components in *Physarum polycephalum*. *Biochim. Biophys. Acta* **287,** 232–235.

Wetmur, J. G., and Davidson, N. (1968). Kinetics of renaturation of DNA. *J. Mol. Biol.* **31,** 349–370.

Williamson, D. H., and Fennel, D. J. (1974). Apparent dispersive replication of yeast mitochondrial DNA as revealed by density labeling experiments. *Mol. Gen. Genet.* **131,** 193–207.

Wintersberger, E. (1968). Synthesis of DNA in isolated yeast mitochondria. *In* "Roundtable Discussion on Biochemical Aspects of Biogenesis of Mitochondria" (E. C. Slater *et al.*, eds.), pp. 189–201. Adriatica Editrice, Bari.

Wolstenholme, D. R., and Dawid, I. B. (1968). A size difference between mitochondrial DNA molecules of urodele and anuran amphibia. *J. Cell Biol.* **39,** 222–228.

Wolstenholme, D. R., and Gross, N. J. (1968). The form and size of mitochondrial DNA of the red bean, *Phaseolus vulgaris. Proc. Natl. Acad. Sci. U.S.A.* **61,** 245–252.

Wolstenholme, D. R., Kirschner, R. G., and Gross, N. J. (1972). Heat denaturation studies of rat liver mitochondrial DNA: A denaturation map and changes in molecular configurations. *J. Cell Biol.* **53,** 393–406.

Wolstenholme, D. R., Koike, K., and Cochran-Fouts, R. (1973a). Single strand-containing replicating molecules of circular mitochondrial DNA. *J. Cell Biol.* **56,** 230–245.

Wolstenholme, D. R., Koike, K., and Cochran-Fouts, R. (1973b). Replication of mitochondrial DNA: Replicative forms of molecules from rat tissues and evidence for discontinuous replication. *Cold Spring Harbor Symp. Quant. Biol.* **38,** 267–280.

Wong, F. Y., and Wildman, S. G. (1972). Simple procedure for isolation of satellite DNA's from tobacco leaves in high yield and demonstration of minicircles. *Biochim. Biophys. Acta* **259,** 5–12.

Wong-Staal, F., Mendelsohn, J., and Goulian, M. (1973). Ribonucleotides in closed circular mitochondrial DNA from HeLa cells. *Biochem. Biophys. Res. Commun.* **53,** 140–148.

Wood, D. D., and Luck, D. J. L. (1969). Hybridization of mitochondrial ribosomal RNA. *J. Mol. Biol.* **41,** 211–224.

Wu, M., Davidson, N., Attardi, G., and Aloni, Y. (1972). Expression of the mitochondrial genome in HeLa cells. XIV. The relative positions of the 4 S RNA genes and of the ribosomal RNA genes in mitochondrial DNA. *J. Mol. Biol.* **71,** 81–93.

Younghusband, H. B., and Inman, R. B. (1974). The electron microscopy of DNA. *Annu. Rev. Biochem.* **43,** 605–619.

5

Mapping the Functional Organization of Eukaryotic Chromosomes

Burke H. Judd

I. INTRODUCTION

Several techniques have proved to be valuable tools for investigating the organization of genetic information in chromosomes. Foremost among these are the mechanisms of recombination. While meiotic recombination gives the greatest resolution and precision for mapping genes, somatic recombination also has provided important information on gene positions, particularly in those eukaryotes in which no sexual phase is known. A third technique utilizes chromosome breakage. Although almost all types

ISBN 0-12-289502-9

of rearrangements are useful for mapping genes, small deletions and duplications serve this purpose particularly well. The value of a rearrangement is enhanced greatly if the points of breakage can be defined cytologically as well as genetically. Other more recently developed techniques, such as *in situ* hybridization, the visualization of denaturation loops or heteroduplex regions in DNA by electron microscopy, and the analysis of cloned recombinant DNA segments using restriction enzymes, now are used to extend genetic and cytological maps to the molecular level. Each of these techniques will be outlined. Views about the nature of genes and their organization in chromosomes that emerge from utilizing these techniques then will be briefly discussed.

II. TECHNIQUES

A. Meiotic Recombination

Meiosis consists of two cell divisions, of which only the first is preceded by chromosome replication (See Chapter 6, Volume 1, of this Treatise). Therefore, the process results in the formation of haploid gametes or spores from diploid meiocytes (Figure 1). Recombination or crossing-over takes place between homologous chromosomes during prophase of the first meiotic division. Replication has taken place prior to prophase so that each chromosome is composed of two identical chromatids that share a centromere. Homologous chromosomes synapse early in prophase (Figure 1), producing sets of paired homologues that are quadripartite. Recombination occurs by the exchange of segments between nonsister chromatids as diagrammed in Figure 2.

Each exchange creates a chiasma which tends to hold paired homologues together. Homologues repel each other late in prophase, causing the chiasmata to terminalize. This produces the characteristic dyad structures seen at diakinesis. When a chiasma forms in a chromosome interval that is flanked by heterozygous loci, two recombinant chromatids are produced; the other two chromatids retain the parental gene combinations (Figure 2). The position of genes in chromosomes can be mapped based on the assumption that the frequency with which an exchange occurs in a given chromosome interval is related directly to the length of that interval. Thus, if exchange produces nonparental gene combinations for two loci in 10% of gametes, those loci are positioned on a map 10 units apart, since, by definition, one map unit equals 1% recombination. It is important to recognize that if an interval is large enough (50 map units or greater) so that there is an average of at least one exchange per meiotic cell in that interval, there will be produced 50% recombinant chromosomes and 50% nonrecombinant chromosomes. This one-to-one

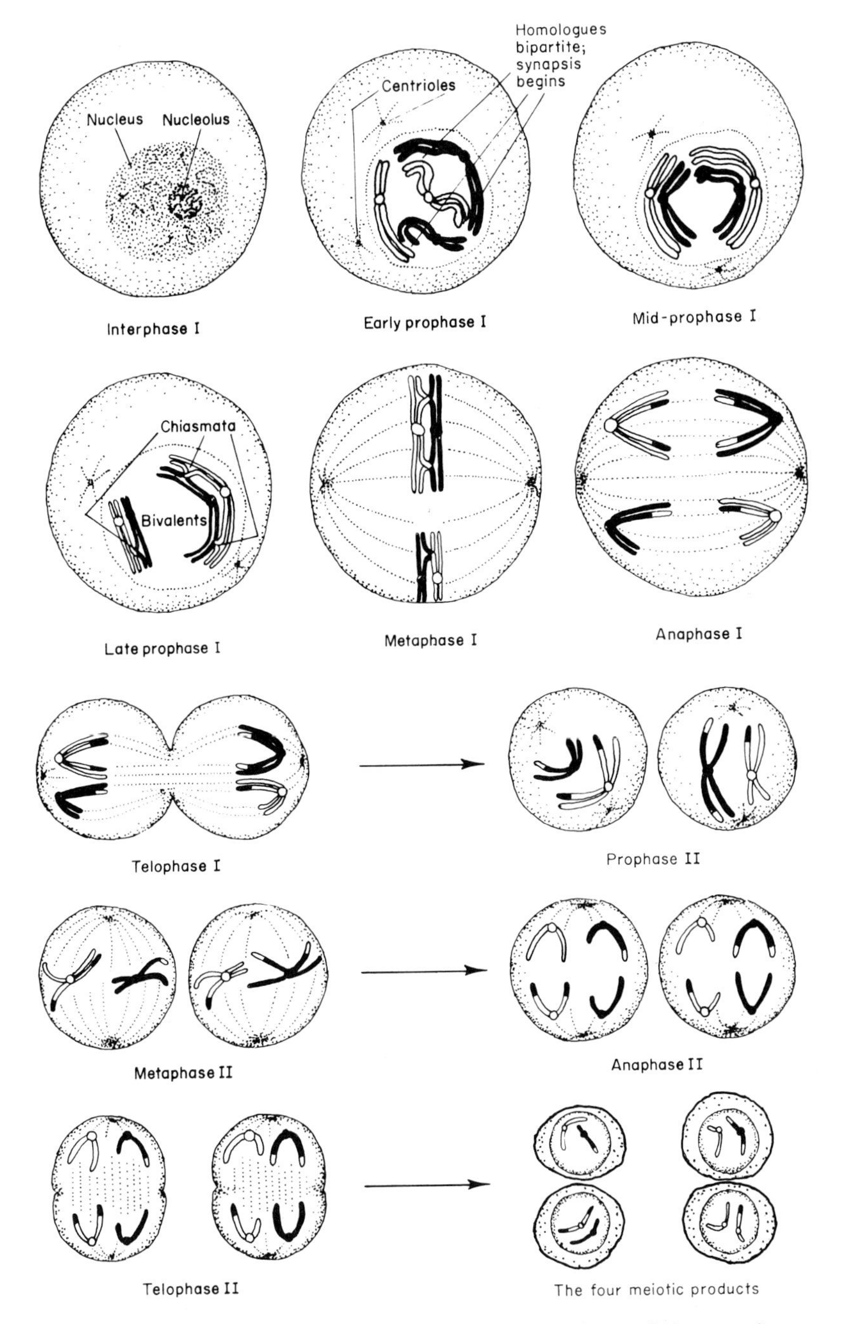

Fig. 1. Selected stages of the process of meiosis are shown for a cell that contains two pairs of chromosomes. Solid chromosomes were contributed by one parent, and open chromosomes were contributed by the other. Following chromosome replication (interphase) and synapsis (early prophase), homologues may exchange parts, a process which results in chiasma formation. After two cell divisions, each of the four products contains one chromosome of each pair in the original cell.

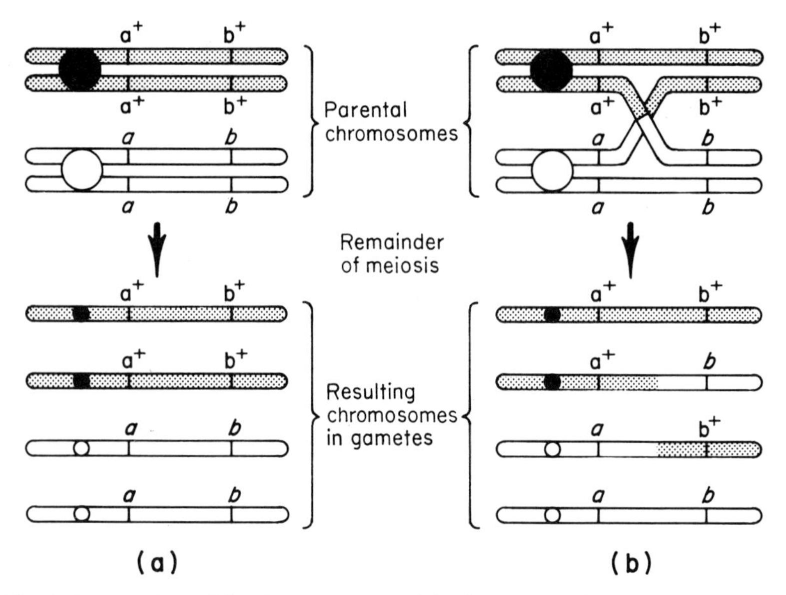

Fig. 2. A comparison of the chromosomes resulting from meiosis in two cells, one of which has formed a chiasma. (a) When no crossover has occurred in the interval between the a and b loci, the four meiotic products will contain nonrecombinant chromosomes (a^+b^+ and ab, like the parental chromosomes). (b) When crossing-over between the loci does occur, two of the cells will contain recombinant chromosomes (a^+b and ab^+) while the other two will contain nonrecombinant chromosomes (a^+b^+ and ab).

ratio of parental and nonparental gene combinations is exactly the same as expected if the loci were not linked at all. Of course, maps of length greater than 50 units can be constructed by mapping intervals of shorter lengths and adding them to complete the map of an entire chromosome.

Because each exchange involves only two of the four chromatids, the theoretical maximum of recombination for two loci is 50%, even if an interval is long enough for multiple exchanges to occur within it. This is illustrated in Figure 3, which shows the four types of double exchanges. If only two of the four chromatids are involved in both exchanges (a two-strand double), parental gene combinations are maintained; thus, all four chromosomes would appear to be nonrecombinant. Three-strand double exchanges, which may occur in two ways, produce 50% recombinant and 50% parental gene combinations. Four-strand double exchanges result in all four chromosomes being recombinant. It follows that, if the ratio of two-strand : three-strand : four-strand double exchanges is 1 : 2 : 1, as expected if strands are involved randomly, then doubles produce 50%

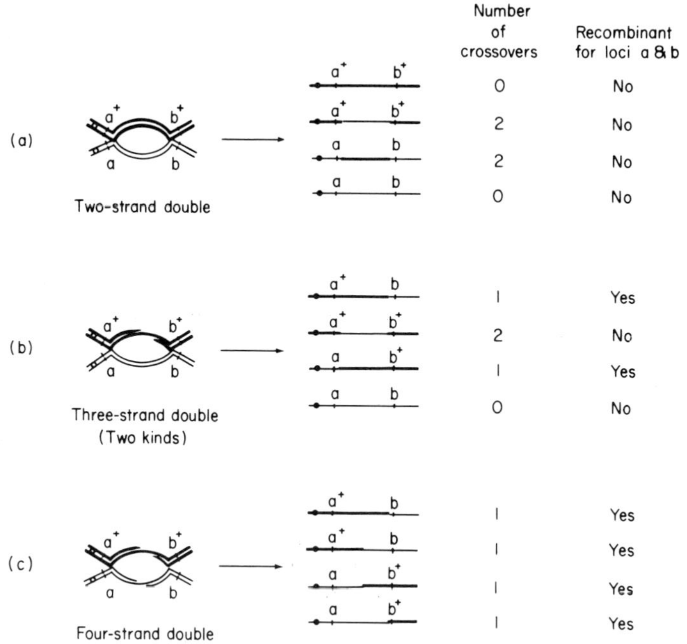

(a) Two-strand double

Chromatid	Number of crossovers	Recombinant for loci a & b
a^+ b^+	0	No
a^+ b^+	2	No
a b	2	No
a b	0	No

(b) Three-strand double (Two kinds)

Chromatid	Number of crossovers	Recombinant for loci a & b
a^+ b	1	Yes
a^+ b^+	2	No
a b^+	1	Yes
a b	0	No

(c) Four-strand double

Chromatid	Number of crossovers	Recombinant for loci a & b
a^+ b	1	Yes
a^+ b	1	Yes
a b^+	1	Yes
a b^+	1	Yes

Fig. 3. Four types of double exchanges can occur in the interval between locus *a* and locus *b*. Part (a) shows a two-strand double exchange, so called because only two of the four chromatids are involved. Notice that only nonrecombinant chromosomes are formed. Part (b) shows one of the two kinds of three-strand double exchanges. Both types generate two recombinant and two nonrecombinant chromosomes. Part (c) depicts a four-strand double in which all four chromatids are involved. Notice that it generates only recombinant chromosomes. If the three types are equally likely, the expected ratio will be 1:2:1 for two-strand: three-strand: four-strand types. This results in an average of 50% nonrecombinant and 50% recombinant chromosomes.

recombinant and 50% parental gene arrangements since the regressive two-strand doubles are equal to the progressive four-strand double exchanges. Higher multiples such as triple or quadruple exchanges, although generally very rare, also, in sum, produce a maximum of 50% recombinant chromosomes (Emerson and Rhoades, 1933; Mather, 1938).

1. Mapping Function

If it is assumed that chiasmata are distributed randomly along the length of paired homologues, the relationship between observed recombination and true map distance is that graphed by line B in Figure 4. For short map distances (less than 15 map units) there is almost a linear relationship between observed recombination frequency and map distance, but for

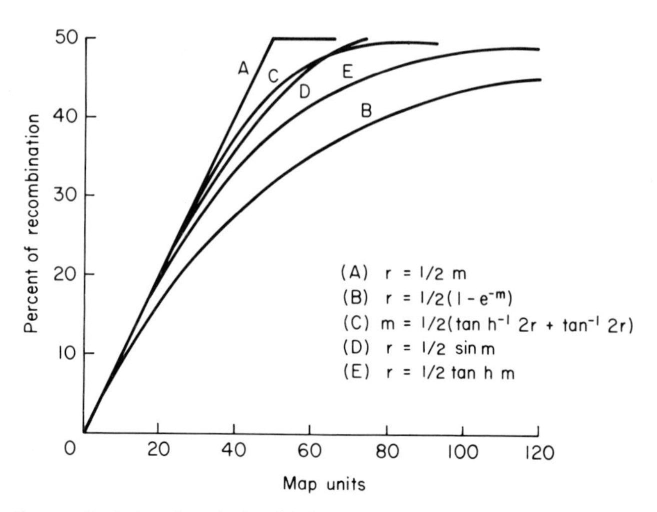

Fig. 4. Curves depicting the relationship between observed recombination frequency and map distance under conditions of: (A) complete interference; (B) zero interference; (C) an empirical function by Carter and Falconer derived from data for the mouse; (D) an empirical function by Ludwig from data for *Drosophila;* (E) an empirical function by Kosambi based on mathematical simplicity (Adapted from Barrett *et al.*, 1954.)

longer distances multiple exchanges occur in numbers large enough to distort the linear relationship significantly. For this reason, it is sometimes useful to use a mapping function for estimating more accurately the true map distance from observed recombination frequencies. Mapping functions may be of only theoretical value when dealing with an organism which is well known genetically and thoroughly mapped, but in cases in which only a few loci are known, mapping functions are of considerable value.

The Poisson distribution may be used to approximate a random distribution of chiasmata within a given chromosome interval for a population of tetrads. The expression

$$P_n = \frac{(Np)^n e^{-Np}}{n!}$$

approximates a binomial distribution when N (the number of trials) is large, but p (the probability of success) is so small that Np (the average number of successes in N trials) is small compared to N. The mean (m) of the Poisson distribution is Np; therefore we may write

$$P_n = \frac{m^n e^{-m}}{n!}$$

The several terms of this expression represent the distribution of tetrads with 0, 1, 2, . . . i exchanges in a given chromosome interval as shown in the following tabulation:

Tetrad rank				
0	1	2	3 . . .	i
$\frac{m^0 e^{-m}}{0!}$	$\frac{m^1 e^{-m}}{1!}$	$\frac{m^2 e^{-m}}{2!}$	$\frac{m^3 e^{-m}}{3!}$	$\frac{m^i e^{-m}}{i!}$

If we actually could measure the mean number of exchanges in a given chromosome interval per meiotic cell, we could calculate the true map length of that interval, because, as we have already seen, all those tetrads of rank 1 or above give 50% recombinant chromosomes. Map distance, then, is $m/2$, i.e., the frequency of exchange tetrads divided by 2. The fraction of tetrads with no exchange is

$$\frac{e^{-m} m^0}{0!} = e^{-m}$$

Therefore, recombination frequency equals (Haldane, 1919)

$$r = \tfrac{1}{2}(1 - e^{-m})$$

This expression is a mapping function which can be used to translate observed recombination frequencies into values for m. More accurate map distances then can be estimated using the relationship $m/2$ = map distance.

2. *Interference*

The mapping function outlined above is based on the assumption that chiasmata are distributed along synapsed chromosomes randomly. This situation usually is not realized, however. There is almost always interference in chiasma placement, and it is generally positive in eukaryotes. Positive chiasma interference shows up as a reduction in the number of double-exchange chromosomes compared to the number expected. For example, from a heterozygote for loci a, b, and c in which map distances are a–b = 10, b–c = 15, a–c = 25, the number of double-exchange chromosomes should total $0.10 \times 0.15 = 1.5\%$ if there is no interference. Interference, when it occurs, usually is expressed as the coefficient of coincidence:

$$C = \text{Observed double exchanges/Expected double exchanges}$$

Interference, then, is: $I = 1 - C$. In those organisms such as *Drosophila* and *Neurospora,* in which chromosomes may be marked well enough for it to be measured accurately, interference is positive within chromosome arms (for reviews, see Mather, 1938; Emerson, 1963). Of course, if interference is complete, then the relationship of recombination to map distance is linear, as represented by line A in Figure 4. In the X chromosome of *Drosophila,* for example, coincidence is very close to 0.2 (Bridges and Olbrycht, 1926; Anderson, 1925).

Several mapping functions have been developed to accommodate measured levels of interference. Empirical functions by Carter and Falconer (1951) and Ludwig (1934) stated to agree with data from the mouse and *Drosophila,* respectively, produce curves (C and D) intermediate between those of complete interference and zero interference, as shown in Figure 4. Also shown in Figure 4 is still another function, line E, proposed by Kosambi (1944), based on mathematical simplicity. Discussions of these and other mapping functions can be found in Owen (1950), Carter and Falconer (1953), and Carter and Robertson (1952).

During the consideration of double exchanges, we noted that the expected ratio of two-, three-, and four-strand doubles is 1 : 2 : 1. This expectation assumes that involvement of nonsister chromatids in the exchange events is random. Any departure from randomness constitutes *chromatid interference*. Positive chromatid interference would result in excess three- and four-strand doubles, while excess two-strand doubles would mean negative chromatid interference. Although examples of chromatid interference are found in the literature, mostly from fungi, it is difficult to make a general statement because of considerable heterogeneity between different experiments. Data compiled from several sources and reviewed by Emerson (1963) indicate a slight excess of two-strand doubles at the expense of four-strand doubles.

3. Tetrad Analysis and Centromere Mapping

In organisms such as the fungi and some of the algae, it is possible to recover all of the products from a meiotic cell. The analysis of such tetrads is particularly useful for investigating the process of recombination. Some basic genetic tenets, such as allelic segregation during meiosis, are testable; and it is possible to explore whether recombination occurs before or after chromosome replication or whether it is causally related to replication. In addition, interference of both the chromosome and chromatid types is open to closer scrutiny from tetrad data than from analysis of random meiotic products. Some double exchanges are recognizable in tetrads that would not be classified as doubles from random product analysis. Calculating map distances from tetrad data is straightforward, but it presents the same type of problems found in single chromosome or random spore data.

From a cross between two fungal strains differing by two loci (*ab* × ++ or *a*+ × +*b*), four spore types will be obtained. If tetrads are collected, they will be of three types: first, the parental ditype (PD) containing two types of spores that are like the two parents; second, the nonparental ditype (NPD) containing two types of spores unlike those of the parent strains; and third, the tetratype (T) that contains all four spore types. These combinations for the cross *ab* × ++ are shown in the tabulation below:

PD	NPD	T
ab	*a*+	*ab*
ab	*a*+	*a*+
++	+*b*	+*b*
++	+*b*	++

In order to map genes using tetrad data, it is important to recognize how the three types are generated during meiosis. Consider first the case in which loci *a* and *b* are not linked. Figure 5 illustrates that parental and nonparental ditypes will be equal in frequency. Each will constitute 50% of the total only if both *a* and *b* were located at the centromeres of their respective chromosomes. Tetratypes are generated when either locus *a* or *b* segregates at second division. Second-division segregation occurs when there is a crossover in the interval between the locus and the centromere. If both *a* and *b* segregate at second division, any of the three types of tetrads may result, with an expected ratio of 1 PD : 2 T : 1 NPD. Depending on the positions of the two loci relative to their respective centromeres, the frequency of tetratypes in the absence of interference will range from 0 to 66.7% The remainder will be parental and nonparental ditypes in equal numbers.

If loci *a* and *b* are linked, parental ditypes will be considerably more numerous than the nonparental ditypes. Nonparental ditypes are formed only by four-strand double exchange in the interval between *a* and *b*. Parental ditypes result from cells in which there is no exchange or from two-strand double exchanges in the *a*–*b* interval. Tetratypes are generated by a single exchange or by three-strand double exchanges in the *a*–*b* interval. To use tetrad data to map the *a* to *b* distance, the relationship is:

$$\text{Map distance} = (\text{NPD} + \tfrac{1}{2}\text{T})/(\text{PD} + \text{NPD} + \text{T}) \times 100$$

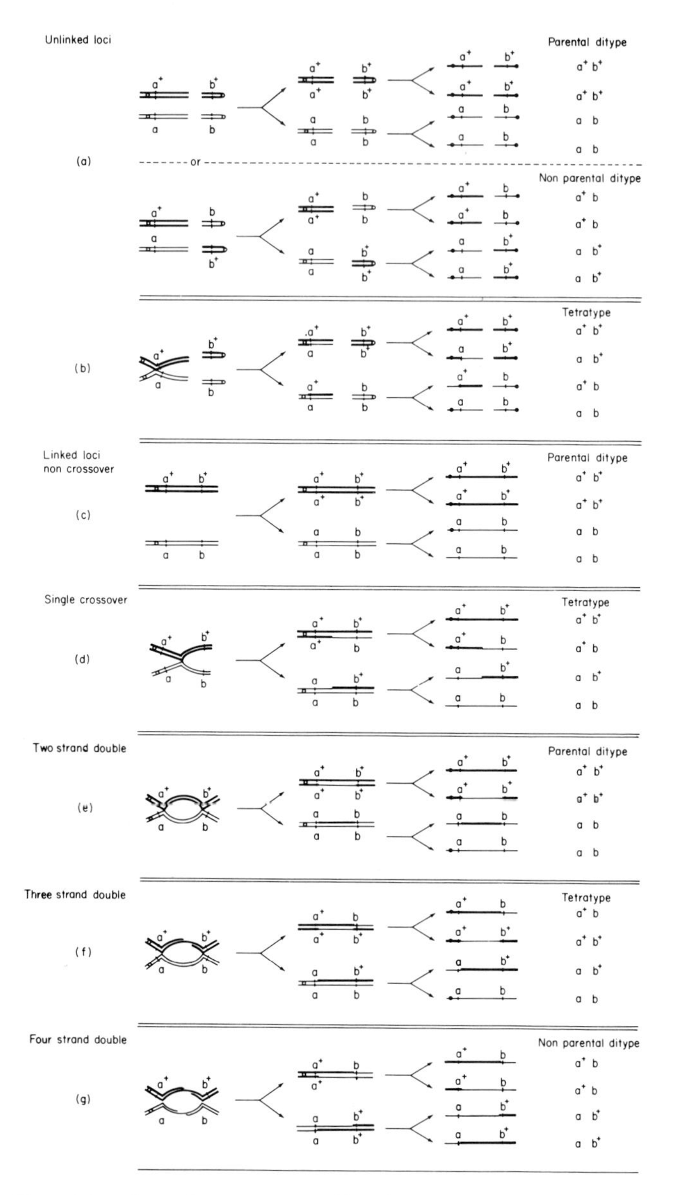

Fig. 5. The origins of tetrad types generated by a heterozygote for two unlinked loci. Part (a) shows independent assortment in bivalents in which no crossing-over has occurred. This results in parental ditype (PD) and nonparental ditype (NPD) tetrads in equal numbers. Part (b) shows one way in which a tetratype (T) tetrad, can be generated. Here, a single exchange in the interval between locus *a* and the centromere gives such a tetrad. Similarly, a single exchange in the interval between locus *b* and the centromere would also produce a tetratype tetrad. For linked genes, PD tetrads predominate, since they are formed by noncrossover cells (part c) and from those having a two-strand double exchange (part e). Single exchanges in the *a* to *b* interval (part d) produce T tetrads as do three-strand double exchanges that occur within the *a* to *b* interval (part f). NPD tetrads are generated only by four-strand double exchanges (Part g).

This formula takes into account the fact that all four chromosomes of the NPD tetrads are recombinant, whereas only two of the four are recombinant in T tetrads. If the map interval is large enough for a significant number of double exchanges to have escaped attention, a more accurate map distance estimate is provided by Perkins (1949):

$$\text{Map distance} = 3\ \text{NPD} + \tfrac{1}{2}\text{T/Total tetrads} \times 100$$

The number of NPD tetrads is multiplied by 3 to take into account the missed double-exchange classes. The logic is that NPD tetrads are formed only by four-strand doubles, and that there should have been an equal number of two-strand doubles classified as PD and twice as many three-strand doubles classified as T. One exchange of each T tetrad generated by a three-strand double is counted among the singles; therefore, the unrecognized exchanges are equal to just twice the NPD class and are simply added to it, with the sum equal to $3 \cdot \text{NPD}$.

It was pointed out above that linkage is often easy to detect from tetrad data because PD $\gg$ NPD. The frequencies of PD and NPD theoretically can approach equality if there is no chiasma interference and if the two loci are sufficiently far apart that multiple exchanges are frequent. Theoretically, both PD and NPD approach 16.7% as the map distance between loci increases. The frequency of T then approaches 66.7%. On occasion, the NPD : T ratio is helpful in detecting linkage. The value of this ratio in the absence of interference is 0.25 to ∞ for unlinked genes and is between 0 and 0.25 for linked genes. Linkage also is indicated if, because of positive interference, tetratype segregations exceed the expected maximum frequency of two-thirds of the total tetrads.

4. *Mapping Centromeres*

Centromere positions on chromosomes can be mapped in those tetrads in which segregation of alleles at the first division of meiosis can be distinguished from their segregation at second division. First versus second division segregation can be scored in *Neurospora,* in which the order of the spores in the ascus sac reflects their history during the two meiotic divisions. Centromeres of homologous chromosomes segregate at the first division of meiosis, and since spindles of the two divisions do not overlap in *Neurospora,* the two pairs of spores at one end of the sac contain those chromatids that migrated to one pole of the spindle while the two pairs at the other end contain those that segregated to the other pole. By scoring the genotype and the position of each spore, it is possible to determine whether a given heterozygous locus segregated alleles at the first or at the second meiotic division. This is illustrated in Figure 6.

Mapping gene–centromere distance can be done by scoring the per-

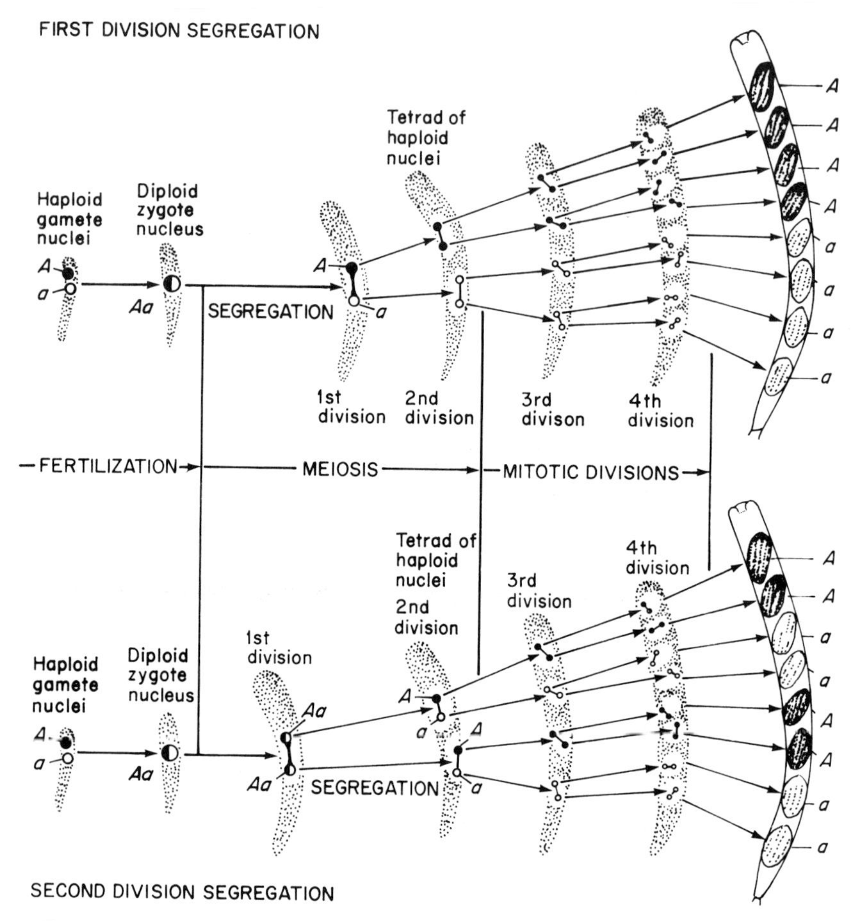

Fig. 6. In *Neurospora crassa,* meiosis follows zygote formation and results in four nuclei, each of which divides twice by mitosis to produce eight haploid binucleate ascospores. Because the spindles of the dividing nuclei do not overlap, the eight spores are ordered in the ascus sac such the segregation of chromosomes during the two meiotic divisions is reflected. Segregation at the first meiotic division is the case for centromeres of homologous chromosomes. Alleles also segregate at first division unless a crossover has occurred in the interval between the locus and the centromere. First and second division segregation patterns are illustrated for the heterozygous mating type locus (*A*/*a*). (Redrawn from Emerson, 1963.)

centage of tetrads that show second division segregation for the locus and dividing that number by 2. Thus

Gene–centromere map distance = percent second division segregation/2

As before, this distance will be shorter than actual map distance if the length of the interval is great enough to allow a significant number of double exchanges. In fact, even greater error is introduced because both two- and four-strand double exchanges are regressive in the sense that they restore first division segregation to a distal locus. In tetrads marked by two heterozygous loci, four-strand double exchanges can be detected.

5. *Mapping Functions for Tetrads*

Because two of the four chromatids of a tetrad are involved in any exchange, map distance (x) with complete interference is equal to the number of tetratype tetrads or second division segregation tetrads (y_t) divided by 2 or

$$y_t = 2x$$

With no interference, the expression developed by Rizet and Englemann (1949) and by Papazian (1952) applies:

$$y_t = \tfrac{2}{3}(1 - e^{-3x})$$

This expression for tetrads is equivalent to the above-discussed function developed by Haldane for single strands with zero interference. To take into consideration the interference which typifies the arms of eukaryotic chromosomes, Barratt *et al.* (1954) developed a mapping function for tetrads. The function is consistent with extensive single-strand data from *Drosophila* and with the more limited tetrad data from *Neurospora*.

B. Chromosome Mapping in Somatic Cells

While several techniques for mapping genes in somatic cells are available, here I shall discuss only mitotic crossing-over. The reader is referred to Chapter 5, Volume 1, of this Treatise for a discussion of the detection of linkage relationships through somatic cell hybridization and chromosome loss as well as for a review of mapping by several chromosome-banding procedures.

Mitotic Crossing-Over

Recombination at mitosis first was described by Stern (1936), examining *Drosophila*. The genetic consequences of mitotic crossing-over are diagrammed in Figure 7. It is assumed that several features of the system are the same as in meiosis, namely, that exchange occurs after chromosomes have replicated, and that it is essentially reciprocal. Unlike meiosis, no synapsis of homologues occurs, nor is there any structure like the synaptonemal complex. Since the centromeres shared by sister chromatids divide and move to opposite poles of the spindle, there are

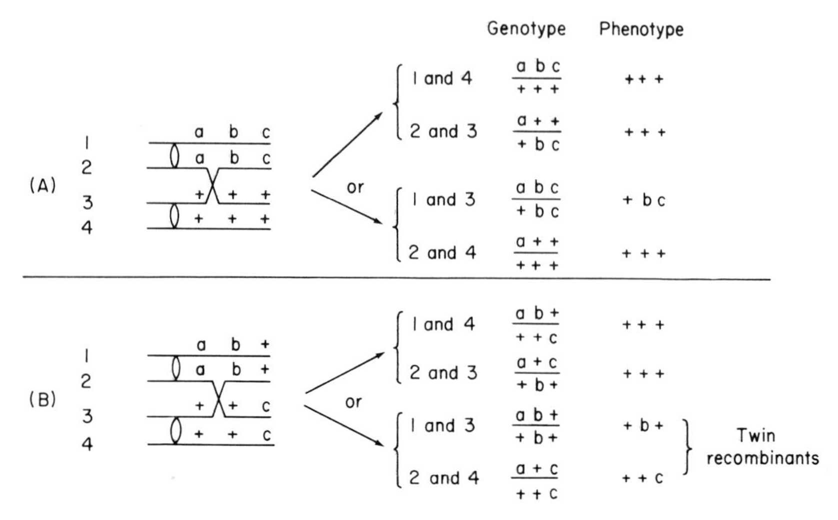

Fig. 7. Mitotic exchange can produce a homozygous condition for loci distal to the point of exchange. The assumption is that the exchange occurs after chromosomes have replicated and that sister chromatids segregate to different daughter nuclei. Example A illustrates results of segregation patterns following an exchange between chromatids with all recessive alleles in cis arrangement. Example B shows a case with recessive alleles distal to the exchange in trans arrangement. (Redrawn from Pritchard, 1963.)

two different segregation patterns following an exchange event. The figure illustrates these patterns for cases in which recessive alleles are in cis (A) and in trans (B) arrangements. In each case there are four genetically different combinations possible, three of which are different from the parental cell. In two cases, all loci distal to the point of exchange become homozygous. This produces one type which is phenotypically different from the parent cell if the markers were in cis. If mutant markers were in trans distal to the exchange, two of the four types will be phenotypically distinguishable, and these appear as complementary twins generated from a single event.

In *Drosophila,* mutations such as yellow body color or singed bristles that change the structures of the integument are useful markers because clones of mutant cells can be scored as patches surrounded by wild-type tissues. Somatic exchange events in *Drosophila* are employed widely in a variety of developmental studies (see Hotta and Benzer, 1972), but they are not very useful for mapping chromosomes because only a relatively few loci produce phenotypes that are readily classifiable in the majority of somatic tissues, and because sectors cannot be recovered and subjected to complete genetic analysis. Despite these handicaps, a sizable store of data on both spontaneous and induced somatic exchange is available for *Drosophila*. Becker (1974) has compiled these data and compared genetic maps derived by somatic exchange to meiotic and cytological maps.

In fungi in which recombinant tissue can be isolated and propagated for study, the technique of somatic exchange has been exploited to produce useful maps (see Pritchard, 1963, for review). In *Aspergillus nidulans*, for example, it is possible to construct mitotic maps and compare them to meiotic maps. In the fungi imperfecti, in which a sexual cycle is unknown, mitotic recombination is the major tool for genetic analysis.

Essential to recombination at mitosis is a stable diploid state. While not a barrier in higher eukaryotes, this is not typical of most fungi. Although certain fungi, such as *Saccharomyces cervisiae*, do have diploid phases, diploidy is found only in the zygote stage or not at all in the majority of fungi. Heterozygous diploid strains of many fungi can be obtained, however, by selection of conidia produced by heterokaryons. Stable diploid cells apparently arise by fusion of two nuclei in a forced heterokaryon. Treatment of heterokaryons with camphor (Pontecorvo and Roper, 1953) or ultraviolet light (Ikeda *et al.*, 1957) greatly increases the frequency with which such diploid cells can be obtained. In sexually reproducing ascomycetes, diploid ascospores are found with a frequency of about 0.1%. Their origin is not entirely understood, but stable diploid lines can be obtained by selecting such spores.

Diploid cells can be recognized by using color markers such as yellow (y) and white (w) in *Aspergillus nidulans*. The heterokaryon that is a mixture of (yw^+) and (y^+w) nuclei usually will produce either yellow or white haploid colonies from plated conidia. The occasional diploid (yw^+/y^+w) colony, however, will be green and therefore easily recognized. Similarly, conidia from heterokaryons with two nuclear types, each having a different nutritional requirement, can be plated on a minimal medium that will allow only diploid cells to grow.

Several schemes are used to detect mitotic recombinant segregants from heterozygous diploid colonies. Visual selection of either white or yellow sectors from a green colony (Pontecorvo *et al.*, 1954) is a scheme easily employed, as is harvesting a red sector in an otherwise white colony of yeast due to the accumulation of precursors behind a blocked step in adenine biosynthesis (Roman, 1956a,b). Segregant conidia can be isolated and grown for analysis of their complete genotypes. Selection for drug resistance (Roper and Kafer, 1957), suppression of a nutritional requirement, or recombination between two mutant sites in the same cistron are some other schemes that have been employed.

The data in Table I illustrate how these selective schemes are used to map a mitotic chromosome in *Aspergillus nidulans*. The parental diploid

TABLE I

Selective Schemes Used to Map a Mitotic Chromosome in *Aspergillus nidulans*

su-ad20	ribo	thi	0	pro	paba	y	ad20
+	+	+		+	+	+	ad20

Segregant type selected	Phenotype	Number
yellow	y ad20	35
	paba y ad20	110
	pro paba y ad20	9
su-ad20 adenine independence	su-ad20	59
	su-ad20 ribo	21
	su-ad20 ribo thi	181

su		ribo		thi		0		pro		paba		y
	22.6		8.0		69.4		5.9		71.4		22.7	

genotype is given in the table. Conidia from it were (1) plated on complete medium and colonies were inspected for yellow sectors, or (2) plated on minimal medium supplemented with all nutrients except adenine. Small colonies develop because ad20 is leaky, and these developed sectors of normal growth in segregants in which suppressor of ad20 had become homozygous. The sectors in both groups were isolated and plated on minimal medium supplemented with various combinations of required nutrients to determine their genotypes.

The frequency of the various classes can be used to construct the mitotic chromosome map shown at the bottom of Table I. Each chromosome arm is set arbitrarily at length 100. Units in the two arms are not comparable because the two selection procedures are different. Figure 8 is a comparison of the meiotic and mitotic maps of two chromosomes of *A. nidulans* drawn by Pritchard (1963) and from data by Pontecorvo and Kafer (1958) and Kafer (1958).

It is clear that, while the order of the loci is the same in both maps, the distances are quite different. The greatest discrepancy is in the regions flanking the centromeres. Becker (1974) points out that both induced and spontaneous mitotic exchange is distributed almost uniformly along the chromosomes of *Drosophila*. Meiotic exchange, on the other hand, is rare or absent in heterochromatic regions flanking the centromeres of

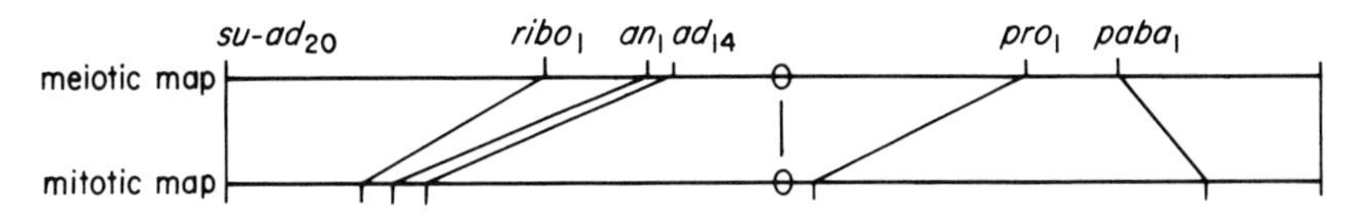

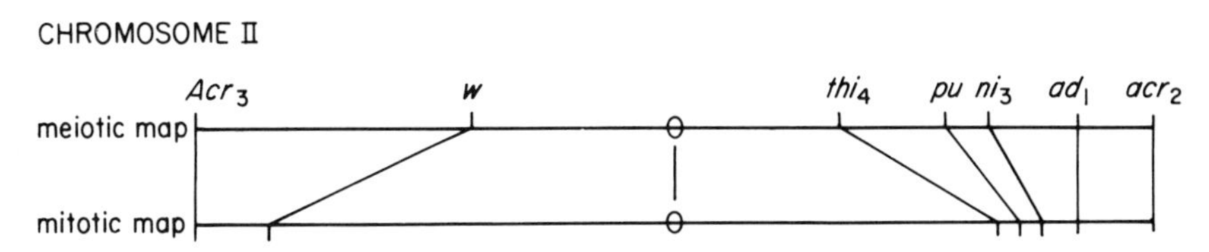

Fig. 8. A comparison of the meiotic and mitotic maps of chromosomes I and II of *Aspergillus nidulans*. Map length of each interval on the meiotic map is calculated as the fraction of the total map length between the centromere and the terminal marker. For the mitotic map, the length of each interval is calculated as the fraction of all crossovers between the centromere and the terminal marker occurring in that interval. *Acr* and *acr,* resistance to acriflavin; *ad,* adenine; *an,* aneurin; *ni,* nitrite; *paba, p*-aminobenzoic acid; *pro,* proline; *pu,* putrescine; *ribo,* riboflavin; *su-ad*20, suppressor of *ad*20; *thi,* thiazole; *w,* white conidia. (Redrawn from Pritchard, 1963.)

Drosophila chromosomes (Baker, 1958). This results in relatively good correspondence between mitotic and cytological maps of metaphase chromosomes, while the meiotic map is greatly constricted near the centromere (Figure 9).

C. Mapping Genes by Chromosomal Rearrangement

Using the breakpoints of chromosome rearrangements for positioning genes provides good linear maps of linked genes. More importantly, however, rearrangements are the tools whereby the physical size and nature of genes can be related to the units defined in genetic terms by operations such as recombination, mutation, and complementation.

The principle involved in gene mapping by chromosome rearrangements is the creation of aneuploid conditions. Deletions and duplications of small segments of chromosomes that survive as heterozygotes are used directly for mapping purposes. Constructing individuals heterozygous for a deletion and a recessive point mutation usually allows a decision about whether the point mutation site is included in the deleted segment. Failure

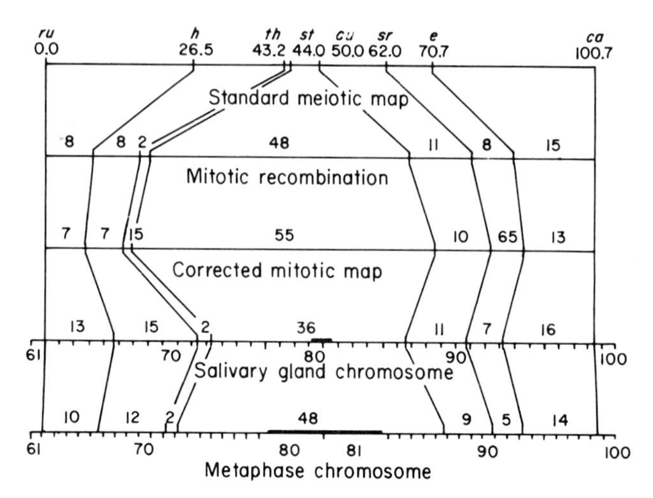

Fig. 9. A comparison of relative map positions for eight loci as the third chromosome of *Drosophila melanogaster*. The standard meiotic map shows distribution of exchange events in diploid females (line 1). Line 2 shows the distribution of 168 X-ray-induced mitotic exchanges in spermatogonia. Line 3 depicts the distribution of the 168 mitotic exchanges corrected by adding 38 more exchanges expected if clusters of exchanges were considered to have been derived from a single event. Line 4 shows the cytological position of the eight loci on the polytene chromosome map. Line 5 relates the map positions of the loci in the metaphase chromosome. Thin line depicts euchromatic positions of the chromosomes; thick line shows heterochromatic segments. *ca*, claret; *cu*, curled; *e*, ebony; *h*, hairy; *ru*, roughoid; *sr*, stripe; *st*, scarlet; *th*, thread. (Adapted from Becker, 1974.)

to complement phenotypically or failure to allow wild-type recombinants is taken to mean overlap of the two lesions. Likewise, the complementation of the homozygous mutant by a duplication indicates inclusion of the point mutant site within the duplicated segment. If series of overlapping deletions, duplications, and point mutants are used in complementation tests, it is possible to generate a map of the chromosome region. An example for a segment of the X chromosome of *Drosophila melanogaster* is shown in Figure 10. A series of point mutations, selected because they fail to complement with one of the deletions, were crossed inter se to define the complementation groups listed across the top of the figure. Members of each group then were crossed to a series of deletion or duplication stocks. The genetic limits of each deletion and duplication are shown in the figure by a line and a double line, respectively. The map was generated by assuming that point mutant sites removed by a deletion or included in a duplication are situated in adjacent positions in the chromosome. In the figure, some of the point mutant complementation groups cannot be placed in unambiguous linear order by this mapping procedure. They are

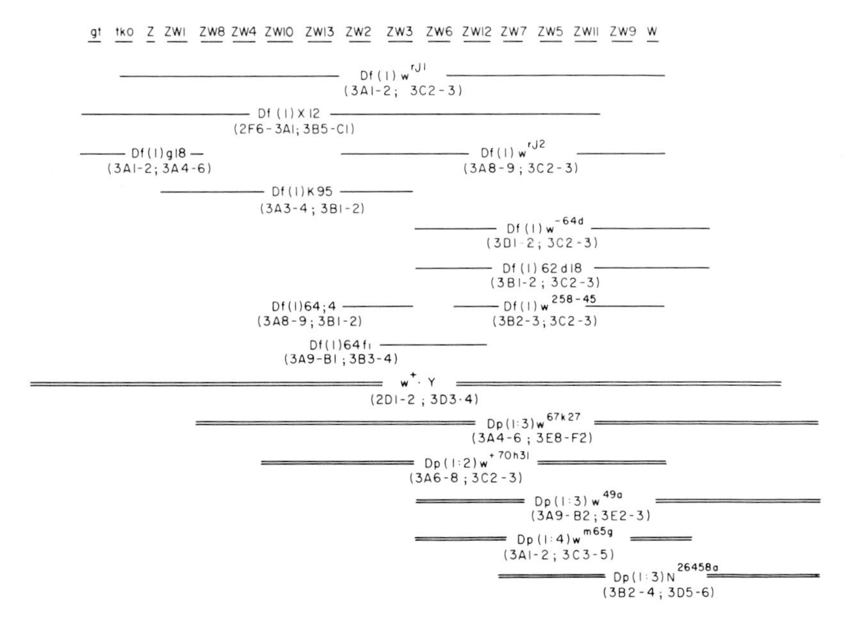

Fig. 10. A complementation map for a position of the X chromosome of *Drosophila melanogaster*, as determined by deletion and duplication mapping. Complementation units are arranged across the top. Deletions are shown as single lines with the extent of each line indicating failure of complementation between a given deletion and the single site mutations in the top row. Duplications are shown by double lines. The extent of each double line shows which loci are included in the duplicated segment. *gt*, giant; *tko*, behavior mutant; *w*, white eye; *z*, zeste eye; *zw 1-13*, lethal mutants.

enclosed in brackets. The ambiguity can be removed as additional breakpoints that separate such clustered groups become available. When deletion–duplication mapping is combined with recombination mapping, these ambiguities also can be resolved. The order of the complementation groups shown in the figure is the correct one, as established by combining the two methods (Judd *et al.*, 1972).

1. Polytene Chromosomes

In most organisms, the cytological determination of the positions of breakpoints in chromosomes is difficult and imprecise at best. However, some organisms, such as *Drosophila* and other Diptera, form polytene chromosomes in the nuclei of some secretory cells. Polytene chromosomes form when closely associated homologues undergo repeated rounds of replication without cell division. In the salivary glands of the third larval instar of *Drosophila melanogaster*, most cells have completed eight to nine

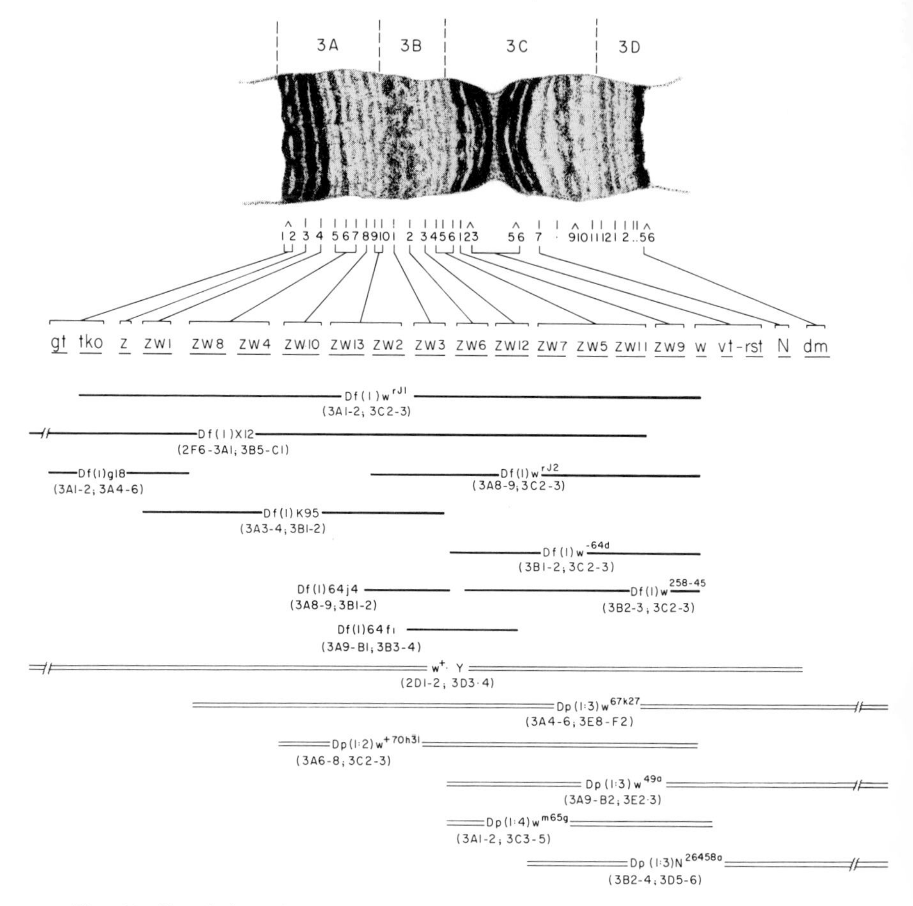

Fig. 11. Correlation of genetic and cytological maps of loci shown in Figure 10. The cytological extent of each deletion and duplication is given, and this information allows positioning of the complementation units to specific band–interband segments of the polytene chromosome.

rounds of replication; thus each polytene chromosome, composed of 512 chromonemata, is closely apposed and aligned with its equally polytene homologue. The resulting structure is viewed easily with light optics. A typical chromosome arm is about 200 μm in length and can be stretched to more than twice that length by squashing and fixing on a glass slide. The most distinctive feature of polytene chromosomes is the pattern of alternating light and dark bands in sequences that are unique for any segment of the chromosomes. These bands can be used for constructing cytological maps of chromosomes and in turn, through the use of rearrangement

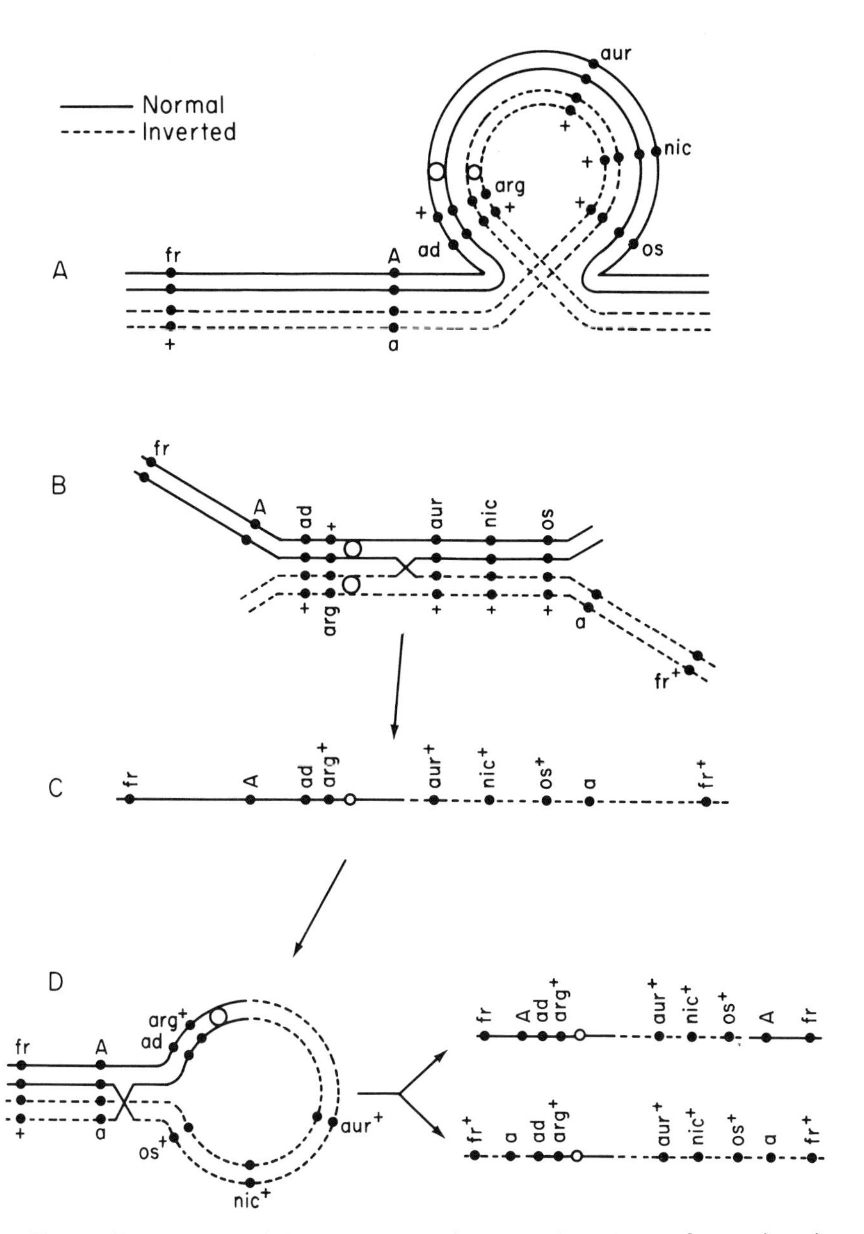

Fig. 12. Chromosome 1 of *Neurospora crassa* in a zygote heterozygous for a pericentric inversion. (A) shows synapsed chromosomes forming an inversion loop. (B) pictures only the inverted segment synapsed with a crossover in the paired region. (C) shows the viable recombinant product of the crossover. It is deficient for the right tip of the chromosome and duplicated for most of the left arm, making it heterozygous for mating type (*A*/*a*) and frost (*fr*$^+$/*fr*). (D) shows how mitotic crossing-over in the duplicated region can produce chromosomes homozygous for the mating type and frost loci. *ad*, adenine; *arg*, arginine; *aur*, aurescent; *nic*, nicotinamide; *os*, osmotic. (Adapted from Newmeyer and Taylor, 1967.)

breakpoints, the cytological position of genes can be correlated with their genetic map positions. This has been done for the segment of the X chromosome of *D. melanogaster* discussed above by determining the cytological extent of each deletion and duplication. This correlation is shown in Figure 11. The number of genes in this region is approximately the same as the number of bands, an interesting point that has been reviewed by Beermann (1972) and Lefever (1974).

2. *Creation of Aneuploidy by the Use of Inversions and Translocations*

Inversions and translocations can be manipulated to create aneuploidy. Simple inversions have two breakpoints in the same chromosome without loss of segments but with the section between the breaks inserted in reverse order. Genes in the inverted segment map in different positions relative to their old neighbors. Homologues of inversion heterozygotes synapse in a loop configuration, and if a crossover occurs within the inverted segment, duplication-deficiency products are formed. Depending on the positions of the breakpoints relative to the centromere and the chromosome tips, these aneuploid products may survive. For example, Newmeyer and Taylor (1967) used a pericentric inversion (one break in each chromosome arm) as illustrated in Figure 12 to create complementary duplication-deficiency products for a segment of chromosome 1 in *Neurospora*. The resulting progeny consisted of one type which was viable that could be used to study the genes in the duplicated segment. An interesting point is that mitotic crossing-over can occur in the duplicated segment to yield products homozygous for originally heterozygous loci.

Paracentric inversions (both breaks in same chromosome arm) in heterozygotes also can generate duplication-deficiency products, but in this case, one is dicentric and the other is acentric. The dicentric condition produces a chromatid bridge at first meiotic anaphase. McClintock (1951) used a chromosome that carried a duplication in inverted order to generate such bridges which break; after replication, the two broken ends may fuse, producing another bridge at the next division. Such a cycle can be used in tissue such as maize endosperm to study chromosome organization and gene expression.

Another method for generating aneuploidy is shown in Figure 13, which illustrates a technique for creating deletion- and duplication-bearing chromosomes by recombination between two chromosomes with similar but not identical inversions. This and similar techniques have been exploited in a variety of organisms, particularly *Drosophila* (Raffel and Muller, 1940; Lefevre and Wilkins, 1966), for gene mapping, including gene fine structure organization, and for investigations of gene dosage effects.

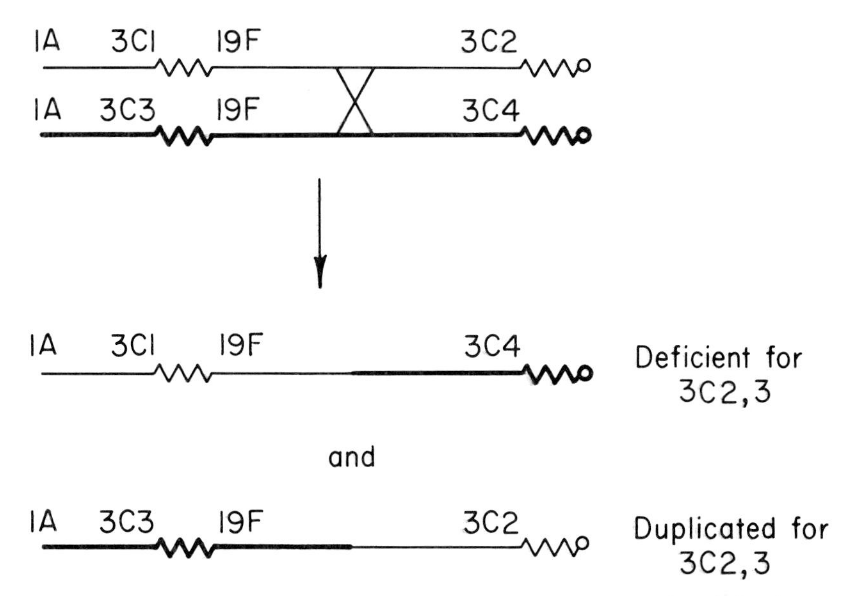

Fig. 13. Pairing and crossing-over between two chromosomes having similar but not identical paracentric inversions. The recombinant products are deficient or duplicated for the regions between the breakpoints of the inversions. Heterochromatin is shown by the wavy line and euchromatin by the straight line.

Translocation breakpoints may be used in ways rather similar to those of inversions. Translocations involve breaks in nonhomologous chromosomes and the switching and reattachment of fragments to form new linkage groups. Translocations can be used in two ways to generate duplication and deficiency situations. Consider the reciprocal transolcation as it would pair with structurally normal homologues in meiosis, as shown in Figure 14. Such a heterozygous individual will generate six different types of gametes as shown in the figure. Only two of the six will contain a complete set of genes for the chromosomes involved in the translocation. The remaining four types are duplicated for some loci and deficient for others. Whether any of these aneuploid types survive will depend on the positions of the transolcation breakpoints and, thus, on the extent of the aneuploidy. For this reason, translocation heterozygotes are usually semisterile. The relative frequency of the three types of disjunction varies with each translocation and depends primarily on the position of breakpoints relative to centromeres. In this respect, it should be noted that the effect of a crossover between centromere and breakpoint in either of the two segments in effect will change the type of disjunction. For example, alternate disjunction following a crossover in the region between the cen-

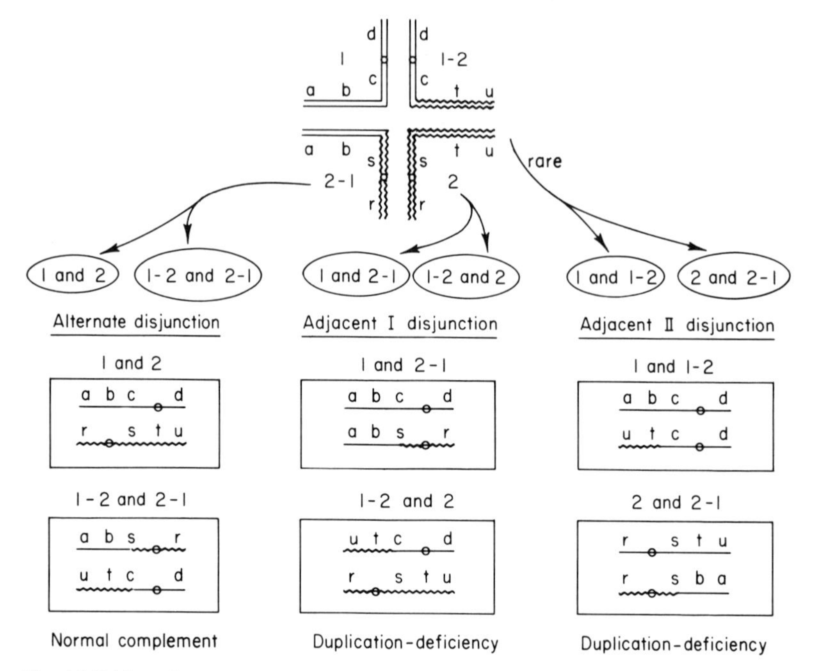

Fig. 14. Disjunction patterns in a meiotic cell heterozygous for a reciprocal translocation. Only alternate disjunction produces cells with euploid genotypes. Both types of adjacent disjunctions produce aneuploid duplication-deficiency types.

tromere and the breakpoint of either chromosome 1 or 2 would produce adjacent I-type gametes.

The usefulness of translocations is enhanced greatly if one of the breakpoints is quasi-terminal so that there is no essential gene in the translocated tip. This situation also may be realized when one of the chromosomes is small or is of a nature that aneuploidy for it is of little consequence. Examples are the tiny chromosome 4 and the Y chromosome in *Drosophila*. Triplo-4 flies are essentially normal in phenotype, and partial duplications for 4 go almost unnoticed. As a result, translocations involving 4 can be used to manipulate the genes of the other chromosome of the translocation without concern for fourth chromosome effects. The Y chromosome or parts of it have no effect in female *Drosophila*, and hyperploidy for segments of it do not interfere with normal development and function of males. Thus, any effect due to Y chromosome involvement can be offset easily.

A very useful way to employ translocations in the mapping of genes is by making combinations of two rearrangements having similar but not identical breakpoints. This technique was exploited by Lindsley and Sandler and associates (Lindsley *et al.*, 1972) to survey the organization of the major autosomes in *Drosophila.* The scheme they employed involved recovery of Y-autosome translocations, determining the autosomal breakpoint cytologically, and then making crosses that would yield aneuploid classes of offspring as diagrammed in Figure 15. This technique can be applied in haploid organisms such as *Neurospora,* in which Perkins (1975) has developed it to create partial diploids. Such aneuploidy is particularly useful for studying gene interactions, for example, the heterokaryon incompatibility genes or regulatory genes such as those controlling alkaline phosphate synthesis (Metzenberg *et al.*, 1974). In partial diploids created by combining different parts of reciprocal translocations, it is possible to construct both cis and trans configurations for two interacting mutants and thereby to determine whether the interaction is dependent on a diffusible product or on a physically adjacent position in the chromosome.

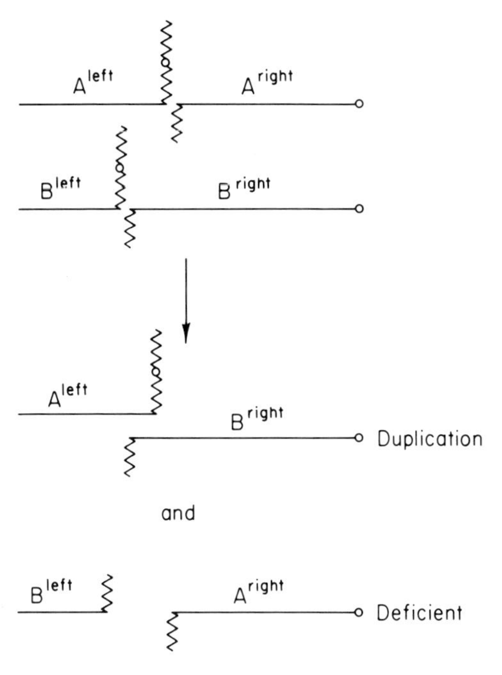

Fig. 15. Two X–Y translocations of *D. melanogaster* with breakpoints in the X chromosomes at different positions. By appropriate crosses, it is possible to construct duplication or deficiency types by recombining the left portion of translocation A with the right portion of translocation B or B-left with A-right. Aneuploidy for a Y chromosome segment also will be created unless Y chromosome breakpoints are identical for the two translocations. This aneuploidy is offset in males by providing another intact Y chromosome. X chromosome is shown by straight line, Y chromosome by wavy line.

D. Molecular Maps of Eukaryotic Chromosomes

In recent years, a number of techniques have been developed which extend the characterization of genes to the level of the DNA itself. A necessary first step to such analyses is the isolation of a homogeneous population of a specific segment of a chromosome. This is accomplished by linking fragments of DNA to a vehicle such as an *E. coli* plasmid or the phage λ, infecting host cells with such hybrid molecules and finally amplifying the fragments through cloning of the transformed cells.

One procedure worked out by Wensink *et al.* (1974), based on methods described by Jackson *et al.* (1972) and by Lobban and Kaiser (1973), is shown in Figure 16. The vehicle is the plasmid pSC101, chosen because it contains a factor that confers tetracycline resistance on the *E. coli* host cell and because it has only one site that is cleaved by the restriction endonuclease *Eco*RI. The plasmid DNA is purified, cleaved by *Eco*RI, and treated with λ-exonuclease, which degrades both chains of the double helix at their 5′-ends. Treatment with terminal transferase plus dATP or

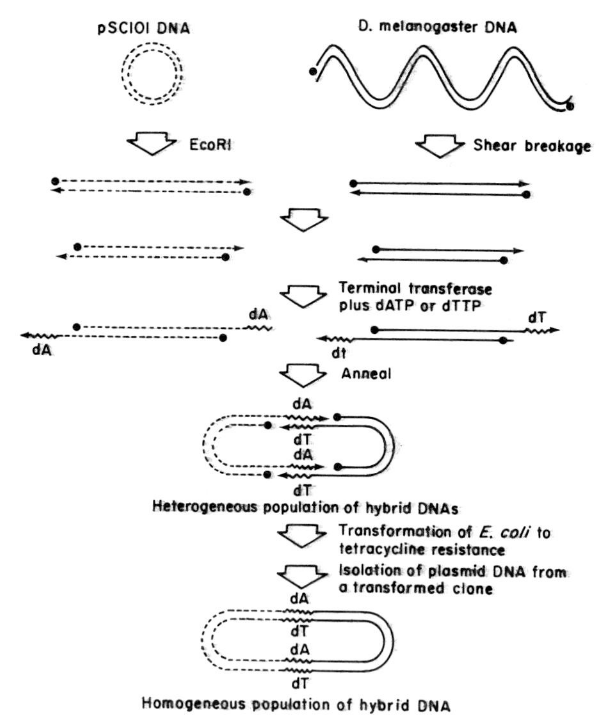

Fig. 16. The construction of hybrid molecules containing DNA from plasmid pSC101 and *D. melanogaster*. See text for explanation of the technique. (Adapted from Wensink *et al.*, 1974.)

dTTP extends the 3′-terminus of both chains with about forty dA or dT nucleotides.

The eukaryotic DNA is sheared to the desired size, treated with the exonuclease, and extended as described for the plasmid DNA. If dATP is used to extend the plasmid, then dTTP is used for the eukaryotic DNA. The two types of DNA are mixed and allowed to form circular hybrid molecules through pairing of dA tails with dT tails.

A technique that uses a restriction endonuclease such as *Eco*RI to cleave both the carrier DNA and the eukaryotic DNA has been described by Mertz and Davis (1972). Endonuclease cuts which are staggered, as is that produced by *Eco*RI, give single-stranded tails that are complementary and identical for all of the DNA molecules of a population. These tails will pair in complementary fashion and can then be ligased to form circular molecules. When this system is used, however, it is important to ascertain that one and only one fragment of foreign DNA is inserted into the plasmid. This method also presents some bias in the recovery of segments because only those segments with endonuclease cleavage sites at their ends will be available for linkage to the vehicle.

For uptake by host cells, the nicked circular molecules are mixed with *E. coli* in the presence of $CaCl_2$. The nicks are closed after uptake by action of the ligase from the host cell. Transformed cells form tetracycline-resistant colonies, and these can be picked and grown to produce many copies of the plasmid–eukaryote DNA hybrid.

The techniques for cloning molecular hybrids between bacteriophage λ and any foreign DNA have been worked out by Thomas *et al.* (1974) and Murray and Murray (1974). This system is useful for selecting specific genes from eukaryotes. The selection is based on the ability of a λ-borne eukaryotic gene to be expressed and to complement a nonreverting bacterial mutation in a gene specifying a known enzymatic function. For example, Struhl *et al.* (1976) isolated a segment of DNA from yeast which, when inserted into the chromosome of an *E. coli* histidine auxotroph, allows the bacterium to grow in the absence of histidine.

The pool from which cloned fragments are made can be from the entire genome by the so-called shotgun method, which allows recovery of random segments of the chromosome, or it can be from DNA fractionated in various ways. Ribosomal cistrons can be enriched by hybridizing to ribosomal RNA, whereupon these can be used in the cloning procedure. If a specific mRNA can be purified, it is possible by a technique developed by Grunstein and Hogness (1975) to screen a very large number of colonies of *E. coli* carrying different hybrid plasmids to determine which contains a piece of DNA that is complementary to that mRNA. The colonies are grown on nitrocellulose filters and, after a replica plate of each set of colonies is made, the cells are lysed and the DNA is denatured and fixed to the filter. Radioactive probe RNA is incubated with the DNA, and hybrid formation is assayed by autoradiography. Colonies carrying the DNA sequence of interest can be identified and picked from the replica reference plate. Another approach is to use the specific RNA to prepare cDNA, which can be replicated to double-stranded DNA using polymerase I. Such DNA then can be used in construction of the hybrid molecule or as a probe to recover cloned segments of the genome which are homologous to it. In this fashion, the sequence encoding the mRNA can be recovered for study. Even more important, segments adjacent to that encoding the message will be found in some fragments, and the analysis of these should provide information about the regulation of transcription and other control mechanisms.

When a homogeneous population has been obtained and purified, it may be characterized by a series of physical and biochemical techniques. For example, buoyant density measurements for the hybrid molecule can be recorded and compared to those values for the vehicle molecule (Champoux and Hogness, 1973). In addition, heteroduplexes between the hybrid molecule and the plasmid vehicle can be prepared as described by Davis *et al.* (1971). When these are viewed by electron microscopy, the plasmid should form a double-stranded ring attached to a single-stranded loop of eukaryotic DNA. The length of the single-strand loop can then be determined. Reassociation kinetics of randomly sheared hybrid DNA yields information about sequence repetition in the cloned fragment. It is also important to know whether the sequences inserted in the plasmid form a homogeneous class with respect to the frequency of their occurrence in the eukaryotic genome from which they were derived.

This information can be related directly to data derived by hybridizing ^3H-labeled RNA copies of the cloned DNA to chromosomes *in situ*. The technique worked out by Gall and Pardue (1969) provides a powerful tool for examining the spatial arrangement of specific DNA sequences in chromosomes. In *Drosophila,* for example (Pardue *et al.,* 1970; Wimber and Steffensen, 1970; Wensink *et al.,* 1974), *in situ* hybridization to polytene chromosomes allows localization of particular sequences to specific bands. Combining this technique with genetic and cytological analyses described earlier allows mapping of genes in a highly precise fashion. It also provides information about the number of copies of a given sequence found in a set of chromosomes and the nature of their relationships to one another.

Cloned fragments of eukaryotic DNA can be mapped in detail and ultimately can be completely sequenced (Maxam and Gilbert, 1977). Molecular mapping takes several forms. First, restriction endonucleases that cleave DNA at specific sites can be used to subdivide a given segment. [For review of the restriction enzymes and their uses, see Nathans and Smith (1975).] A pSC101–eukaryotic DNA hybrid digested with *Eco* RI, for example, will break the eukaryotic DNA at every

$$\begin{array}{l}\downarrow\\ \mathrm{GAATTC}\\ \mathrm{CTTAAG}\\ \uparrow\end{array}$$

sequence. The plasmid DNA, since it contained only one such site which was destroyed by treatment with λ exonuclease and the addition of poly(dA) or poly(dT), is not cleaved by this treatment. Displaying digested DNA on gels shows how many sites for a given nuclease are present and also gives information about the size of each fragment. Maps of restriction enzyme cleavage sites can be constructed by isolating fragments and using each one to form heteroduplexes with the original cloned segment for viewing with electron microscopy. The double-stranded region thus is visible, and its position in the original segment and its size can be determined.

Further, those pieces which encode an mRNA can be hybridized with such a message, and the position of that region within the cloned segment can be deduced. Another technique for mapping regions encoding specific RNA's has been worked out by White and Hogness (1976). R-loop mapping is done by mixing duplex DNA with single-stranded RNA in the presence of formamide. At high temperature, RNA which is homologous to a DNA sequence will pair with the complementary DNA strand to form an RNA/DNA duplex with a loop of single-stranded DNA. Such R-loops are visualized readily with electron microscopy, and the relative position of the RNA encoding region can be worked out by measuring the position of the R-loop within the entire fragment. An example of R-loop figures and the arrangement of the coding sequences for 18 and 28 S ribosomal RNA's as described by White and Hogness are shown in Figure 17.

DNA Sequence Diversity and Arrangement

A hallmark of eukaryotic genomes is the repetition of some of their DNA sequences; some, in fact, exist in as many as 10^6 copies per genome. The bulk of the repetitious sequences for most organisms range from 10^2 to 10^4 copies [see Davidson and Britten (1973) for review]. Studies of how these repetitious sequences are distributed throughout the genome have produced some interesting results that are relevant to gene organization and the mechanism of gene regulation. The mapping of repetitious sequences thus becomes very important. At the present time, several techniques can be used to determine the arrangements of sequences in the chromosomes. The method developed by Britten and his associates [see

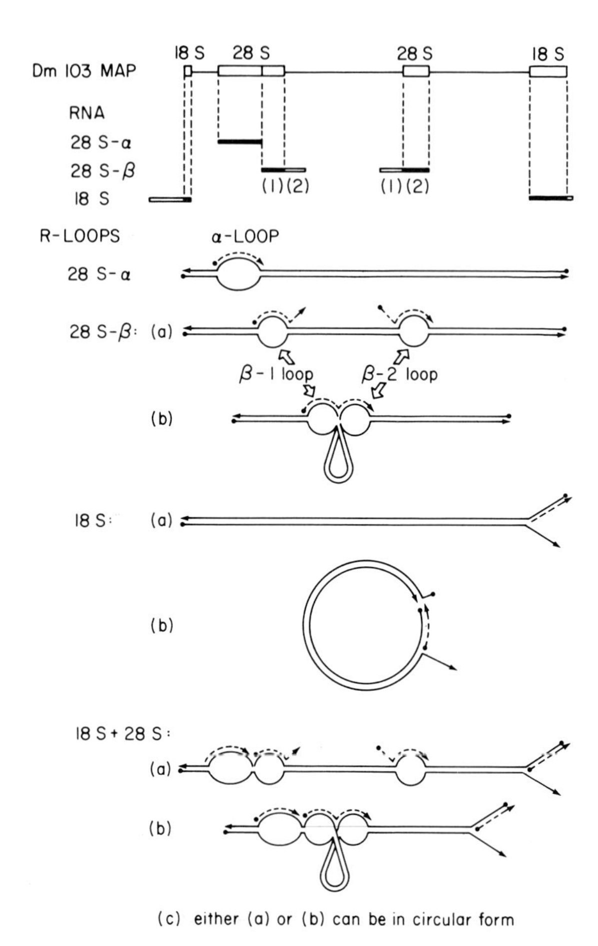

Fig. 17. One of the repeated ribosomal RNA cistrons cloned in Dm 103 has been mapped by hybridizing fragments of the DNA, created by restriction enzyme digests, to purified ribosomal RNA. The purified RNA's are 28 and 18 S molecules with the 28 S cleaved into two parts of about equal sizes. The map is shown on the top line with the expected sequence homologies between DM 103 and the rRNA's indicated by the filled portions of the bars that represent the RNA's. The rRNA gene was cloned from fragments created by Eco RI digest which cleaves in the 18 S sequence. Therefore, the cloned fragment begins and ends with the 18 S region. The 28 S sequence is interrupted in this type of rRNA gene by a spacer sequence. When R-loop mapping is done using 28 S, the predicted configuration viewed by electron microscopy is a single loop. 28 S RNA should produce two loops separated by the spacer sequence or a twisted configuration produced by looping out the spacer DNA. 18 S RNA should hybridize at one end to give a fork, or possibly at both ends to produce a circle. Mixtures of 28 and 18 S will give the configurations shown in the two bottom lines. The RNA is shown as dashed lines, the dot and arrowhead mark the 5′- and 3′-termini, respectively. (From White and Hogness, 1977.)

Davidson and Britten (1973) for review] depends on the kinetics of reassociation of denatured DNA fragments. When long labeled fragments of calf DNA were reassociated in the presence of a great excess of short unlabeled fragments (driver DNA), the results indicated that three-quarters of the fragments 4000 nucleotides long contained repetitive sequences, and that most of these also had regions of unique sequence (Britten, 1969; Britten and Smith, 1970).

Davidson *et al.* (1973) described a pattern of sequence interspersion in *Xenopus* mainly characterized by repetitive sequences of a few hundred nucleotides interspersed with unique single-copy sequences ranging in length from one thousand to several thousand nucleotides. Similar patterns typify a majority of animals studied. Examples include the calf (Britten and Smith, 1970), rat (Bonner *et al.*, 1973), sea urchin (Graham *et al.*, 1974), seven invertebrate species of diverse phylogenetic groups (Goldberg *et al.*, 1975; Davidson *et al.*, 1975; Angerer *et al.*, 1975), and a slime mold (Firtel and Kindle, 1975). In all of these, the majority of repetitive sequences are short, averaging about 300 nucleotides that alternate with single-copy sequences, the majority of which average about 1000 nucleotides in length.

A very different pattern was described for *Drosophila melanogaster* by Manning *et al.* (1975) using renaturation kinetics and viewing renatured duplexes of DNA with electron microscopy. They studied DNA fragments 4200, 11,600, and 17,400 nucleotides long that renatured at intermediate C_0t values and thus contained moderately repetitious sequences. They examined the type of duplexes that were formed and measured the lengths of the double-stranded regions. They found that most of the repeat sequences are more than 5000 nucleotide pairs long, and that these are interspersed with single-copy stretches most of which, although too long to measure, average more than 5000 nucleotide pairs and range well over 10,000 nucleotide pairs. Only a small fraction of the repeats are short and interspersed with unique sequence stretches of one to a few thousand nucleotide pairs. A similar pattern of organization is seen in the honeybee (Crain *et al.*, 1976); however, another insect, *Musca domestica* (housefly), has the short-period interspersion pattern typical of *Xenopus* and other organisms.

The analysis of cloned fragments of DNA by restriction enzyme digestion and *in situ* hybridization also gives some information about the repetitious sequences and how they are distributed in chromosomes. Preliminary indication from analyses carried out by Hogness and his associates (Finnegan *et al.*, 1977) is that repeat sequences in some fragments are made up of a set of elements, some of which are found repeated in other parts of the genome. The multiunit repeat may be important in orchestrating the action of genes which are not clustered together in a chromosome.

All of the techniques for mapping genes outlined here, from the rough determination of linkage groups to the most detailed fine structure mapping, leading eventually to the sequencing of the DNA, bring into focus the importance of knowing the construct of the gene and how it relates to its neighbors in the organization of chromosomes. How eukaryotic genes are regulated is a central, still unanswered, set of questions that only can be understood when the spatial relationships of sequences encoding proteins, sequences that are transcribed but not translated, and finally sequences that are not even transcribed are mapped out in detail. Only then will the function and the control of genes become evident. [More detailed discussion of DNA sequence arrangement in eukaryotic genomes will be found in Chapter 1 of Volume 3 of this Treatise.]

III. THE NATURE OF EUKARYOTIC GENETIC SYSTEMS

The information which emerges from the many studies applying the techniques outlined in the first part of this chapter serves to focus our attention on several aspects of eukaryotic genetic systems. These points all deal with genome, chromosome, and gene organization and with the regulation of gene activity. Questions of interest involve estimates of gene size and number, the details of gene fine structure, the nature of the units that are transcribed, and the mechanisms for controlling transcription. Despite the tools available for mapping genes and determining their structure, organization, and function, we still do not have answers to these fundamental questions that lead to understanding coordinate control of gene activity in eukaryotes. It is useful, therefore, to summarize briefly selected points about the nature of the eukaryotic genetic systems as we view them today.

A. Genome Complexity

Complexity is the key word in the organization of eukaryotic genomes. This complexity is expressed in several different ways. First of all, the packaging of eukaryotic DNA's into several chromosomes represents a level of organization above that seen in prokaryotes. The chromosome represents one of the units of function. Many examples of the inactivation of chromosomes or even entire genomes, through condensation to heterochromatic states or the selective elimination of genomes or chromosomes from certain somatic cells by chromosome diminution, testify to the effectiveness of this type of gene regulation (review by Sager and Kitchin, 1975; Brown and Chandra, Chapter 4, Volume 1, this Treatise).

Second, the arrangement of DNA sequences within chromosomes represents an organization on a regional basis that clearly has a functional significance. The very highly repeated simple sequence satellite DNA's in general are found in blocks which are heterochromatic and which are located near centromeres. The pattern of interspersion of intermediately repeated sequences alternating with unique sequences has attracted considerable attention, and it has been speculated that this arrangement plays a major role in the mechanism of gene regulation (Britten and Davidson, 1969; Davidson *et al.*, 1977).

A third level of complexity is represented by families of genes such as the ribosomal RNA cistrons and the histone genes that are clustered, presumably to facilitate their regulation and possibly to maintain their fidelity. Even the simplest single cistron in eukaryotes may be considerably more complex than the DNA sequence encoding a polypeptide plus a few control signals. While the mechanism for coordinately regulating gene function is not known, some clues may become available by examining the paradox which emerges when attention is directed to gene numbers and sizes relative to genome sizes.

B. Gene Number and Size

There is enough DNA in the average mammalian genome to encode more than one and one-half million different mRNA's each 1.5 kb long. Even when the repetitious DNA sequences are removed from consideration on the basis that they are not transcribed, the amount of DNA in most eukaryotes is very large compared to estimates of gene numbers. This apparent discrepancy is based on several approaches to this question. For example, Muller and Altenberg (1919) and Muller (1929) tried to estimate the number of mutable loci in the X chromosome of *Drosophila* by scoring mutation frequency at a few loci. The principle employed was to observe how often mutation events could occur without striking a locus which had been mutated previously. It was assumed that the total number of mutational events that could occur without striking a locus for a second or third time should be related to the total number of mutable loci. Alikhanian (1937) extended this approach by measuring the lethal mutation rates at a few loci and comparing those rates to the lethal mutation rate for an entire chromosome. In this manner, he estimated that 968 genes would be found in the X chromosome. Since the X chromosome represents about one-fifth of the total genome, based on cytological observations he estimated that there should be about 5000 genes per set of chromosomes of *Drosophila*. Good agreement with this figure is provided by the work of Mukai and associates (Mukai, 1964; Mukai *et al.*, 1972), who estimated, by dividing mutation rate per genome per generation by mutation rate per gene per generation, that there are between 5000 and 10,000 genes in *Drosophila*.

Ohta and Kimura (1971) concluded that a large proportion of the DNA of eukaryotes must be noninformational, in the sense that mutations in such sequences are of neutral effect. They reasoned, from data about the average rate of amino acid substitution in nine different proteins, that the true mutation rate at the molecular level is ca. 8.3×10^{-9} per amino acid site per year. The human genome, with 3×10^{9} nucleotide pairs, has the potential to encode 10^{9} amino acid sites. Therefore, the total mutation rate due to nucleotide changes per genome per year is $8.3 \times 10^{-9} \times 10^{9}$ or about 8 substitutions per genome per year. This, they assert, is far higher than could be tolerated by a species because of selection, if a majority of mutations are deleterious, thereby arguing that most of the nucleotide sequences do not code for amino acid sequences.

Do calculations such as these indicate that genes encoding proteins may be interspersed with spacer sequences within which mutation events may have little effect? The 18 and 28 S subunit sequences in the cistrons encoding ribosomal RNA are known to be cotranscribed as a 45 S precursor molecule even though they are separated by a sizeable transcribed spacer (Miller and Beatty, 1969; Brown and Weber, 1968). In turn, each transcriptional unit is separated from the next in this repeating array by a nontranscribed spacer. Spacers also are known to exist between 5 S RNA genes (Brown *et al.*, 1971) and between histone genes (Birnstiel *et al.*, 1974; Shutt and Kedes, 1974).

Evidence for spacer DNA between genes of ciliated protozoa of the genus *Oxytricha* comes from the work of Prescott and his associates (1971; Prescott and Murti, 1973). A most remarkable series of events occurs in the chromosomes of the macronucleus of this organism. Polytene chromosomes are formed by rapid DNA synthesis in the micronucleus that is destined to develop into the macronucleus. After polytene chromosomes are synthesized, partitions are formed by membranes that separate the DNA of each band into individual vesicles. Next, most of the DNA is degraded, and evidence indicates that there is a reduction in sequence diversity (Prescott, 1977). The interpretation is that spacer sequences have been degraded, leaving only the polytene copies of the gene sequences.

Possibly the most straightforward approach to gene number, size, and organization is the cytogenetic studies on *Drosophila* that have attempted to saturate selected regions of chromosomes mutationally, and then map each region genetically and cytologically. The purpose is to identify every functional unit in a given segment of chromosome and to determine their positions and functions. The results from these investigations carried out primarily by Judd *et al.* (1972), Hochman (1973), and Lefevre (1973) are summarized in reviews by Beermann (1972) and Lefevre (1974). In general, there is good correspondence between the number of functional units and the number of bands and interbands contained in a given segment of the polytene chromosomes. This means that a gene of average size is about 25 kb, if all of the DNA encodes information that is useful in the development and function of the organism. There are exceptions to the one band : one function relationship (Young and Judd, 1977), but there is strong evidence that the number of genes in *Drosophila* is no greater than about 10,000 if the chromosome regions that have been studied are representative of the entire genome. One of the central questions, then, concerns the organization of genes in the chromosomes and just how much of the DNA is informational.

C. RNA Transcripts

Support for the idea that genes are considerably larger than the mRNA's they encode comes from two types of evidence. First, the comparison of the size and sequence diversity of RNA molecules found in the nucleus (hnRNA) with those found in polysomes in the cytoplasm (mRNA) shows that a very large proportion of the transcribed sequences is not represented in mRNA, and that the average size of mRNA molecules is about an order of magnitude smaller than the average size of hnRNA. Kinetic analyses of RNA synthesis and turnover show that only a small fraction of nuclear RNA is transported to the cytoplasm [for review, see Lewin (1976a,b) and Davidson and Birtten (1973)]. It is possible but unlikely that mRNA's are produced only from the smaller size classes of hnRNA.

A second line of evidence about gene size and number comes from viewing transcription units by electron microscopy. McKnight and Miller (1976) measured lengths of actively transcribing segments of *Drosophila* embryo cell chromatin. Nonribosomal transcription units averaged 2.93 μm in length. Laird and Chooi (1976) also observed such units, which averaged 3.8 μm in length. The packing ratio of this chromatin is such that these lengths correspond to about 15 to 20 kb of DNA. This is similar to the average DNA content per band of the polytene chromosomes when corrected to a haploid value.

To reconcile the measurements of transcription units with the sizes of mRNA molecules in polysomes and the sizes of polypeptides, we must assume some processing of the original transcripts or that those transcription units which produce mRNA's are small and underrepresented in electron microscope preparations. In this respect, the studies of the puffing of specific bands of polytene chromosomes provide evidence that a puff represents an actively transcribing segment of the chromosome, and that each puff has its origin in a single band (Beermann, 1972; Ashburner, 1972). In specific cases such as *Chironomous,* the DNA of essentially the entire band is transcribed (Daneholt, 1975). If each band of the polytene chromosomes is a unit of transcription, as these cytological studies indicate, then the number of such units in *Drosophila* is between 7000 and 10,000.

If transcription units are large and contain only one structural gene sequence, what purpose, if any, is served by the other information in the transcripts? It would appear, in fact, that if transcripts do contain important sequences in addition to that encoding a polypeptide, such information must be used primarily in the nucleus itself because a major fraction of nuclear RNA is not transported to the cytoplasm. The answers to these and related questions may soon be available, as specific genes and their RNA products are isolated and analyzed. For the present, however, we must turn to what is known about the fine structure organization of genes for clues to the understanding of these problems.

D. Multigenic Families, Complex Loci, Allelic Complementation, and Genetic Fine Structure

The genetic analysis of selected loci in a variety of eukaryotic organisms identifies at least two categories of loci. These are (1) members of multigenic families as contrasted with (2) single-cistron loci.

Hood *et al.* (1975) and Tartof (1975) recently have reviewed multigenic families, their organization, expression, and evolution. Multigenic families include the tandemly repetitious genes encoding the 18, 28, and 5 S RNA's of the ribosomes and the similarly arranged transfer RNA cistrons and the histone genes. Another type of multigenic family is composed of individual loci which have related functions and appear to have evolved from a common ancestral gene. Examples of these are the antibody genes and the loci for the hemoglobin chains.

A complex locus is one which by genetic analysis is composed of several complementation groups but which behaves as a single cis-acting unit; that is, there are alleles which map as single-site mutations which fail to complement with all members of all complementation units in the complex. Some complex loci appear to belong to the latter type of multigenic family, while others may qualify as single cistron loci when more information is available about the gene product(s) encoded by the locus. To illustrate this point, consider the rudimentary locus in *Drosophila melanogaster.* Mutant *r* strains lack activity for one or more of the enzymes which catalyze the first three steps in pyrimidine biosynthesis. These are carbamyl phosphate synthetase (CPSase), aspartate transcarbamylase (ATCase), and dihydroorotase (DHOase). The complementation pattern for rudimentary mutants is complex (Carlson, 1971), but in general three units can be identified which correspond to each of the three enzyme activities. Mutations lacking only ATCase activity generally will complement those that lack CPSase and/or DHOase, and so on. A most interesting group is composed of *r* alleles which map as single-site lesions but which fail to complement with all other alleles.

There are at least two models which can account for such a complementation pattern. One model assumes that rudimentary is a multicistronic complex that encodes several polypeptides and is regulated in much the same way as an operon in a prokaryote. Complementing alleles would be those in different structural genes of the complex. The totally noncomplementing alleles might be operator, promotor, or other polar-type mutations. The other model assumes that the locus encodes a single polypeptide which has three active sites, and that complementation can occur between two mutant polypeptides which differ in the active sites affected by the mutation. All three enzyme activities affected by *r* mutations can be recovered in a single complex (Brothers *et al.*, 1977), but as yet it is not clear whether one or several polypeptides is involved.

There are numerous other examples of so-called complex loci for which a great deal of genetic data are available, particularly in *Drosophila* (Judd, 1976, for review). By definition, complex loci show allelic complementation and usually some pleiotropism. It is surprising, however, just how many loci in eukaryotes may qualify as complex. A survey of loci with four or more allelic representatives shows about one-third of them exhibiting allelic complementation and pleiotropism (Judd, 1977). This is expected if the units of function are polycistronic operons or if the gene products function as multimeric complexes. However, in eukaryotes there are no clear-cut cases of polycistronic units, and the incidence of allelic complementation seems much too high to be accounted for by the formation of multimers. If an allelic complementation cannot be explained by the interaction of mutant polypeptides in a multimeric complex, can some of it be attributed to the organization and function of the regulatory elements of eukaryotic genes?

Fine structure maps give some clues to the solution of this problem. Maps have been developed for a number of eukaryotic loci, particularly in *Drosophila* (reviewed by Finnerty, 1976). The best developed example is the rosy locus in *Drosophila* that encodes the enzyme xanthine dehydrogenase. Chovnick *et al.* (1977) have recently reviewed the work that they have done over a number of years. They have mapped a variety of null mutants and electrophoretic variants to establish the genetic limits of the structural gene. Another variant which causes increased production of the enzyme and which thus is an excellent candidate for a regulatory mutation has been mapped well outside of the structural gene limits. Most significant is the observation that this variant acts in cis configuration to increase the amount of the particular structural gene product encoded by the linked sequence. Estimates of the structural element size based on the relationship of map distances spanning the structural unit to the molecular weight of the polypeptide product allow estimates of the size of the control element as well. The structural element is about 4.4 kb and spans 0.005 map units. The regulatory site maps about 0.0035 units beyond the closest structural gene mutation site, a length which corresponds to about 3.0 kb. A minimum size for the entire gene, then, is about 7.4 kb. Most important is the discovery that the regulatory sequence is linked in close proximity to the structural element it controls, and at least some of the control is in a cis-functioning mode.

In summary, the concept of eukaryotic genes that is emerging from the several techniques of mapping outlined here points up several basic characteristics. It appears that there is a spectrum of types of genetic units ranging from clusters of tandemly repetitious sequences to the simplest single cistron. Even with the simplest unit, however, the size of the gene appears to exceed considerably that which is necessary to encode the polypeptide gene product. Although evidence for polycistronic transcription units is lacking for eukaryotes, there is increasing evidence to indicate that a great deal of DNA of eukaryotes is important in gene control and coordination.

REFERENCES

Alikhanian, S. I. (1937). A study of the lethal mutations in the left end of the sex chromosome in *Drosophila melanogaster. Zool. Zh.* **16,** 247–279 (in Russian with English summary).

Anderson, E. G. (1925). Crossing-over in a case of attached X chromosomes in *Drosophila melanogaster. Genetics* **10,** 403–417.

Angerer, R. C., Davidson, E. H., and Britten, R. J. (1975). DNA sequence organization in the mollusc *Aplysia californica. Cell* **6,** 29–39.

Ashburner, M. (1972). Puffing patterns in *Drosophila melanogaster,* and related species. *Results Probl. Cell Differ.* **4,** 101–151.

Baker, W. K. (1958). Crossing-over in heterochromatin. *Am. Nat.* **92,** 59–60.

Barratt, R. W., Newmeyer, D., Perkins, D. D., and Garnjobst, L. (1954). Map construction in *Neurospora crassa. Adv. Genet.* **6,** 1–93.

Becker, H. J. (1974). Mitotic recombination maps in *Drosophila melanogaster. Naturwissenschaften* **61,** 441–448.

Beermann, W. (1972). Chromomeres and genes. *Results Probl. Cell Differ.* **4,** 1–33.

Birnstiel, M., Telford, J., Weinberg, E., and Stafford, D. (1974). Isolation and some properties of the genes coding for histone proteins. *Proc. Natl. Acad. Sci. U.S.A.* **71,** 2900–2904.

Bonner, J., Garrard, W. T., Gottesfeld, J., Holmes, D. S., Sevall, J. S., and Wilkes, M. (1973). Functional organization of the mammalian genome. *Cold Spring Harbor Symp. Quant. Biol.* **38,** 303–310.

Bridges, C. B., and Olbrycht, T. M. (1926). The multiple stock "Xple" and its use. *Genetics* **11,** 41–56.

Britten, R. J. (1969). Repeated DNA and transcription. *In* "Problems in Biology" (E. W. Hanly, ed.), pp. 187–216. Univ. of Utah Press, Salt Lake City.

Britten, R. J., and Davidson, E. H. (1969). Gene regulation for higher cells: A theory. *Science* **165,** 349–357.

Britten, R. J., and Smith, J. (1970). A bovine genome. *Carnegie Inst. Washington, Yearb.* **68,** 379–386.

Brothers, V. M., Tsubota, S. I., Germeraad, S. E., and Fristrom, J. W. (1978). The rudimentary locus of *Drosophila melanogaster:* Partial purification of a carbamyl phosphate synthase–aspartate transcarbamylase–dihydroorotase complex. *Biochem. Genet.* **16,** 321–332.

Brown, D. D., and Weber, C. S. (1968). Gene linkage by RNA–DNA hybridization. II. Arrangement of the redundant gene sequences for 28 and 18 S ribosomal RNA. *J. Mol. Biol.* **34,** 681–697.

Brown, D. D., Wensink, P. C., and Jordan, E. (1971). Purification and some characteristics of 5 S DNA from *Xenopus laevis. Proc. Natl. Acad. Sci. U.S.A.* **68,** 3175–3179.

Carlson, P. (1971). A genetic analysis of the rudimentary locus of *Drosophila melanogaster. Genet. Res.* **17,** 53–81.

Carter, T. C., and Falconer, D. S. (1951). Stocks for detecting linkage in the mouse, and the theory of their design. *J. Genet.* **50,** 307–323.

Carter, T. C., and Falconer, D. S. (1953). Independence of linkage groups I, II and XI in the house mouse. *J. Genet.* **51,** 373–374.

Carter, T. C., and Robertson, A. (1952). A mathematical treatment of genetic recombination, using a four-strand model. *Proc. R. Soc. London, Ser. B* **139,** 410–426.

Champoux, J. J., and Hogness, D. S. (1973). The topography of Lambda DNA: Polyriboguanylic acid binding sites and base composition. *J. Mol. Biol.* **71,** 383–405.

Chovnick, A., Gelbart, W., and McCarron, M. (1977). Organization of the rosy locus in *Drosophila melanogaster. Cell* **11,** 1–10.

Crain, W. R., Davidson, E. H., and Britten, R. J. (1976). Contrasting patterns of DNA sequence arrangement in *Apis mellifera* (honeybee) and *Musca domestica* (housefly). *Chromosoma* **59,** 1–12.

Daneholt, B. (1975). Transcription in polytene chromosomes. *Cell* **4,** 1–9.

Davidson, E. H., and Britten, R. J. (1973). Organization, transcription, and regulation in the animal genome. *Q. Rev. Biol.* **48,** 565–613.

Davidson, E. H., Hough, B. R., Amenson, C. S., and Britten, R. J. (1973). General interspersion of repetitive with nonrepetitive sequence elements in the DNA of *Xenopus. J. Mol. Biol.* **77,** 1–23.

Davidson, E. H., Galau, G. A., Angerer, R. C., and Britten, R. J. (1975). Comparative aspects of DNA sequence organization in metazoa. *Chromosoma* **51,** 253–359.

Davidson, E. H., Klein, W. H., and Britten, R. J. (1977). Sequence organization in animal DNA and a speculation on hnRNA as a coordinate regulatory transcript. *Dev. Biol.* **55,** 69–84.

Davis, R., Simon, M., and Davidson, N. (1971). Electron microscope heteroduplex methods for mapping regions of base sequence homology in nucleic acids. *In* "Methods in Enzymology" (L. Grossman and K. Moldave, eds.), Vol. 21, pp. 413–428. Academic Press, New York.

Emerson, R. A., and Rhoades, M. (1933). Relation of chromatid crossing-over to the upper limit of recombination percentages. *Am. Nat.* **67,** 374–377.

Emerson, S. (1963). Meiotic recombination in fungi with special reference to tetrad analysis. *In* "Methodology in Basic Genetics" (W. J. Burdette, ed.), pp. 167–208. Holden-Day, San Francisco, California.

Finnegan, D. J., Rubin, G. M., Young, M., and Hogness, D. S. (1977). Multigene families in the genome of *Drosophila melanogaster. Cold Spring Harbor Symp. Quant. Biol.* (in press).

Finnerty, V. (1976). Genetic units of Drosophila—simple cistrons. *In* "The Genetics and Biology of Drosophila" (M. Ashburner and E. Novitski, eds.), Vol. 1B, pp. 721–765. Academic Press, New York.

Firtel, R. A., and Kindle, K. (1975). Structural organization of the genome of the cellular slime mold *Dictyostelium discoideum:* Interspersion of repetitive and single-copy DNA sequences. *Cell* **5,** 401–411.

Gall, J. G., and Pardue, M. L. (1969). Formation and detection of RNA–DNA hybrid molecules in cytological preparations. *Proc. Natl. Acad. Sci. U.S.A.* **63,** 378–383.

Goldberg, R. B., Crain, W. R., Ruderman, J. V., Moore, G. P., Barnett, T. R., Higgins, R. C., Gelfand, R. A., Galau, G. A., Britten, R. J., and Davidson, E. H. (1975). DNA sequence organization in the genomes of five marine invertebrates. *Chromosoma* **51,** 225–251.

Graham, D. E., Neufeld, B. R., Davidson, E. H., and Britten, R. J. (1974). Interspersion of repetitive and nonrepetitive DNA sequences in the sea urchin genome. *Cell* **1,** 127–138.

Grunstein, M., and Hogness, D. S. (1975). Colony hybridization: A method for the isolation of cloned DNA's that contain a specific gene. *Proc. Natl. Acad. Sci. U.S.A.* **72,** 3961–3965.

Haldane, J. B. S. (1919). The combination of linkage values, and the calculation of distances between the loci of linked factors. *J. Genet.* **8,** 299–309.

Hochman, B. (1973). Analysis of a whole chromosome in *Drosophila. Cold Spring Harbor Symp. Quant. Biol.* **38,** 581–589.

Hood, L., Campbell, J. H., and Elgin, S. C. R. (1975). The organization, expression and evolution of antibody genes and other multigene families. *Annu. Rev. Genet.* **9,** 305–353.

Hotta, Y., and Benzer, S. (1972). Mapping of behaviour in *Drosophila* mosaics. *Nature (London)* **240,** 527–535.

Ikeda, Y., Ishitani, C., and Nakamura, K. (1957). A high frequency of heterozygous diploids and somatic recombination induced in imperfect fungi by ultraviolet light. *J. Gen. Appl. Microbiol.* **3,** 1–11.

Jackson, D., Symons, R., and Berg, P. (1972). Biochemical methods for inserting new genetic information into DNA of Simian Virus 40: Circular SV40 DNA molecules containing Lambda phage genes and the galactose operon of *Escherichia coli. Proc. Natl. Acad. Sci. U.S.A.* **69,** 2904–2909.

Judd, B. H. (1976). Genetic units of Drosophila—complex loci. *In* "The Genetics and Biology of Drosophila" (M. Ashburner and E. Novitski, eds.), Vol. 1B, pp. 767–799. Academic Press, New York.

Judd, B. H. (1977). The nature of the module of genetic function in *Drosophila. In* "The Organization and Expression of the Eukaryotic Genome" (E. M. Bradbury and K. Javaherian, eds.), pp. 469–483. Academic Press, New York.

Judd, B. H., Shen, M. W., and Kaufman, T. C. (1972). The anatomy and function of a segment of the X chromosome of *Drosophila melanogaster*. *Genetics* **71,** 139–156.

Kafer, E. (1958). An 8-chromosome map of *Aspergillus nidulans*. *Adv. Genet.* **9,** 105–145.

Kosambi, D. D. (1944). The estimation of map distances from recombination values. *Ann. Eugen., London* **12,** 172–175.

Laird, C. D., and Chooi, W. Y. (1976). Morphology of transcription units in *Drosophila melanogaster*. *Chromosoma* **58,** 193–218.

Lefevre, G., Jr. (1973). The one band–one gene hypothesis: Evidence from a cytogenetic analysis of mutant and nonmutant rearrangement breakpoints in *Drosophila melanogaster*. *Cold Spring Harbor Symp. Quant. Biol.* **38,** 591–599.

Lefevre, G., Jr. (1974). The relationship between genes and polytene chromosome bands. *Annu. Rev. Genet.* **8,** 51–62.

Lefevre, G., Jr., and Wilkins, M. D. (1966). Cytogenetic studies on the white locus in *Drosophila melanogaster*. *Genetics* **53,** 175–187.

Lewin, B. (1976a). Units of transcription and translation: The relationship between heterogeneous nuclear RNA and messenger RNA. *Cell* **4,** 11–20.

Lewin, B. (1976b). Units of transcription and translation: Sequence components of heterogeneous nuclear RNA and messenger RNA. *Cell* **4,** 77–93.

Lindsley, D. L., Sandler, L., Baker, B. S., Carpenter, A. T. C., Denell, R. E., Hall, J. C., Jacobs, P. A., Miklos, G. L. G., Davis, B. K., Gethmann, R. C., Hardy, R. W., Hessler, A., Miller, S. M., Nozawa, H., Parry, D. M., and Gould-Somero, M. (1972). Segmental aneuploidy and the genetic gross structure of the *Drosophila* genome. *Genetics* **71,** 157–184.

Lobban, P. E., and Kaiser, A. D. (1973). Enzymatic end-to-end joining of DNA molecules. *J. Mol. Biol.* **78,** 453–471.

Ludwig, W. (1934). Über numerische Beziehungen der Crossover-Werte untereinander. *Z. Indukt. Abstamm. Vererbungsl.* **67,** 58–95.

McClintock, B. (1951). Chromosome organization and genic expression. *Cold Spring Harbor Symp. Quant. Biol.* **16,** 13–47.

McKnight, S. L., and Miller, O. L., Jr. (1976). Ultrastructural patterns of RNA synthesis during early embryogenesis of *Drosophila melanogaster*. *Cell* **8,** 305–319.

Manning, J. E., Schmid, C. W., and Davidson, N. (1975). Interspersion of repetitive and nonrepetitive DNA sequences in the *Drosophila melanogaster* genome. *Cell* **4,** 141–155.

Mather, K. (1938). Crossing-over. *Biol. Rev. Cambridge Philos. Soc.* **13,** 252–292.

Maxam, A. M., and Gilbert, W. (1977). A new method for sequencing DNA. *Proc. Natl. Acad. Sci. U.S.A.* **74,** 560–564.

Mertz, J. E., and Davis, R. W. (1972). Cleavage of DNA by RI restriction endonuclease generates cohesive ends. *Proc. Natl. Acad. Sci. U.S.A.* **69,** 3370–3374.

Metzenberg, R. L., Gleason, M. K., and Littlewood, B. S. (1974). Genetic control of alkaline phosphate synthesis in *Neurospora:* The use of partial diploids in dominance studies. *Genetics* **77,** 25–43.

Miller, O. L., Jr., and Beatty, B. R. (1969). Visualization of nucleolar genes. *Science* **164,** 955–957.

Mukai, T. (1964). The genetic structure of natural populations of *Drosophila melanogaster*. I. Spontaneous mutation rate of polygenes controlling viability. *Genetics* **50,** 1–19.

Mukai, T., Chigusa, S. I., Mettler, L. E., and Crow, J. F. (1972). Mutation rate and dominance of genes affecting viability in *Drosophila melanogaster*. *Genetics* **72,** 335–355.

Muller, H. J. (1929). The gene as the basis of life. *Proc. Int. Congr. Plant Sci., 1926* Vol. 1, pp. 897–921.

Muller, H. J., and Altenberg, E. (1919). The rate of change of hereditary factors in *Drosophila*. *Proc. Soc. Exp. Biol. Med.* **17,** 10–14.

Murray, N. E., and Murray, K. (1974). Manipulation of restriction targets in phage λ to form receptor chromosomes for DNA fragments. *Nature* (*London*) **251,** 476–481.

Nathans, D., and Smith, H. O. (1975). Restriction endonucleases in the analysis and restructuring of DNA molecules. *Annu. Rev. Biochem.* **44,** 273–293.

Newmeyer, D., and Taylor, C. W. (1967). A pericentric inversion in *Neurospora,* with unstable duplication progeny. *Genetics* **56,** 771–791.

Ohta, T., and Kimura, M. (1971). Functional organization of genetic material as a product of molecular evolution. *Nature* (*London*) **233,** 118–119.

Owen, A. R. G. (1950). The theory of genetical recombination. *Adv. Genet.* **3,** 117–157.

Papazian, H. P. (1952). The analysis of tetrad data. *Genetics* **37,** 175–188.

Pardue, M. L., Gerbi, S. A., Eckhardt, R. A., and Gall, J. G. (1970). Cytological localization of DNA complementary to ribosomal RNA in polytene chromosomes of Diptera. *Chromosoma* **29,** 268–290.

Perkins, D. D. (1949). Biochemical mutants in the smut fungus *Ustilago maydis*. *Genetics* **34,** 607–626.

Perkins, D. D. (1975). The use of duplication-generating rearrangements for studying heterokaryon incompatability genes in *Neurospora*. *Genetics* **80,** 87–105.

Pontecorvo, G., and Kafer, E. (1958). Genetic analysis based on mitotic recombination. *Adv. Genet.* **8,** 71–104.

Pontecorvo, G., and Roper, J. A. (1953). The genetics of *Aspergillus nidulans*. VII. Diploids and mitotic recombination. *Adv. Genet.* **5,** 218–238.

Pontecorvo, G., Tarr Gloor, E., and Forbes, E. (1954). Analysis of mitotic recombination in *Aspergillus nidulans*. *J. Genet.* **53,** 226–237.

Prescott, D. M. (1977). Genetic organization of eukaryotic chromosomes. *In* "Chromosomes: From Simple to Complex" (P. A. Roberts, ed.), pp. 55–78. Oregon State Univ. Press, Corvallis.

Prescott, D. M., and Murti, K. G. (1973). Chromosome structure in ciliated protozoans. *Cold Spring Harbor Symp. Quant. Biol.* **38,** 609–618.

Prescott, D. M., Bostock, C. J., Murti, K. G., Lauth, M. R., and Gamow, E. (1971). DNA of ciliated protozoa. I. Electron microscopic and sedimentation analyses of macronuclear and micronuclear DNA of *Stylonychia mytilus*. *Chromosoma* **34,** 355–366.

Pritchard, R. H. (1963). Mitotic recombination in fungi. *In* "Methodology in Basic Genetics" (W. J. Burdett, ed.), pp. 228–246. Holden-Day, San Francisco, California.

Raffel, D., and Muller, H. J. (1940). Position effect and gene divisibility considered in connection with three strikingly similar scute mutations. *Genetics* **25,** 541–583.

Rizet, G., and Englemann, C. (1949). Contribution à l'étude génétique d'un Ascomycete tétrasporé. *Rev. Cytol. Biol. Veg.* **11,** 201–304.

Roman, H. (1956a). Studies of gene mutation in *Saccharomyces*. *Cold Spring Harb. Symp. Quant. Biol.* **21,** 175–185.

Roman, H. (1956b). A system selective for mutations affecting the synthesis of adenine in yeast. *C. R. Tran. Lab. Carlsberg, Ser. Physiol.* **36,** 299–314.

Roper, J. A., and Kafer, E. (1957). Acriflavin-resistant mutants of *Aspergillus nidulans*. *J. Gen. Microbiol.* **16,** 660–667.

Sager, R., and Kitchin, R. (1975). Selective silencing of eukaryotic DNA. *Science* **189,** 426–433.

Shutt, R. H., and Kedes, L. H. (1974). Synthesis of histone mRNA sequences in isolated nuclei of cleavage stage sea urchin embryos. *Cell* **3,** 283–290.

Stern, C. (1936). Somatic crossing-over and segregation in *Drosophila melanogaster*. *Genetics* **21,** 625–730.

Struhl, K., Cameron, J. R., and Davis, R. W. (1976). Functional genetic expression of eukaryotic DNA in *Escherichia coli. Proc. Natl. Acad. Sci. U.S.A.* **73,** 1471–1475.

Tartof, K. D. (1975). Redundant genes. *Annu. Rev. Genet.* **9,** 355–385.

Thomas, M., Cameron, J. R., and Davis, R. W. (1974). Viable molecular hybrids of bacteriophage lambda and eukaryotic DNA. *Proc. Natl. Acad. Sci. U.S.A.* **71,** 4579–4583.

Wensink, P. C., Finnegan, D. J., Donelson, J. E., and Hogness, D. S. (1974). A system for mapping DNA sequences in the chromosomes of *Drosophila melanogaster. Cell* **3,** 315–325.

White, R. L., and Hogness, D. S. (1977). R-loop mapping of the 18S and 28S sequences in the long and short repeating units of *Drosophila melanogaster* rDNA. *Cell* **10,** 177–192.

Wimber, D. E., and Steffensen, D. M. (1970). Localization of 5S RNA genes on *Drosophila* chromosomes by RNA–DNA hybridization. *Science* **170,** 639–641.

Young, M. W., and Judd, B. H. (1978). Nonessential sequences, genes and the polytene chromosome bands of *Drosophila melanogaster. Genetics* **88,** 723–742.

6

Chromosome Structure and Levels of Chromosome Organization

Hans Ris and Julie Korenberg

ISBN 0-12-289502-9

I. GENERAL ORGANIZATION OF THE EUKARYOTIC GENOME

Chromosomes long have fascinated biologists by their exact, balletlike movements during nuclear division and their mysterious appearance and disappearance during the cell cycle. Based on the longitudinal splitting of chromosomes and distribution of the halves into opposite daughter cells, first described by Flemming, Strasburger, and van Beneden in the 1880s, Roux suggested that they must be the bearers of hereditary factors. This was confirmed after the rediscovery of Mendel's laws in 1900, when further studies on chromosome behavior laid the foundation for the "chromosome theory of inheritance." The number of chromosomes was found to be constant for a species; chromosomes were shown to be qualitatively distinct, each one corresponding to a genetic linkage group. Their persistence as individual structures through the cell cycle was demonstrated, despite their apparent disappearance during interphase. The behavior during meiosis explained the independent assortment of linkage groups, and genetic recombination was shown to result from breakage and reunion between paternal and maternal chromosomes (Chapters 1 and 6, Volume 1, of this treatise and Chapter 5, this volume). Quantitative studies of recombination established the linear order of genes on chromosomes. In the 1940s and 1950s, this association of genes with chromosomes was extended to a chemical relationship: The hereditary material turned out to be the deoxyribonucleic acid molecule (DNA), and the linear order of genes could be expressed as a linear order of nucleotide sequences.

The rapid successes of molecular genetics depended on the simplicity of the genetic system in prokaryotes. The early tendency simply to transfer the knowledge gained with viruses and bacteria to the vastly more complex eukaryote genetic system now is being replaced by biochemical analysis of the properties unique to the eukaryotic chromosome. One striking characteristic of eukaryotes is their large amount of DNA, including not only unique nucleotide sequences coding for proteins but also simple or complex sequences that are repeated from a few hundred- to a millionfold, some of which are never transcribed. This mass of DNA requires an elaborate packing order in which specific proteins (histones) play a dominant role. This packing also is related to the control of gene function, with genes turned off in the compacted state and activation related to the physical unraveling of specific regions. Further condensation leads to the tightly packed mitotic chromosomes, which can be handled by the mitotic apparatus during nuclear division without becoming tangled. With the exception of the rare organisms with polytene chromosomes, it is only in this compacted state during mitosis and meiosis that the characteristic number, size, and shape of chromosomes can be determined. This number and the DNA content of the haploid chromosome set (C value) are a basic characterization of the genome in eukaryotes. Table I provides representative examples of DNA values and chromosome numbers. (For more detailed lists of DNA values, see Rees and Jones, 1972; Shapiro, 1976; Sparrow *et al.,* 1972; for chromosome numbers, Altman and Dittmer, 1962; Hsu and Benirschke, 1967–1974.) While DNA content tends to increase with the evolutionary complexity of organisms, this relationship is far from straightforward. Diverse groups of both plants and animals are notable for their high C value (e.g., amphibians and lilies). Other groups surprise us by the enormous spread of genome size among members (e.g., amphibians) (Britten and Davidson, 1971). Mammals show minimal differences between species. The significance of these variations in DNA content is still little understood.

The haploid chromosome number varies from two to many hundreds, but in most species lies between 6 and 30. Even in mammals with minimal variation in DNA content the numbers vary from 6 (*Muntjac*) to 92 (the rodent *Auctomys leander*).

Chromosome volume generally is proportional to the DNA content. The yeast *Saccharomyces cerevisiae* has the smallest chromosomes, judging from their DNA content (average about 300 μm), although—as in some other fungi—they do not condense in mitosis and thus are not visible individually. Minute chromosomes are known in duck weeds (Lemnaceae), in which they measure 0.1×0.2 μm. Perhaps the largest metaphase chromosomes are found in lilies, measuring 2×23 μm in a *Haemanthus* species. Besides DNA content, other factors influence the size of chromosomes as shown by variations in volume of chromosomes from different cell types in the same organism. During early cleavage stages, chromosome volume often is considerably larger than in somatic mitosis (Schrader and Hughes-Schrader, 1926). In haploid salamanders, chromosomes are larger than in normal diploids (Fankhauser, 1934). Simi-

TABLE I

Haploid Chromosome Number and Corresponding DNA Content for Selected Plants and Animals[a]

	Number of chromosomes (haploid)	DNA content[b] (pg)
Bacteria		
Mycoplasma	1	0.000840
Escherichia coli	1	0.004
Algae, Protozoa		
Amphidinium carteri	25	3.0
Euglena gracilis	45	3.0
Ameba proteus	Ca. 250	40.0
Plasmodium berghei	—	0.06
Fungi		
Saccharomyces cerevisiae	17	0.024
Neurospora crassa	7	0.04
Aspergillus nidulans	4	0.044
Flowering plants		
Lilium longiflorum	12	50.0
Allium cepa	8	20.0
Crepis capillaris	6	2.1
Sponges		
Tube sponge	—	0.06
Echinoderms		
Strongylocentrotus purpuratus	18	0.9
Molluscs		
Aplysia californica	Ca. 16	1.0
Octopus vulgaris	—	5.0
Insects		
Chironomus tentans	4	0.25
Drosophila melanogaster	4	0.09
Melanoplus differentialis (grasshopper)	12	8.0
Tunicates		
Cionia intestinalis	9	0.14
Fish		
Scyllium (shark)	Ca. 40	3.3
Amia (bowfin)	Ca. 23	1.0
Protopterus (lungfish)	17	50.0
Esox (pike)	9	0.8
Cyprinus (carp)	52	1.6

TABLE I (*Continued*)

	Number of chromosomes (haploid)	DNA content[b] (pg)
Amphibians		
Amphiuma means	12	100.0
Notophthalmus viridescens (=*Triturus*)	12	22.0
Bufo bufo (toad)	11	6.0
Rana pipiens (frog)	13	7.0
Reptiles		
Chelonia (turtle)	28	2.7
Anolis (lizard)	18	2.0
Constrictor (boa constrictor)	18	1.8
Birds		
Gallus domestica (chicken)	39	1.2
Passer domesticus (sparrow)	23	1.9
Columba (pigeon)	40	2.0
Mammals		
Human	23	3.0
Cattle	30	3.2
Dog	39	2.8
Cat	19	3.5
Microtus oregoni (vole)	9	2.7
Perameles nasuta (bardicoot)	7	4.6
Trichosurus vulpecula (possum)	10	3.0
Tachyglossus aculeatus (spiny anteater)	32	2.9

[a] For references, see Sparrow *et al.* (1972), Rees and Jones (1972), Shapiro (1976), and Altman and Dittmer (1962).

[b] 1 μm of double-standard DNA = 3000 bp or 2×10^6 daltons or 3.1×10^{-6} pg.

lar variation during development and in tumors were reported by Biesele (1947). Pierce (1937) found a fourfold increase of chromosome volume in root tips of violets when grown in phosphate-rich medium as compared with tap water. While the chromosomal component responsible for these differences is not known, it is certainly not DNA.

The genome of a species can be characterized further by the chromosome morphology that includes the position of the *centromere* (primary constriction or point of spindle attachment), which determines the relative length of "short" and "long" arm of a chromosome, the position of other constrictions ("secondary constrictions"), the location of "nucleolar organizers," the pattern of "prophase chromomeres," and metaphase "banding patterns" revealed by fluorescent dyes or Giemsa staining. The

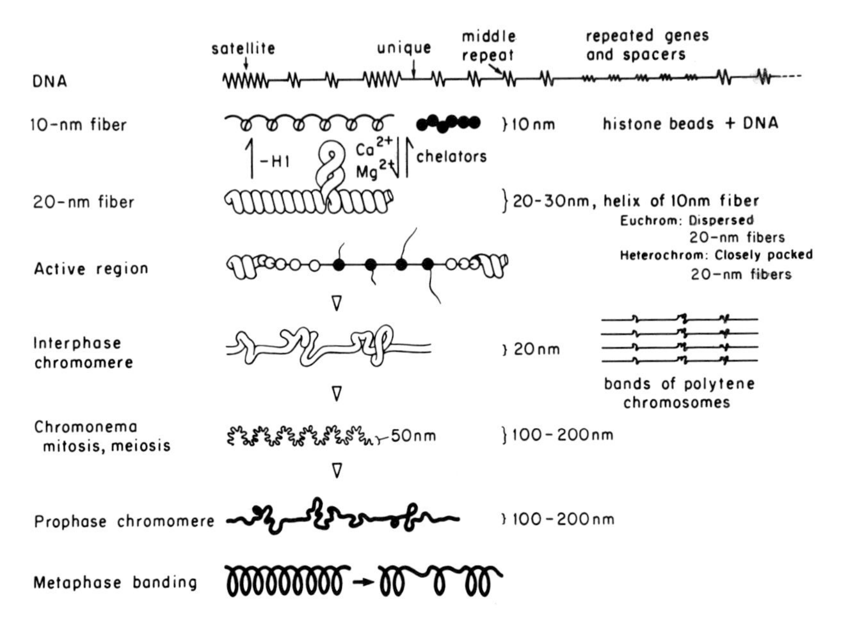

Fig. 1. This diagram summarizes the present knowledge on the orders of packing of DNA in chromosomes. The triangle indicates a change in magnification.

mitotic morphology of all the chromosomes taken together is called the karyotype of a species (idiogram if it is a diagram representing many karyotypes). In a few species, a more detailed chromosome morphology is visible in "interphase chromomere" patterns. This occurs when successive chromosome replications have resulted in "polytene" (multistranded) chromosomes in which chromomeres are lined up in register, producing a characteristic banding pattern (for example, the salivary gland chromosomes of *Drosophila*). These detailed morphological markers have been of great importance for the establishment of gene maps, relating genes to specific chromosome abnormalities, and evolutionary changes caused by chromosome alterations.

Light microscope studies had suggested that the longitudinal differentiation of the chromosome (chromomeres, bands), as well as the structural changes during the cell cycle, are caused by a differential packing or condensation of a basic chromosome thread. Mitotic condensation, for example, was shown to involve a helical coiling of the smallest unit resolved by the light microscope, the *chromonema*. Certain chromosome segments do not undergo a cyclic behavior and remain condensed during interphase. Heitz (1928) called these regions *heterochromatin*. Today we recognize two major types of heterochromatin: (1) "constitutive heterochromatin," which is condensed in every cell and (2) "facultative heterochromatin," which remains condensed in some cells but not in others (see Chapter 4, Volume 1, of this treatise). The first type contains DNA that never is transcribed, such as the highly repetitive simple sequence DNA (satellite DNA). The inactive sex chromosome in mammalian females (sex chromatin or Barr body) is an example of the second type and represents chromosome regions carrying normal genes that have been turned off during early development.

The electron microscope has revealed that the chromonema represents a state of condensation of a still thinner thread. Indeed, it now is possible to recognize a whole hierarchy of organizational levels in the chromosome from the DNA molecule to the condensed metaphase chromosome. This is diagrammatically illustrated in Figure 1. The following discussion will be organized according to this sequence of increasingly complex structures, in which the DNA is folded and coiled at various stages of the cell cycle.

II. THE ORGANIZATION OF DNA IN EUKARYOTES

A. The Arrangement of DNA in Chromosomes *

Once DNA was established as the bearer of genetic information, the arrangement of this extraordinarily long polymer became of immediate concern. The effect of DNase, RNase, and proteases on the longitudinal integrity of chromosomes established that DNA formed the longitudinal backbone of chromosomes. DNase breaks chromosomes, while RNase and proteases do not (Callan and Macgregor, 1958; Gall, 1963a; Lezzi, 1965). The enormous amount of DNA in an average eukaryotic chromosome, on the order of 6×10^{10} MW (ca. 10^8 base pairs), and previous microscopic observations had led early workers to hypothesize that eukaryotic chromosomes consisted of side-by-side arrays of a number of DNA helices (multinemy) and, further, that each longitudinal member itself was held together at intervals by protein linkers. Although mitigating the difficulties associated with long molecules, this model necessitated elaborate mechanisms for strand segregation after replication and complicated the interpretation of mutation theory. Electron microscopic observations and the kinetics of DNase breakage of lampbrush chromosomes in amphibian oocytes supported the presence of a single DNA molecule per chromatid (uninemy). The demonstration in two organisms of DNA

* See also Chapter 1, Volume 3, of this series.

molecules corresponding to the DNA content of chromosomes (see below) and the presence in all species studied of nucleotide sequences that occur only once in the DNA complement of a chromosome (Laird, 1971) make it highly probable that uninemy is the rule.

Three aspects of chromosome structure have been established firmly in both *Drosophila melanogaster* and *Saccharomyces cerevisiae*. First, in its simplest form, one chromosome contains one molecule of DNA. This, then, implies the second, that the DNA is continuous through the centromere, and the third, that protein linkers do not maintain the continuity of the chromosomal axis. In *Drosophila*, Kavenoff and Zimm (1973) utilized a viscoelastometric method that is uniquely suited to the measurement of long molecules. Briefly, in this method, a rotatable cylinder is placed inside a fixed cylinder that contains a solution of detergent-lysed cells. The inner cylinder then is rotated through a given arc and released. The long molecules, which have been stretched by the rotation, now return to their relaxed state of random coils and, in doing so, rotate the inner cylinder back toward its initial position. The rate of this recoil movement is proportional to the molecular weight and number of the largest molecules in the solution, viz., the DNA. Thus, they could estimate both the number of similar-sized large chromosomes per cell and the approximate molecular weight. Although burdened with a 30% uncertainty, their absolute measurements fell within a few percent of previous cytophotometric measurements of the DNA content of similar chromosomes. Further, with respect to continuity through the centromeric region, they observed that molecular weight estimates for a given chromosome were constant regardless of alteration of centromere position by pericentric inversion.

The very small chromosomes of yeast (6.2×10^8 MW) are more accessible to measurement by both electron microscopy and sedimentation velocity analysis. Chromosome-sized DNA molecules have been demonstrated by both Petes *et al.* (1973) and Cryer *et al.* (1973). Although the longest yeast molecules observed are shorter than predicted, perhaps due to shear or to residual nuclease activity, these nonetheless are too long to be compatible with a multistranded model.

It is worth mentioning that, although the yeast chromosomes have been shown to be single linear molecules resistant to digestion by RNase, the *Drosophila* work has not excluded the possibility of RNA or other protease resistant linkers. Neither can either group exclude the possibility of a slightly branched network, or even the existence of a single super DNA molecule that is susceptible to precise breakage comprising the entire genome (DuPraw, 1968; Lauer and Klotz, 1975). However, they have made this possibility exceedingly unlikely.

As is usual in biology, the exceptions to this have been known far longer than the rule. For example, in the endosperm of certain plants chromosomes clearly appear double at anaphase (Holm and Bajer, 1966). In *Drosophila melanogaster*, third instar larval ganglionic metaphase chromosomes contain two and four times the normal amount of DNA and appear to be multinemic (Gay *et al.*, 1970). Multinemic mitotic chromosomes thus do exist occasionally in certain tissues. Finally, this model most clearly does not apply to the polytene chromosomes found in dipteran larval tissues (Beermann, 1962), in salivary glands of Collembola (Cassagnau, 1968, 1971), in macronuclear anlagen of hypotrich ciliates (Ammermann, 1971), and in certain plant tissues (Nagl, 1962).

Some form of stable branching does occur in polytene chromosomes. This may be inferred from the observation (described in detail later) that the satellite DNA and rDNA cistrons are underreplicated to different degrees relative to the euchromatic arms of *Drosophila melanogaster* salivary gland chromosomes. Whereas the rDNA undergoes five replication cycles, it is flanked on both sides by satellites which undergo only one to two cycles (Hennig and Meer, 1971; Spear and Gall, 1973). These findings suggest the existence of stable specific branchpoints or discontinuities at the borders between satellite and rDNA and satellite and euchromatin.

B. Unique and Repeated Nucleotide Sequences, Genome Size, and Genome Complexity

It now is well established that, in contrast to prokaryotes, the large amount of DNA in the genomes of higher organisms is not simply a collection of unrelated unique sequences. To the contrary, it consists of sequences whose lengths vary from a few to 2000 base pairs and whose frequencies vary from one to one million. In this section, we shall describe methods used to determine the number and organization of DNA sequences in a genome and shall discuss the organization of the genome that emerges and its relationship to the substructure of specific genes. The most commonly used method for determining the numbers of genes depends on principles clearly stated by Britten and Kohne (1968) for DNA and by Bishop (1972) for RNA. In essence, the method first requires the preparation of total DNA from the organism in question, followed by shearing it to reasonably small pieces and melting it. It then is allowed to reassociate under specific conditions, such as cation concentration, temperature of incubation, DNA concentration, and DNA fragment size. The temperature must be sufficiently high to weaken intrastrand secondary structure and to allow free movement of the DNA chain. The optimum temperature for reassociation is about 25°C below the temperature required for dissociation of the double strands. The time must be sufficient to permit an adequate number of collisions so that the DNA can reassociate. The size of the nucleotide chains must be controlled, because one collision between two long strands will result in more hybridized DNA than one event between two short fragments; however, the relationship is not straightforward.

Reassociation has been measured in a variety of ways, each of which depends on the different properties of single- and double-stranded DNA. For example, double-stranded DNA sticks preferentially to nitrocellulose filters and to calcium phosphate (hydroxyapatite) columns. Reassociation reactions can be followed by incubating the mixture on the column and then eluting at a phosphate buffer concentration that removes only single-stranded DNA. The remaining reassociated DNA then may be eluted by raising the phosphate concentration. The amount of DNA in each fraction can be determined by ultraviolet (uv) absorption or by the inclusion of a radioactive label. To characterize the proportion in a genome of sequences of a given order of repetition, total genomic DNA is used. To determine the numbers of a specific gene, a small amount of highly labeled single-stranded probe for this gene or sequence (RNA or DNA) is added to an excess of the cold single-stranded DNA. As the time of incubation (t) increases, more and more of the probe hybridizes to the cold DNA or, in the former case, the cold DNA to itself. Since the association of single strands to form a hybrid requires a bimolecular collision, the rate of this association is proportional to the square of the concentration and may be expressed in the following way:

$$dC/dt = -kC^2$$

where C = concentration of single-stranded DNA at time t, dC/dt = rate at which single-stranded DNA is reassociating to form a duplex, and k = rate constant for the rate-limiting (collision or nucleation) step. One also can change the degree of hybridization obtained in a given time by increasing the total DNA plus probe concentration (C_0). In general, the degree of hybridization proves to be related in an intuitively reasonable way to the product of the total DNA concentration times the time of incubation:

$$C/C_0 = 1/(1 + kC_0t)$$

where C, C_0, k, and t are as defined above.

With relatively simple systems, such as bacteriophage lambda DNA being hybridized either to itself or to traces of complementary RNA (cRNA) made from the same lambda DNA, a smooth S-shaped curve is obtained. The extent of hybridization is related to the log of C_0t as seen in

Figure 2. This curve is characterized by a single parameter, ($C_0t_{1/2}$), which is the C_0t value at 50% hybridization. The actual value of ($C_0t_{1/2}$) depends on the various conditions used (temperature of incubation, salt concentration, DNA fragment size, and DNA concentration), but under a defined set of conditions has a very important relationship to the sequence complexity. In lower organisms, the complexity is equal to the genome size; in higher organisms in which a fraction of the sequences are repeated, the complexity (expressed in base pairs) is the sum of base pairs of each different sequence in the genome, i.e., the total number of base pairs of single-copy DNA plus the sum of base pairs of one representative of each family of repeated sequences. Now, lambda cRNA or DNA hybridized to lambda DNA gives a $C_0t_{1/2}$ of approximately 10^{-1} (moles/liter) times seconds while *E. coli* DNA hybridized to itself gives a $C_0t_{1/2}$ of approximately 10^1; this means that *E. coli* DNA hybridizes one hundred times slower. This also is the number of times by which the *E. coli* genome is larger than the lambda genome. In general, the relationship found is that the less

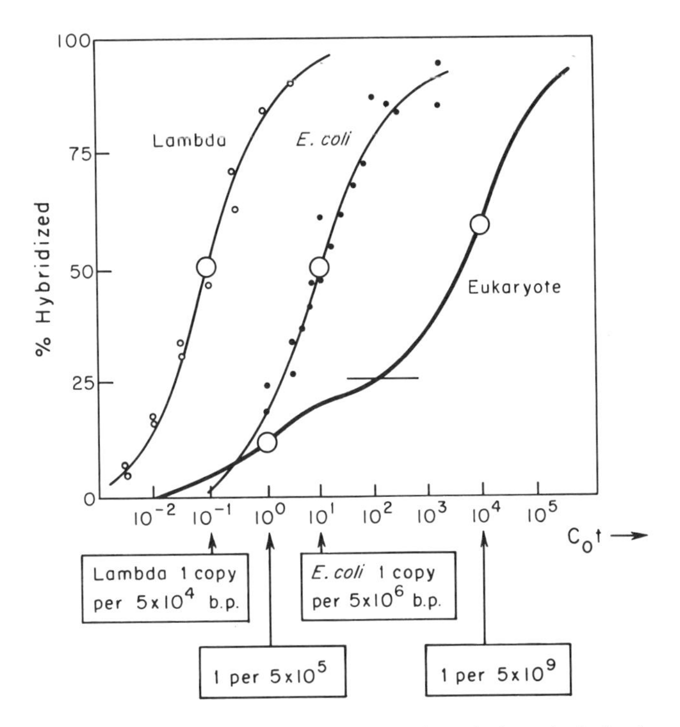

Fig. 2. Examples of C_0t curves: lambda phage, *E. coli*, and a hypothetical eukaryote. See text for explanation. Redrawn after original from G. Smith and O. Smithies, Department of Genetics, University of Wisconsin, Madison, Wisconsin.

complex the DNA sequence, the less time it will take the complementary DNA (or RNA) to form a hybrid. The principle is simple. If one has ten copies of a DNA all with a given sequence, it is ten times easier for complementary RNA to form a hybrid with the DNA than if one has one copy of a DNA with a sequence length ten times longer. Thus, for simple DNA's, whose complexity equals the genome size, the value of $C_0t_{1/2}$ is directly porportional to the size of the genome.

When DNA (or cRNA) from eukaryotic organisms is hybridized to the DNA from the same organism, simple S-shaped curves no longer are found, because the proportionality between the C_0t required for half-reassociation of the DNA and the genome size is true only in the absence of repeated sequences. Part of the mixture may hybridize at a faster rate than *E. coli* and part much more slowly. Figure 2 shows the curves obtained from *E. coli* and bacteriophage lambda along with a hypothetical version of what one might expect to see from the hybridization of a eukaryotic cell. These curves now may be interpreted in terms of the number of base pairs in the genome. Curves such as these must be calibrated by hybridizing DNA of known size under the same conditions. The lambda genome is approximately 5×10^4 base pairs long; *E. coli* is one hundred times larger, with approximately 5×10^6 base pairs. For the hypothetical eukaryote, we assume a genome size a thousand times larger than *E. coli,* or 5×10^9 base pairs. Consequently, the $C_0t_{1/2}$ of 10^{-1} seen for lambda is that observed for a gene present as one copy per 5×10^4 base pairs (the genome size of lambda). Similarly, the $C_0t_{1/2}$ of *E. coli* corresponds to a gene present in one copy per 5×10^6 base pairs (the genome size of *E. coli*). The hypothetical eukaryote shows two components of the hybridization curve. The first, about 25% of the total, hybridizes at a $C_0t_{1/2}$ equal to 10^0, that is, to a sequence present at one copy per 5×10^5 base pairs. Since the genome size of the eukaryote is 10^4 times larger than this, the number of copies of this sequence in the genome is 10^4. The second component of the curve, representing 75% of the total, hybridizes at a $C_0t_{1/2}$ of 10^4, 1000 times slower than that of *E. coli*. This component, then, represents DNA present 1000 times less frequently than that in *E. coli,* and therefore corresponds to one copy in 5×10^9 base pairs—the assumed size of the eukaryotic genome. Thus we can infer from these data that 25% of the hypothetical genome consists of DNA reiterated 10,000 times in the genome, and that the other 75% is single-copy DNA.

In practice, curves are not always so good, and the calibrations are not always so precise. One often observes some apparent hybridization before the incubation period (zero-time binding) and some incomplete hybridization at high C_0t values. The former most likely is due to the foldback of a given single-stranded DNA upon itself (intrachain hybridization) so that an insufficient amount is available to hybridize with the probe. When the numbers of genes are determined with a specific labeled probe, the presence of other labeled species in the probe can affect the estimate greatly. Early methods for determining gene numbers were not kinetic, but, rather, dependent on saturation of hybridization. They consisted of adding increasing amounts of labeled RNA to small amounts of cold DNA until a saturation level was reached. Then, knowing the specific activity of the RNA, one could calculate the proportion of the DNA that was complementary to the RNA. These methods were limited to the analysis of repeated sequences.

Although some sequences consist of perfectly matched duplexes, many do not. In general, the stability of the helix depends on the number of correctly matched base pairs and thus can be used as a measure of mismatching. The temperature at which purified double-stranded DNA dissociates into single strands depends in part on the number of correctly paired bases and therefore is a measure of stability of the helix. The proportion of single-stranded DNA at a given temperature can be measured by the change in uv absorbancy at 260 nm. A solution of double-stranded DNA gains 35% more absorbance when completely melted to single strands. The degree of mismatch measured in this way can be used to determine the homogeneity of a family of sequences within an individual, or to determine the relatedness of putatively similar or identical sequences in different individuals or species. In this connection, the greater the evolutionary divergence, the greater the degree of mismatch that is expected, an expectation that has been proved true between species. Within an individual, the more perfectly matched sequences after hybridization are those most likely to have evolved more recently.

In general, then, the DNA of prokaryotes consists of a collection of unique sequences. In fact, certain sequences must derive from the same ancestral gene, but these have diverged sufficiently to now be considered unique under the specific conditions of reassociation. One notable exception is the fivefold reiteration of the rDNA genes in *E. coli*. These are not contiguous but are separated by tRNA genes, called the "spacer tRNA's" (Lund *et al.*, 1976). This suggests that the ancestral sequence containing both rDNA and tRNA genes was duplicated and the reiterated sequence conserved during evolution. The DNA of eukaryotes, on the other hand, consists of both unique DNA and families of repeated sequences with varying degrees of similarity. Nucleic acid hybridization studies have established a general scheme of sequence organization in eukaryotes. A certain percentage of the genome (about 20–80%, depending on the species) consists of genes present in one or at most a few copies; another fraction consists of sequences reiterated from ten to a few thousand times; and the remainder of the DNA represents sequences that are reiterated from a few thousand to a few million times.

1. Unique Sequences

While most structural genes are found among the unique sequences, not all unique sequences are transcribed, and not all transcripts reach the cytoplasm to be translated into proteins. Until recently, the investigation of unique DNA has been hampered by the lack of sufficient amounts of relatively pure sequences. The development of nucleic acid cloning techniques has removed this barrier (see Chapter 5, this volume). Studies in the next few years should begin to reveal the detailed arrangement of sequences neighboring specific unique DNA's (see Chapter 1, Volume 3) and thereby also reveal the role these sequences play in gene control.

2. Intermediate Repetitive Sequences

A certain fraction of the genome consists of sequences that are repeated from a few times to several thousandfold. Although we know next to nothing about the function of most of these sequences, in this group there are certain repeated genes whose products are required in large numbers, such as the genes for 5 S RNA, the ribosomal RNA's, the histones, and the immunoglobulin genes. There undoubtedly are other families of related sequences whose functions are distinct albeit related, and families whose sequences have derived from a common ancestor but whose functions now are quite different. The occurrence of multigene families and their evolution are discussed by Hood *et al.* (1975).

Although the scheme varies considerably among gene families and within gene families of various organisms, a general organization of middle repetitive genes has begun to emerge. This is well illustrated by the genes for ribosomal RNA (rDNA) (Figure 3). Eukaryote ribosomes contain three species of RNA, the 5, 18, and 28 S molecules. The genes for 5 S RNA are linked to those for 18 and 28 S in bacteria, in yeast, and in the slime mold *Dictyostelium discoideum* (Maizels, 1976) but not in the higher eukaryotes studied thus far. One sequence each for the 18 and 28 S RNA is organized into a larger unit in which these are separated by a spacer region which is transcribed along with them to form the single precursor RNA. From 100 to 500 such transcriptional units are arranged in tandem, separated by nontranscribed spacer regions.

In general, the 18 and 28 S sequences are highly conserved among species, whereas both the transcribed and nontranscribed spacer regions are not. The transcript of the transcribed spacer is cleaved during maturation of the precursor rRNA molecule. Although the transcribed segment is much larger in birds and mammals (4.1×10^6) than in lower vertebrates and invertebrates (2.5×10^6 MW), the difference is entirely due to variation in the size of the spacer regions (Wellauer *et al.*, 1974). The spacer

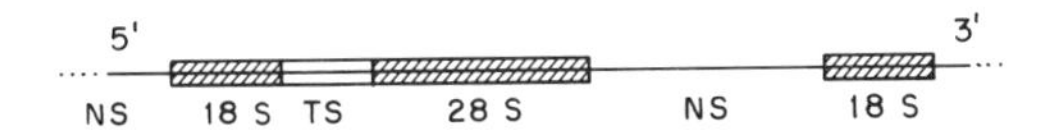

Fig. 3. Arrangement of sequences in a transcription unit coding for ribosomal RNA (rDNA) that often are repeated a few hundred times in a cluster. The sequence for the 18 S RNA is transcribed first followed by a spacer (TS) and the sequence for 28 S RNA. Many such units are arranged in tandem separated by spacer DNA (NS) that is not transcribed. Most of the transcribed spacer (TS) is discarded during maturation of ribosomal RNA's (for a recent reference, see Pellegrini *et al.*, 1977).

regions of repeated genes are of particular interest. They appear to be composed of internally repetitious DNA, whose size may vary even in neighboring units, (e.g., as for the 5 S genes in *Xenopus laevis*). In *Drosophila melanogaster,* in which rDNA is located on both the X and Y chromosomes, the sequence of the nontranscribed spacer DNA differs considerably in the two locations but is homogeneous within either. (C. D. Laird and W. Y. Chooi, personal communication). Because of the relatively short repeat unit within a spacer region, the base composition may be extreme and consist of either GC- or AT-rich regions. Clearly, these would significantly alter the average base composition of a region, as spacers may be from one to six times the size of the genes that they separate. Although their function at present is unknown, these and other neighboring sequences are being scoured for clues to their possible role in regulating the genes they border.

The genes for 5 S RNA have been sequenced completely in a number of organisms (Tartof, 1975). They are analyzed best in *Xenopus laevis* and *X. mulleri*. They are reiterated about 9,000–24,000 times in the former and are located at the telomere region of 15 of the 18 chromosomes (Pardue *et al.*, 1973). They are of particular interest in that ovaries synthesize four types of 5 S sequences, whereas kidney cells synthesize only one of the four. The distribution of the genes for the four types with respect to specific chromosomes is not known. It appears likely that in this 5 S system and perhaps in other single genes and families, the luxury afforded by the large amount of DNA in eukaryotes allows for duplicate genes whose functions are nearly identical but whose activity is specific for a given tissue. Although the 40 tRNA's in *Xenopus laevis* each are reiterated about 200 times and separated by AT-rich spacer (Clarkson *et al.*, 1973), they are reiterated only 8–10 times in *Drosophila*. As yet, it is not known whether the expression of any of these genes is tissue specific.

Among repeated genes, the histone genes are of special interest because they represent repetitive DNA coding for proteins (reviewed by Kedes, 1976). In various sea urchin species, the histone genes are repeated 300–1000 times and in *Xenopus* and man 10–20 times. Histones are synthesized only during chromosome replication in the S phase of the cell cycle, are required in large amounts at that time, and in specific relative proportions for the five types of histones. The recent work in the laboratories of Kedes and Birnstiel has given some exciting insight into the organization of these genes in two species of sea urchins (*Strongylocentrotus purpuratus* and *Psammechinus miliaris*). During cleavage, about 70% of polysome-associated mRNA is 9 S histone message. This allowed purification of histone mRNA and more recently the preparation of pure mRNA for each of the five histones. By means of a wheat germ cell-free translation system, each of the mRNA fractions could be assigned to one of the histones (H1, H2A, H2B, H3, H5) (Gross *et al.,* 1976).

The histone DNA is relatively GC-rich and could be greatly enriched as a satellite on actinomycine–CsCl gradients. Pure histone DNA also was obtained by cloning in *E. coli* through incorporation into plasmids (Kedes *et al.,* 1975). Restriction endonucleases were used to analyze further the histone DNA. Hind III, for instance, cleaves histone DNA into uniform fragments about 6000–7000 base pairs (6–7 kilobases or kb) long, and all histone mRNA's hybridize to this fragment, suggesting that the genes for the different histones are clustered into a repeat unit about 6–7 kb long. Further subdivision with restriction nucleases of both purified and cloned DNA produced pieces to which only one specific histone mRNA hybridizes (Cohn *et al.,* 1976; Schaffner *et al.,* 1976). Since the arrangement of the restriction enzyme fragments on the 7 kb repeat unit and the histone coded by each fragment were known, the sequence of the histone genes on the repeat unit could be determined. It was further found that the mRNA's all are transcribed from the same strand. Strand-specific exonuclease treatment revealed the polarity of the coding strand with regard to the gene sequence; accordingly, transcription can be said to proceed in the direction of H4 → H2B → H3 → H2A → H1 on the gene cluster. This order of the histone genes is identical in both species of sea urchins that have been studied in detail.

Of the 6–7 kb in the repeat unit, only 2 kb are required to code for the five histones. This and the presence of about 50% AT-rich sequences in the histone DNA indicated by its melting profile (Schaffner *et al.,* 1976) suggest the presence of spacer sequences between the histone genes. Electron microscope studies now have given direct evidence in both species for such spacers. Wu *et al.* (1976) used the T4 gene 32 protein which binds to single-strand DNA to visualize single-strand segments in histone DNA hybridized with total histone mRNA. Portmann *et al.* (1976) employed partial denaturation mapping to visualize AT-rich regions. The results show that each of the five histone coding segments is present only once on the 7 kb unit and that they are separated by AT-rich spacers of variable length.

The "immunoglobulin genes" comprise another interesting multigene system with properties somewhat different from the previously discussed examples. The essence here is not uniformity of the repeat units but diversity, which allows for the enormous variability of the immune response. (For a recent review, see Hood *et al.,* 1975.) Antibody molecules consist of pairs of two different polypeptide chains, the small light chain and longer heavy chain. Both polypeptides contain regions of constant amino acid sequences and regions that are variable and determine the great diversity of these molecules. In mammals, there are at least three different multigene families located on different autosomes. In the germ line, the variable and constant regions of the polypeptide chains are coded by separate genes. Since an entire light chain is known to be translated from a single mRNA, its variable and constant regions must be joined in some unknown way at the RNA or DNA level. A small number of genes code for the constant regions but the variable regions seem to be coded by multiple gene families having from one hundred to several thousand copies. Since an antibody-producing cell secretes a single specific antibody protein, the differentiation of these cells requires some interesting control mechanisms that select and combine a few genes out of the hundreds of possibilities in the multigene families for expression in a particular cell.

3. *Highly Repeated Single Sequence DNA*

Families of highly repeated sequences (10^6–10^8 copies per haploid genome) occur in all eukaryotes; their investigation has provided critical information on the evolution and organization of chromosomes. Since repeated genes have been thoroughly reviewed recently (Tartof, 1975), we will summarize the major points only briefly. Satellite DNA originally was found as DNA that banded apart from the main band in cesium chloride density gradient centrifugation. Later work showed that satellite DNA consisted of clusters of tandemly repeating units 2–10 base pairs in length (Southern, 1970; Tartof, 1975). Detailed sequence analysis of a number of satellites with restriction endonucleases has shown that there is a longer range repeat unit spanning hundreds to thousands of base pairs superimposed on the shorter repeat unit. (Sequences of several satellites are found in Tartof, 1975.) In some species, different satellites can be related to one another by a single base change (Gall and Atherton, 1974). Often, satellites contain high concentrations of the unusual base 5-hydroxymethylcytosine (Salomon *et al.,* 1969; Miller *et al.,* 1974). Satellite DNA's occur in long blocks at few sites in chromosomes. Using *in situ* hybridization, most, but not all, satellite DNA's have been localized in the centromeric heterochromatin (see Figure 33), a fact which has suggested to many that they might function in meiotic pairing and therefore correct chromosome disjunction. Although it is well known that heterochromatin regions tend to associate, it is speculative whether the association depends on the presence of repetitive sequences in heterochromatin.

In humans, three major blocks of satellite DNA are concentrated at the three secondary constrictions on chromosomes 1, 9, and 16 (Jones and Corneo, 1971). A fourth block is located on the distal portion of the long arm of the Y chromosome (Evans *et al.,* 1974). This brilliantly fluorescing region is found in gorillas and in all races of man and, although its absence is compatible with normal male development, it clearly must be of some selective advantage. More than one satellite may be located at a given position on a chromosome, although it is not known how they are arranged. This is true in *Drosophila melanogaster* (Peacock *et al.,* 1973), *Dipodomys ordii* (Bostock and Christie, 1975), and man (Gosden *et al.,* 1975). Related species may have similar satellites [for instance, satellite III in chimpanzee and man (Jones *et al.,* 1973)] in addition to satellites unique to each species.

The amounts of satellites clearly can vary within a species. In man, for example, the regions containing satellite DNA's vary visibly in size and may, without affecting the phenotype, occur in twice the usual amount or not at all. The length of the Y chromosome, for example, varies greatly within the population and a lack of the quinacrine brilliant portion (50% of the chromosome) is compatible with normal development. No specific function for this simple sequence, highly repeated DNA has been discovered as yet. All we know is that they are not transcribed (Flamm *et al.,* 1969).

4. Arrangement of Sequences

We have seen that, for repetitive genes in which gene sequences are amenable to detailed study, the transcribed regions are separated from one another by nontranscribed spacer DNA. These spacers resemble satellite DNA in that they are composed of short repeated sequences and are not conserved in evolution. The arrangement of unique sequences has been investigated with a series of clever DNA reassociation techniques, including a comparison of reassociation kinetics of short (360) and long (1830 nucleotide) segments combined with the use of single-strand-specific nuclease and measurements of the hyperchromicity of the reassociated repetitive fragments as a function of length (reviewed in Crain *et al.,* 1976). Of the more than twenty higher organisms now studied, most have more than 50% of their DNA arranged in a pattern of alternating short repeated sequences of about 300 nucleotides and longer single-copy sequences of about 1000 nucleotides. Two species, *Drosophila melanogaster* and *Apis mellifera* (honey bee), seem to be exceptions to this general pattern (Crain *et al.,* 1976). In *D. melanogaster,* for instance, the single-copy sequences are on the average 10,000 bp long with intermediate repetitive sequences of about 5600 bp. At present, there is no understanding of the significance of these patterns.

In eukaryotes, the most striking dilemma has been to discern the function of the great excess of DNA over what would be required to code for the 5000–10,000 genes estimated to be present. Recent findings in prokaryotes suggest that the situation may be even more complex. Bacteriophage ϕX174 and the RNA phage QB contain too little DNA to account for the number of genes in some areas of the gene map. In fact, neighboring genes overlap; the DNA is shared, and the polypeptides for the two genes are read in different reading frames from the same DNA. Such genetic arrangement also is suspected in bacteria, and it is entirely reasonable to assume that the pressures which led to this arrangement in prokaryotes also may exist in specific regions of eukaryotes. With the increasing use of DNA cloning techniques, we can expect rapid progress in the understanding of the arrangement and functional significance of the various types of nucleotide sequences observed in eukaryote chromosomes.

5. *Changes in Gene Repetition*

Several mechanisms are known which change the degree of repetition of specific sequences. "Gene amplification" refers to the differential synthesis of specific sequences in a cell that increases the number of copies beyond that in the basic complement. The best known example is the increase in the number of ribosomal RNA genes in the oocytes of many animals. During meiotic prophase (pachytene), differential DNA synthesis of the nucleolar organizer forms a microscopically visible mass of DNA that is later dispersed to form the many extrachromosomal accessory nucleoli of the oocyte (Gall, 1968). In the toad *Xenopus laevis,* for instance, the rDNA is increased 20- to 30-fold over that of a somatic cell. The present evidence indicates that amplification originates from chromosomal DNA and is accomplished by the rolling circle process starting with monocistronic circles which double in size with each replication step (Rochaix *et al.,* 1974). This amplification may be related to the large number of ribosomes that must be assembled during oogenesis to provide for protein synthesis during early development.

"Gene compensation" is an increase in the number of copies of a gene on a chromosome when the corresponding locus on the homologous chromosome is missing. In X/O males of X/X_{NO}^- females of *Drosophila,* the single nucleolar organizer (NO) present increases its rDNA multiplicity by 60–70% (Tartof, 1971). A similar phenomenon has been observed for the 5 S ribosomal RNA gene, which in *D. melanogaster* is located in a single euchromatic site. When opposite a deficiency in this region of the homologous chromosome, the number of copies of 5 S genes in the site present is increased from about 165 to 265 (Procunier and Tartof, 1975).

Magnification and reduction first were described for the rDNA in *D. melanogaster* (for references, see Tartof, 1975). It is a heritable increase or decrease of rDNA copies in the germ line. The bobbed (*bb*) mutation is the result of a partial deficiency in the nucleolar organizer of the X chromosome. If this deficient X chromosome is placed with a Y chromosome deficient for rDNA, the number of rDNA copies in the X is increased to the number of the wild type, and concomitantly the *bb* mutation reverts to wild type. The opposite (reduction) cases, in which bb^+ changes to *bb*, also has been observed, but only in flies that also exhibit magnification. The mechanism for magnification reduction appears to be unequal sister strand exchange (Tartof, 1975).

6. *Maintenance of Homogeneity in Repeated Genes*

One characteristic property of multigene families (except antibody genes) is the homogeneity of the repeated sequences. On first inspection, this is surprising, since one would expect mutation to result in differences between the repeats and therefore to induce heterogeneity. It appears that there exists some mechanism that corrects for differences arising from mutation. Several models have been proposed for such a correction mechanism. They are reviewed in Tartof (1975) and in Hood *et al.* (1975). So far, the most satisfactory proposal is the unequal crossing-over model (Smith, 1976). Out-of-register pairing and crossing-over produce both deletions and tandem duplications. This process not only can establish new tandem repeats and eliminate old ones, but it also leads to homogeneity within the repeated segments if crossover is sufficiently frequent relative to mutation.

III. THE 10-NM CHROMATIN FIBER

A. The First Level of DNA Packing

It is characteristic of eukaryote chromosomes that only a small fraction of the DNA is transcribed in a particular nucleus while the bulk of it is silent. This inactivation is related to the compacting of DNA by specific proteins. We shall describe first the nature of this compaction in "inactive chromatin'' and then consider the changes which lead to activation of specific DNA sequences.

It has been known since the pioneering work of Miescher and Kossel almost 100 years ago that the DNA in chromosomes is associated with characteristic basic proteins (protamines and histones). However, the structural nature of this nucleoprotein is becoming understood only now. With recent technical advances and increased interest in the biochemistry of gene action, research on chromatin has become one of the most active fields of molecular biology. Rapid progress is being made in the chemical analysis of the components of chromatin, in the details of their interaction, and in the reconstitution of functional chromatin from its components.

From the late 1950s, electron microscope studies have shown that a fiber about 10 nm thick is a universal component of inactive chromatin. We shall call it the ``10-nm fiber'' (Figure 4). It appears to be the smallest unit obtainable without removing or dislocating the histones. In the next section, we shall provide evidence that, although in the native nucleus this fiber usually is folded or coiled into higher order structures, it becomes unraveled when treated with metal binding agents or after removal of the very lysine-rich histones (H1). Isolation of nuclei or chromatin generally involves chelators (EDTA, EGTA, citrate) or metal-binding buffers (Tris, phosphate buffer). Therefore, chromatin dispersed for biochemical or biophysical studies consists of 10-nm fibers (Ris, 1961). These fibers also are obtained when cells are spread by hypotonic shock on a hypophase containing chelators (Ris, 1967) and seen in thin sections if cells are fixed in the presence of metal binding buffers. Pronase digestion of 10-nm fibers leaves a DNase-sensitive, 2-nm thick fiber, suggesting that it contains a single DNA double helix (Ris, 1967).

Small-angle X-ray scattering analysis of chromatin in solution, either fixed or unfixed, confirmed the electron microscopic observations by indicating the presence of cylindrical structures with an average diameter of 10 nm (Bram and Ris, 1971). More detailed information came from X-ray diffraction studies on partially oriented chromatin gels (Pardon *et al.*, 1967). The 10-nm fiber was found to have a periodic structure with repeat periods of 11, 5.5, 3.5, and 2.7 nm. This regular structure is preserved by fixing in formaldehyde (Pardon *et al.*, 1973) or glutaraldehyde (D. E. Olins and Olins, 1972), but collapses upon dehydration. The X-ray diffraction pattern (see Figure 3 in Richards and Pardon, 1970) is an important indicator for the native configuration of a nucleohistone preparation. It first was thought to result from a supercoil of the DNA–histone complex, but recent studies suggest a different interpretation.

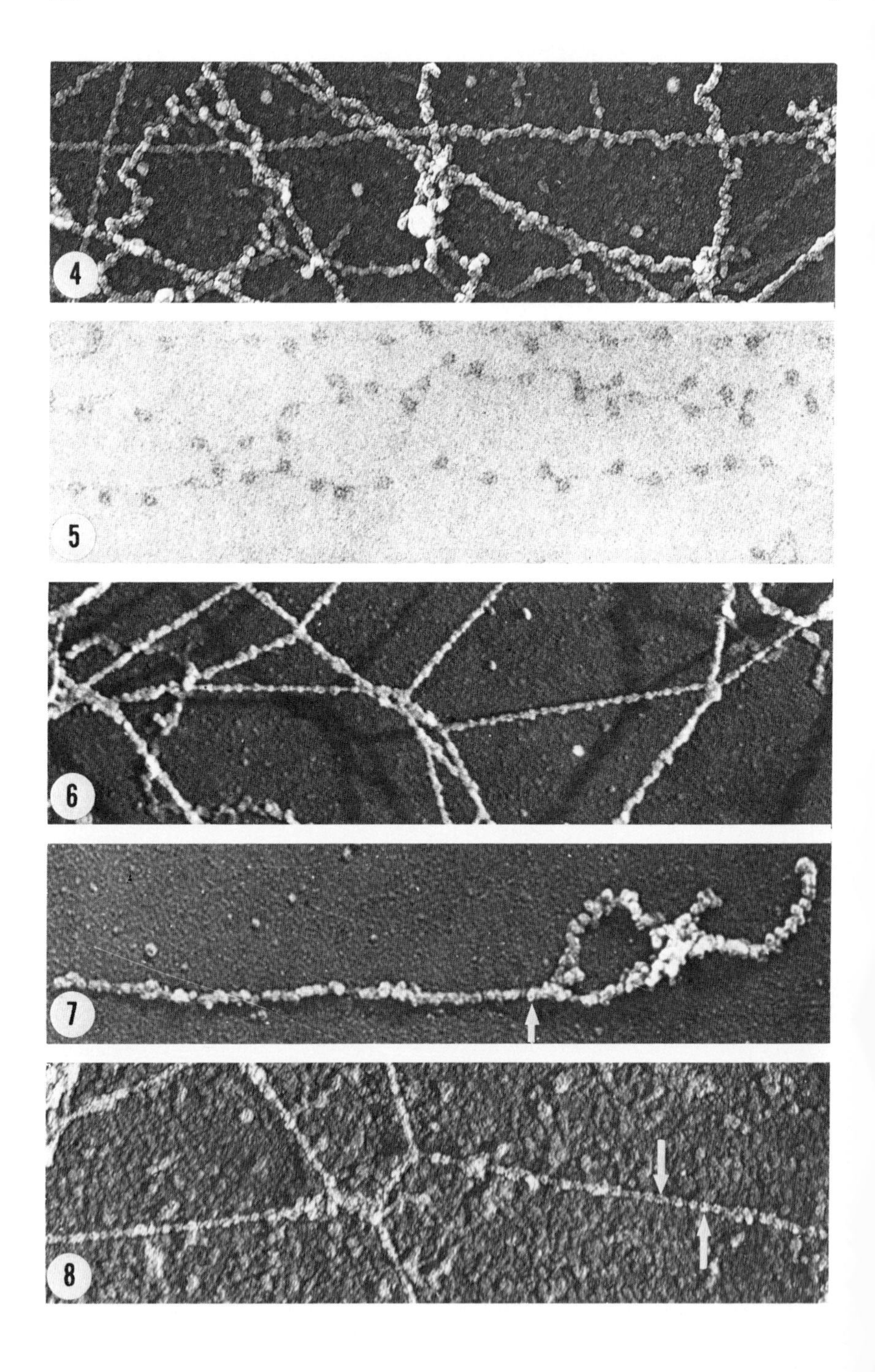
4
5
6
7
8

B. The 10-nm Fiber—A Closely Packed, Flexible String of Beads

Recent work has established conclusively that the periodicity along the elementary fiber represents a tandem array of chromatin subunits 11 nm in diameter. These particles are assemblies of histones strung together by a DNA double helix which is wrapped around each particle. We first shall review the evidence for these chromatin beads and then review the nature of the histone assemblies and their association with DNA.

1. Electron Microscopy

Olins and Olins (1973, 1974; Olins *et al.,* 1975) and independently Woodcock (1973; Woodstock *et al.,* 1976) first showed a regular beaded structure of chromatin in electron micrographs. Olins and Olins called the beads "ν (nu) bodies" and suggested that they were regular histone aggregates (Figure 5). This beaded structure appears in chromatin prepared with a method originally developed to demonstrate actively transcribing chromosome segments (Miller and Beatty, 1969a; Miller and Bakken, 1972). This involves lysing cells with a nonionic detergent at a relatively high pH (8.5–9) and low ionic strength, centrifugation on a grid, and air-drying from the detergent Photo-flo.

This treatment, particularly the high pH and low ionic strength, change the structure of the 10-nm fiber, allowing the DNA connecting the beads to unwind (Figure 6: H. Ris, unpublished observations). In chromatin

---

Fig. 4. Ten-nanometer fibers from erythrocytes of the newt *Notophthalmus* (=*Triturus*) *viridescens* spread on 0.5 m*M* Na citrate. Fixed in 2% formaldehyde, critical point-dried, and shadowed with carbon–platinum. Uniform 10-nm fibers. × 72,000.

Fig. 5. String of chromatin beads streaming from a chicken erythrocyte nucleus ruptured in a hypotonic medium, centrifuged on a grid, air-dried from Photo-Flo, and stained with 5 m*M* uranylacetate. × 230,000. (Photograph by A. L. Olins, Oak Ridge National Laboratory, Oak Ridge, Tennessee.)

Fig. 6. Chromatin fiber from erythrocyte of *Notophthalmus viridescens* spread on water and treated for 10 min at 4°C with 10 m*M* sodium borate, pH 9. Fixed in 2% formaldehyde, critical point-dried, and shadowed with carbon–platinum. At this pH, the 20-nm fiber becomes unwound into a string of beads. × 72,000.

Fig. 7. Chromatin fiber from chicken erythrocyte nucleus lysed in 1 m*M* EDTA, fixed in 1% formaldehyde, centrifuged on grid, critical point-dried, and shadowed with carbon–platinum. Chromatin beads with central hole or depression (arrow). (From H. Ris, 1976, *Electron Microsc. Proc. Eur. Congr. Electron Microsc., 6th 1976*, Vol. II, pp. 21–25. Copyright © 1976 by INTERTAL International Publishing Co. Ltd., Givatayim, Israel.)

Fig. 8. Chromatin fiber from erythrocyte of *Notophthalmus viridescens* spread on water. H1 was extracted with 0.6 *M* NaCl at 4°C for 10 min. Fixation in 2% formaldehyde, critical point-dried, and shadowed with carbon–platinum. Two aspects of chromatin beads are visible: (1) (right arrow) front view with central depression, (2) (left arrow) side view appearing as two lines which may represent the two tetramers (compare with Figure 11). × 72,000.

spread on a hypophase and critical point-dried, beads are not visible, because they normally are closely packed. Beads become clearly visible with these techniques if the fibers are centrifuged (Figure 7), treated with high pH (Figure 4), or treated to remove the very lysine-rich histone (H1), e.g., by 0.6 *M* NaCl (Figure 8). These treatments unwind the DNA which is not bound tightly to the histone beads. In well-preserved chromatin, however, the beads remain as closely packed as they are in the native state (Figure 4) (see also Finch *et al.,* 1975). Although the wide spacing of beads obtained with Miller's method therefore clearly is an artifact, it is an "informative artifact" revealing a substructure which was not recognizable in better preserved chromatin.

Oudet *et al.* (1975) described similar beads in the chromatin from a number of vertebrate tissues and introduced the term "nucleosome." Their preparatory methods included removal of the very lysine-rich histones, exposure to Tris buffer at pH 8 and Triton X-100, and the use of positively charged carbon films. Recently, the beaded structure also was demonstrated in mitotic chromosomes using the procedures listed above (Howze *et al.,* 1976; Rattner *et al.,* 1975). In mouse spermatids, the beaded fiber is replaced by a smooth fiber at the time when histones are replaced by more basic proteins, demonstrating that histones are related to the presence of beads (Kierszenbaum and Tres, 1975).

2. *Stoichiometry of Histones in Chromatin and Specificity of Histone Complexes*

With few exceptions, eukaryote chromosomes contain five types of histones with a total mass about equal to that of the DNA. One of these histones (H1) has a relatively high lysine content (Lys/Arg 22), two of them (H2A, H2B) are moderately lysine-rich (Lys/Arg 1.2 and 2.5), and two are rich in arginine (H3, H4) (Lys/Arg 0.7 and 0.8). Quantitative studies have established an interesting stoichiometry: Histones 2A, 2B, 3, and 4 are present in equimolar concentrations (Johns, 1967; Panyim and Chalkley, 1969; Oliver and Chalkley, 1972; Fambrough *et al.,* 1968) and H1 at exactly half this concentration. In 1974, Kornberg and Thomas suggested that this stoichiometry was not coincidental. Using salt-extracted histones and reversible cross-linking agents, they found that histones in solution form very specific aggregates. The arginine-rich histones H3 and H4 associate as a tetramer $(H3)_2$ $(H4)_2$ and the slightly lysine-rich histones H2A and H2B seem to occur as dimers. When these histone complexes were added to DNA, X-ray patterns characteristic of original chromatin were observed. Based on these observations, Kornberg proposed the following new model for the elementary chromatin fiber. Repeating units composed of two of each of the histones 2A, 2B, 3, and 4 (the four core histones) are wrapped with DNA and joined into a flexible chain with about 200 base pairs of DNA (equal weight to the four histones) per repeat unit. Although H1 is not necessary for this basic structure, one molecule per repeat unit is present in chromatin. This idea of a regular histone complex as the basic repeat unit of chromatin has been confirmed by other investigators. The arrangement of histones in the beads and their interaction with DNA will be discussed in Section III,C in more detail.

3. *Neutron Diffraction*

Neutron diffraction has the great advantage over X-ray diffraction in that the scattering by DNA and by protein can be assessed independently. The scatter from protein or DNA can be "contrast-matched" by immersion in different ratios of D_2O and H_2O (analogous to the invisibility of a glass rod immersed in a medium of the same refractive index). In a solution of about 60% D_2O, the neutron scatter of DNA is matched, and the observed scattering mainly is due to proteins. When the D_2O concentration is around 40%, the observed scatter is mainly from DNA. Using this method, Baldwin *et al.* (1975) showed that the 10-nm periodicity of calf thymus chromatin largely is due to proteins, while the 5-5 and 2-7 nm peaks are caused by DNA. The 10-nm repeat, therefore, corresponds to the spacing of the linearly arranged chromatin beads. Neutron scattering data also give information on the relative location of DNA and protein in the chromatin particles. When scattering by DNA predominates, the calculated radius of gyration is larger than when scattering is due to protein. It follows that the protein is mainly on the inside and the DNA on the outside of the chromatin particle.

4. *DNase Digestion Experiments*

The presence of a regular subunit structure in chromatin also is indicated by the action of deoxyribonucleases (DNase). When pure DNA is digested with such enzymes, random fragments are produced. With chromatin, however, the DNA fragments are of specific lengths, suggesting that DNase-sensitive regions alternate at regular intervals with regions that are protected, presumably through association with histones. This first was shown by Hewish and Burgoyne (1973) by and Burgoyne *et al.* (1974), who made use of the endogenous Ca–Mg activated endonuclease in rat liver nuclei. Isolated nuclei were digested, and the DNA was isolated and analyzed on polyacrylamide gels. Regular bands were produced on the gels, and the relative position of the bands indicated that they represented pieces of DNA that were multiples of a basic unit about 200 bp long. Similar results were obtained by digesting nuclei or isolated chromatin with *Staphylococcus* (*Micrococcus*) nuclease (Noll, 1974a; Sahasrabuddhe and Van Holde, 1974; Sollner-Webb and Felsenfeld,

1975). DNase I (pancreatic) (Noll, 1974b), and DNase II (spleen) (Oosterhof *et al.,* 1975). If chromatin is used, shear must be avoided, since this distorts the native chromatin structure and digestion no longer produces regular DNA fragments (Noll *et al.,* 1975). At least 85% of the DNA from rat liver nuclei exists as multiples of the basic subunit (Noll, 1974a). When digestion is prolonged, more nucleotides become acid soluble and a plateau in the size of DNA pieces about 140 bp long is reached. How these protected and exposed DNA regions fit into the chromatin beads will be discussed next.

5. *Isolation of Chromatin Beads*

Sucrose gradient analysis of chromatin digested with nuclease reveals a particle which sediments at 11–12 S and its multimers (Sahasrabudde and Van Holde, 1974). Further digestion reduces the multimers to monomers. The monomer contains the repeat unit length of DNA (about 200 bp in rat liver) or, after longer digestion, the DNA closely associated with the histone beads, about 140 bp long. These particles contain the same histone fractions as whole chromatin and in the same proportions. Similar particles were isolated after sonication of fixed, water-swollen nuclei (Senior *et al.,* 1975). In electron micrographs, these particles appear identical to the nu bodies and nucleosomes described in lysed nuclei and isolated chromatin. Reassociation analysis of the DNA present in chromatin beads shows that all fractions of DNA, including that transcribed, are associated with histone beads (Lacy and Axel, 1975).

6. *Viral Minichromosomes*

The SV40 virus, which replicates in monkey cell cultures, can be isolated as a nucleohistone complex. Griffith (1975) demonstrated that this complex resembles chromatin fibers in the histone/DNA ratio, the equimolar concentration of histones 2A, 2B, 3, and 4, its appearance in electron micrographs. In 0.15 *M* NaCl, this complex forms a ring with a fiber thickness of about 10 nm and contour length of 210 nm (Figure 9). Since the deproteinized SV40 DNA has a contour length of 1480 nm, the packing ratio in the nucleohistone fiber is about 7 : 1, similar to that in the 10-nm chromatin fiber (Kornberg and Thomas, 1974). Griffith, therefore, called the SV40 nucleohistone complex a "minichromosome." When such minichromosomes are suspended at low ionic strength, the ring expands and appears as a string of 21 beads 10 nm in diameter connected by thin bridges 13.4 nm long (Figure 10). The uniform length of the viral DNA beautifully demonstrates that histone complexes 10 nm in diameter are associated with 200 bp of DNA, and of these, about 30 bp extend at low ionic strength and form bridges between adjacent histone beads.

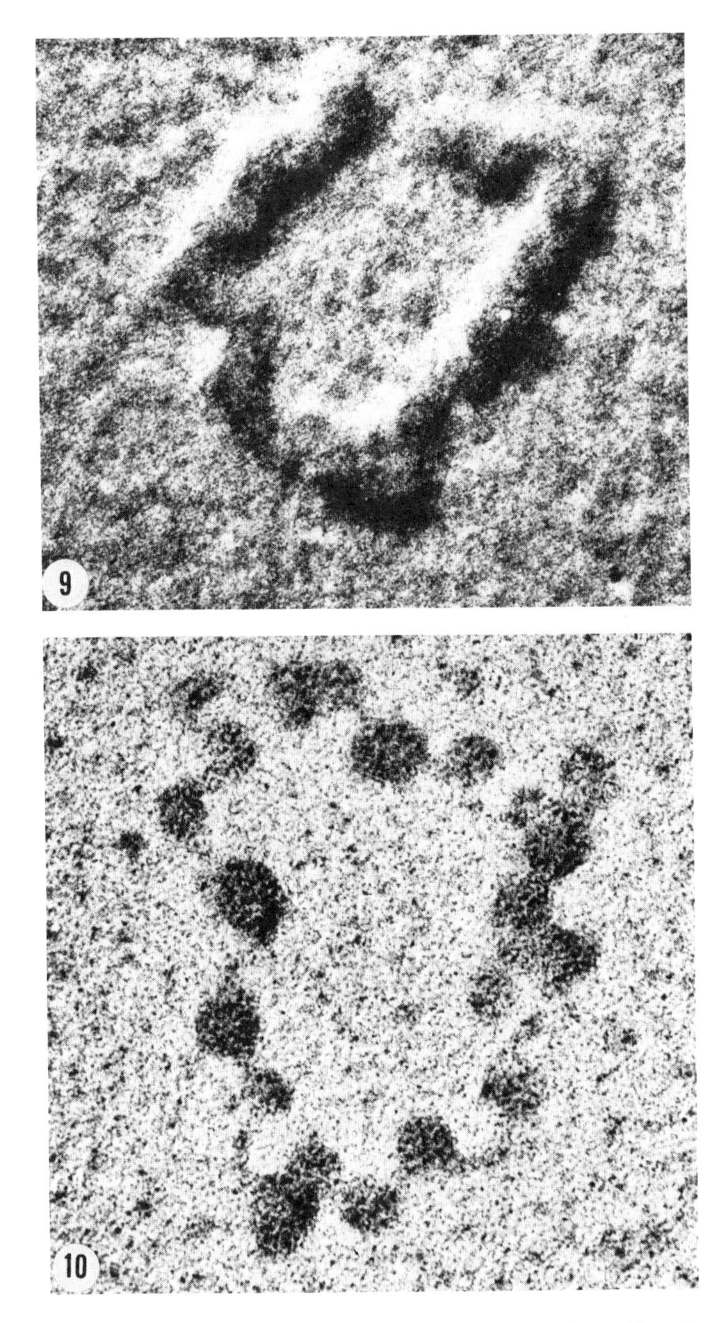

Fig. 9. SV40 minichromosome in 0.15 *M* NaCl. Ring of nucleohistone fiber 11 nm thick and 210 nm long. (From J. D. Griffith, 1975, *Science* **187,** 1202–1203. Copyright 1975 by the American Association for the Advancement of Science.) × 80,000.

Fig. 10. SV40 minichromosome in 0.015 *M* NaCl. Twenty-one beads connected by DNA about 13 nm long have become visible. The beaded string is 510 nm long. After removal of histones, a ring of DNA with a contour length of 1500 nm remains (From J. D. Griffith, 1975, *Science* **187,** 1202–1203. Copyright 1975 by the American Association for the Advancement of Science.) × 80,000.

7. *Reconstruction of Chromatin Beads*

We have seen that there are three methods which reveal a repeated subunit structure in chromatin: (1) electron microscopy, imaging a string of 10-nm beads; (2) X-ray diffraction displaying 11, 5.5, 3.7, 2.7 nm reflections; and (3) brief DNase digestion, producing characteristic DNA fragments. All three of these probes for chromatin structure have demonstrated that when DNA and equimolar amount of the four core histones 2A, 2B, 3, and 4 are mixed in high salt, and the salt concentration is gradually reduced, a nucleohistone complex forms which is identical to the orginal chromatin:

1. Electron microscopy. Oudet *et al.* (1975) dissociated chicken erythrocyte chromatin (H1 depleted) in 2 *M* NaCl. Slow decrease in salt concentration resulted in reassociation into strings of beads as in the original chromatin. Even more interesting was the combination of vertebrate histones with viral DNA's. Calf thymus or chicken erythrocyte core histones added to adenovirus-2 or λ phage DNA resulted in strings of beads with 200 bp of DNA per bead and a sevenfold packing of DNA. Thus, histone beads will pack any DNA, irrespective of origin or base composition, into a typical 10-nm fiber, demonstrating that the *information for the structure formed resides in the histone molecules.*

2. Richards and Pardon (1970) and Kornberg and Thomas (1974) demonstrated that the four core histones were necessary and sufficient to reconstitute chromatin that gives X-ray diffraction patterns typical of chromatin fibers.

3. Axel *et al.* (1974) reconstituted nucleohistone with DNA from various sources (including bacterial and λ phage) and histones from chicken reticulocytes. The same nuclease limit digests were obtained as with original vertebrate chromatin. These studies demonstrate that under appropriate conditions chromatin is a self-assembly system.

C. The Structure of Histones

Knowledge of the structure and behavior of the histone beads is basic to an understanding of the role of histones in the control of gene activity. Since the properties of the beads are determined by the histones and their mutual interactions, we must review briefly what is known about these proteins. (For recent reviews and more details, see Hnilica, 1972; DeLange and Smith, 1975; Elgin and Weintraub, 1975.) Histones are small basic proteins restricted to chromatin of eukaryotes. Five major types have been identified, and these show very little variation during evolution of the eukaryotes. They are coded by gene sequences that are repeated

(400 copies in one species of sea urchin) and the genes for the different histones are clustered into a unit (see Section II). Their transcription takes place primarily during the DNA synthetic period in the cell cycle when new chromatin is being assembled, with very little turnover taking place at other times. The nomenclature used here is that proposed in Ciba Foundation Symposium 28 (Bradbury, 1975a).

1. *Primary Structure*

a. The Very Lysine-Rich Histone, H1. This is the largest histone with a molecular weight of 21,500, about 215 amino acid residues, and a lysine/arginine ratio of 22. It is also the most divergent histone both within and between species. Within one species, several subfractions are present, and different tissues vary in the number and type of subfractions expressed. One subfraction from trout and one from rabbits have been largely sequenced (cf. Elgin and Weintraub, 1975). The distribution of charge is strikingly asymmetric; the basic amino acids occur predominantly at the carboxyl end of the protein; a less basic region occurs near the amino terminal, and the central region contains a high concentration of hydrophobic amino acids. The behavior of H1 differs in many ways from that of the other histones. For example, it is removed most easily from chromatin by weak acids or by 0.5 *M* NaCl (Ohlenbush *et al.*, 1967). Its antigenic groups are more exposed in native chromatin (Goldblatt and Bustin, 1975), it is more accessible to proteases (Weintraub and Van Lente, 1974), and most easily exchanges to added nucleic acids (Ilyin *et al.*, 1971). Its function obviously is quite different from that of the other histones. We shall return to this subject later in this chapter.

b. The Slightly Lysine-Rich Histones, H2A and H2B. These are proteins with molecular weights of 14,000 and a lysine/arginine ratio of 1.17 (2A) and 2.5 (2B), respectively. They have been sequenced completely in several species. H2A contains 129 residues and H2B 125 (see Elgin and Weintraub, 1975; Sautière *et al.*, 1975). They have been highly conserved during evolution. A comparison of the primary structure of H2A in calf, rat, trout, and sea urchin shows only a small number of amino acid changes, most of them conservative; that is, they do not affect charge or secondary structure. This conservatism suggests that the entire molecule is involved in the function of these proteins. There again is an asymmetry in the distribution of basic amino acids. Their highest concentration is at the amino terminal with a smaller cluster at the carboxy terminal. The center is rich in hydrophobic and acidic amino acids.

c. The Arginine-Rich Histones, H3 and H4. These histones have a lysine/arginine ratio of 0.72–0.79, MW 15,000 (H3) and 11,000 (H4) with 135 (H3) and 102 (H4) amino acid residues. They have been sequenced completely for several species (DeLange and Smith, 1975). A comparison of their primary structure shows amazing evolutionary conservation. H3 of calf and pea differ only in 3% of the residues, H4 in 2%. This corresponds to a mutation rate of 0.006 per 100 amino acid residues per 100 million years. Obviously, the complete sequence must be essential for the function of these proteins. As in H2A abd H2B, the basic amino acids are clustered at the amino end with a smaller basic region at the carboxyl end. H3 is the only histone that contains cysteine—one residue in plants, invertebrates, and lower vertebrates, and two residues in mammals.

2. *Secondary Structure of Histones*

Histones are highly charged molecules and show little secondary structure in aqueous solutions at low ionic strength. However, considerable α-helical conformation appears if the charge is neutralized by salt. Helical structure also is induced by DNA, and it must be assumed that in chromatin the core histones contain considerable α-helix (Tuan and Bonner, 1969). This is shown to be true from both theoretical considerations and direct measurements. The location of those residues, which either favor or hinder α-helix formation, allows predictions of which parts of the molecule are likely to be helical and which extended. In practice, the proton magnetic resonance spectra agree well with these theoretical predictions, that is, the arginine-rich histones (3,4) are coiled at the carboxyl half while the basic amino half is extended. The slightly lysine-rich histones (2A,2B) have an extended basic region at the amino terminal, a shorter one at the carboxyl terminal, and a helical central region (reviewed in Bradbury, 1975b). H1 shows a large coiled segment in the amino-terminal half and a long extended basic region at the carboxyl-terminal region (Chapman *et al.,* 1976). This secondary structure of histones most likely determines important aspects of their function; it has been shown that the extended basic regions bind to DNA and the coiled hydrophobic segments interact with other histones or with nonhistone proteins.

3. *Tissue-Specific Histones*

H1 generally occurs as several subfractions, and variations have been observed during development and in adult tissues (references in Elgin and Weintraub, 1975; Areci *et al.,* 1976). Improved methods of separation recently have demonstrated variation in H2A and H2B during development and adult tissue cells (Blankstein and Levy, 1976; Cohen *et al.,* 1975). Tissue-specific histones are present in cell types in which chromatin is inactivated completely and highly condensed, e.g., erythrocytes of vertebrates (except of mammals, which are enucleate) and many spermatozoa. Best known are the lysine-rich histones H5, which partially replace H1 in chicken erythrocytes and sea urchin sperm (reference in Hnilica, 1972). Replacement of the typical histones by more basic (arginine-rich) proteins is common in the sperm of many animals. In mammalian sperm, an arginine-rich, keratinlike protein replaces the somatic histones during spermiogenesis (Coelingh *et al.,* 1972). In many fish species, histones are replaced during sperm maturation by protamines. These are arginine-rich polypeptides with a molecular weight of about 5000. Protamines from several fish species have been sequenced completely (reference in Hnilica, 1972).

4. Histone Modification

Despite the small number of histone types and the extreme conservation of their primary structure, histones do vary according to their side-chain modifications. These postsynthetic changes involve acetylation and methylation at the 6-amino group in lysines and O-phosphorylation and N-phosphorylation of serine and threonine (reviewed in Dixon *et al.,* 1975).

These alterations can change the charge, size, and shape of the histones and may specifically change their interaction with each other and with DNA. That these modifications are not trivial is indicated by their specificity. They affect only a fraction of histones in a particular cell and occur at specific sites which differ with cell type and functional state. Enzymes which either add or remove acetylmethyl or phosphoryl groups are highly specific; some are activated by cyclic AMP, while others are not. The modifications also show correlations with stage in the cell cycle, or specific functions such as gene activation or replication. Three examples shall be mentioned which suggest that histone modifications may be specific controlling factors in chromosome functions.

a. Replacement of Histones with Protamine in Trout Testis. During sperm maturation, histones are replaced by protamine. Candido and Dixon (reference in Dixon *et al.,* 1975) found that the histones, before they leave the DNA, become heavily acetylated in their basic halves, i.e., the part linked to DNA. Since acetylation of lysine reduced the positive charge, this process may facilitate histone removal. The protamines become phosphorylated after they enter the nucleus. As the chromatin condenses, dephosphorylation of the protamines takes place, which leads to increased charge and binding strength to DNA. These protein modifications may facilitate the ordered replacement of histones by protamines and the concomitant condensation of DNA in the sperm nucleus.

b. Assembly of the New 10-nm Elementary Fiber during DNA Replication. The fate of histone beads during DNA replication is still unresolved. Whether they remain with the template strand or are distributed between old and new strands remains to be determined; but, in any case, new beads must be assembled in the process. Histones are synthesized on cytoplasmic ribosomes and then moved to the assembly sites in the nucleus. During this process, interesting modifications take place (Ruiz-Carillo *et al.*, 1975). After synthesis, H4 molecules are acetylated at a lysine, and the N-terminal serine residue is both acetylated and phosphorylated. After the histone enters the nucleus, specific enzymes remove the acetyl from the lysine and the phosphate from the terminal serine. Later, acetylation of four specific lysines takes place. The phosphorylation occurs only on the newly synthesized H4 molecules. These modifications controlling charge and conformation of the histones may facilitate the assembly process of new chromatin beads.

c. Phosphorylation of H1 during the Cell Cycle. Extensive phosphorylation of H1 has been noted at specific times of the cell cycle. In synchronized mammalian cells, up to 85% of H1 becomes phosphorylated at from one to four sites just prior to S phase with a peak during S. All the phosphorylated sites are in the basic carboxy-terminal half, and these modifications which lower the overall charge could control the interaction of H1 with DNA. Bradbury and his group (1975b) have studied H1 phosphorylation in the highly synchronous divisions of the slime mold *Physarum polycephalum*. They found a peak in G_2, just before chromosome condensation and suggested that this H1 modification is related to the mitotic condensation of chromosomes.

D. The Detailed Structure of Chromatin Beads

1. *Bead and Tail (Spacer) of the Chromatin Subunit*

Many observations indicate that the role of the very lysine-rich histones H1 and H5 is quite different from that of the other histones. These histones are bound more weakly to chromatin and show greater evolutionary diversity, and many subfractions exist within a species. Their antigenic groups are more exposed than those of other histones in chromatin (Goldblatt and Bustin, 1975). They are not required for the basic subunit structure of chromatin as revealed by the methods described in Section II,B. Cross-linking experiments show that in chromatin they are close to each other, but not to the other histones (Bonner and Pollard, 1975; Hardison *et al.*, 1975). Digestion of DNA with micrococcal DNase distinguishes between a more easily accessible region of about 40–60 bp and a more resistant segment of 140 bp that is closely bound to the four bead histones. H1 appears to be associated with the less protected 40–60 bp DNA which connects adjacent beads. If H1 is removed from chromatin with 0.6 *M* NaCl (which also removes some non-histone proteins), this

DNA is digested five times more rapidly than in the original chromatin (Whitlock and Simpson, 1976). After brief nuclease digestion, two different monomer beads can be distinguished on sucrose gradients. One carries H1 and contains 200 bp of DNA, while the other is free of H1 and contains about 170 bp. Similarly, three kinds of dimers were observed, containing either one, two, or no H1 molecules (Olins *et al.*, 1976; Varshavsky *et al.*, 1976).

The presence or absence of H1 profoundly affects the appearance of the elementary fiber as seen in electron micrographs. Oudet *et al.* (1975) published electron micrographs of H1-depleted chromatin from various sources. They found that histone beads were separated by variable lengths of naked DNA. Figure 8 shows newt erythrocyte chromatin extracted with 0.6 *M* NaCl to remove H1: The beads are separated by thin strands about 15 nm long. In the presence of H1, the beads are packed closely even at low ionic strength, and no connecting bridges are visible (see Figure 4). Thus, a chromatin subunit consists of about 140 bp of DNA tightly bound to the bead histones 2A, 2B, 3, and 4 and a less protected segment which is associated with H1.

2. *The Core Histone Particle*

We now must consider how the bead (or core) histones are arranged in the particle and how they interact with DNA. Although this problem has not yet been solved in detail, enough now is known to build a reasonable model.

Hydrodynamic studies (Olins *et al.*, 1976) and electron microscopy (Olins *et al.*, 1977) suggest that the particle is nearly spherical. The molecular weight for histones of about 100,000 (Shaw *et al.*, 1976; Olins *et al.*, 1976) and the equimolar ratio of histones require the presence of two molecules of each core histone. The sharp peaks on sucrose gradients for the monomer particles, the conserved primary structure of the histones, and the antibody precipitation studies (Simpson and Bustin, 1976) support the view that these particles are quite uniform. The particles are highly hydrated (analogous to ribosomes) with a diameter of 11.0 nm calculated from hydrodynamic studies (Olins *et al.*, 1976), in agreement with X-ray data. The corresponding anhydrous sphere would have a diameter of about 8 nm, close to the values reported from electron microscopy (Olins *et al.*, 1976).

The secondary structure of the histones predicts that the hydrophobic, coiled C-terminal halves are involved in the histone–histone binding and are located inside the particle while the extended basic amino-terminal regions stick out from the surface to bind the DNA. This arrangement is supported by exhaustive trypsin digestion, which removes only 20–30 amino acids from the histones, but leaves the hydrophobic core intact (Weintraub and Van Lente, 1974). ^{13}C nuclear magnetic resonance studies (Bradbury and Crane-Robinson, 1971; Lilley *et al.,* 1976) of H3–H4 tetramers provide further evidence for such random coil N-terminal tails projecting from histone aggregates.

The relative position of histones within the complex has been analyzed with cross-linking agents (reviewed in Hardison *et al.,* 1975). From these studies, it follows that in solution there is strong pairwise association between the "homotypic" (Weintraub *et al.,* 1975) pairs 2A and 2B and 3 and 4 (D'Anna and Isenberg, 1974) and the heterotypic dimer 2B–4. Cross-linking of 2B and 4 also has been observed in native chromatin (Martinson and McCarthy, 1975; Van Lente *et al.,* 1975). Thomas and Kornberg (1975) observed an octamer in solution at ionic strength 2 and pH 9 (conditions of diminished charge) with properties similar to native particles. Weintraub *et al.* (1975) described a "heterotypic tetramer" (2A, 2B–3, 4) in 2 *M* NaCl. This tetramer reacted to trypsin digestion like native chromatin, suggesting that the native histone bead consists of two heterotypic tetramers (Weintraub *et al.,* 1976). A very attractive model with interesting implications for chromatin functions was developed on this concept (Worcel and Benyajati, 1977). The transition from open string of beads to 10-nm fiber is brought about by supercoiling of the spacer DNA between the beads. This brings adjacent beads into contact so that tetramer-to-tetramer bonding can occur between the beads analogous to the bonding within each bead. The result is a stable stack of beads which is stabilized further by H1 molecules that link adjacent beads (Figure 11).

3. *Interaction of the Histone Particle with DNA*

Reconstitution experiments have shown that DNA is not required for the proper association of histones into beads. The information for their assembly resides in the histones themselves. According to the major

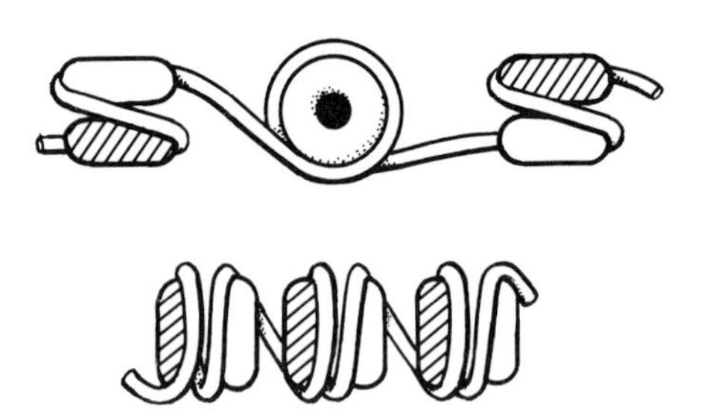

Fig. 11. Diagram illustrating a possible arrangement of the histone beads (pairs of tetramers of the histones 2A, 2B, 3, 4) in the open string of beads form (H1 removed) and in the closely packed form (10-nm fiber). (After Worcel and Benyajati, 1977.)

probes for native chromatin structure (X-ray diffraction, neutron diffraction, electron microscopy, DNase digestion, trypsin digestion), all four bead histones are required to organize DNA into native chromatin, although H3 and H4 play the major role in this process (Sollner-Webb *et al.*, 1976). The base sequence of DNA is irrelevant for the formation of chromatin beads; any nucleotide sequence (Polisky and McCarthy, 1975) or even poly(dG–dC) poly(G–dC) (Sollner-Webb *et al.*, 1976) is compacted properly with no regard to specific sequences.

The closed circular DNA of SV40 provided some important information on how DNA is wrapped around the histone beads (Germond *et al.*, 1975). These authors analyzed the relationship of histone beads to the number of supercoils on the DNA ring. This number can be determined by agarose polyacrylamide gel electrophoresis. In a series of experiments, it was shown that each histone bead is associated with a left-handed supercoil on the DNA.

Neutron diffraction indicates a radius of gyration of 50 Å for the DNA (Pardon *et al.*, 1975), which corresponds to a circle of about 95 bp. The four basic "fingers" on each tetramer protect 45 additional bp (Weintraub *et al.*, 1976), which add up to the observed 140 bp protected from nuclease digestion by each histone particle (Sollner-Webb and Felsenfeld, 1975; Axel, 1975; Shaw *et al.*, 1976).

The use of DNases which can identify protected and exposed regions of the DNA has provided more detailed insight into the packing of DNA in the beads. Clark and Felsenfeld (1971) first showed that about half of the DNA in chromatin is protected against DNase attack. We know now that this is the DNA intimately coiled around each histone bead. Digestion with DNase I (pancreatic) has dissected this DNA further. It produces single-strand pieces which are multiples of 10 bp (Noll, 1974b). This is interesting, because in the B form of DNA 10 bp corresponds to one turn of the double helix. Perhaps each basic finger of the core histones wraps around one turn of the helix, protecting it against DNase action. By labeling the 5′-ends of the monomer DNA with ^{32}P, Simpson and Whitlock (1976) were able to map the DNase-sensitive sites with respect to fixed reference points (the 5′-ends). They found sites 20, 40, 50, 100, 120, 130 nucleotides from the 5′-end most susceptible, while site 80 never was cleaved. Thus, the susceptible sites appear clustered near the ends of the DNA loop, where the basic "fingers" of the histones make contact with the DNA (Simpson, 1976).

Recently, crystals have been prepared from purified chromatin beads. This opens exciting new possibilities for the analysis of their molecular organization. A preliminary low resolution X-ray diffraction study has been published (Finch *et al.*, 1977). It indicates that chromatin beads are

TABLE II

Interspecies and Tissue Variation in DNA Repeat Length

Cell type	Repeat length (bp)	Length of core DNA (bp)	References
Saccharomyces cerevisiae (yeast)	161	142	Lohr *et al.*, 1977
Neurospora crassa	170	140	Noll, 1976
Aspergillus nidulans	154	—	Morris, 1976a
Physarum polycephalum	171	—	Compton *et al.*, 1976
Sea urchin, sperm	241	140	Spadafora *et al.*, 1976
Sea urchin, gastrula	218	140	Spadafora *et al.*, 1976
Chicken erythrocyte	211	—	Morris, 1976b
Chicken erythrocyte	197	139	Lohr *et al.*, 1977
Chicken liver	202	—	Morris, 1976b
Chicken oviduct	196	—	Compton *et al.*, 1976
Rat bone marrow	192	—	Compton *et al.*, 1976
Rat kidney	196	—	Compton *et al.*, 1976
Rat liver	200	—	Morris, 1976b
Rat liver	198	140	Noll, 1976
Rat liver	196	—	Compton *et al.*, 1976
CHO cell line (Chinese hamster)	178	—	Compton *et al.*, 1976
HeLa cell line	188	—	Compton *et al.*, 1976
HeLa cell line	182	135	Lohr *et al.*, 1977

disk-shaped, 11 nm in diameter, and 5.7 nm high. Electron micrographs of negatively stained preparations in early stages of crystallization show wavy columns of stacked bipartite particles. The DNA seems to be arranged in a helix of about $1\frac{3}{4}$ turns with a pitch of 28 Å and an average diameter of 90 Å around the rim of the particle. As yet, no details are available on the exact arrangement of the histones in the particle. Electron micrographs of chromatin beads prepared in a variety of ways (negative staining, carbon-platinum shadowing, and dark-field EM) consistently show a central hole or depression as illustrated in Figure 8 (see also Finch *et al.*, 1977; Langmore and Wooley, 1975).

4. Variation in Length of the DNA Repeat Unit

Recently, more accurate measurements on the length of the DNA fragments after minimal (micrococcal) DNase digestion have shown that the basic repeat unit can vary in different tissues of one organism and between organisms (Table II). These differences are due to variations in the "spacer" which is associated with H1. On the other hand, the 140 bp piece combined with the core particle is the same in all cases.

These variations most likely are related to differences in H1. It may be significant that the longest DNA subunits are found in sea urchin sperm and chicken erythrocytes, which contain the very lysine-rich H5 in addition to H1 and are characterized by highly condensed and transcriptionally inactive chromatin.

IV. THE 20-NM CHROMATIN FIBER

A. The Unit of Inactive Chromatin in the Intact Cell

In most cells, the 10-nm fiber is compacted further into a fiber 20–30 nm thick. This structure first was described in water-extracted chromatin from sea urchin sperm (Bernstein and Mazia, 1953) and later in squashes and thin sections of many plant and animal nuclei (Ris, 1956). The widespread occurrence of this chromatin unit became recognized with the introduction of the surface-spreading method by Gall (1963b). In this procedure, cells and nuclei are lysed by osmotic shock on a clean water surface. Proteins form a monolayer on the surface from which dispersed chromatin and other cell components are suspended into the hypophase. This surface film can be picked up by touching a Formvar–carbon-filmed grid to it. After fixation in aldehydes, the material may be negatively stained or dried with the critical point method. Interphase chromatin and mitotic or meiotic chromosomes prepared in this way from a large variety of plant and animal cells were found to consist mainly of 20–30 nm fibers (Figure 12; for references, see Wolfe, 1969; Ris and Kubai, 1970; Davies and Haynes, 1976). Similar thick fibers are found in condensed chromatin after freeze-etching (Ris, 1969; see Figure 6). These results were questioned, because in thin sections the chromatin units were frequently 10 nm thick, and it was suggested that the fibers spread on water were thicker because of contamination with cell proteins (Wolfe and Grim, 1967). The relationship between these thin and thick fibers was resolved when the effect of cations on fiber thickness was recognized.

B. The Effect of Cations on Chromatin Fibers

The nucleus contains a relatively high concentration of cations, much of it bound to chromatin (Langendorf *et al.*, 1961; Clark and Ackerman, 1971; Steffensen, 1961). According to Naora *et al.* (1961), Mg^{2+} is largely bound to DNA and Ca^{2+} to chromosomal proteins. Using ^{45}Ca, Steffensen and Bergeron (1959) found that Ca^{2+} remains bound to chromosomes through the cell cycle. Removing cations from chromosomes with chelat-

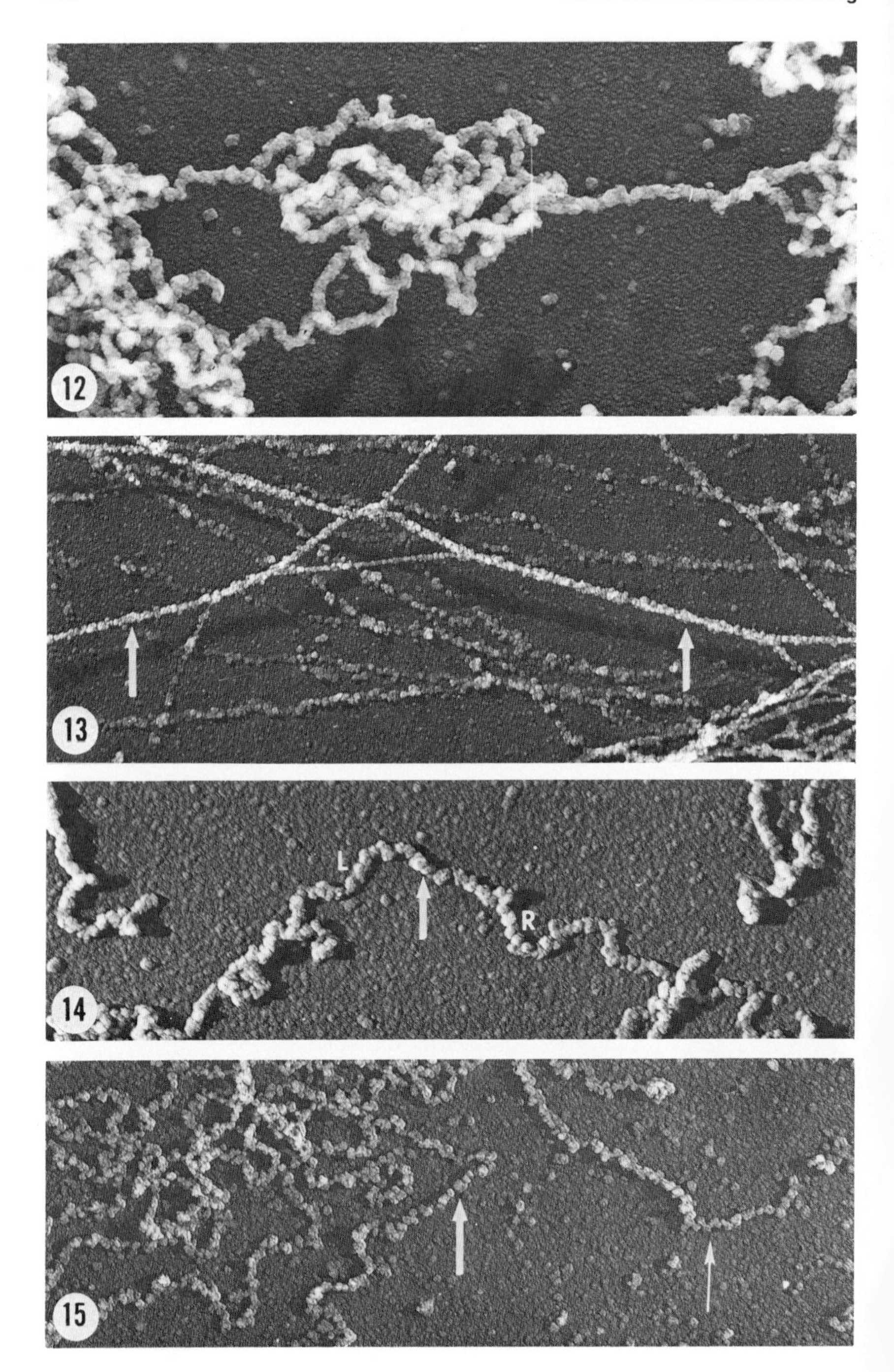
12
13
14
L
R
15

ing agents (e.g., EDTA) causes them to disperse and lose structure as seen by light microscopy (Mazia, 1954). On the electron microscope level, Ris (1967) demonstrated that chelating agents unwind the 20-nm to give 10-nm fibers. Since the buffers generally used in fixatives and in nuclear isolating media (Veronal–acetate, phosphate, cacodylate, Tris–HCl) are metal binding, it became clear that the 10-nm fibers seen in sections and isolated nuclei were produced by the buffers used and did not represent the native state (Ris, 1968). Where fixatives with nonmetal binding buffers (Good *et al.*, 1966) or unbuffered formaldehyde are used, the fibers in sections are generally 20–30 nm thick. Calcium and magnesium therefore seem to be necessary for the maintenance of the 20–30-nm fiber. Potassium pyroantimonate has been used for localization of cations with the electron microscope (Clark and Ackerman, 1971; Simson and Spicer, 1975). When 20-nm chromatin fibers are exposed to this reagent, a fine precipitate appears along the fiber. At the same time, the fiber width is reduced to 10 nm since the reaction removes the cation from the fiber. No precipitate ever forms over 10-nm fibers (Ris, 1975b).

Several authors have studied the effect of ions on fiber thickness in sections of isolated nuclei (Brasch *et al.*, 1971; D. E. Olins and Olins, 1972). At physiological concentrations (0.1–0.2 *M* NaCl or KCl), the fibers are about 20 nm thick. At low ionic strength (distilled water or 0.02 *M* KCl), the diameter is 8–10 nm. It must be remembered that the nuclear isolating media used in these studies contained chelating agents, and these authors in fact described the effect of monovalent cation concentration on 10-nm fibers! Chromatin that was not exposed to chelators remains as 20-nm fibers even in distilled water, and the fiber thickness is reduced only after treatment with chelating agents. Pooley *et al.* (1974) studied the effect of ions on calf thymus chromatin isolated in the presence of EDTA.

Fig. 12. Twenty-nanometer chromatin fibers from erythrocytes of *Notophthalmus viridescens* spread on water, fixed in 2% formaldehyde, critical point-dried, and carbon–platinum shadowed. × 72,000

Fig. 13. Chromatin of erythrocytes from *Notophthalmus viridescens* was spread on 5 m*M* Na citrate (10-nm fibers), picked up on a Formvar–carbon-coated grid, and floated on 1 m*M* $CaCl_2$. The 10-nm fiber coils up into a 20-nm fiber (arrow). × 72,000.

Fig. 14. Twenty-nanometer chromatin fiber from erythrocyte of *Notophthalmus viridescens* spread on water, breifly treated with 5 m*M* citrate, fixed in 2% formaldehyde, critical point-dried, and shadowed with carbon–platinum. A left-handed coil (L) turns into a right-handed coil (R) at arrow, suggesting that the direction of coiling along the 20-nm fiber might be random. × 72,000.

Fig. 15. Twenty-nanometer chromatin fibers from erythrocytes of *Notophthalmus viridescens,* spread on water, fixed in 2% formaldehyde, treated with 24 m*M* EDTA in 140 m*M* NaCl (pH 6.4), and fixed in 1% glutaraldehyde in 1 m*M* Pipes buffer. In addition to a coiled 10-nm fiber (thin arrow), one sees "super beads" (thick arrow; see also Hozier *et al.*, 1977). × 40,000.

At low ionic strength, the fibers were 11 nm thick. In the presence of 1 m*M* Mg^{2+} or Ca^{2+}, the fibers became 25 nm thick. Na^+ at 20–50 m*M* concentration had the same effect. Na-treated fibers became reduced in thickness in distilled water, but fibers with divalent cations required EDTA to return them to 10 nm.

This reversible effect of cations also can be demonstrated with chromatin spread on a hypophase. When the 10-nm fibers (spread on 5 m*M* Na citrate) are treated with 0.5–1 m*M* $CaCl_2$, the fibers become 25–30 nm thick (Figure 13). These studies show that, in the presence of cations, especially Ca^{2+} and Mg^{2+}, the 10-nm fiber is compacted into a 20–30-nm fiber. To preserve this native fiber during isolation or during fixation, metal binding agents must be avoided.

C. The 20-nm Fiber—A Helical Coil of the 10-nm Fiber

Several models have been proposed for the structure of the 20–30-nm fiber. Water-spread chromatin fibers often show side branches which were shown to be foldbacks of the 10-nm fiber on itself (Ris, 1969). However, it is unlikely that such foldbacks are responsible for long stretches of 20-nm fibers. DuPraw (1965), Gall (1966), and Lampert (1971) have suggested, without presenting convincing evidence, that the 20-nm fiber is a helical supercoil of the DNA–histone fiber. Two recent publications (Filip *et al.*, 1975; Finch and Klug, 1976) provide micrographs compatible with a helical coil, but other explanations are still possible. More convincing pictures can be obtained by exposing newt erythrocyte chromatin spread on water to a brief treatment with chelating agents (H. Ris, unpublished observations). Regions which are unraveled to various extents can be recognized (see Figure 16). In some areas, the direction of the gyres can be determined. In Figure 14, one can recognize both right-handed (R) and left-handed (L) coils and regions where the direction changes (Figure 14 arrow). The coils are preserved especially well in negatively stained preparations (Figure 17). The native 20-nm fiber thus appears to be a supercoil of the elementary fiber with the direction of coiling apparently random. Cations (especially Ca^{2+} and Mg^{2+}) play a role in the initiation and maintenance of the supercoil. Since the compaction of DNA in the elementary fiber is about 7 to 1, the tight coiling into the 20-nm fiber leads to a final packing of DNA of about 40 to 1 in this second-order chromatin fiber, which represents the unit of inactive chromatin in most cells.

A different structure for the 20-nm fiber was proposed by Kiryanov *et al.* (1976) and by Hozier *et al.* (1977). They found that under certain conditions the 20-nm fiber appears as a chain of 20-nm particles ("super beads"). These can be isolated on sucrose gradients as 40 S particles after brief micrococcal nuclease digestion. The beaded appearance is not necessarily in conflict with a helical structure. A periodic discontinuity in the 10-nm beaded string could produce a flexible joint at two gyre intervals throughout the helix. This would give the appearance of "super beads" and allow for greater flexibility in the folding of the 20-nm fiber (see Worcel and Benyajati, 1977). Figure 15 lends some support to this interpretation. The same material which shows the helical organization (Figure 14 and 16) appears as a string of 20-nm beads if treated with EDTA *after* formaldehyde fixation, suggesting a different binding at two gyre intervals.

D. The Role of the Very Lysine-Rich Histones (H1, H5) in Maintaining the 20-nm Fiber

In the previous section, we pointed out that the very lysine-rich histone, H1, is located between the beads and that cross-linking agents bind an H1 to another H1. This suggests that H1 could be a factor in folding the elementary fiber into a coil. There now is some evidence that removal of H1 does indeed unravel the native fiber. Brasch *et al.* (1972) and Brasch (1976) studied the structure of condensed chromatin in chicken erythrocytes and isolated liver nuclei after sequential extraction of histone types. H1 and H5 were extracted selectively, either with 0.1 *M* critic acid or 0.65 *M* NaCl. These treatments changed the fiber width in condensed chromatin from 20 to 10 nm, indicating that H1 may play a role in organizing the 20-nm fiber. Thin sections through condensed chromatin are not the best way to demonstrate such changes in fiber thickness, however, because individual fibers are not resolved easily. Clearer pictures are obtained with erythrocyte chromatin spread on water (H. Ris, unpublished). Figure 8 shows what happens to the 20-nm fibers when extracted with 0.6 *M* NaCl (which removes H1 but no other histones). Not only are the native fibers unraveled, but the beads become clearly visible and separated by a 15-nm thread, presumably DNA. Both H1 and cations thus are involved in coiling the 10-nm fiber into the 20-nm unit. This begins to make sense with the demonstration that H1 binds cations and is responsible for the condensation and precipitation of chromatin in the presence of cations (Billett and Barry, 1974; Chapman *et al.,* 1976; Olins *et al.,* 1976; Varsharsky *et al.,* 1976). Worcel and Benyajati (1977) have constructed a space-filling model which incorporates the existing chemical and structural observations. It is based on four major assumptions: (1) symmetrical chromatin beads containing 140 bp DNA; (2) intra- and interbead DNA exists as uniform left-handed supercoil with 90 bp per turn and 4.7 nm average pitch; (3) basic symmetry principles of any helix; and (4) inter-

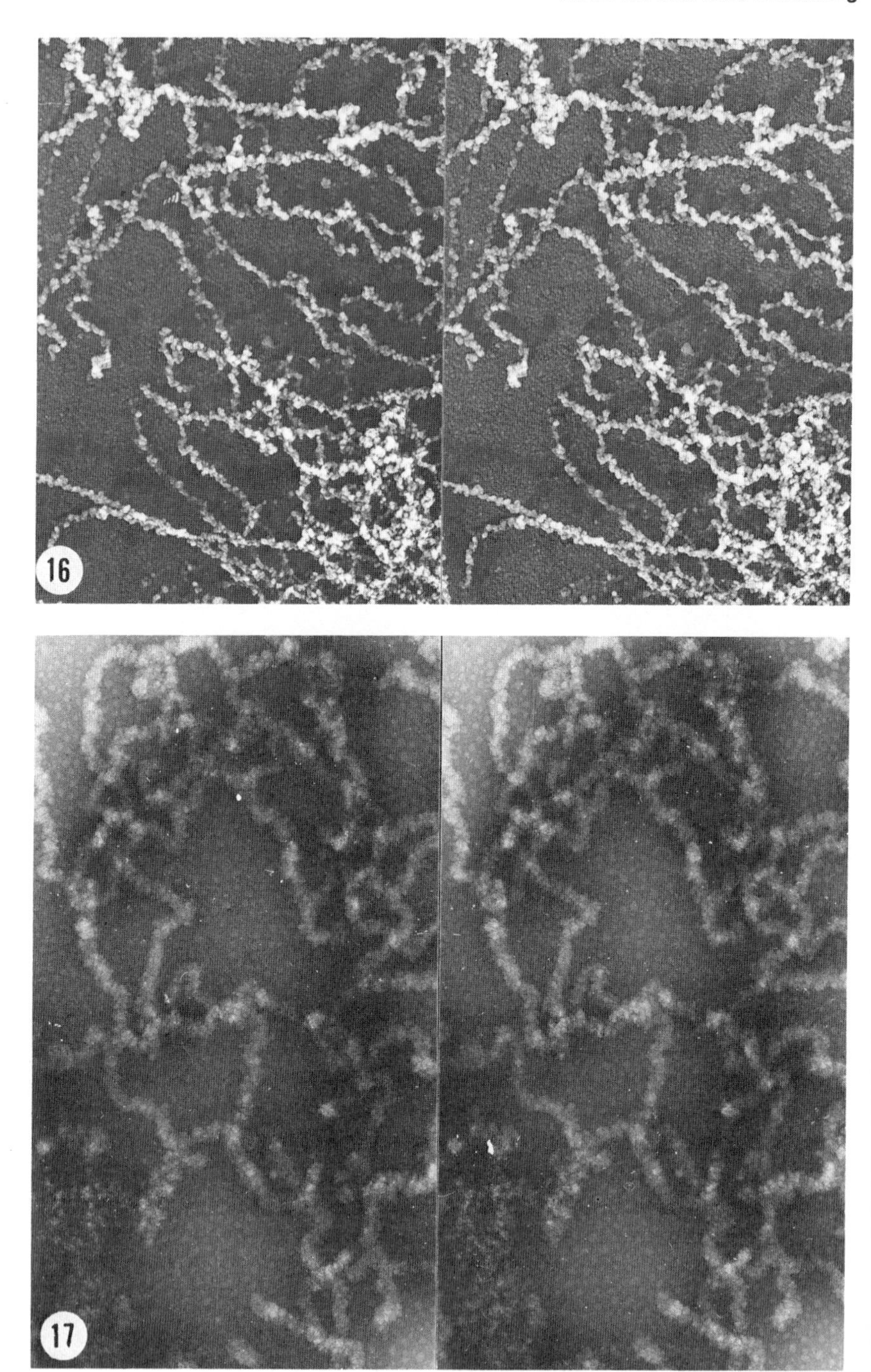
16
17

bead DNA faces the outside of the helix (Figure 18). The helix is stabilized by the interactions between the beads from one gyre to the next and cation binding by H1, which bridges the gap between beads.

E. Exceptions

In *Neurospora crassa,* chromatin consists of 10-nm fibers both in intact nuclei and after spreading on water (H. Ris, unpublished observations). H1 is present in this species but has a low lysine/arginine ratio compared with other H1. It is possible that the unusual H1 in this species is responsible both for the short DNA repeat unit seen by DNase digestion (Noll, 1976) and for the failure of the elementary fiber to coil into a 20-nm fiber.

Johmann and Gorovsky (1976) have reported that H1 is missing from the micronucleus of *Tetrahymena* while present in the macronucleus. It will be interesting to see how this affects chromatin structure.

V. ACTIVE CHROMATIN

A. The Active Fraction of Chromation

The two major functions of chromosomes are replication and transcription. We define active chromatin as those segments of chromosomes that are being transcribed. A chromosome region may be inactive because the DNA is never transcribed (such as simple sequence, highly repeated DNA), because it is not expressed in a particular differentiated cell type (such as globin genes in a hepatic cell), or because it is inactive at a certain physiological state of a cell or part of the cell cycle (probably all genes at mitotic metaphase). Genes which are expressed in a cell type but are not transcribed at a particular time will be called "potentially active." We shall see later that there are some interesting differences between this potentially active chromatin and regions never expressed in a specific cell type. Even where a cell is fully functional, only a small part of its DNA is

---

Fig. 16. Stereomicrograph of 20-nm chromatin fibers from erythrocytes of *Notophthalmus viridescens* spread on water and briefly treated with Na citrate, fixed in 2% formaldehyde, critical point-dried, and shadowed with carbon–platinum. The coil of the 20-nm fiber is partially unwound after this treatment. × 50,000.

Fig. 17. Stereomicrograph of 20-nm chromatin fibers from erythrocytes of *Notophthalmus viridescens,* spread on water, briefly treated with Na citrate, and fixed in 2% formaldehyde. The fibers were embedded in a thick layer (about 100 nm) of 2% uranylacetate (deep negative stain) and photographed at 1 MeV with the High Voltage Electron Microscope Facility at the University of Wisconsin, Madison. Twenty-nanometer fiber is a helical coil. × 80,000.

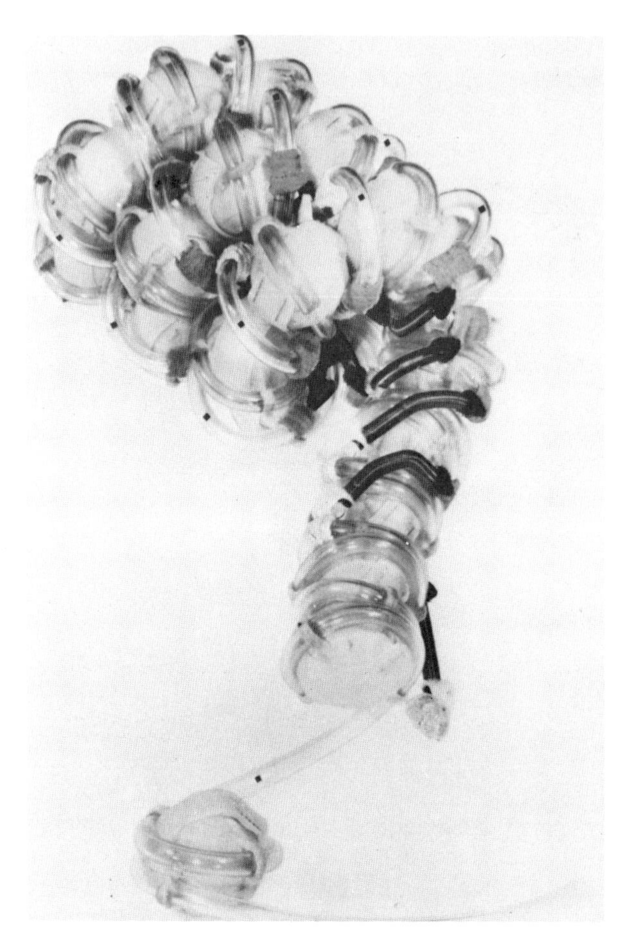

Fig. 18. Scale model illustrating the arrangement of chromatin beads in the 20-nm fiber. The dark structures bridging adjacent beads represent H1. (Photo by A. Worcel, Princeton University.)

being transcribed. This has been determined by measuring the reassociation kinetics of total nuclear RNA of a tissue against single-stranded DNA of the species (see McCarthy *et al.,* 1973). In most eukaryotic cells, the transcribed RNA corresponds to only 2–15% of the single-stranded DNA. Since generally only one strand is transcribed, it means that between 4 and 30% of the DNA in a particular cell is transcribed. In contrast, practically all the DNA in bacteria is transcribed.

B. Transcription and Chromosome Decondensation

In most eukaryote nuclei, a large fraction of chromatin is highly condensed and, in general, localized along the nuclear membrane or around the nucleolus (see Section VIII on heterochromatin). A variable proportion of total chromatin is unraveled and dispersed. Autoradiography after incubation with labeled RNA precursors has demonstrated that RNA synthesis occurs only in the dispersed chromatin fraction (Littau *et al.*, 1964; Granboulan and Granboulan, 1965; Noorduyn and de Man, 1966). In nuclei in which all chromatin is highly condensed (mature erythrocyte nuclei of vertebrates, sperm nuclei), no RNA synthesis occurs. Stimulation of transcription in such nuclei is associated with swelling and dispersal of chromatin (Ringertz *et al.*, 1972). However, it must be emphasized here that *not all dispersed chromatin* (*or euchromatin*) *is in an active state*, and it therefore is wrong to equate euchromatin with active chromatin. Decondensation is a precondition for activation, but further structural changes are required to allow transcription, as will be illustrated later.

Usually, individual chromosomes are not visible when transcription takes place. Therefore, the exceptions are particularly useful and have been studied intensively. They are the "polytene chromosomes" (Beermann, 1972) and chromosomes of meiotic prophase, "the lampbrush chromosomes" (Hess, 1971).

Polytene chromosomes are bundles of up to a thousand chromatids that are formed by endomitosis (replication without following disjunction). They occur in larval tissues of dipteran insects (Beermann, 1962), in salivary glands of Collembola (Cassagnau, 1968, 1971), in developing macronuclei of hypotrich ciliates (Ammermann, 1971), and in certain plant tissues (Nagl, 1962). These chromosomes appear banded because alternating condensed segments (chromomeres) and unfolded segments of the chromatin fibers are in register. Cytogenetic studies in *Drosophila* have identified each band with specific genetic loci. Active transcription in a band, indicated by autoradiography after administration of [^{3}H]uridine or staining for RNA, often is associated with "puffing" of the band, i.e., swelling and dispersion of chromatin presumably due to unraveling of the DNA. The pattern of puffed bands in a cell indicates which genes are turned on in that cell. In salivary glands of *Chironomus tentans,* about 300 of the 2000 bands are puffed. Most of these puffs are present in all tissues, and only a small percentage are cell specific. *In situ* hybridization, with radioactive RNA or cDNA complementary to it, is a sensitive way to analyze which bands are active (Spradling *et al.*, 1975). Puffs thus show again that active transcription is related to structural change, mainly an unraveling of the compacted DNA, although the details of these changes have not been worked out in this material.

Lampbrush chromosomes also testify to the unraveling of DNA in active regions (Hess, 1971). These are chromosomes in the diplotene stage of meiosis, in spermatocytes or oocytes, and consist of the two homologous chromosomes held together by chiasmata. Each chromosome axis shows a string of dense bodies of different sizes (prophase chromomeres). In oocytes of certain amphibia, there are about 5000 such chromomeres per haploid set. Two loops project laterally from each chromomere (one from each chromatid). These loops contain unraveled DNA many microns long which is covered by transcription products forming a fibrous or granular "matrix." The significance of this active transcription in meiotic prophase is not yet understood. The actual percentage of DNA transcribed may be rather small. The total length of all the loops represents about 5% of the DNA (Gall and Callan, 1962). This agrees well with RNA/DNA hybridization studies (Davidson *et al.*, 1966), suggesting that about 5% of the DNA actually is transcribed.

C. Electron Microscopy of Active Chromatin

Attempts to identify active chromatin in thin sections have utilized electron microscopic autoradiography after labeling with [^{3}H]uridine, specific staining of ribonucleoproteins (especially the EDTA technique of Bernhard, 1969), ribonuclease digestion, and specific transcription inhibitors. These investigations recently were reviewed by Bouteille *et al.* (1974). Characteristic RNA-containing structures, such as *interchromatin granules, perichromatin fibrils,* and *perichromatin granules* were described. However, it has not been possible to relate these structures to a specific conformation of transcribed chromatin or transcription products. A more informative visualization of active chromatin was achieved by O. Miller and collaborators (see Miller and Bakken, 1972). For large cells such as amphibian oocytes, cells and nuclei were disrupted by hand or by lysis with a nonionic detergent ("Joy," Triton X-100, Nonidet 40). The chromatin then was dispersed in hypotonic media at pH 8–9 (0.01 *M* borate buffer) and centrifuged through 1% formaldhyde in 0.1 *M* sucrose onto an electron microscope grid in a plastic chamber. The grid was immersed in Photo-flo, air-dried, and stained with ethanolic PTA. This method has been spectacularly successful for visualizing actively transcribed DNA segments, but it must be recognized that it definitely alters the structure of *inactive* chromatin and also might change active segments. The high pH unravels the 20-nm fiber characteristic of inactive chromatin into the familiar string of beads, perhaps by removing H1 or by altering its association with DNA. To distinguish inactive fibers clearly from actively transcribing fibers, one needs a method that displays the active region with its transcripts but does not alter the natural state of inactive segments.

The method of Miller now has been applied to many different cell types and organisms. Two kinds of active regions have been recognized, the nucleolar, ribosomal RNA coding regions (rDNA) and nonnucleolar cistrons.

1. *Nucleolar Transcription Units*

Active rDNA cistrons first were described in accessory nucleoli of amphibian oocytes (Miller and Beatty, 1969a). A transcription unit includes a DNA-containing axis with laterally projecting ribonucleoprotein fibrils. The fibrils originate from a 125-nm granule (presumably RNA polymerase) on the axis and form a gradient of 80–100 short to long fibrils ("Christmas tree," or matrix unit) (Figure 19). Many such transcription units are arranged in tandem, separated by nontranscribed spacer DNA (Figure 20). In general, the gradients have the same polarity, although occasionally adjacent units may face in opposite direction. The original transcript is known to be a 40 S precursor rRNA containing the sequences for 18 and 28 S ribosomal RNA and transcribed spacers. Since the longest fibril is much shorter than a 40 S RNA, it is assumed that the transcripts rapidly become compacted by proteins. This folding of transcripts into a compact granule begins at their free ends and ultimately involves the entire fibril.

In amphibians, the 40 S RNA has a molecular weight of 2.6×10^6. The length of DNA corresponding to this transcript would be about $5.2 \times 10^6/1.92 \times 10^6 = 2.7$ μm (1.92×10^6 daltons = 1 μm of B-form DNA). The length of the transcribed regions has been determined for several amphibians (e.g., Miller and Hamkalo, 1972; Scheer *et al.*, 1973) and varies from 2.3 to 2.8 μm. This indicates that the DNA is extended almost completely in active rDNA transcription units. It therefore is not surprising that histone beads are absent (Scheer *et al.*, 1977). This does not prove that the histones themselves are missing, since they could be bound to DNA in a different way.

Between the "Christmas trees" are spacers which are generally free of transcripts, although a few short RNP fibrils have been reported in some cases (e.g., Scheer *et al.*, 1973). The spacers usually are shorter than the transcribed regions and more variable in length (cf. Berger and Schweiger, 1975). They carry histone beads (Foe *et al.*, 1976; Scheer *et al.*, 1977) characteristic of inactive chromatin that has been unraveled at pH 9 in low ionic strength. In the intact nucleus, these regions may be

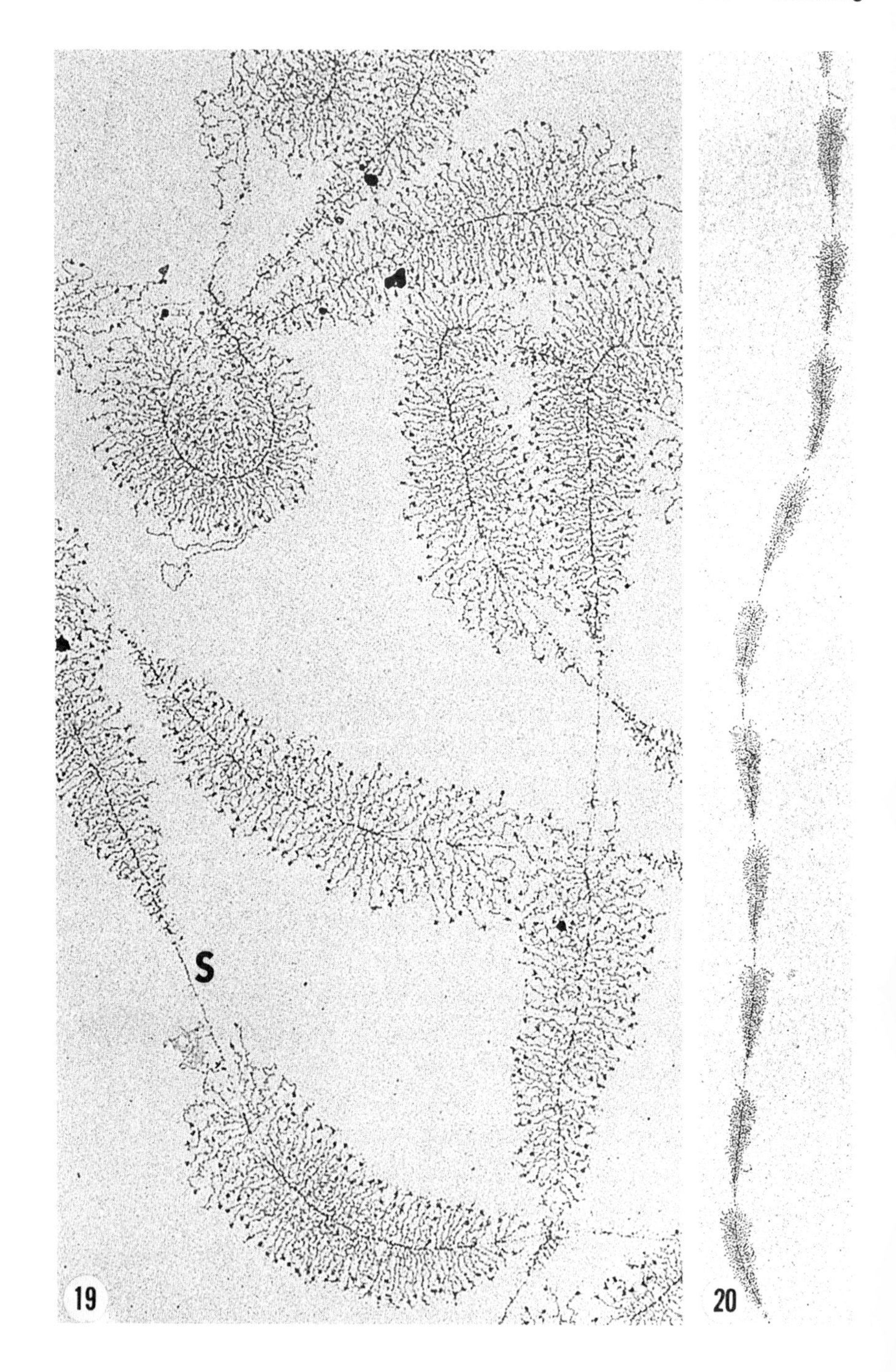
S
19
20

unraveled or coiled into 20-nm fibers. Spacers have been described for the majority of rDNA regions both in nucleoli and in amplified regions. Interesting exceptions are nucleoli of certain green algae in which spacers are either absent or extremely short (Berger and Schweiger, 1975).

2. Nonnucleolar Transcription Units

In addition to the rDNA "Christmas trees," nuclei spread with Miller's technique usually show other transcribed regions which are assumed to be active nonnucleolar segments. These generally occur as single transcription units with relatively few fibers. The units are spaced more widely than in rDNA, but frequently are arranged in a regular gradient as well (Foe *et al.*, 1976; Laird *et al.*, 1976; Laird and Chooi, 1976) (Figures 21 and 22). In contrast to rDNA, histone beads are present between the ribonucleoprotein fibrils (Foe *et al.*, 1976; Laird and Chooi, 1976). The beads, however, are spaced more widely than in nontranscribed regions. The presence of histone beads in transcribed regions appears related to the density of RNA polymerases. Where these are closely packed, the DNA is unraveled completely and beads are missing, as in rDNA. When polymerases are spaced more widely, as in non-rDNA, histone beads are present although more separated than in inactive chromatin. The packing ratio of DNA is about 2 instead of 7 as in the 10-nm fiber (Laird *et al.*, 1976).

3. Lampbrush Chromosomes

Lampbrush chromosome loops from amphibian oocytes show gradients of RNP fibrils ("Christmas trees") with closely packed polymerases similar to those in rDNA (Miller and Bakken, 1972). A loop may contain several such gradients separated by spacers and with either parallel or opposite polarity (Angelier and Lacroix, 1975). The transcribed RNA is complexed with specific proteins (Sommerville and Hill, 1973) into characteristic particles which are released from the chromosomes.

Lampbrushlike structures also have been described for the Y chromosome in spermatocytes of *Drosophila*. This localized transcriptional activity on the Y chromosome is required for normal differentiation of sper-

Fig. 19. Ribosomal RNA transcription units from nucleoli in oocytes of *Notophthalmus viridescens*. RNA polymerases are packed closely on the rDNA with transcripts (rRNA complexed with protein) projecting laterally ("Christmas trees"). The transcripts are short near the origin and become increasingly longer toward the terminus of the transcription unit. Many such transcription units are arranged in tandem separated by nontranscribed spacers (S). See also Figure 3. (From O. L. Miller, Jr., and B. R. Beatty, 1969, *Science* **164,** 955–957. Cover, May 1969. Copyright 1969 by the American Association for the Advancement of Science.) × 27,500.

Fig. 20. Ribosomal RNA transcription units from nucleolar chromatin of *Acetabularia mediterranea* (From Spring *et al.*, 1976.) × 6000.

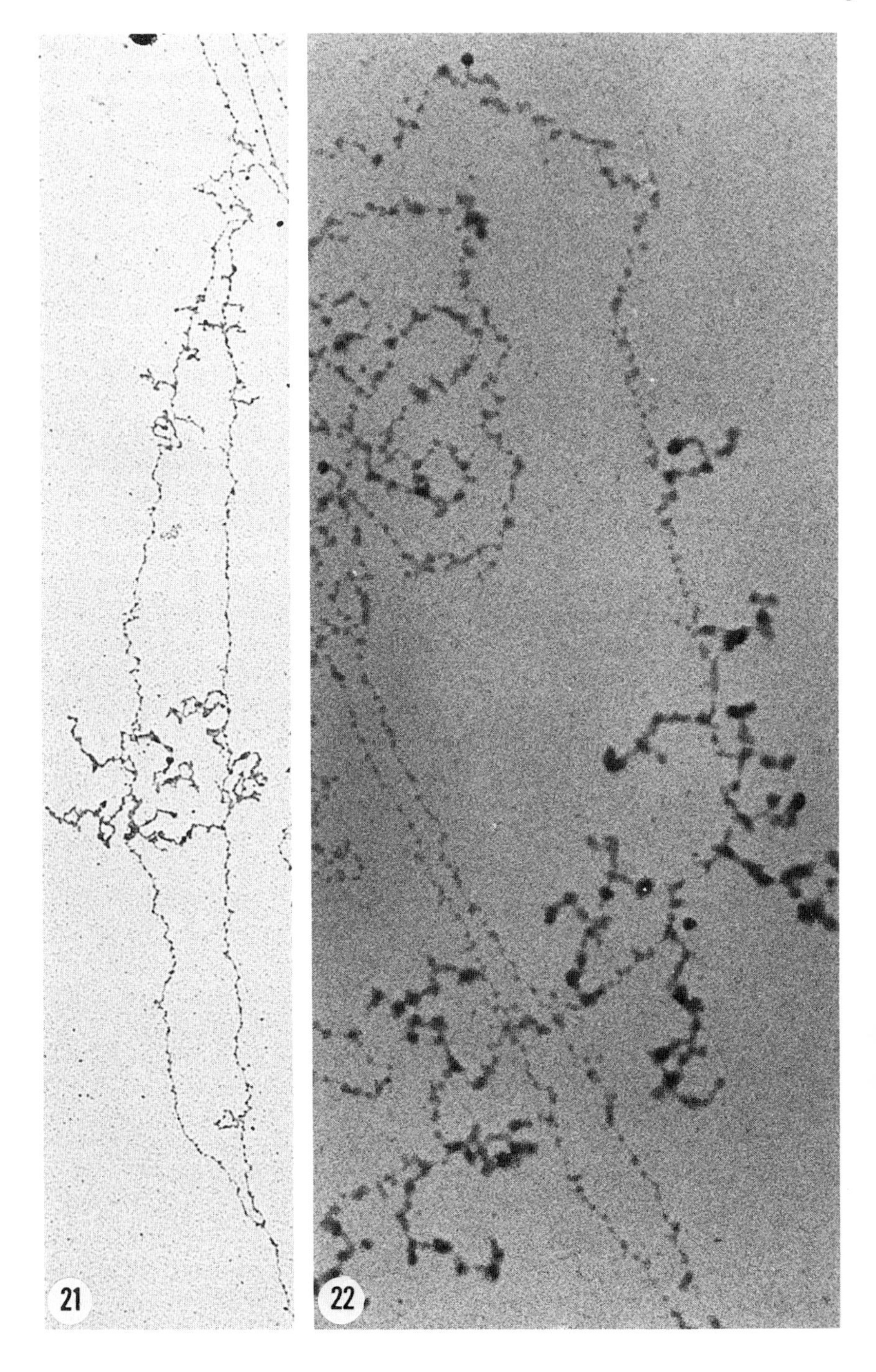
21
22

matids (Hess and Meyer, 1968; Hess, 1970). Henning *et al.* (1973) have adapted the Miller technique to *Drosophila* spermatocytes and have shown that the loops are typical transcription units separated by spacers. The distance between RNA polymerases varies considerably for different regions of the Y chromosome. The RNA molecules produced are extremely large, perhaps exceeding 10 μm. The transcriptional activity in spermatocytes is not restricted to the Y chromosome. Using spermatocytes of XO males, Glätzer (1975) described non-rDNA transcription units with long transcripts widely spaced and chromatin beads on the axis between the RNA polymerases. Similar active transcription may be a general phenomenon of the diplotene state in meiotic prophase.

Lampbrush chromosomes recently have been found in the large primary nucleus of the green alga *Acetabularia* (Spring *et al.*, 1975). The loops of these chromosomes contain gradients of closely spaced transcripts which are not separated by spacers.

D. Biochemical Studies

1. *Histones in Active Chromatin*

DNA of active chromatin is associated with "histone beads." This was shown by preparing highly labeled complementary DNA (cDNA) to nuclear RNA or mRNA of a tissue and hybridizing this probe with DNA extracted from chromatin monomers or with total DNA. The cDNA hybridizes to the unique sequence fraction, and the kinetics of reassociation is the same in total DNA and monomer DNA (Lacy and Axel, 1975; Kuo *et al.*, 1976). Therefore, DNA transcribed in a particular cell also is complexed with histone beads, in agreement with electron micrographs of active segments. Association with beads, therefore, is not the controlling factor in determining whether certain DNA sequences will be transcribed. But there is evidence that the association of beads with DNA may be different in *potentially active* chromatin and chromatin never expressed in a cell type (Weintraub and Groudine, 1976; Garel and Axel, 1976). DNase I (which attacks single-stranded DNA) preferentially digests potentially active DNA sequences. For instance, globin sequences are lost rapidly by

---

Fig. 21. Transcription unit near replication fork in chromatin isolated from cellular blastoderm stage of *Drosophila melanogaster*. (Photo by S. L. McKnight and O. L. Miller, Department of Biology, University of Virginia, Charlottesville.) × 40,000.

Fig. 22. Nonnucleolar transcription unit from embryonic chromatin of *Drosophila melanogaster* (From Laird *et al.*, 1976.) × 70,000.

DNase I treatment of reticulocyte chromatin but not from that of fibroblasts. This preferential sensitivity also is found in differentiated cells after transcription has stopped, as in *mature* erythrocytes. Perhaps association with nonhistone proteins, which are known to be the factors controlling cell specific transcription, effects this change in DNA–histone association.

Certain experiments suggest that the *very lysine-rich histones* (H1) may play a role in the control of transcription. We have seen above that active chromatin always is unwound and that removal of H1 leads to unwinding of the 20-nm fiber. Bacheler and Smith (1976) recently have shown that extraction with 0.6 *M* NaCl, which removes H1 and some nonhistone proteins, eliminates the restriction of transcription in unextracted chromatin as compared with pure DNA. Several authors have reported the absence or at least decreased amounts of H1 in chromatin fractions enriched for active chromatin (Berkovitz and Doty, 1975; Lau and Ruddon, 1977; Reeck *et al.*, 1972; Simpson and Reeck, 1973; Bonner *et al.*, 1973). However, others find H1 present in such fractions (McCarthy *et al.*, 1973; Gottesfeld *et al.*, 1975).

2. *Isolation of Active Chromatin Fraction*

Studies of biochemical properties of active chromatin rely on methods used to purify actively transcribed chromatin. Various methods based on structural differences between active and inactive chromatin have been tried. Shear and differential centrifugation lead to some enrichment of the active fraction. Thus far, the most successful method is based on the increased sensitivity of active chromatin to DNase II (Billing and Bonner, 1972). Gottesfeld *et al.* (1975) have used this approach to isolate chromatin subunits from the active fraction. In this fraction, the resistant DNA is associated with 14 and 19 S particles. All histone fractions are present, but the protein/DNA ratio is reduced compared with inactive fractions. These particles also contain nonhistone protein and RNA.

While biochemical analysis of purified active chromatin is still in the early stages, it has contributed information on the structure of active chromatin that complements electron microscopy as follows. (1) Histone beads are present in active chromatin, although relatively less concentrated than in the inactive chromatin. Particles of active regions also contain nonhistone protein and RNA. Wether H1 is missing from active regions is not yet resolved. (2) DNA of the active fraction is more accessible to DNase digestion between the beads as well as within the beads than is the inactive chromatin. Thermal denaturation studies also indicate that DNA is extended more and protected less in active regions (Berkowitz and Doty, 1975).

VI. CHROMOMERES

With the light microscope, chromosomes of mitosis and meiosis appear beaded until metaphase, when they resemble uniform rods. Wilson (1896) introduced the term "chromomere" for these darkly staining beads which were especially evident in meiotic prophase. Wenrich (1917), studying meiotic prophase chromosomes in male grasshoppers, noticed the remarkable constancy of the chromomere pattern and the side-by-side alignment of homologous chromomeres during meiotic pairing of chromosomes. After the linear order of genes was established, it was tempting to see its structural equivalent in the chromomere pattern. Belling (1928) counted about 1500–2500 "ultimate chromomeres" in early meiotic prophase (leptotene to pachytene) of several plants and proposed that chromomeres were equivalent to genes since their number appeared reasonable for the total number of genes. The discovery of the banded polytene chromosomes in dipteran larvae and their detailed cytogenetic analysis strengthened this concept. It generally is assumed that the "interphase chromomeres" in polytene chromosomes and the meiotic "prophase chromomeres" are comparable structural and functional units, and that the larger and less numerous chromomeres of late prophase are aggregates of the "ultimate chromomeres." However, it has not been shown either genetically or structurally that interphase chromomeres and leptotene chromomeres are identical units. Differences between these recently have been emphasized by Lima-de-Faria (1975). Our electron microscopic studies also suggest that the structures of interphase and of prophase chromomeres are quite different, and we therefore shall discuss them separately.

A. Interphase Chromomeres

In the normal interphase nucleus, it is practically impossible to follow a single 20-nm fiber to establish its arrangement. The polytene chromosomes in dipteran larvae, interphase chromosomes which have become individually distinct because of their polyteny, seem to offer a solution. The longitudinal organization of the chromosome is made visible by the close packing of the chromatids, whose structural variations are all in register. The constant pattern of thicker or thinner bands visible in the light microscope represents the elusive interphase chromomere pattern amplified. Although there is no evidence that the organization of chromatids in polytene chromosomes is characteristic of all interphase chromosomes, the genetic, structural, and biochemical analysis possible in *Drosophila* has raised some provocative questions and has afforded some interesting insights concerning interphase chromosome organization.

1. Chromomeres as Structural Equivalents of Genes

Recent studies have strengthened the long-held belief of *Drosophila* cytogeneticists that bands of polytene chromosomes, i.e., chromomeres, correspond to gene loci (for reviews, see Chapter 5, this volume; Beermann, 1972; Lefevre, 1974). With X rays and chemical mutagens, Judd *et al.* (1972) recovered over one hundred lethal and semilethal mutants in a restricted area of the X chromosome of *Drosophila melanogaster*. These mutants could be arranged into 13 complementation groups. Thus, thirteen different gene functions are present in this region of the X chromosome which cytologically shows 13 bands. Similar correspondence between complementation group number and number of invisible bands has been observed by Hochmann (1973) for the small fourth chromosome of *Drosophila melanogaster*. Another approach relating genes to bands uses chromosome rearrangements to determine the location of specific genes. The position of the breakpoints associated with a particular mutation pinpoints the location of the gene on the salivary gland map. Thus far, each band has been found associated with only one mutant gene. In conclusion, interphase chromomeres as seen in salivary gland chromosomes are the sites of specific genes.

2. DNA Content of Chromomeres

The large size of chromosomes in salivary glands of Diptera makes it possible to determine the DNA content of bands by cytophotometry and thus to calculate the amount of DNA per chromomere if the degree of polyteny is known. Polyteny can be determined by comparing the DNA content with that of a sperm nucleus. A correction must be made here for the chromosome regions that are underreplicated in polytene chromosomes. For instance, the DNA in centric heterochromatin (chromocenter) remains unchanged, or, at most, is doubled during polytenization, in which the bulk of the DNA may have increased two thousandfold (Rudkin, 1972; Mulder *et al.*, 1968). From such data, it was calculated that in *Drosophila melanogaster* the DNA per chromomere ranges from 3×10^6 daltons [1.5 μm or 5.1 kilobases (kb)] to 365×10^6 daltons (80 μm or 240 kb) (Rudkin, 1972). An average band contains about 17×10^6 daltons (9 μm or 27 kb). This is many times the DNA required to code for an average protein. If a chromomere corresponds to a single genetic unit, what then is the function of this variable amount of excessive DNA? Two basic models have been proposed: (1) Chromomeres include a linear repetition of functionally equivalent units (Bonner and Wu, 1973). (2) Each chromomere contains only one or a few copies of a structural gene plus a large amount of regulating DNA (Britten and Davidson, 1969; Georgiev, 1969; Crick, 1971; Paul, 1972). (For a more detailed discussion of these models of chromomeres, see Beermann, 1972.)

The polytene chromosomes in the developing macronucleus of hypotrich ciliates represent an interesting special case of chromomere organization (Prescott and Murti, 1973; Prescott *et al.*, 1973). During macronuclear development, banded polytene chromosomes appear. The subsequent breakdown of these is accompanied by loss of about 93% of DNA from each band. In the mature macronucleus, the DNA consists of small pieces 0.75 or 2 μm (rDNA) in length. Since 99% of the cellular RNA is transcribed from macronuclear DNA, one must conclude that less than 10% of the DNA sequences in this organism code for proteins. If the bands are units of genetic function as in *Drosophila,* then each unit or chromomere consists of a small piece that is transcribed and a large segment of "spacer" DNA with unknown function.

3. *Electron Microscopy*

Polytene chromosomes have been studied in thin sections of intact glands and of squashed chromosomes, by spreading on a water surface followed by critical point drying, and as whole mounts obtained by squashing of cells fixed *in situ*. Because of the complex twisting and folding of the chromatin fiber, thin sections of intact nuclei are not very informative. Swift (1962) and Sorsa and Sorsa (1967) made longitudinal sections of squashed chromosomes in which the longitudinal organization into bands and interbands still was visible. Both reported fibrils 40–60 Å thick in interbands. However, these thin interband fibers are certainly artifacts due to the use of metal chelating buffers and the extreme stretching during the squashing. Sections of chromosomes fixed in aldehydes without chelating buffers show 20-nm fibers both in interbands and bands (H. Ris, unpublished). Similar fibers also are present after spreading on water (Rae, 1966) and in whole mounts squashed after fixation *in situ* and photographed with the million volt EM (Ris, 1975a, 1976). In inactive regions of polytene chromosomes, as in other interphase nuclei, a 20-nm fiber is the visible structural unit. Although the fibers cannot be followed through the bands, it appears that the 20-nm fiber, which represents a single chromatid, runs more or less parallel to the chromosome axis in interbands and is highly folded in bands (Figure 23). In bands, several fibers associate to form tight knots with different fibers coming together from band to band, creating an interconnected network.

While electron micrographs suggest that interphase chromomeres originate essentially by a tight folding of the 20-nm fiber, the structure of

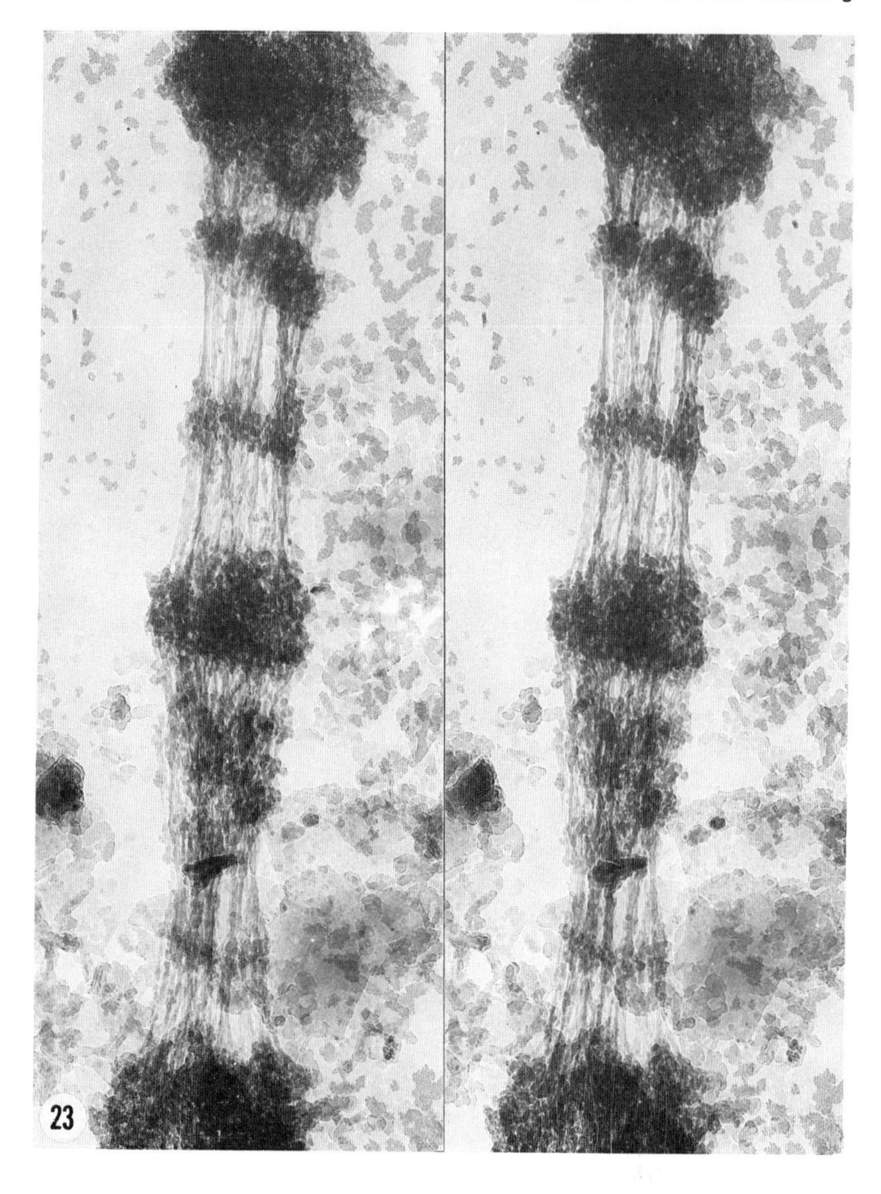

Fig. 23. Stereomicrograph of polytene chromosome from fat cell of *Drosophila melanogaster*. Fixation methanol–acetic acid (3 : 1) squashed in 50% acetic acid. The 20-nm fibers are straight in the interbands and compacted to form interphase chromomeres in the bands. Photographed at 1 MeV, High Voltage Electron Microscope Facility, Madison, Wisconsin. (From H. Ris, 1976, *Electron Microsc. Proc. Eur. Congr. Electron Microsc., 6th, 1976,* Vol. II, pp. 21–25. Copyright © 1976 by INTERTAL International Publishing Co. Ltd., Givatayim, Israel.) × 40,000.

bands in highly polytenized chromosomes might be more complex. C. D. Laird (personal communication) has compared estimates of the DNA packing ratio in the 20-nm fiber with the relative DNA concentration in bands and interbands. Equal degrees of polyteny in bands and interbands predict more DNA in interbands than is actually found by cytophotometry. This suggests that interbands might be underreplicated with respect to band regions.

4. *Chromomere-Sized Domains in Nonpolytene Interphase Chromosomes*

Benyajati and Worcel (1976) obtained evidence for chromomere-sized domains in the chromosome fibers of *Drosophila* tissue culture cells. Cells were lysed gently in 0.9*M* NaCl and the liberated chromatin was analyzed in a sucrose gradient at 4°C. Under these conditions, the chromosome fiber is a simple string of histone beads with the DNA highly supercoiled because of the high salt and low temperature. Single-strand nicks or ethidium bromide relax the supercoiling in distinct domains about 85 kb long. This restricted unwinding suggests the presence of "clamps" (probably nonhistone proteins) which subdivide the chromatin fiber into chromomere-sized domains. It is possible that these domains correspond to the differentially folded regions of the 20-nm fiber that represent the chromomeres of polytene chromosomes.

B. Prophase Chromomeres

1. *Variability of the Number and Arrangement of Prophase Chromomeres*

The relationship of interphase chromomeres to units of genetic function is supported well by the constancy of chromomere patterns in different tissues and their correspondence to units of mutation and of transcription. The situation is much more complex for prophase chromomeres. The relationship of prophase chromomeres to interphase chromomeres is not known, because organisms with polytene chromosomes do not have analyzable prophase chromomeres. There is no evidence for a one-to-one relationship of prophase chromomeres to cistrons. In maize, at least one gene is localized in the region between two pachytene chromomeres and several genes within a chromomere (McClintock, 1944). One important aspect of prophase chromomeres is that the chromomere pattern is different in mitosis and meiosis and in prophases of different cell types. As chromosomes condense during prophase, the number of chromomeres decreases. Small chromomeres aggregate specifically to form larger chromomeres (Yunis, 1976; Luciani *et al.*, 1976).

Variation in chromomere patterns in different cell types was studied in detail by Lima-de-Faria (1975). He found that the chromomere number varied drastically in prophase of the first meiotic division, second meiotic division, and pollen mitosis of several plant species. However, size and spacing of chromomeres remained the same, indicating that the smaller number was not due to fusion of chromomeres but represented a totally different chromosome organization. The chromomeres of lampbrush chromosomes in amphibian oocytes present another interesting situation. They frequently are compared with bands of polytene chromosomes. However, a recent study of lampbrush chromosomes in closely related species of *Plethodon* makes a one-to-one relationship of lampbrush chromomeres to genes highly unlikely (Vlad and Macgregor, 1975). Three species were compared, all having the same chromosome number and arm ratio. However, one of them (*P. cinereus*) has only about half the genome size of the other two (*P. dunni* and *P. vehiculum*). The difference in DNA content is due almost entirely to the moderately repetitive fraction which is dispersed throughout the genome. In the species with higher DNA content, the chromosomes are longer and have about 60% more chromomeres. The average distance between chromomeres is the same in all species (about 2 μm). The number of chromomeres, therefore, has no direct relationship to the number of unique sequences (which presumably represent most structural genes) but is proportional to total DNA content.

2. *Significance of Prophase Chromomeres*

What is the functional significance of prophase chromomeres? The constant pattern for the same stage in the same cell type indicates that chromomeres are not random structures but correspond to definite structural and functional units of the chromosome. But chromomeres are not all homologous and can be understood only in relation to the specific physiology of the cell in which they occur. A possible functional relationship can be inferred for the pachytene chromomeres of mammalian spermatocytes. These prophase chromomeres seem to be identical to Giemsa bands of mitotic chromosomes (see Section IX on metaphase chromosome bands). Stubblefield (1975) has shown that these chromosome regions are units of DNA replication. The DNA within each unit is replicated within the same 2-hr period of S phase. Whatever controls the initiation of DNA replication acts on all initiation points within such a chromosome segment at the same time.

3. *Prophase Chromomeres–Tightly Packed Regions of the 100-nm Chromonema*

Prophase chromomeres are aspects of mitotic chromosome condensation, a topic which will be discussed later. The study of prophase chromosomes by electron microscopy has been very difficult because they are embedded in a fibrous protein material which resembles chromatin fibers (see Berezney and Coffey, 1974; Comings and Okada, 1976; Wunderlich and Herlan, 1977). Squashes of meiotic prophase chromosomes have been more informative than those of mitotic chromosomes, because less of this material is present. Leptotene and pachytene chromosomes of plants and animals exhibit a structural unit about 100 nm thick which seems to correspond to a chromatid in the paired chromosomes. This structure is packed more highly in chromomeres than in interchromomere regions (See Figure 31). Thus, prophase chromomeres differ fundamentally from interphase chromomeres in ultrastructure; they are units of packing of a higher order chromatin unit (the 100-nm fiber), whereas the interphase chromomere of polytene chromosomes is a unit of folding of the 20-nm fiber.

VII. MITOTIC CONDENSATION OF CHROMOSOMES

A. Light Microscopy of Mitotic Chromosomes

During nuclear division, the long chromosomes of interphase condense into much shorter rods which may be moved about by the spindle apparatus. After reaching the poles, they begin to decondense to the interphase structure characteristic for a cell type. There are only few exceptions to this cyclic condensation of chromosomes, and they are generally restricted to species with small chromosomes (low haploid DNA content), such as the yeast *Saccharomyces cerevisiae* (see Peterson and Ris, 1976).

Mitotic condensation on the light microscopic level was intensively investigated between 1930 and 1950. An excellent review is found in Manton (1950). Although earlier studies mainly were confined to a few plant species with especially large chromosomes, they showed mitotic condensation essentially to be a helical coiling of the chromonema, the 0.2-μm thick chromosome fiber resolved by light microscopy. More recently, Ohnuki (1968) has documented beautifully that the same process applies to the smaller human chromosomes and therefore probably to all chromosomes. Since recent electron microscopic studies largely have ignored these findings, the main results will be summarized briefly here.

1. The metaphase chromosome is a helical coil of a fiber (the chromonema) with a diameter of about 0.2 μm, close to the resolution limit of the light microscope. In a few instances, this helix is visible in living cells (Figure 24) (*Haemanthus* endosperm, Bajer, 1966; Holm and Bajer, 1966). Generally, however, the gyres are packed tightly so that the living or well-fixed chromosome has a uniformly solid appearance. A

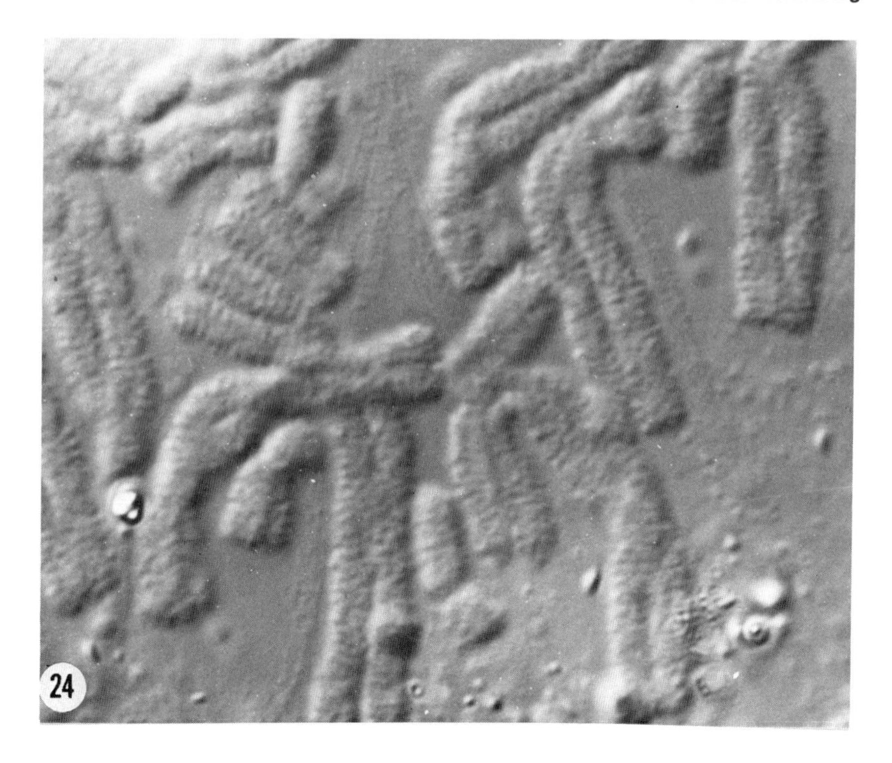

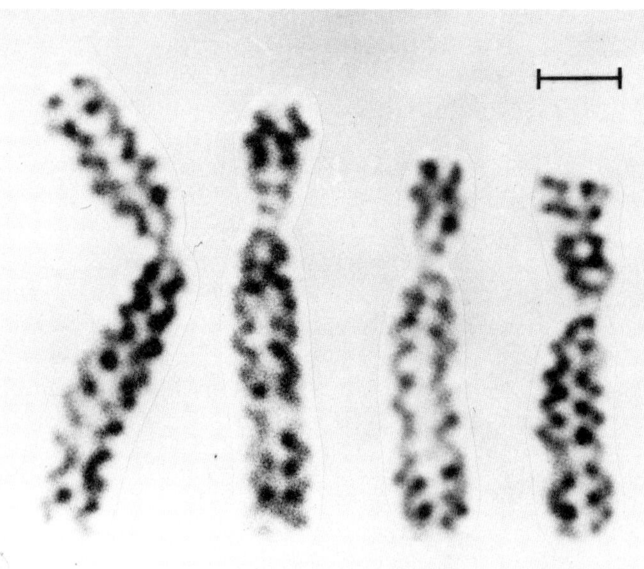

number of prefixation treatments have been described which make the coil visibly clearer. Treatments with hypotonic salt solutions are most successful (Figure 25, Ohnuki, 1968). In tissue culture, cells exposed to 0.07 *M* KCl or to 0.04 *M* phosphate buffer (pH 7.0) for 20–90 min give highly reproducible results. An interesting case was described by Therman (1972). In cells treated with 1-methyl-2-benzylhydrazine, the helix is clearly visible in one or two chromosomes while the others remain tightly condensed. In pollen mother cells of *Tradescantia,* the coil of meiotic metaphase is clearly visible without treatment if anthers are excised in late prophase and cultured in nutrient medium. These cells complete meiosis normally (Taylor, 1949).

2. The direction of coiling can be right- or left-handed and may change either at the centromere or within a chromosome arm (Ohnuki, 1968).

3. In the first meiotic metaphase, a double coil usually is present. The few large gyres of the "major coil" show evidence of many small gyres of a "minor coil" (Coleman and Hillary, 1941). In some species, this double coil persists into the second division. In others, the coiling of the second division resembles that in somatic cells ("standard coil"). The number of gyres in a chromosome is constant for a cell type under the same external conditions. The width and number of gyres determine the relative diameter and length of a particular chromosome.

4. The increasing condensation of chromosomes during prophase is accomplished by a process of gyre elimination (Swanson, 1943). Early prophase chromosomes show a large number of very small gyres. As prophase progresses, the width of the gyres increases while their number decreases. These changes have been observed in living cells (Holm and Bajer, 1966).

B. Electron Microscopy of Mitotic Chromosomes

In thin sections, metaphase chromosomes have the same structure as inactive condensed chromatin in interphase. When properly fixed, they are composed of typical 20–30-nm fibers. However, the specific arrangement of the 20-nm fiber in mitotic chromosomes cannot be determined by this method. Therefore, many attempts have been made to prepare intact

---

Fig. 24. Chromosome coiling in a living endosperm cell of *Haemanthus katherinae* (Nomarski interference contrast.) (From Bajer, 1966.) × 2000.

Fig. 25. Photomicrograph of human metaphase chromosomes treated with hypotonic salt solutions to unwind the coiled chromonema partially. (From Ohnuki, 1968.) × 1000.

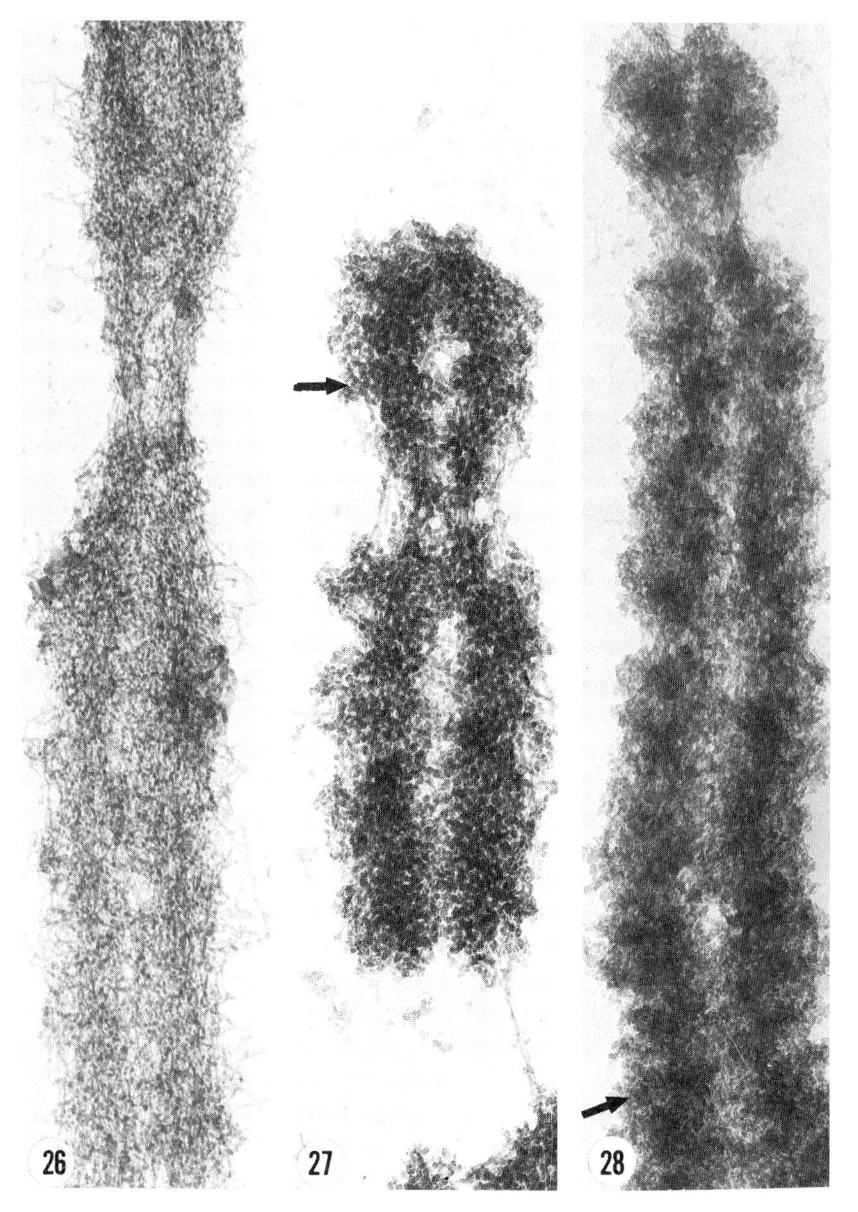

Fig. 26. Metaphase chromosome from Chinese hamster (CHO cell line), spread on water, fixed in 4% formaldehyde, and critical point-dried. The component fiber is 20 nm thick. Photographed at 1 MeV, HVEM Facility, Madison, Wisconsin. × 20,000.

Fig. 27. Metaphase chromosome (CHO cell line). Cells were pretreated in hypotonic phosphate buffer and chromosomes were isolated in 0.1 m*M* Pipes (pH 6.7), 0.5 m*M* $CaCl_2$, 1 *M* hexylene glycol (Wray and Stubblefield, 1970), fixed in methanol–acetic acid (3 : 1),

chromosomes (while mounts) for analysis with the electron microscope. We can distinguish three major approaches.

1. *Isolation by Osmotic Shock*

Colcemid-arrested metaphase cells are spread on a water surface. Chromosomes prepared in this manner often are very distorted. Where the overall shape is preserved, they appear as a dense tangle of 20-nm threads (Figure 26) with no obvious organization into a coiled chromonema as seen with light microscopy (for other examples, see DuPraw, 1970). On the basis of such preparations, DuPraw (1970) and Bahr (1975) have proposed a folded fiber model of metaphase chromosome structure according to which the 20-nm fiber is woven into an intricate lateral and longitudinal pathway as it threads its way back and forth through the chromatid. As we shall see later, this method drastically alters the organization of metaphase chromosomes and is useless in determining the arrangement of the 20-nm fiber.

2. *Bulk Isolation of Chromosomes*

Various methods have been developed to isolate chromosomes for chemical analysis and for the ultimate separation of specific chromosomes (for reviews, see Wray, 1973; Hanson, 1975; Mendelsohn, 1974). For example, hexylene glycol preserves chromosome morphology in media at neutral pH (Wray and Stubblefield, 1970) and allows isolation of metaphase chromosomes relatively free of cytoplasmic contaminants (Stubblefield and Wray, 1971). Chromosomes isolated in this medium are rather highly condensed and consist of a fiber about 50 nm thick (Figure 27; see also Daskal *et al.*, 1976). When placed in water or 1 m*M* Pipes buffer (pH 7.0), they become less dense, and the unit fiber is now 20 nm thick. The dense 50-nm fiber is caused by the presence of 0.5 m*M* Ca^{2+} in the medium and not the hexylene glycol, since such a structure reappears when chromosomes are placed in the isolation buffer without hexylene glycol. Is this 50 nm a natural unit of mitotic chromosomes? It seems rather unlikely that 0.5 m*M* Ca^{2+} would create anew such a fiber of uni-

---

and squashed in 50% acetic acid. The component fiber is 50 nm thick under these conditions. It appears to be coiled into a 200-nm unit (arrow) which corresponds to the chromonema seen by light microscopy. Photographed at 1 MeV, HVEM Facility, Madison, Wisconsin. × 20,000.

Fig. 28. Metaphase chromosome (CHO cell line). Cells are pretreated in 0.07 *M* KCl for 10 min, fixed in methanol–acetic acid (3 : 1), and squashed in 50% acetic acid. The coiled chromonema (arrow) is visible at the end of the long arm. Photographed at 1 MeV, HVEM Facility, Madison, Wisconsin. (From H. Ris, 1976, *Electron Microsc. Proc. Eur. Congr. Electron Microsc., 6th, 1976,* Vol. II, pp. 21–25. Copyright © 1976 by INTERTAL International Publishing Co. Ltd., Givatayim, Israel.) × 20,000.

form thickness. More probably, it tightens an existing structure of that order of magnitude. Indeed, we shall present below preliminary evidence that the 50-nm fiber is a regular component of mitotic and meiotic chromosomes and is formed by helical coiling of the 20-nm fiber.

3. *Squash Preparation of Chromosomes Fixed* in Situ

With this method, the chromosomes first are fixed in the intact cell and separated from the cells only afterward (Ris, 1961, 1976, 1978). Unfortunately, because aldehyde fixation hardens the cytoplasm and prevents successful squashing, cells therefore must be fixed with methanol–acetic acid (3 : 1). Despite the fact that this fixative extracts some of the histones, the 20-nm fiber, although it is somewhat thinner, remains intact (Ris, 1978). After fixation, the cells are squashed between slide and coverslip, and selected chromosomes are transferred to EM grids (Ris, 1961). Preparations of whole chromosomes are rather too thick for conventional 100-kV electron microscopes. Better results are obtained with a million-volt electron microscope (Ris, 1976). Figure 28 shows a chromosome from Chinese hamster (CHO) cells treated for 30 min in hypotonic phosphate buffer (0.04 *M*), pH 7.0. The 20-nm fibers in such preparations are organized into a thicker unit, 100–200 nm in diameter. In stereoscopic photographs, this unit is seen to be helically coiled and unquestionably represents the chromonema known from light microscopy. At the edge of the chromosomes, one often sees a 20-nm fiber coiled into a unit 50–70 nm thick (Figure 29). In turn, this fiber is coiled to form the 100–200-nm-thick chromonema. Therefore, two orders of coiling seem to be present between the 20-nm fiber and the chromonema known from light microscopy, which in turn is arranged helically in the condensed chromatid of metaphase. The degree of compaction of the chromonema varies somewhat from chromosome to chromosome. It seems that the hypotonic treatment loosens to variable degrees the coiling in the 50-nm and 100-nm fiber and thus swells their diameters (compare Figures 26–29). More extensive swelling, as for instance in distilled water, disorganizes the chromonema completely, and only an irregularly twisted mass of 20-nm fibers remains. This explains why the chromonema is not visible in water-spread chromosomes. It is preserved only if the hypotonic swelling is slight and if the chromosome is fixed before it is isolated from the cell. The hypotonic spreading method and bulk isolation techniques therefore are not suitable to study the various levels of organization in mitotic chromosomes.

4. *Prophase Chromosomes*

To understand mitotic chromosome condensation, it is necessary to follow chromosome organization through prophase. This has been difficult because the nuclear envelope is not dispersed easily, and in the intact nucleus the chromosomes are embedded in a fibrous "matrix" (Wunderlich and Herlan, 1977; Berezney and Coffey, 1974) which complicates their isolation. Meiotic prophase chromosomes are prepared more easily, and we have obtained squashes of pachytene chromosomes from plant

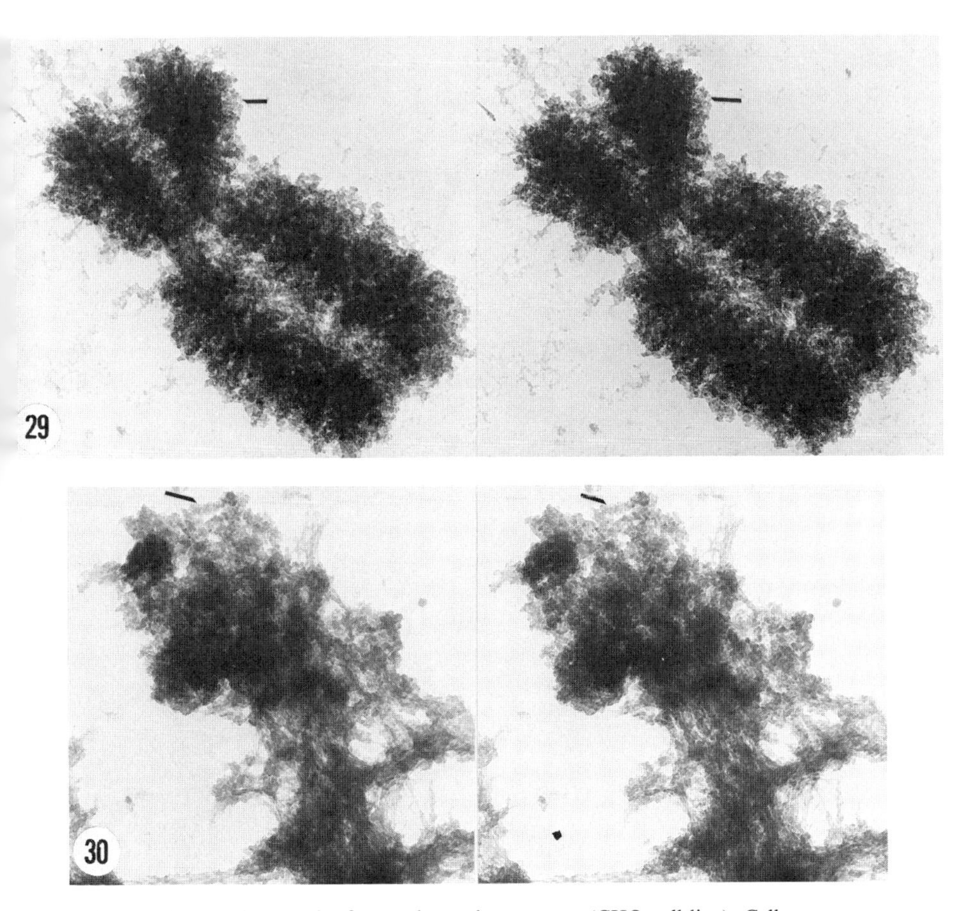

Fig. 29. Stereomicrograph of metaphase chromosome (CHO cell line). Cells were pretreated in 0.07 *M* KCl, fixed in methanol–acetic acid (3 : 1), and squashed in 50% acetic acid. Note the coiled chromonema. At the edge of the chromosome one sees short stretches of the 50-nm fiber which appear as helical coils of the 20-nm fiber (arrow). Photographed at 1 MeV, HVEM Facility, Madison, Wisconsin. (From H. Ris, 1976, *Electron Microsc. Proc. Eur. Congr. Electron Microsc., 6th, 1976,* Vol. II, pp. 21–25. Copyright © 1976 by INTERTAL International Publishing Co. Ltd., Givatayim, Israel.) × 20,000.

Fig. 30. Stereomicrograph of mouse pachytene chromosome. Testis fixed in methanol–acetic acid (3 : 1), squashed in 50% acetic acid, and critical point-dried. Twenty-nanometer fibers are coiled to form 50-nm fibers (arrows). Photographed at 1 MeV, HVEM Facility, Madison, Wisconsin. × 20,000.

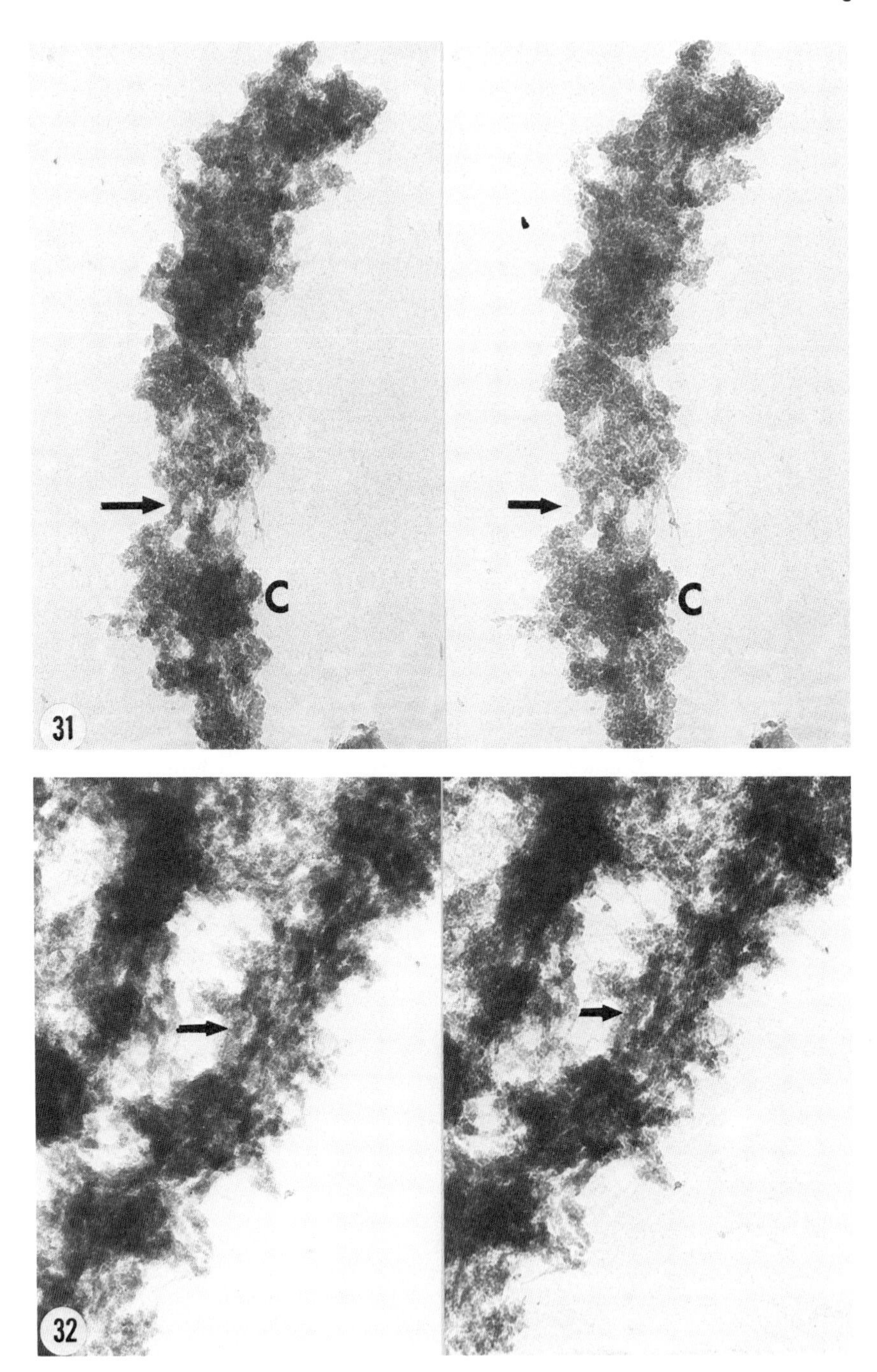
C
C
31
32

anthers and insect and mouse testes. Figure 30 shows the 50-nm fiber in a mouse pachytene chromosome. It consists of a coiled 20-nm fiber. Figure 31 illustrates the 100-nm component in a leptotene chromosome of the grasshopper *Chorthippus curtipennis*. It is folded tightly in the prophase chromomeres. Two side-by-side 100-nm fibers are visible in a region that might be the centromere (arrow): The 100-nm fiber therefore corresponds to the chromatids in this leptotene chromosome. Figure 32 represents a pachytene chromosome in the grasshopper *Dissosteira carolina*. Four 100-nm fibers corresponding to the four chromatids are visible in the interchromomere region. The 50-nm and 100–200-nm components seem to be units of organization in both meiotic and mitotic chromosomes.

VIII. HETEROCHROMATIN

A. Definitions

Certain chromosome segments or even entire chromosomes do not uncoil at the end of mitosis and remain condensed in interphase and form "chromocenters." Entire chromosomes remaining condensed were called "heterochromosomes" by Montgomery (1904). In 1928, Heitz discovered that specific and constant chromosome regions were responsible for the chromocenters of interphase, and he called these segments "heterochromatin" in contrast to the "euchromatin" which disperses in interphase. Heterochromatin abstains from the normal condensation cycle; it is "allocyclic" (Darlington and LaCour, 1940) or "heteropycnotic" (Gutherz, 1907). Under some conditions, chromosome regions may be undercondensed and lighter-staining in metaphase; in this case, they are said to be "negatively heteropycnotic" (White, 1935) in contrast to the "positive heteropycnosis" of heterochromatin in interphase. A further refinement of the concept of heterochromatin was introduced by Brown (1966). He showed that we must differentiate between chromosome regions that always are condensed in both homologues of the diploid set

---

Fig. 31. Stereomicrograph of a grasshopper (*Chorthippus curtipennis*) leptotene chromosome. Testis fixed in methanol–acetic acid, squashed in 50% acetic acid, and critical point-dried. The chromatid is a fiber about 100 nm thick (arrow) which is compacted in the prophase chromomeres (C). Photographed at 1 MeV, HVEM Facility, Madison, Wisconsin. × 20,000.

Fig. 32. Stereomicrograph of a pachytene chromosome from the grasshopper *Dissosteira carolina*. Testis fixed in methanol–acetic acid (3 : 1), squashed in 50% acetic acid, and critical point-dried. Four 100-nm chromatids are visible (arrow) which are more tightly compacted in the prophase chromosomeres (C). Photographed at 1 MeV, HVEM Facility, Madison, Wisconsin. × 20,000.

("constitutive heterochromatin") and regions which are condensed only under special conditions and usually only in one homologue ("facultative heterochromatin"). More recently, various new methods of studying chromosome composition and behavior have made it clear that allocycly is a common behavior of a variety of differential chromosome segments, and additional criteria must be used to subdivide structurally and functionally diverse chromatin further. Such characterizations will remain rather crude until we can describe the functionally diverse chromosome segments in terms of nucleotide sequences, histone and nonhistone protein composition, and functional specializations. In the meantime, constitutive heterochromatin has been differentiated further into "Y-chromosome type" (brilliant quinacrine fluorescence), "centric heterochromatin" (C-banding methods), and "intercalary heterochromatin" (Pätau, 1973; Comings, 1974).

Finally, it must be emphasized that not all condensed chromatin of interphase nuclei is "heterochromatin." In certain cells, a fraction or all of the euchromatic chromosome regions condense in connection with temporary and reversible inactivation of transcription. A good example are the nucleated erythrocytes of vertebrates. During their maturation, all chromatin becomes condensed (tightly packed 20-nm fibers), and transcription ceases. Such cells can be reactivated by fusion with tissue culture cells leading to dispersal of condensed chromatin (Ringertz *et al.*, 1972). Another example of decondensation of condensed euchromatin is the activation of peripheral lymphocytes by phytohemagglutinin (Killander and Rigler, 1965). While condensation always is related to inactivation, decondensation into dispersed 20-nm fibers by itself is not activation. Neither "euchromatin" nor "decondensed chromatin" is synonymous with active chromatin. As we have pointed out in Section V on active chromatin, transcription involves unwinding of the 20-nm fiber and perhaps even removal of H1 to produce an open string of beads.

At the present state of our knowledge, it is useful to distinguish the following major categories of heteropycnotic chromatin: (1) constitutive heterochromatin, always heteropycnotic in both homologues, genetically inert, not transcribed; (2) facultative heterochromatin, entire chromosomes or large chromosome blocks inactivated for many cell generations, rarely reversed in somatic cells; and (3) temporarily inactivated chromatin reversible within the same cell generation.

B. Properties of Heterochromatin

1. General Properties

a. Condensation. Allocycly or heteropycnosis is a general property of all types of heterochromatin. The detailed nature of this condensation, however, is not really understood. We have seen that mitotic condensation is not merely a closer packing of the 20-nm fiber but involves higher orders of arrangement that as yet are known imperfectly. Is heteropycnosis a persistence of mitotic orders of packing, a different but specific ordering of the 20-nm fiber, or a more or less random aggregation of the 20-nm fiber? There is little specific information on this question, and its analysis will depend on the development of new techniques which preserve and display such higher orders of arrangement of the 20-nm fiber. At present, we can stress only that condensation affecting the general spacing of the 20-nm fiber must be distinguished from higher orders of coiling or folding. In the living nuclei, little structural differentiation is visible except for nucleoli. This state can be preserved by fixation in aldehydes or OsO_4 in neutral buffers (Ris and Mirsky, 1949). Cell injury or fixation at low pH produces a distinctive pattern of condensed chromatin. In isolated nuclei, low ionic strength gives a diffuse appearance, while at physiological ion concentration one finds the differentiation into condensed and dispersed chromatin (Ris and Mirsky, 1949; D. E. Olins and Olins, 1972). The difference between these states, as seen in electron micrographs, is mainly in the average distance between the 20-nm fibers (Ris, 1961), presumably depending on the charge on the nucleoprotein fibers.

This contraction of condensed chromatin in the presence of cations seems to be mediated by very lysine-rich histones, namely, H1 in rat liver nuclei and H1 plus H5 in erythrocyte nuclei (Billett and Barry, 1974). The role of these histones in the formation of the 20-nm fiber was discussed earlier. In addition to maintaining the 20 nm supercoil, they also determine the closeness of packing of the fibers in the presence of cations.

The swelling of nuclei at low ionic strength, however, does not abolish the basic nuclear organization. Upon addition of cations, the same patterns reappear. This is true not only for condensed chromatin patterns of interphase (Brasch *et al.*, 1971), but even for the more specific chromomere patterns in polytene chromosomes (H. Ris, unpublished observations). This suggests that the ion-induced, and perhaps also H1-associated, condensation only affects the relative distance between the 20-nm fibers but not the basic allocycly or heteropycnosis, which most probably is related to higher orders of folding or coiling of the 20-nm fibers.

b. Late Replication. Both constitutive and facultative heterochromatin tend to replicate late in S phase (Lima-de-Faria, 1969), and late replication sometimes has been used as a criterion for heterochromatin. The situation, however, is more complex, since in a number of organisms, chromosome segments which behave like constitutive heterochromatin replicate

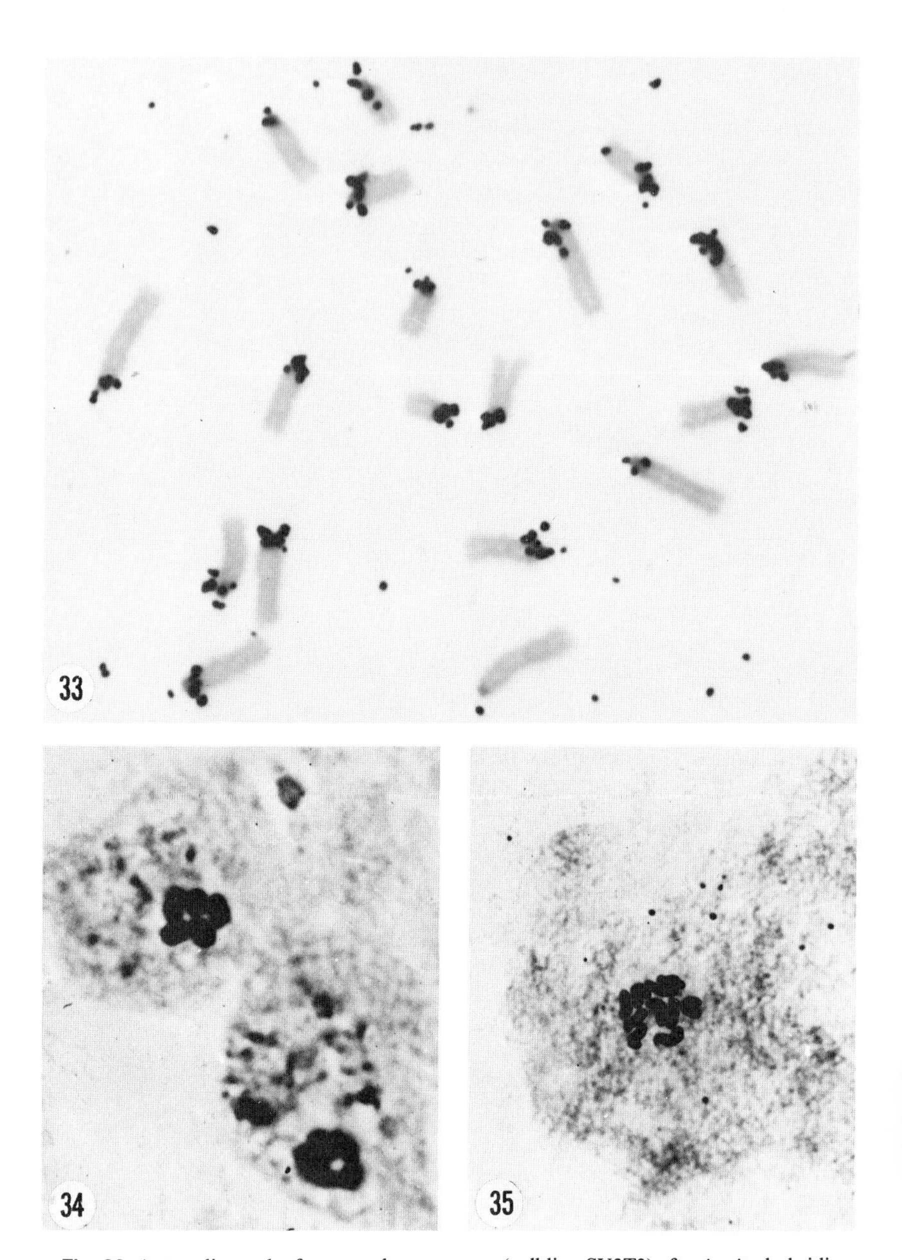

Fig. 33. Autoradiograph of mouse chromosomes (cell line SV3T3) after *in situ* hybridization with radioactive RNA copied *in vitro* from mouse satellite DNA. The grains are localized over the centromeric constitutive heterochromatin, indicating the location of the satellite sequences on the chromosomes. (From M. L. Pardue and J. G. Gall, 1970, *Science* **168**, 1356–1358. Copyright 1970 by the American Association for the Advancement of Science.)

Fig. 34. Somatic cell interphase of male mealybug: the five paternal chromosomes are totally heterochromatic. (Photo by Dr. M. Sabour, Brown University. From S. W. Brown,

early or at least before many euchromatic segments (Bostock and Christie, 1975; Hsu and Arrighi, 1971; Sharma and Dhaliwal, 1974).

c. Absence of Transcription. All condensed chromatin, including all types of heterochromatin, is transcriptionally inactive as shown, for instance, by autoradiography after labeling with [H^3]uridine (Sieger *et al.*, 1970). In lampbrush chromosomes, in which the loops on the chromomeres are a visible expression of transcription, it has been observed that constitutive heterochromatin carries no loops (Batistoni *et al.*, 1974).

2. Properties of Constitutive Heterochromatin

a. Genetic Inertness. Early cytogenetic studies already demonstrated that certain heterochromatic regions (now recognized as constitutive heterochromatin) were genetically inert in that they contained few mutable genes and could be lost or duplicated with little effect on the organism (Bridges, 1916; Muller and Painter, 1932).

b. Location of Highly Repeated Simple Sequence DNA. The discovery that satellite, or highly repeated simple sequence, DNA is localized in constitutive heterochromatin was an important step in the evolution of the heterochromatin concept, since it demonstrated that some heterochromatin is chemically and functionally distinct from the rest of the chromatin. Two methods showed this association of satellite DNA with heterochromatin. First, fractionation of condensed and dispersed chromatin revealed that the bulk of satellite DNA was in the condensed chromatin (Yasmineh and Yunis, 1970; Yunis and Yasmineh, 1972). Second, *in situ* hybridization allows cytological localization of specific satellite DNA's in constitutive heterochromatin of individual chromosomes (Figure 33; Pardue and Gall, 1970). These highly repeated DNA fractions are located mostly in the centric heterochromatin, more rarely in telomeres or in intercalary heterochromatin.

c. Polymorphism within a Species. The amount of constitutive heterochromatin can vary considerably between individuals of a species. In humans, for instance, a high frequency of heterochromatin polymorphism has been found in the long arm of the Y chromosome and centric heterochromatin of six autosomes (Craig–Holmes and Shaw, 1971). Similar polymorphism in centric heterochromatin also was described in the mouse (Forejt, 1973).

d. Variation between Related Species. The DNA content per chromosome set often varies manyfold between species of a genus. This variation

---

1966, *Science* **151,** 417–425. Copyright 1966 by the American Association for the Advancement of Science.)

Fig. 35. Metaphase cell of male mealybug. All chromosomes are equally condensed during mitosis. (Photo by Dr. M. Sabour, Brown University. From S. W. Brown, 1966, *Science* **151,** 417–425. Copyright 1966 by the American Association for the Advancement of Science.)

usually is due to an increase or decrease in the repetitive DNA fraction and is related directly to changes in constitutive heterochromatin content (Narayan and Rees, 1976; Mizuno and Macgregor, 1974; Hatch *et al.*, 1976).

e. Nonhomologous Association. Heitz (1934) noticed the tendency of heterochromatin of nonhomologous chromosomes to associate. In *Drosophila,* for instance, the centric heterochromatin of all chromosomes fuses into the chromocenter in interphase. In a similar fashion, telomeric heterochromatin may result in fusion of chromosome ends, or nucleolar heterochromatin in fusion of nucleolar organizers (for references, see Yunis and Yasmineh, 1971). Specific pairing between nonhomologous bands has been described for polytene chromosomes in *Drosophila* (*ectopic pairing*). Such association is especially prominent for bands that are very bright after quinacrine staining and thus presumably contain AT-rich DNA. Mayfield and Ellison (1975) have proposed that this association is based on similarity in base sequences which are recognized by specific proteins which then bind such regions together. Perhaps related to this are the recent findings of Musich *et al.* (1977) that, in constitutive heterochromatin of several mammals, nucleosomes are arranged in a regular way with regard to restriction enzyme sites on the simple sequence repetitive DNA so that specific sequences are exposed in the spacers between nucleosomes. H1 is replaced here by specific nonhistone proteins which recognize these sequences and are bound tightly to the DNA.

f. Underreplication in Polytene Chromosomes. In polytene chromosomes of *Drosophila,* constitutive heterochromatin remains at the diploid or at most tetraploid level even when the euchromatic regions are replicated several hundred fold (for references, see Section II on DNA organization).

g. Reduced Recombination. Meiotic recombination is reduced greatly in constitutive heterochromatin as compared to euchromatin (Cooper, 1959; Roberts, 1965; Brown, 1966). A similar relative decrease also has been demonstrated for sister strand exchanges (Carrano and Wolff, 1975). However, an increase in both types of exchanges is found at the border of heterochromatin and euchromatin (Carrano and Wolff, 1975).

3. *Properties of Facultative Heterochromatin*

Facultative heterochromatin refers to chromosome segments or entire chromosomes which become inactivated, condensed, and late-replicating during early development and, except for rare cases of reversal, remain inactive throughout the life of an individual. The best known examples are the inactive X chromosome of mammalian females (Barr body, sex chromatin) and the paternal chromosome set in mealybugs (see Chapter 4 in Volume 1 of this treatise).

a. Mammalian X Chromosome. In somatic cells of females, only one X chromosome is active. Early in embryogenesis, one of the X chromosomes becomes heterochromatic and inactive and remains in that state in all descendants of that particular cell. In marsupials, it is always the paternal X that is inactive (Cooper *et al.*, 1971), while in Eutheria it may be either X in any one cell (Lyon, 1972). Heterochromatization thus represents a long-term inactivation of chromosomes, ostensibly as a means of dosage compensation in the female cells.

One property characteristic of facultative heterochromatin is *reactivation,* either as a normal process during development or under special experimental conditions. During differentiation of female gametes, long before meiosis, both X chromosomes are found to be active and euchromatic (Ohno, 1963; Migeon and Jelalian, 1977). Localized activation of a gene on an inactive X was observed in human–mouse hybrid cells (Kahan and DeMars, 1975).

b. The Mealybug (Lecanoid) System.* In males of these insects, the paternal set of chromosomes becomes heterochromatic early in embryogenesis (Figures 34 and 35). Concomitantly, paternal chromosomes are genetically inactive, males expressing only genes received from their mothers (Brown and Nur, 1964). Autoradiography after treatment with tritiated uridine reveals grains only over the euchromatic chromosome set. Heterochromatization thus prevents transcription of the paternal chromosomes (Berlowitz, 1965).

Reversal of heterochromatization has also been observed in the mealybug system (Nur, 1967). In certain tissues, the heterochromatic set becomes euchromatic at a specific time in development. This can explain the residual genetic activity of the paternal chromosome set suggested by experiments involving heavy irradiation of the father and from species crosses (Brown and Nur, 1964). It thus is probable that in these tissues certain paternal genes are expressed after euchromatization.

IX. METAPHASE CHROMOSOME BANDING

A. Methods That Differentiate Chromosome Regions

Metaphase chromosomes are condensed uniformly along their length. As a result, in organisms lacking polytene chromosomes, cytogenetic analysis largely was restricted to meiotic prophase chromomere patterns. This changed radically in 1968, when Caspersson found that certain fluorochromes produced a chromosome-specific pattern of bright and dark bands (Caspersson *et al.*, 1968). This soon was followed by the development of banding methods that combined pretreatments of chro-

* See Chapter 4, Volume 1, for more details.

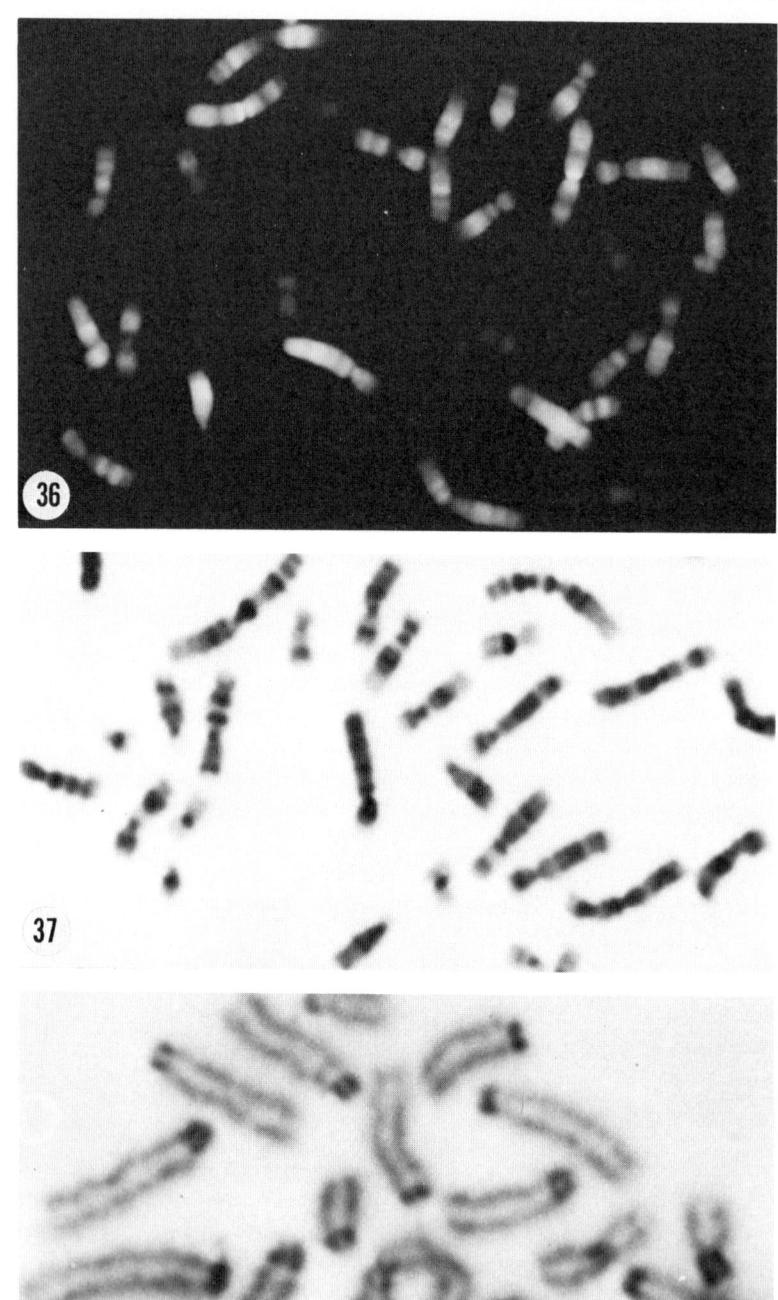
36
37
38

mosomes with Giemsa staining. These methods greatly increased the power of cytogenetic analysis and provided new insights into the structural and functional differentiation of chromosomes. The major types of chromosome regions defined by these methods are listed below.

1. *Quinacrine Bands (Q)/Giemsa Bands (G)*

Fluorescence staining with quinacrine shows that each chromosome is subdivided into a unique pattern of bright and dark bands (for examples, see Caspersson and Zech, 1973). With few exceptions, the bright segments stain darkly with Giemsa stain after certain pretreatments. These include treatment with alkali, phosphate and citrate buffers, and trypsin digestion. In general, Q bands and G bands define similar chromosome subunits (Figures 36 and 37).

2. *C Bands*

The segments stained correspond to constitutive heterochromatin which contains simple sequence repetitive DNA (Figures 33 and 38). These methods derived originally from combining Giemsa staining with *in situ* hybridization techniques (Pardue and Gall, 1970). They include treatment with alkali at 60°C or extraction with 0.2 *N* HCl at room temperature, followed by Giemsa staining (Arrhigi and Hsu, 1971).

3. *Centromere Dots, Cd Bands*

This modified Giemsa method (Eiberg, 1974) specifically stains two dots of uniform size at the centromere region of human chromosomes.

4. *Human Chromosome 9 Centric Chromatin (G-11 Bands)*

The secondary constriction of chromosome 9, a subgroup of C bands, is stained differentially with Giemsa at pH 11 (Bobrow *et al.*, 1972; Gagné and Laberge, 1972). This staining also is observed in meiotic prophase and in interphase.

5. *Telomere Staining, T Bands*

Staining by acridine orange or Giemsa is limited to small dots at most chromosome ends when chromosomes are exposed to certain buffers at 87°C (Dutrillanx, 1973).

Fig. 36. Human metaphase chromosomes (leukocyte culture) stained with quinacrine: Q banding.

Fig. 37. Human metaphase chromosomes (leukocyte culture). Trypsin–Giemsa technique: G banding.

Fig. 38. Mouse metaphase chromosomes (A9 cell line) stained with C banding method. Only centric constitutive heterochromatin is stained.

6. *Nucleolar Organizer, N Bands*

Two very different methods which appear to be specific for nucleolar organizers have been described: (1) heating (96°C) in 1 *M* NaH_2PO_4, adjusted to pH 4.2, followed by Giemsa staining, specifically stains the nucleolar organizer in plants and animals (Funaki *et al.*, 1975); (2) a modified ammoniacal silver staining method (Ag–As, Goodpasture and Bloom, 1975; Denton *et al.*, 1976) selectively stains the active nucleolar organizer.

7. *Inactive X Staining*

Kanda (1973) developed a method involving prefixation treatment in hypotonic KCl at 50°C, which, in the mouse, preferentially stains the inactive X chromosome at metaphase when the X chromosomes usually are indistinguishable.

8. *Differential Staining in Somatic Cell Hybrids (G-11 Technique)*

In Chinese hamster–human or mouse–chimpanzee hybrid cells, alkaline Giemsa stain (pH 11.3) differentiates chromosomes according to species. Chromosome pieces retain the staining characteristic of the species even in interspecific chromosome translocations (Friend *et al.*, 1976).

B. Chemical Basis for Metaphase Banding

With certain exceptions, the brightness of chromosome segments stained with quinacrine is related to the AT/GC ratio (AT richness), with the reservation that interspersion with G decreases brightness. The length of uninterrupted T sequences thus is an important factor in this staining. This interpretation is supported by a variety of observations. *In vitro* studies on the interaction of DNA with quinacrine demonstrated that AT enhances fluorescence but that QC quenches it (Weisblum and de Haseth, 1972; Pachman and Rigler, 1972). Actinomycin D and chromomycin A_3 (CMA), on the other hand, bind specifically to guanine and give bright fluorescence to the quinacrine-dark bands (Schweizer, 1976). Fluorescence-labeled antibodies to adenosine give essentially the same pattern as quinacrine (Dev *et al.*, 1972). If anticytosine antibodies are used, however, the quinacrine dark bands become bright (Schreck *et al.*, 1973). Perhaps the strongest evidence comes from a direct comparison of base ratios established from autoradiographic experiments. After specific labeling with T or with C, autoradiographic counts were used to determine AT/GC ratios in 75 segments of the human karyotype. A significant correlation was found between the AT/GC ratio thus obtained and the brightness of quinacrine fluorescence (Korenberg, 1976). However, it is clear that other factors, including the arrangement of bases, also are important.

Very little is known about the chemical bases of the Giemsa staining techniques. The evidence indicates that the specificity probably is due to proteins, especially nonhistone proteins. However, very little has been learned about chemical differences between chromosome segments from the many attempts to identify compounds that are specifically extracted by the banding techniques and the nature of interaction between Giemsa stain and chromosome components.

C. Structural Basis for Metaphase Bands

Several authors have suggested that banding patterns are caused by basic structural differences along chromosomes rather than differential extractions that are part of the banding procedures. G bands, according to this view, correspond to regions of chromosomes that already are condensed relatively highly in untreated chromosomes. For instance, G bands often are seen by phase microscopy in chromosomes that did not receive special treatments (McKay, 1973) and in electron micrographs of water-spread whole mounts of metaphase chromosomes (Golomb and Bahr, 1974; Green and Bahr, 1975). What these authors ignore, however, is that phase microscopy does not show such banding *in vivo* and metaphase chromosomes fixed *in situ* appear uniformly dense in electron micrographs. Where "structurally banded" chromosomes have been observed, the chromosomes were exposed to hypotonic salt solutions, distilled water, or even chelating agents prior to fixation. These treatments very well could affect chromosome regions differentially. This suggests that banding may well be structurally based, i.e., only concentration dependent. However, underlying chemical differences determine how specific chromosome regions react toward the pre- or postfixation treatments.

We have approached this problem by asking how the treatments affect the various levels of metaphase chromosome organization as we described them in this review (Ris and Korenberg, 1977). The successful G banding methods include treatment with hypotonic saline, high pH buffers, chelating agents, and trypsin, all of which are known to unwind the higher orders or chromatin structure, for example, by extracting H1. We therefore exposed mouse (A9 cell line) chromosomes isolated by the method of Wray and Stubblefield (1970) to agents that are known to extract H1 (0.6 *M* NaCl, borate buffer at pH 9, tRNA, and brief trypsin treatment). With each of these treatments, we obtained first G banding and, with longer treatment, C banding (Figure 39). At higher magnification, it is obvious that in interbands the fibers are 10 nm thick or strings of 10-nm beads, while in the dense bands they are still 50 or 20 nm thick, as in untreated chromosomes (Figure 40). In the dense bands, the chromatin fiber remains supercoiled, either because it contains a more tightly bound H1 subfraction or perhaps nonhistone proteins which stabilize the supercoil. A recent paper by Musich *et al.* (1977) may provide the explanation; they were able to isolate nucleosomes from constitutive heterochromatin of African green monkey cells (CV-1) that contains highly repetitive satellite DNA and to analyze the proteins associated with these nucleosomes. Most significantly, the H1 proteins are depleted or perhaps absent and replaced with several nonhistone proteins of similar molecular weight which are bound tightly to DNA.

These observations suggest a new explanation for metaphase banding. The replacement of H1 with more tightly bound nonhistone proteins in specific chromosome regions makes the chromatin superstructure more resistant to the treatments used. In interbands, the chromatin is unraveled to a string of nucleosomes, and loss of DNA also may occur subsequently. The chemical basis of banding resides first in the nature of nucleotide sequences and second in the proteins which stabilize the supercoiling of the nucleosome chain. Differential resistance to the banding techniques leads to structural differentiation which is visible by phase microscopy, electron microscopy, and staining.

D. Functional Significance of Metaphase Bands

Metaphase banding reminds us that chromosomes are not homogeneous structures but are organized into blocks with different functional properties. C bands contain the simple sequence highly repeated DNA, which apparently never is transcribed. Their higher-order structure is unusually stable, perhaps because of unique nonhistone proteins. This certainly is related to properties of constitutive heterochromatin. G bands also show greater structural stability, although little is known about the chemical basis. In general, they also are correlated to constitutive (intercalary) heterochromatin and correspond to the regions that are visible in meiotic prophase as pachytene chromomeres (Ferguson-Smith and Page, 1973; Okada and Comings, 1974; Luciani *et al.,* 1976). These regions also behave as units with regard to replication. According to Stubblefield (1975), all the DNA within a band replicates within a small interval of the S period. In general, Q and G bands replicate late in S (Ganner and Evans, 1971). Breakage induced by X ray occurs predominantly in the Q-dark regions (Seabright, 1973; Holmberg and Jonasson, 1973). Somatic chias-

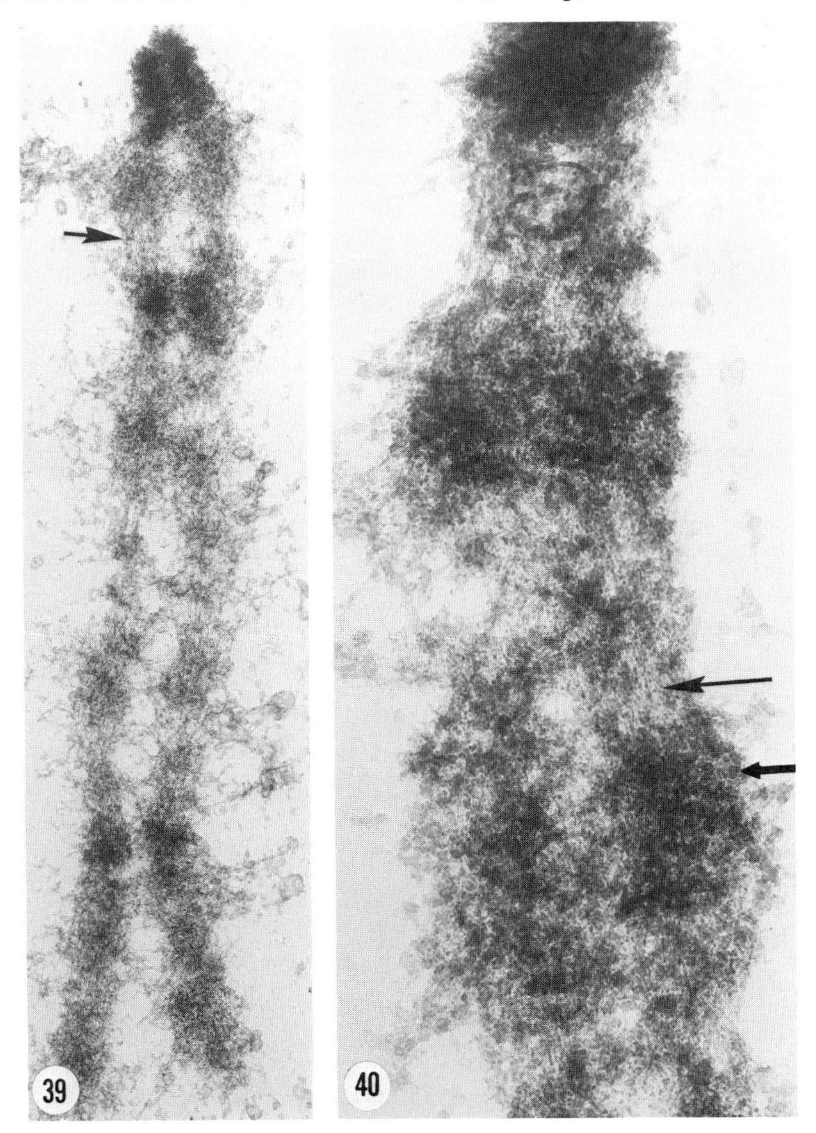

Fig. 39. Mouse (A9 cell line) metaphase chromosome isolated with the Wray–Stubblefield (pH 6.7) method (see also Figure 26), spread on water, picked up on a Formvar–carbon-coated grid, treated with 0.6 *M* NaCl (10 min at 4°C) to extract H1, fixed in 2% formaldehyde, and critical point-dried. In the bands, the chromatin remains compacted as 20- or 50-nm fibers, while in the interbands the fibers are unwound into 10-nm fibers or beaded strings (arrow). × 13,000.

Fig. 40. Mouse (A9 cell line) metaphase (metacentric) chromosome, prepared as in Figure 39. Centric heterochromatin contains 50-nm fibers (thick arrow) while in interbands the fibers are unwound into beaded strings (thin arrow). × 40,000.

mata also occur predominantly in the dark regions (Kuhn, 1975). A comparison of the gene density map on human chromosomes (McKusick and Ruddle, 1977) with Q and G banding suggests that regions with prominent Q or G bands are relatively poor in the number of known genes. The presence of segments along the chromosome that differ qualitatively in nucleotide sequences and protein content, that have different structural properties and transcriptional activity and relative density of genes, is well established. But much still is to be learned about the functional significance of this organization into units which involve tremendous stretches of DNA.

ACKNOWLEDGMENTS

Supported by research career program award (K-6-GM 21,948) (H.R.) and a research grant (GM 04738) from the National Institutes of Health. The High Voltage Electron Microscope Facility—Madison—is supported by a grant from the Biotechnology Resources Program, NIH (RR570).

REFERENCES

Altman, P. L., and Dittmer, D. S., eds. (1962). "Growth," Biol. Handb., pp. 1–70. Fed. Am. Soc. Exp. Biol., Washington, D.C.

Ammermann, D. (1971). Morphology and development of the macronuclei of the ciliates *Stylonychia mytilus* and *Euplotes aediculatus*. *Chromosoma* **33,** 209–238.

Angelier, N., and Lacroix, J. C. (1975). Complèxes de transcription d'origines nucléolaire et chromosomique d'ovocytes de *Pleurodeles waltlii* et *P. poireti* (Amphibiens, Urodèles). *Chromosoma* **51,** 323–335.

Areci, R., Senger, D. R., and Gross, P. R. (1976). The programmed switch in lysine-rich histone synthesis at gastrulation. *Cell* **9,** 171–178.

Arrighi, F. E., and Hsu, T. C. (1971). Localization of heterochromatin in human chromosomes. *Cytogenetics* **10,** 81–86.

Axel, R. (1975). Cleavage of DNA in nuclei and chromatin with staphylococcal nuclease. *Biochemistry* **14,** 2921–2925.

Axel, R., Melchior, W., Jr., Sollner-Webb, B., and Felsenfeld, G. (1974). Specific sites of interaction between histones and DNA in chromatin. *Proc. Natl. Acad. Sci. U.S.A.* **71,** 4101–4105.

Bacheler, L. T., and Smith, K. D. (1976). Transcription of isolated mouse liver chromatin. *Biochemistry* **15,** 3281–3290.

Bahr, G. F. (1975). The fibrous structure of human chromosomes in relation to rearrangements and aberrations; a theoretical consideration. *Fed. Proc., Fed. Am. Soc. Exp. Biol.* **34,** 2209–2217.

Bajer, A. (1966). Morphological aspects of normal and abnormal mitosis. *In* "Probleme der biologischen Reduplikation" (P. Sitte, ed.), pp. 91–119. Springer-Verlag, Berlin and New York.

Baldwin, J. P., Boseley, P. G., Bradbury, E. M., and Ibel, K. (1975). The subunit structure of the eukaryotic chromosome. *Nature (London)* **253,** 245–249.

Batistoni, R., Nardi, I., and Barsacchi Pilone, G. (1974). Banding patterns on lampbrush chromosomes of *Triturus marmoratus* (Amphibia Urodela) by the Giemsa stain. *Chromosoma* **49**, 121–134.

Beermann, W. (1962). *Riesenchromosomen. Protoplasmatologia* **6**, 1–161.

Beermann, W. (1972). Chromomeres and genes. *Results Probl. Cell Differ.* **4**, 1–33.

Belling, J. (1928). The ultimate chromomeres of *Lilium* and *Aloë* with regard to the numbers of genes. *Univ. Calif., Berkeley, Publ. Bot.* **14**, 307–318.

Benyajati, C., and Worcel, A. (1976). Isolation, characterization, and structure of the folded interphase genome of *Drosophila melanogaster. Cell* **9**, 393–407.

Berezney, R., and Coffey, D. S. (1974). Identification of a nuclear protein matrix. *Biochem. Biophys. Res. Commun.* **60**, 1410–1417.

Berger, S., and Schweiger, H.-G. (1975). An apparent lack of nontranscribed spacers in rDNA of a green alga. *Mol. Gen. Genet.* **139**, 269–275.

Berkowitz, E. M., and Doty, P. (1975). Chemical and physical properties of fractionated chromatin. *Proc. Natl. Acad. Sci. U.S.A.* **72**, 3328–3332.

Berlowitz, L. (1965). Correlation of genetic activity, heterochromatization, and RNA metabolism. *Proc. Natl. Acad. Sci. U.S.A.* **53**, 68–73.

Bernhard, W. (1969). A new staining procedure for electron microscopical cytology. *J. Ultrastruct. Res.* **27**, 250–265.

Bernstein, M. H., and Mazia, D. (1953). The desoxyribonucleoprotein of sea urchin sperm. 1. Isolation and analysis. *Biochim. Biophys. Acta* **10**, 600–612.

Biesele, J. J. (1947). Chromosomes in lymphatic leukemia of C58 mice. *Cancer Res.* **7**, 70–77.

Billett, M. A., and Barry, J. M. (1974). Role of histones in chromatin condensation. *Eur. J. Biochem.* **49**, 477–484.

Billing, R. J., and Bonner, J. (1972). The structure of chromatin as revealed by deoxyribonuclease digestion studies. *Biochim. Biophys. Acta* **281**, 453–462.

Bishop, J. O. (1972). Molecular hybridization of ribonucleic acid with a large excess of deoxyribonucleic acid. *Biochem. J.* **126**, 171–185.

Blankstein, L. A., and Levy, S. B. (1976). Changes in histone f2a2 associated with proliferation of Friend leukaemic cells. *Nature (London)* **260**, 638–640.

Bobrow, M., Madan, K., and Pearson, P. L. (1972). Staining of some specific regions of human chromosomes, particularly the secondary constriction of No. 9. *Nature (London), New Biol.* **238**, 122–124.

Bonner, J., and Wu, J.-R. (1973). A proposal for the structure of the *Drosophila* genome. *Proc. Natl. Acad. Sci. U.S.A.* **70**, 535–537.

Bonner, J., Garrard, W. T., Gottesfeld, J., Holmes, D. S., Sevall, J. S., and Wilkes, M. (1973). Functional organization of the mammalian genome. *Cold Spring Harbor Symp. Quant. Biol.* **38**, 303–310.

Bonner, W. M., and Pollard, H. B. (1975). The presence of F3-F2a1 dimers and F1 oligomers in chromatin. *Biochem. Biophys. Res. Commun.* **64**, 282–288.

Bostock, C. J., and Christie, S. (1975). Chromosomes of a cell line of *Dipodomys panamintinus* (kangaroo rat)—a banding and autoradiographic study. *Chromosoma* **51**, 25–34.

Bouteille, M., Laval, M., and Dupuy-Coin, A. M. (1974). Localization of nuclear functions as revealed by ultrastructural autoradiography and cytochemistry. *In* "The Cell Nucleus" (H. Busch, ed.), Vol. 1, pp. 3–71. Academic Press, New York.

Bradbury, E. M. (1975a). Forword: Histone nomenclature. *Ciba Found. Symp.* **28** (New Ser.), 1–4.

Bradbury, E. M. (1975b). Histones in chromosomal structure and control of cell division. *Ciba Found. Symp.* **28** (New Ser.), 131–148.

Bradbury, E. M., and Crane-Robinson, C. (1971). Physical and conformational studies of histones and nucleohistones. *In* "Histones and Nucleohistones" (D.-M. P. Phillips, ed.), pp. 85–134. Plenum, New York.

Bram, S., and Ris, H. (1971). On the structure of nucleohistone. *J. Mol. Biol.* **55,** 325–336.

Brasch, K. (1976). Studies on the role of histones H1 (f1) and H5 (f2c) in chromatin structure. *Exp. Cell Res.* **101,** 396–410.

Brasch, K., Seligy, V. L., and Setterfield, G. (1971). Effects of low salt concentration on structural organization and template activity of chromatin in chicken erythrocyte nuclei. *Exp. Cell Res.* **65,** 61–72.

Brasch, K., Setterfield, G., and Neelin, J. M. (1972). Effects of sequential extraction of histone proteins on structural organization of avian erythrocyte and liver nuclei. *Exp. Cell Res.* **74,** 27–41.

Bridges, C. B. (1916). Nondisjunction as proof of the chromosome theory of heredity. *Genetics* **1,** 1–52.

Britten, R. J., and Davidson, E. H. (1969). Gene regulation for higher cells: A theory. *Science* **165,** 349–357.

Britten, R. J., and Davidson, E. H. (1971). Repetitive and nonrepetitive DNA sequences and a speculation on the origins of evolutionary novelty. *Q. Rev. Biol.* **46,** 111–138.

Britten, R. J., and Kohne, D. E. (1968). Repeated sequences in DNA. *Science* **161,** 529–540.

Brown, S. W. (1966). Heterochromatin. *Science* **151,** 417–425.

Brown, S. W., and Nur, U. (1964). Heterochromatic chromosomes in the coccids. *Science* **145,** 130–136.

Burgoyne, L. A., Hewish, D. R., and Mobbs, J. (1974). Mammalian chromatin substructure studies with the calcium–magnesium endonuclease and two-dimensional polyacrylamide gel electrophoresis. *Biochem. J.* **143,** 67–72.

Callan, H. G., and Macgregor, H. C. (1958). Action of deoxyribonuclease on lampbrush chromosomes. *Nature (London)* **181,** 1479–1480.

Carrano, A. V., and Wolff, S. (1975). Distribution of sister chromatid exchanges in the euchromatin and heterochromatin of the Indian muntjac. *Chromosoma* **53,** 361–369.

Caspersson, T., and Zech, L., eds. (1973). "Chromosome Identification," Nobel Symp. No. 23. Academic Press, New York.

Caspersson, T., Farber, S., Foley, G. E., Kudynowski, J., Modest, E. J., Simonsson, E., Wagh, U., and Zech, L. (1968). Chemical differentiation along metaphase chromosomes. *Exp. Cell Res.* **49,** 219–222.

Cassagnau, P. (1968). Sur la structure des chromosomes salivaires de *Bilobella massoudi* Cassagnau (*Collembola: Neanuridae*). *Chromosoma* **24,** 42–58.

Cassagnau, P. (1971). Les chromosomes salivaires polytènes chez *Bilobella grassei* (Denis) (*Collemboles: Neanuridae*). *Chromosoma* **35,** 57–83.

Chamberlin, M. E., Britten, R. J., and Davidson, E. H. (1975). Sequence organization in *Xenopus* DNA studied by the electron microscope. *J. Mol. Biol.* **96,** 317–333.

Chapman, G. E., Hartman, P. G., and Bradbury, E. M. (1976). Studies on the role and mode of operation of the very-lysine-rich histone H1 in eukaryote chromatin. *Eur. J. Biochem.* **61,** 69–75.

Clark, M. A., and Ackerman, G. A. (1971). A histochemical evaluation of the pyroantimonate–osmium reaction. *J. Histochem. Cytochem.* **19,** 727–737.

Clark, R. J., and Felsenfeld, G. (1971). Structure of chromatin. *Nature (London), New Biol.* **229,** 101–106.

Clarkson, S. G., Birnstiel, M. L., and Serra, V. (1973). Reiterated transfer RNA genes of *Xenopus laevis*. *J. Mol. Biol.* **79,** 391–410.

Coelingh, J. P., Monfoort, C. H., Rozijn, T. H., Leuven, J. A. G., Schiphof, R., Steyn-Parvé, E. P., Braunitzer, G., Schrank, B., and Ruhfus, A. (1972). The complete amino acid sequence of the basic nuclear protein of bull spermatozoa. *Biochim. Biophys. Acta* **285,** 1–14.

Cohen, L. H., Newrock, K. M., and Zweidler, A. (1975). Stage-specific switches in histone synthesis during embryogenesis of the sea urchin. *Science* **190,** 994–997.

Cohn, R. H., Lowry, J. C., and Kedes, L. H. (1976). Histone genes of the sea urchin (*S. purpuratus*) cloned in *E. coli:* Order, polarity, and strandedness of the five histone-coding and spacer regions. *Cell* **9,** 147–161.

Coleman, L. C., and Hillary, B. B. (1941). The minor coil in meiotic chromosomes and associated phenomena as revealed by the Feulgen technique. *Am. J. Bot.* **28,** 464–469.

Comings, D. E. (1974). The role of heterochromatin. *In* "Birth Defects, Proceedings of the Fourth International Conference" (A. G. Motulsky, W. Lentz, and F. J. G. Ebling, eds.), Int. Cong. Ser. No. 310, pp. 44–52. Excerpta Med. Found. Amsterdam.

Comings, D. E., and Okada, T. A. (1976). The fibrillar nature of the nuclear matrix. *J. Cell Biol.* **70,** 119a.

Compton, J. L., Bellard, M., and Chambon, P. (1976). Biochemical evidence of variability in the DNA repeat length in the chromatin of higher eukaryotes. *Proc. Natl. Acad. Sci. U.S.A.* **73,** 4382–4386.

Cooper, D. W., VandeBerg, J. L., Sharman, G. B., and Poole, W. E. (1971). Phosphoglycerate kinase polymorphism in kangaroos provides further evidence for paternal X inactivation. *Nature (London), New Biol.* **230,** 155–157.

Cooper, K. W. (1959). Cytogenetic analysis of major heterochromatic elements (especially Xh and Y) in *Drosophila melanogaster,* and the theory of 'heterochromatin.' *Chromosoma* **10,** 535–588.

Craig-Holmes, A. P., and Shaw, M. W. (1971). Polymorphism of human constitutive heterochromatin. *Science* **174,** 702–704.

Crain, W. R., Davidson, E. H., and Britten, R. J. (1976). Contrasting patterns of DNA sequence arrangement in *Apis mellifera* (Honeybee) and *Musca domestica* (Housefly). *Chromosoma* **59,** 1–12.

Crick, F. (1971). General model for the chromosomes of higher organisms. *Nature (London)* **234,** 25–27.

Cryer, D. R., Goldthwaite, C. D., Zinker, S., Lam, K.-B., Storm, E., Hirschberg, R., Blamire, J., Finkelstein, D. B., and Marmur, J. (1973). Studies on nuclear and mitochondrial DNA of *Saccharomyces cerevisiae. Cold Spring Harbor Symp. Quant. Biol.* **38,** 17–29.

D'Anna, J. A., Jr., and Isenberg, I. (1974). Interactions of histone LAK (f2a2) with histones KAS (f2b) and GRK (f2a1). *Biochemistry* **13,** 2098–2104.

Darlington, C. D., and LaCour, L. (1940). Nucleic acid starvation of chromosomes in *Trillium. J. Genet.* **40,** 185–213.

Daskal, Y., Mace, M. L., Jr., Wray, W., and Busch, H. (1976). Use of direct current sputtering for improved visualization of chromosome topology by scanning electron microscopy. *Exp. Cell Res.* **100,** 204–212.

Davidson, E. H., Crippa, M., Kramer, F. R., and Mirsky, A. E. (1966). Genomic function during the lampbrush chromosome stage of amphibian oögenesis. *Proc. Natl. Acad. Sci. U.S.A.* **56,** 856–863.

Davies, H. G., and Haynes, M. E. (1976). Electron microscope observations on cell nuclei in various tissues of a teleost fish: The nucleolus-associated monolayer of chromatin structural units. *J. Cell Sci.* **21,** 315–327.

DeLange, R. J., and Smith, E. L. (1975). Histone function and evolution as viewed by sequence studies. *Ciba Found. Symp.* **28** (New Ser.), 59–70.

Denton, T. E., Howell, W. M., and Barrett, J. V. (1976). Human nucleolar organizer chromosomes: Satellite association. *Chromosoma* **55,** 81–84.

Dev, V. G., Warburton, D., Miller, O. J., Miller, D. A., Erlanger, B. F., and Beiser, S. M. (1972). Consistent pattern of binding of antiadenosine antibodies to human metaphase chromosomes. *Exp. Cell Res.* **74,** 288–293.

Dixon, G. H., Candido, E. P. M., Honda, B. M., Louie, A. J., Macleod, A. R., and Sung, M. T. (1975). The biological roles of postsynthetic modifications of basic nuclear proteins. *Ciba Found. Symp.* **28** (New Ser.), 229–250.

DuPraw, E. J. (1965). Macromolecular organization of nuclei and chromosomes: A folded fibre model based on whole-mount electron microscopy. *Nature (London)* **206,** 338–343.

DuPraw, E. J. (1968). "Cell and Molecular Biology." Academic Press, New York.

DuPraw, E. J. (1970). "DNA and Chromosomes." Holt, New York.

Dutrillaux, B. (1973). Nouveau système de marquage chromosomique: Les bandes T. *Chromosoma* **41,** 395–402.

Eiberg, H. (1974). New selective Giemsa technique for human chromosomes, C_d staining. *Nature (London)* **248,** 55.

Elgin, S. C. R., and Weintraub, H. (1975). Chromosomal proteins and chromatin structure. *Annu. Rev. Biochem.* **44,** 725–774.

Evans, H. J., Gosden, J. R., Mitchell, A. R., and Buckland, R. A. (1974). Location of human satellite DNA's on the Y chromosome. *Nature (London)* **251,** 346–347.

Fambrough, D. M., Fujimura, F., and Bonner, J. (1968). Quantitative distribution of histone components in the pea plant. *Biochemistry* **7,** 575–584.

Frankhauser, G. (1934). Cytological studies on egg fragments of the salamander Triton. V. Chromosome number and chromosome individuality in the cleavage mitoses of merogonic fragments. *J. Exp. Zool.* **68,** 1–57.

Ferguson-Smith, M. A., and Page, B. M. (1973). Pachytene analysis in a human reciprocal (10;11) translocation. *J. Med. Genet.* **10,** 282–287.

Filip, D. A., Gilly, C., and Mouriquand, C. (1975). Metaphase chromosome ultrastructure. II. Helical organization of the basic chromosome fiber as revealed by acute angle metal deposition. *Humangenetik* **30,** 155–165.

Finch, J. T., and Klug, A. (1976). Solenoidal model for superstructure in chromatin. *Proc. Natl. Acad. Sci. U.S.A.* **73,** 1897–1901.

Finch, J. T., Noll, M., and Kornberg, R. D. (1975). Electron microscopy of defined lengths of chromatin. *Proc. Natl. Acad. Sci. U.S.A.* **72,** 3320–3322.

Finch, J. T., Lutter, L. C., Rhodes, D., Brown, R. S., Rushton, B., Levitt, M., and Klug, A. (1977). Structure of nucleosome core particles of chromatin. *Nature (London)* **269,** 29–36.

Flamm, W. G., Walker, P. M. B., and McCallum, M. (1969). Some properties of the single strands isolated from DNA of the nuclear satellite of the mouse (*Mus musculus*). *J. Mol. Biol.* **40,** 423–443.

Foe, V. E., Wilkinson, L. E., and Laird, C. D. (1976). Comparative organization of active transcription units in *Oncopeltus fasciatus*. *Cell* **9,** 131–146.

Forejt, J. (1973). Centrometric heterochromatin polymorphism in the house mouse: Evidence from inbred strains and natural populations. *Chromosoma* **43,** 187–201.

Friend, K. K., Chen, S., and Ruddle, F. H. (1976). Differential staining of interspecific chromosomes in somatic cell hybrids by alkaline Giemsa stain. *Somat. Cell Genet.* **2,** 183–188.

Funaki, K., Matsui, S., and Sasaki, M. (1975). Location of nucleolar organizers in animal and plant chromosomes by means of an improved N-banding technique. *Chromosoma* **49,** 357–370.

Gagné, R., and Laberge, C. (1972). Specific cytological recognition of the heterochromatic segment of number 9 chromosome in man. *Exp. Cell Res.* **73**, 239–242.

Gall, J. G. (1963a). Kinetics of deoxyribonuclease action on chromosomes. *Nature* (*London*) **198**, 36–38.

Gall, J. (1963b). Chromosome fibers from an interphase nucleus. *Science* **139**, 120–121.

Gall, J. G. (1966). Chromosome fibers studied by a spreading technique. *Chromosoma* **20**, 221–233.

Gall, J. G. (1968). Differential synthesis of the genes for ribosomal RNA during amphibian oogenesis. *Proc. Natl. Acad. Sci. U.S.A.* **60**, 553–560.

Gall, J. G., and Atherton, D. D. (1974). Satellite DNA sequences in *Drosophila virilis. J. Mol. Biol.* **85**, 633–664.

Gall, J. G., and Callan, Y. G. (1962). H^3-uridine incorporation in lampbrush chromosomes. *Proc. Natl. Acad. Sci. U.S.A.* **48**, 562–570.

Ganner, E., and Evans, H. J. (1971). "The relationship between patterns of DNA replication and of quinacrine fluorescence in the human chromosome complement. *Chromosoma* **35**, 326–341.

Garel, A., and Axel, R. (1976). Selective digestion of transcriptionally active ovalbumin genes from oviduct nuclei. *Proc. Natl. Acad. Sci. U.S.A.* **73**, 3966–3970.

Gay, H., Das, C. C., Forward, K., and Kaufmann, B. P. (1970). DNA content of mitotically-active condensed chromosomes of *Drosophila melanogaster. Chromosoma* **32**, 213–223.

Georgiev, G. P. (1969). On the structural organization of operon and the regulation of RNA synthesis in animal cells. *J. Theor. Biol.* **25**, 473–490.

Germond, J. E., Hirt, B., Oudet, P., Gross-Bellard, M., and Chambon, P. (1975). Folding of the DNA double helix in chromatin-like structures from simian virus 40. *Proc. Natl. Acad. Sci. U.S.A.* **72**, 1843–1847.

Glätzer, K. H. (1975). Visualization of gene transcription in spermatocytes of *Drosophila hydei. Chromosoma* **53**, 371–379.

Goldblatt, D., and Bustin, M. (1975). Exposure of histone antigenic determinants in chromatin. *Biochemistry* **14**, 1689–1695.

Golomb, H. M., and Bahr, G. F. (1974). Correlation of the fluorescent banding pattern and ultrastructure of a human chromosome. *Exp. Cell Res.* **84**, 121–126.

Good, N. E., Winget, G. D., Winter, W., Connolly, T. N., Izawa, S., and Singh, R. M. M. (1966). Hydrogen ion buffers for biological research. *Biochemistry* **5**, 467–477.

Goodpasture, C., and Bloom, S. E. (1975). Visualization of nucleolar organizer regions in mammalian chromosomes using silver staining. *Chromosoma* **53**, 37–50.

Gosden, J. R., Buckland, R. A., Clayton, R. P., and Evans, H. J. (1975). Chromosomal localisation of DNA sequences in condensed and dispersed human chromatin. *Exp. Cell Res.* **92**, 138–147.

Gottesfeld, J. M., Murphy, R. F., and Bonner, J. (1975). Structure of transcriptionally active chromatin. *Proc. Natl. Acad. Sci. U.S.A.* **72**, 4404–4408.

Granboulan, N., and Granboulan, P. (1965). Cytochimie ultrastructurale du nucléole. II. Etude des sites de synthèse du RNA dans le nucléole et le noyau. *Exp. Cell Res.* **38**, 604–619.

Green, R. J., and Bahr, G. F. (1975). Comparison of G-, Q-, and EM-banding patterns exhibited by the chromosome complement of the Indian muntjac, *Muntiacus muntjak*, with reference to nuclear DNA content and chromatin ultrastructure. *Chromosoma* **50**, 53–67.

Griffith, J. D. (1975). Chromatin structure: Deduced from a minichromosome. *Science* **187**, 1202–1203.

Gross, K., Probst, E., Schaffner, W., and Birnstiel, M. (1976). Molecular analysis of the histone gene cluster of *Psammechinus miliaris*. 1. Fractionation and identification of five individual histone mRNA's. *Cell* **8,** 455–469.

Gutherz, S. (1907). Zur Kenntnis der Heterochromosomen. *Arch. Mikrosk. Anat. Entwicklungsmech.* **69,** 491.

Hanson, C. V. (1975). Techniques in the isolation and fractionation of eukaryotic chromosomes. *In* "New Techniques in Biophysics and Cell Biology" (R. H. Pain and B. J. Smith, eds.), Vol. 2, pp. 43–83. Wiley, New York.

Hardison, R. C., Eichner, M. E., and Chalkley, R. (1975). An approach to histone nearest neighbors in extended chromatin. *Nucleic Acids Res.* **2,** 1751–1770.

Hatch, F. T., Bodner, A. J., Mazrimas, J. A., and Moore, D. H. (1976). Satellite DNA and cytogenetic evolution: DNA quantity, satellite DNA and karyotypic variations in kangaroo rats (*Genus Dipodomys*). *Chromosoma* **58,** 155–168.

Heitz, E. (1928). Das Heterochromatin der Moose I. *Jahrb. Wiss. Bot.* **69,** 762–818.

Heitz, E. (1934). Uber α- und β-Heterochromatin sowie Konstanz und Bau der Chromomeren bei *Drosophila*. *Biol. Zentralbl.* **54,** 588–609.

Hennig, W., and Meer, B. (1971). Reduced polyteny of ribosomal RNA cistrons in giant chromosomes of *Drosophila hydei*. *Nature* (*London*) *New Biol.* **233,** 70–72.

Hennig, W., Meyer, G. F., Hennig, I., and Leoncini, O. (1973). Structure and function of the Y chromosome of *Drosophila hydei*. *Cold Spring Harbor Symp. Quant. Biol.* **38,** 673–691.

Hess, O. (1970). Genetic function correlated with unfolding of lampbrush loops by the Y chromosome in spermatocytes of *Drosophila hydei*. *Mol. Gen. Genet.* **106,** 328–346.

Hess, O. (1971). Lampenbürstenchromosomen. *Handb. Allg. Pathol.* **2,** Part 2, pp. 215–281.

Hess, O., and Meyer, G. F. (1968). Genetic activities of the Y chromosome in *Drosophila* during spermatogenesis. *Adv. Genet.* **14,** 171–223.

Hewish, D. R., and Burgoyne, L. A. (1973). Chromatin substructure: The digestion of chromatin DNA at regularly spaced sites by a nuclear deoxyribonuclease. *Biochem. Biophys. Res. Commun.* **52,** 504–510.

Hnilica, L. S. (1972). "The Structure and Biological Function of Histones." CRC Press, Cleveland, Ohio.

Hochmann, B. (1973). Analysis of a whole chromosome in *Drosophila*. *Cold Spring Harbor Symp. Quant. Biol.* **38,** 581–589.

Holm, G., and Bajer, A. (1966). Cine micrographic studies on mitotic spiralization cycle. *Hereditas* **54,** 356–375.

Holmberg, M., and Jonasson, J. (1973). Preferential location of X-ray induced chromosome breakage in the R-bands of human chromosomes. *Hereditas* **74,** 57–68.

Hood, L., Campbell, J. H., and Elgin, S. C. R. (1975). The organization, expression, and evolution of antibody genes and other multigene families. *Annu. Rev. Genet.* **9,** 305–353.

Howze, G. B., Hsie, A. W., and Olins, A. L. (1976). ν-bodies in mitotic chromatin. *Exp. Cell Res.* **100,** 424–428.

Hozier, J., Renz, M., and Nehls, P. (1977). The chromosome fiber: Evidence for an ordered superstructure of nucleosomes. *Chromosoma* **62,** 301–317.

Hsu, T. C., and Arrighi, F. E. (1971). Distribution of constitutive heterochromatin in mammalian chromosomes. *Chromosoma* **34,** 243–253.

Hsu, T. C., and Benirschke, K. (1967–1974). "An Atlas of Mammalian Chromosomes," Vols. 1–8. Springer-Verlag, Berlin and New York.

Ilyin, Y. V., Varshavsky, A. Ya., Mickelsaar, U. N., and Georgiev, G. P. (1971). Studies on deoxyribonucleoprotein structure: Redistribution of proteins in mixtures of deoxyribonucleoproteins, DNA and RNA. *Eur. J. Biochem.* **22,** 235–245.

Johmann, C. A., and Gorovsky, M. A. (1976). Immunofluorescence evidence for the absence of histone H1 in a mitotically dividing, genetically inactive nucleus. *J. Cell Biol.* **71,** 89–95.

Johns, E. W. (1967). The electrophoresis of histones in polyacrylamide gel and their quantitative determination. *Biochem. J.* **104,** 78–82.

Jones, K. W., and Corneo, G. (1971). Location of satellite and homogeneous DNA sequences on human chromosomes. *Nature (London) New Biol.* **233,** 268–271.

Jones, K. W., Prosser, J., Corneo, G., Girelli, E., and Bobrow, M. (1973). Constitutive heterochromatin in man. *Symp. Med. Hoechst.* **6,** 45–61.

Judd, B. H., Shen, M. W., and Kaufman, T. C. (1972). The anatomy and function of a segment of the X chromosome of *Drosophila melanogaster. Genetics* **71,** 139–156.

Kahan, B., and DeMars, R. (1975). Localized derepression on the human inactive X chromosome in mouse–human cell hybrids. *Proc. Natl. Acad. Sci. U.S.A.* **72,** 1510–1514.

Kanda, N. (1973). A new differential technique for staining the heteropycnotic X chromosome in female mice. *Exp. Cell. Res.* **80,** 463–467.

Kavenoff, R., and Zimm, B. H. (1973). Chromosome-sized DNA molecules from *Drosophilia. Chromosoma* **41,** 1–27.

Kedes, L. H. (1976). Histone messenger and histone genes. *Cell* **8,** 321–331.

Kedes, L. H., Chang, A. C. Y., Houseman, D., and Cohen, S. N. (1975). Isolation of histone genes from unfractionated sea urchin DNA by subculture cloning in *E. coli. Nature (London)* **255,** 533–538.

Kierszenbaum, A. L., and Tres, L. L. (1975). Structural and transcriptional features of the mouse spermatid genome. *J. Cell Biol.* **65,** 258–270.

Killander, D., and Rigler, R. (1965). Initial changes of deoxyribonucleoprotein and synthesis of nucleic acid in phytohemagglutinine-stimulated human leucocytes *in vitro. Exp. Cell. Res.* **39,** 701–704.

Kiryanov, G. I., Manamshjan, T. A., Polyakov, V. Yu., Fais, D., and Chentsov, Ju. S. (1976). Levels of granular organization of chromatin fibres. *FEBS Lett.* **67,** 323–327.

Korenberg, J. (1976). Human chromosome structure: DNA content, base ratio, quinacrine fluorescence. Ph. D. Thesis, University of Wisconsin, Madison.

Kornberg, R. D., and Thomas, J. O. (1974). Chromatin structure: Oligomers of the histones. *Science* **184,** 865–871.

Kuhn, E. M. (1975). Mitotic chiasmata in Bloom's syndrome. Ph.D. Thesis, University of Wisconsin, Madison.

Kuo, M. T., Sahasrabuddhe, C. G., and Saunders, G. F. (1976). Presence of messenger specifying sequences in the DNA of chromatin subunits. *Proc. Natl. Acad. Sci. U.S.A.* **73,** 1572–1575.

Lacy, E., and Axel, R. (1975). Analysis of DNA of isolated chromatin subunits. *Proc. Natl. Acad. Sci. U.S.A.* **72,** 3978–3982.

Laird, C. D. (1971). Chromatid structure: Relationship between DNA content and nucleotide sequence diversity. *Chromosoma* **32,** 378–406.

Laird, C. D., and Chooi, W. Y. (1976). Morphology of transcription units in *Drosophila melanogaster. Chromosoma* **58,** 193–218.

Laird, C. D., Wilkinson, L. E., Foe, V. E., and Chooi, W. Y. (1976). Analysis of chromatin-associated fiber arrays. *Chromosoma* **58,** 169–192.

Lampert, F. (1971). Coiled supercoiled DNA in critical point dried and thin sectioned human chromosome fibres. *Nature (London), New Biol.* **234,** 187–188.

Langendorf, H., Siebert, G., Lorenz, I., Hannover, R., and Beyer, R. (1961). Kationenverteilung in Zellkern und Cytoplasma der Rattenleber. *Biochem. Z.* **335,** 273–284.

Langmore, J. P., and Wooley, J. C. (1975). Chromatin architecture: Investigation of a subunit of chromatin by dark field electron microscopy. *Proc. Natl. Acad. Sci. U.S.A.* **72,** 2691–2695.

Lau, A. F., and Ruddon, R. W. (1977). Proteins of transcriptionally active and inactive chromatin from Friend erythroleukemia cells. *Exp. Cell Res.* **107,** 35–46.

Lauer, G. D., and Klotz, L. C. (1975). Determination of the molecular weight of *Saccharomyces cerevisiae* nuclear DNA. *J. Mol. Biol.* **95,** 309–326.

Lefevre, G., Jr. (1974). The relationship between genes and polytene chromosome bands. *Annu. Rev. Genet.* **8,** 51–62.

Lezzi, M. (1965). Die Wirkung von DNase auf Isolierte Polytän-Chromosomen. *Exp. Cell Res.* **39,** 289–292.

Lilley, D. M. J., Howarth, O. W., Clark, V. M., Pardon, J. F., and Richards, B. M. (1976). The existence of random coil N-terminal peptides—'tails'—in native histone complexes. *FEBS Lett.* **62,** 7–10.

Lima-de-Faria, A. (1969). DNA replication and gene amplification in heterochromatin. *In* "Handbook of Molecular Cytology" (A. Lima-de-Faria, ed.), pp. 277–325. North-Holland Publ., Amsterdam.

Lima-de-Faria, A. (1975). The relation between chromomeres, replicons, operons, transcription units, genes, viruses, and palindromes. *Hereditas* **81,** 249–284.

Littau, V. C., Allfrey, V. G., Frenster, J. H., and Mirsky, A. E. (1964). Active and inactive regions of nuclear chromatin as revealed by electron microscope autoradiography. *Proc. Natl. Acad. Sci. U.S.A.* **52,** 93–100.

Lohr, D., Corden, J., Tatchell, K., Kovacic, R. T., and Von Holde, K. E. (1977). Comparative subunit structure of *HeLa,* yeast, and chicken erythrocyte chromatin. *Proc. Natl. Acad. U.S.A.* **74,** 79–83.

Luciani, J. M., Devictor, M., Morazzani, M.-R., and Stahl, A. (1976). Meiosis of trisomy 21 in the human pachytene oocyte. *Chromosoma* **57,** 155–163.

Lund, E., Dahlberg, J. E., Lindahl, L., Jaskunas, S. R., Dennis, P. P., and Nomura, M. (1976). Transfer RNA genes between 16 S and 23 S rRNA genes in rRNA transcription units of *E. coli. Cell* **7,** 165–177.

Lyon, M. F. (1972). X chromosome inactivation and developmental patterns in mammals. *Biol. Rev. Cambridge Philos. Soc.* **47,** 1–35.

McCarthy, B. J., Nishiura, J. T., Doenecke, D., Nasser, D. S., and Johnson, C. B. (1973). Transcription and chromatin structure. *Cold Spring Harbor Symp. Quant. Biol.* **38,** 763–771.

McClintock, B. (1944). The relation of homozygous deficiencies to mutations and allelic series in maize. *Genetics* **29,** 478–502.

McKay, R. D. G. (1973). The mechanism of G and C banding in mammalian metaphase chromosomes. *Chromosoma* **44,** 1–14.

McKusick, V. A., and Ruddle, F. H. (1977). The status of the gene map of the human chromosomes. *Science* **196,** 390–405.

Maizels, N. (1976). Dictyostelium 17 S, 25 S, and 5 S rDNA's lie within a 38,000 base pair repeated unit. *Cell* **9,** 431–438.

Mazia, D. (1954). The particulate organization of the chromosome. *Proc. Natl. Acad. Sci. U.S.A.* **40,** 521–527.

Manton, I. (1950). The spiral structure of chromosomes. *Biol. Rev. Cambridge Philos. Soc.* **25,** 486–508.

Martinson, H. G., and McCarthy, B. J. (1975). Histone–histone associations within chromatin. Cross-linking studies using tetranitromethane. *Biochemistry* **14,** 1073–1078.

Mayfield, J. E., and Ellison, J. R. (1975). The organization of interphase chromatin in Drosophildae: The self adhesion of chromatin containing the same DNA sequences. *Chromosoma* **52**, 37–48.

Mendelsohn, J. (1974). Studies of isolated mammalian metaphase chromosomes. *In* "The Cell Nucleus" (H. Busch, ed.), Vol. 2, pp. 123–147. Academic Press, New York.

Migeon, B. R., and Jelalian, K. (1977). Evidence for two active X chromosomes in germ cells of female before meiotic entry. *Nature (London)* **269**, 242–243.

Miller, O. L., and Bakken, A. H. (1972). Morphological studies of transcription. *Acta Endocrinol. (Copenhagen), Suppl.* **168**, 155–173.

Miller, O. L., and Beatty, B. R. (1969a). Visualization of nucleolar genes. *Science* **164**, 955–957.

Miller, O. L., and Beatty, B. R. (1969b). Portrait of a gene. *J. Cell. Physiol.* **74**, 225–232.

Miller, O. L., and Hamkalo, B. A. (1972). Visualization of RNA synthesis on chromosomes. *Int. Rev. Cytol.* **33**, 1–25.

Miller, O. J., Schnedl, W., Allen, J., and Erlanger, B. F. (1974). 5-methylcytosine localized in mammalian constitutive heterochromatin. *Nature (London)* **251**, 636–637.

Mizuno, S., and Macgregor, H. C. (1974). Chromosomes, DNA sequences, and evolution in salamanders of the genus *Plethodon*. *Chromosoma* **48**, 239–296.

Montgomery, T. H. (1904). Some observations and considerations upon the maturation phenomena of the germ cells. *Biol. Bull. (Woods Hole, Mass.)* **6**, 137–157.

Morris, N. R. (1976a). Nucleosome structure in *Aspergillus nidulans*. *Cell* **8**, 357–363.

Morris, N. R. (1976b). A comparison of thc structure of chicken erythrocyte and chicken liver chromatin. *Cell* **9**, 627–632.

Mulder, M. P., van Duijn, P., and Gloor, H. J. (1968). The replicative organization of DNA in polytene chromsomes of *Drosophila hydei*. *Genetica* **39**, 385–428.

Muller, H. J., and Painter, T. S. (1932). The differentiation of the sex chromosomes of *Drosophila* into genetically active and inert regions. *Z. Indukt. Abstamm. Vererbungs.* **62**, 316–365.

Musich, P. R., Brown, F. C., and Maio, J. J. (1977). Subunit structure of chromatin and the organization of eukaryotic highly repetitive DNA: Nucleosomal proteins associated with a highly repetitive mammalian DNA. *Proc. Natl. Acad. Sci. U.S.A.* **74**, 3297–3301.

Nagl, W. (1962). 4096-Ploidie und "Riesenchromosomen" in suspensor von *Phaseolus coccineus*. *Naturwissenschaften* **49**, 261–262.

Naora, H., Naora, H., Mirsky, A. E., and Allfrey, V. G. (1961). Magnesium and calcium in isolated cell nuclei. *J. Gen. Physiol.* **44**, 713–742.

Narayan, R. K. J., and Rees, H. (1976). Nuclear DNA variation in *Lathyrus*. *Chromosoma* **54**, 141–154.

Noll, M. (1974a). Subunit structure of chromatin. *Nature (London)* **251**, 249–251.

Noll, M. (1974b). Internal structure of the chromatin subunit. *Nucleic Acids Res.* **1**, 1573–1578.

Noll, M. (1976). Differences and similarities in chromatin structure of *Neurospora crassa* and higher eucaryotes. *Cell* **8**, 349–355.

Noll, M., Thomas, J. O., and Kornberg, R. D. (1975). Preparation of native chromatin and damage caused by shearing. *Science* **187**, 1203–1206.

Noorduyn, N. J. A., and de Man, J. C. H. (1966). RNA synthesis in rat and mouse hepatic cells as studied with light and electron microscope radioautography. *J. Cell Biol.* **30**, 655–660.

Nur, U. (1967). Reversal of heterochromatization and the activity of the paternal chromosome set in the male mealy bug. *Genetics* **56**, 375–389.

Ohlenbush, H. H., Olivera, B. M., Tuan, D., and Davidson, N. (1967). Selective dissociation of histones from calf thymus nucleoprotein. *J. Mol. Biol.* **25,** 299–315.

Ohno, S. (1963). Life history of female germ cells in mammals. *In* "Congential Malformations," pp. 36–40. National Foundation, New York.

Ohnuki, Y. (1968). Structure of chromosomes. I. Morphological studies of the spiral structure of human somatic chromosomes. *Chromosoma* **25,** 401–428.

Okada, T. A., and Comings, D. E. (1974). Mechanisms of chromosome banding. III. Similarity between G-bands of mitotic chromosomes and chromomeres of meiotic chromosomes. *Chromosoma* **48,** 65–71.

Olins, A. L., and Olins, D. E. (1973). Spheroid chromatin units (ν bodies). *J. Cell Biol.* **59,** 252a.

Olins, A. L., and Olins, D. E. (1974). Spheroid chromatin units (ν bodies). *Science* **183,** 330–332.

Olins, A. L., Carlson, R. D., and Olins, D. E. (1975). Visualization of chromatin substructure: ν bodies. *J. Cell Biol.* **64,** 528–537.

Olins, A. L., Carlson, R. D., Wright, E. B., and Olins, D. E. (1976). Chromatin ν bodies: Isolation, subfractionation and physical characterization. *Nucleic Acids Res.* **3,** 3271–3291.

Olins, A. L., Breillatt, J. P., Carlson, R. D., Senior, M. B., Wright, E. B., and Olins, D. E. (1977). On nu models for chromatin structure. *In* "Molecular Biology of the Mammalian Genetic Apparatus" (P. O. P. Ts'o, ed.), pp. 211–237. Elsevier, Amsterdam

Olins, D. E., and Olins, A. L. (1972). Physical studies of isolated eukaryotic nuclei. *J. Cell Biol.* **53,** 715–736.

Oliver, D., and Chalkley, R. (1972). An electrophoretic analysis of *Drosophila* histones. *Exp. Cell Res.* **73,** 295–302.

Oosterhof, D., Hozier, J. C., and Rill, R. L. (1975). Nuclease action on chromatin: Evidence for discrete, repeated nucleoprotein units along chromatin fibrils. *Proc. Natl. Acad. Sci. U.S.A.* **72,** 633–637.

Oudet, P., Gross-Bellard, M., and Chambon, P. (1975). Electron microscopic and biochemical evidence that chromatin structure is a repeating unit. *Cell* **4,** 281–300.

Pachman, U., and Rigler, R. (1972). Quantum yield of acridines interacting with DNA of defined base sequence. *Exp. Cell Res.* **72,** 602–608.

Panyim, S., and Chalkley, R. (1969). The heterogeneity of histones. I. A quantitative analysis of calf histones in very long polyacrylamide gels. *Biochemistry* **8,** 3972–3979.

Pardon, J. F., Wilkins, M. H. F., and Richards, B. M. (1967). Super-helical model for nucleohistone. *Nature (London)* **215,** 508–509.

Pardon, J. F., Richards, B. M., and Cotter, R. I. (1973). X-ray diffraction studies on oriented nucleohistone gels. *Cold Spring Harbor Symp. Quant. Biol.* **38,** 75–81.

Pardon, H. J., Worcester, D. L., Wooley, J. C., Tatchell, K., Van Holde, K. E., and Richards, B. M. (1975). Low-angle neutron scattering from chromatin subunit particles. *Nucleic Acids Res.* **2,** 2163–2176.

Pardue, M. L., and Gall, J. G. (1970). chromosomal localization of mouse satellite DNA. *Science* **168,** 1356–1358.

Pardue, M. L., Brown, D. D., and Birnstiel, M. L. (1973). Location of the genes for 5 S ribosomal RNA in *Xenopus laevis*. *Chromosoma* **42,** 191–203.

Pätau, K. (1973). Three main classes of constitutive heterochromatin in man: Intercalary, Y-type and centric. *Chromosomes Today* **4,** 1–430.

Paul, J. (1972). General theory of chromosome structure and gene activation in eukaryotes. *Nature (London)* **238,** 444–446.

Peacock, W. J., Brutlag, D., Goldring, E., Appels, R., Hinton, C. W., and Lindsley, D. L.

(1973). The organization of highly repeated DNA sequences in *Drosophila melanogaster* chromosomes. *Cold Spring Harbor Symp. Quant. Biol.* **38,** 405–416.

Pellegrini, M., Manning, J., and Davidson, N. (1977). Sequence arrangement of the rDNA of *Drosophila melanogaster. Cell* **10,** 213–224.

Peterson, J. B., and Ris, H. (1976). Electron microscopic study of the spindle and chromosome movement in the yeast *Saccharomyces cerevisiae. J. Cell Sci.* **22,** 219–242.

Petes, T. D., Byers, B., and Fangman, W. L. (1973). Size and structure of yeast chromosomal DNA. *Proc. Nalt. Acad. Sci. U.S.A.* **70,** 3072–3076.

Pierce, W. P. (1937). The effect of phosphorus on chromosome and nuclear volume in a violet species. *Bull. Torrey Bot. Club* **64,** 345–355.

Polisky, B., and McCarthy, B. (1975). Location of histones on simian virus 40 DNA. *Proc. Natl. Acad. Sci. U.S.A.* **72,** 2895–2899.

Pooley, A. S., Pardon, J. F., and Richards, B. M. (1974). The relation between the unit thread of chromosomes and isolated nucleohistone. *J. Mol. Biol.* **85,** 533–549.

Portmann, R., Schaffner, W., and Birnstiel, M. (1976). Partial denaturation mapping of cloned histone DNA from the sea urchin *Psammechinus miliaris. Nature* (*London*) **264,** 31–34.

Prescott, D. M., and Murti, K. G. (1973). Chromosome structure in ciliated protozoans. *Cold Spring Harbor Symp. Quant. Biol.* **38,** 609–618.

Prescott, D. M., Murti, K. G., and Bostock, C. J. (1973). Genetic appartus of *Stylonychia sp. Nature* (*London*) **242,** 576–600.

Procunier, J. C., and Tartof, K. D. (1975). Genetic analysis of the 5 S RNA genes in *Drosophila melanogaster. Genetics* **81,** 515–523.

Rae, P. M. M. (1966). Whole mount electron microscopy of *Drosophila* salivary chromosomes. *Nature* (*London*) **212,** 139–142.

Rattner, J. B., Branch, A., and Hamkalo, B. A. (1975). Electron microscopy of whole mount metaphase chromosomes. *Chromosoma* **52,** 329–338.

Reeck, G. R., Simpson, R. T., and Sober, H. A. (1972). Resolution of a spectrum of nucleoprotein species in sonicated chromatin. *Proc. Natl. Acad. Sci. U.S.A.* **69,** 2317–2321.

Rees, H., and Jones, R. N. (1972). The origin of the wide species variation in nuclear DNA content. *Int. Rev. Cytol.* **32,** 53–92.

Richards, B. M., and Pardon, J. F. (1970). The molecular structure of nucleohistone (DNH). *Exp. Cell Res.* **62,** 184–196.

Ringertz, N. R., Carlsson, S. A., and Savage, R. E. (1972). Nucleocytoplasmic interactions and the control of nuclear activity. *Adv. Biosci.* **8,** 219–236.

Ris, H. (1956). A study of chromosomes with the electron microscope. *J. Biophys. Biochem. Cytol.* **2,** 385–392.

Ris, H. (1961). The annual invitation lecture: Ultrastructure and molecular organization of genetic systems. *Can. J. Genet. Cytol.* **3,** 95–120.

Ris, H. (1967). Ultrastructure of the animal chromosome. *In* "Regulation of Nucleic Acid and Protein Biosynthesis" (V. V. Koningsberger and C. Bosch, eds.), pp. 11–21. Elsevier, Amsterdam.

Ris, H. (1968). Effect of fixation on the dimension of nucleohistone fibers. *J. Cell Biol.* **38,** 158a–159a.

Ris, H. (1969). The molecular organization of chromosomes. *In* "Handbook of Molecular Cytology" (A. Lima-de-Faris, ed.), pp. 221–250. North-Holland Publ., Amsterdam.

Ris, H. (1975a). High voltage electron microscopy in the analysis of chromosome organization. *Electron Microsc., Proc. Int. Congr., 8th, 1974* Vol. 2, pp. 250–251.

Ris, H. (1975b). Chromosomal structure as seen by electron microscopy. *Ciba Found. Symp.* **28** (New Ser.), 7–23.

Ris, H. (1976). Levels of chromosome organization. *Electron Microsc. Proc. Eur. Congr. Electron Microsc., 6th, 1976*, Vol. II, pp. 21–25.
Ris, H. (1978). Preparation of chromatin and chromosomes for electron microscopy. *Methods Cell Biol.* **18,** 229–246.
Ris, H., and Korenberg, J. (1977). From beaded strings to metaphase chromosomes. *In* "Helsinki Chromosome Conference Abstracts," p. 54. Painovalmiste, Helsinki.
Ris, H., and Kubai, D. F. (1970). Chromosome structure. *Annu. Rev. Genet.* **4,** 263–294.
Ris, H., and Mirsky, A. E. (1949). The state of chromosomes in the interphase nucleus. *J. Gen. Physiol.* **32,** 489–502.
Roberts, P. A. (1965). Difference in the behavior of eu- and heterochromatin: Crossing-over. *Nature (London)* **205,** 725–726.
Rochaix, J.-D., Bird, A., and Bakken, A. (1974). Ribosomal RNA gene amplification by rolling circles. *J. Mol. Biol.* **87,** 473–487.
Rudkin, G. T. (1972). Replication in polytene chromosomes. *In* "Developmental Studies on Giant Chromosomes" (W. Beermann, ed.), pp. 59–85. Springer-Verlag, Berlin and New York.
Ruiz-Carrillo, A., Wangh, L. J., and Allfrey, V. G. (1975). Processing of newly synthesized histone molecules. *Science* **190,** 117–128.
Sahasrabuddhe, C. G., and Van Holde, K. E. (1974). The effect of trypsin on nuclease-resistant chromatin fragments. *J. Biol. Chem.* **249,** 152–156.
Salomon, R., Kaye, A. M., and Herzberg, M. (1969). Mouse nuclear satellite DNA: 5-methylcytosine content, pyrimidine isoplith distribution and electron microscopic appearance. *J. Mol. Biol.* **43,** 581–592.
Sautière, D., Wouters-Tyrou, D., Laine, B., and Biserte, G. (1975). Structure of histone H2A (histone ALK, IIb1 or F2a2). *Ciba Found. Symp.* **28** (New Ser.), 77–88.
Schaffner, W., Gross, K., Telford, J., and Birnstiel, M. (1976). Molecular analysis of the histone gene cluster of *Psammechinus miliaris*. II. The arrangement of the five histone-coding and spacer sequences. *Cell* **8,** 471–478.
Scheer, U., Trendelenburg, M. F., and Franke, W. W. (1973). Transcription of ribosomal RNA cistrons. *Exp. Cell. Res.* **80,** 175–190.
Scheer, U., Trendelenburg, M., Krohne, G., and Franke, W. W. (1977). Lengths and patterns of transcriptional units in the amplified nucleoli of oocytes of *Xenopus laevis*. *Chromosoma* **60,** 147–167.
Schrader, F., and Hughes-Schrader, S. (1926). Haploidy in *Icerya purchasi*. *Z. Wiss. Zool.* **128,** 182–200.
Schreck, R. R., Warburton, D., Miller, O. J., Beiser, S. M., and Erlanger, B. F. (1973). Chromosome structure as revealed by a combined chemical and immunochemical procedure. *Proc. Natl. Acad. Sci. U.S.A.* **70,** 804–807.
Schweizer, D. (1976). Reverse fluorescent chromosome banding with chromomycin and DAPI. *Chromosoma* **58,** 307–324.
Seabright, M. (1973). High resolution studies on the pattern of induced exchanges in the human karyotype. *Chromosoma* **40,** 333–346.
Senior, M. B., Olins, A. L., and Olins, D. E. (1975). Chromatin fragments resembling ν bodies. *Science* **187,** 173–175.
Shapiro, H. S. (1976). DNA content of chordate cell nuclei. *In* "Cell Biology" (P. L. Altman and D. D. Katz, eds.), pp. 367–378. Fed. Am. Soc. Exp. Biol., Bethesda, Maryland.
Sharma, T., and Dhaliwal, M. K. (1974). Relationship between patterns of late S DNA synthesis and C- and G-banding in muntjac chromosomes. *Exp. Cell Res.* **87,** 394–397.
Shaw, B. R., Herman, T. M., Kovacic, R. T., Beaudreau, G. S., and Van Holde, K. E.

(1976). Analysis of subunit organization in chicken erythrocyte chromatin. *Proc. Natl. Acad. Sci. U.S.A.* **73,** 505–509.

Sieger, M., Pera, F., and Schwarzacher, H. G. (1970). Genetic inactivity of heterochromatin and heteropycnosis in *Microtus agrestis. Chromosoma* **29,** 349–364.

Simpson, R. T. (1976). Histones H3 and H4 interact with the ends of nucleosome DNA. *Proc. Natl. Acad. Sci. U.S.A.* **73,** 4400–4404.

Simpson, R. T., and Bustin, M. (1976). Histone composition of chromatin subunits studied by immunosedimentation. *Biochemistry* **15,** 4305–4312.

Simpson, R. T., and Reeck, G. R. (1973). A comparison of the proteins of condensed and extended chromatin fractions of rabbit liver and calf thymus. *Biochemistry* **12,** 3853–3858.

Simpson, R. T., and Whitlock, J. R., Jr. (1976). Mapping DNAase 1-susceptible sites in nucleosomes labeled at the 5′ ends. *Cell* **9,** 347–353.

Simson, J. A. V., and Spicer, S. S. (1975). Selective subcellular localization of cations with variants of the potassium (pyro) antimonate technique. *J. Histochem. Cytochem.* **23,** 575–598.

Smith, G. P. (1976). Evolution of repeated DNA sequences by unequal crossover. *Science* **191,** 528–535.

Sollner-Webb, B., and Felsenfeld, G. (1975). A comparison of the digestion of nuclei and chromatin by staphylococcal nuclease. *Biochemistry* **14,** 2915–2920.

Sollner-Webb, B., Camerini-Otero, R. D., and Felsenfeld, G. (1976). Chromatin structure as probed by nucleases and proteases: Evidence for the central role of histones H3 and H4. *Cell* **9,** 179–193.

Sommerville, J., and Hill, R. J. (1973). Proteins associated with heterogeneous nuclear RNA of newt oocytes. *Nature (London), New Biol.* **245,** 104 106.

Sorsa, M., and Sorsa, V. (1967). Electron microscopic observations on interband fibrils in *Drosophila* salivary chromosomes. *Chromosoma* **22,** 32–41.

Southern, E. M. (1970). Base sequence and evolution of guinea pig α-satellite DNA. *Nature (London)* **227,** 794–798.

Spadafora, C., Bellard, M., Compton, J. L., and Chambon, P. (1976). The DNA repeat lengths in chromatins from sea urchin sperm and gastrula cells are markedly different. *FEBS Lett.* **69,** 281–285.

Sparrow, A. H., Price, H. J., and Underbrink, A. G. (1972). A survey of DNA content per cell and per chromosome of prokaryotic and eukaryotic organisms. Some evolutionary considerations. *Brookhaven Symp. Biol.* **23,** 451–494.

Spear, B., and Gall, J. G. (1973). Independent control of ribosomal gene replication in polytene chromosomes of *Drosophila melanogaster. Proc. Natl. Acad. Sci. U.S.A.* **70,** 1359–1363.

Spradling, A., Penman, S., and Pardue, M. L. (1975). Analysis of *Drosophila* mRNA by *in situ* hybridization: Sequences transcribed in normal and heat shocked cultured cells. *Cell* **4,** 395–404.

Spring, H., Scheer, U., Franke, W. W., and Trendelenburg, M. F. (1975). Lampbrush-type chromosomes in the primary nucleus of the green alga *Acetabularia mediterranea. Chromosoma* **50,** 25–43.

Spring, H., Krohne, W., Franke, W. W., Scheer, U., and Trendelenburg, M. F. (1976). Homogeneity and heterogeneity of sizes of transcriptional units and spacer regions in nucleolar genes of *Acetabularia. J. Microsc. Biol. Cell.* **25,** 107–116.

Steffensen, D., and Bergeron, J. A. (1959). Autoradiographs of pollen tube nuclei with calcium-45. *J. Biophys. Biochem. Cytol.* **6,** 339–342.

Steffenson, D. (1961). Chromosome structure with special reference to the role of metal ions. *Int. Rev. Cytol.* **12,** 163–197.

Stubblefield, E. (1975). Analysis of the replication pattern of Chinese hamster chromosomes using 5-bromodeoxyuridine suppression of 33258 Hoechst fluorescence. *Chromosoma* **53,** 209–221.

Stubblefield, E., and Wray, W. (1971). Architecture of the Chinese hamster metaphase chromosome. *Chromosoma* **32,** 262–294.

Swanson, C. P. (1943). The behavior of meiotic prophase chromosomes as revealed through the use of high temperatures. *Am. J. Bot.* **30,** 422–428.

Swift, H. (1962). Nucleic acids and cell morphology in dipteran salivary glands. *In* "The Molecular Control of Cellular Activity" (J. M. Allen, ed.), pp. 73–125. McGraw-Hill, New York.

Tartof, K. D. (1971). Increasing the multiplicity of ribosomal RNA genes in *Drosophila melanogaster*. *Science* **171,** 294–297.

Tartof, K. D. (1975). Redundant genes. *Annu. Rev. Genet.* **9,** 355–385.

Taylor, J. H. (1949). Chromosomes from cultures of excised anthers. *J. Hered.* **40,** 86–88.

Therman, E. (1972). Chromosome breakage by 1-methyl-2-benzylhydrazine in mouse cancer cells. *Cancer Res.* **32,** 1133–1136.

Thomas, J. O., and Kornberg, R. D. (1975). An octamer of histones in chromatin and free in solution. *Proc. Natl. Acad. Sci. U.S.A.* **72,** 2626–2630.

Tuan, D. Y. H., and Bonner, J. (1969). Optical absorbance and optical rotatory dispersion studies on calf thymus nucleohistone. *J. Mol. Biol.* **45,** 59–76.

Van Lente, F., Jackson, J. F., and Weintraub, H. (1975). Identification of specific crosslinked histones after treatment of chromatin with formaldehyde. *Cell* **5,** 45–50.

Varshavsky, A. J., Bakayev, V. V., and Georgiev, G. P. (1976). Heterogeneity of chromatin subunits *in vitro* and location of histone H1. *Nucleic Acids Res.* **3,** 477–492.

Vlad, M., and Macgregor, H. C. (1975). Chromomere number and its genetic significance in lampbrush chromosomes. *Chromosoma* **50,** 327–347.

Weintraub, H., and Groudine, M. (1976). Chromosomal subunits in active genes have an altered conformation. *Science* **193,** 848–856.

Weintraub, H., and Van Lente, F. (1974). Dissection of chromosome structure with trypsin and nucleases. *Proc. Natl. Acad. Sci. U.S.A.* **71,** 4249–4253.

Weintraub, H., Palter, K., and Van Lente, F. (1975). Histones H2a, H2b, H3, and H4 form a tetrameric complex in solutions of high salt. *Cell* **6,** 85–110.

Weintraub, H., Worcel, A., and Alberts, B. (1976). A model for chromatin based upon two symmetrically paired half-nucleosomes. *Cell* **9,** 409–417.

Weisblum, B., and de Haseth, P. L. (1972). Quinacrine, a chromosome stain specific for deoxyadenylate–deoxythymidylate-rich regions in DNA. *Proc. Natl. Acad. Sci. U.S.A.* **69,** 629–632.

Wellauer, P. K., Dawid, I. B., Kelley, D. E., and Perry, R. P. (1974). Secondary structure maps of ribosomal RNA. II. Processing of mouse L-cell ribosomal RNA and variations in the processing pathway. *J. Mol. Biol.* **89,** 397–407.

Wenrich, D. H. (1917). Synapsis and chromosome organization in *Chorthippus* (*Stenobothrus*) *curtipennis* and *Trimerotropis suffusa* (Orthoptera). *J. Morphol.* **29,** 471–516.

White, M. J. D. (1935). The effects of X-rays on mitosis in the spermatogonial divisions of *Locusta migratoria* L. *Proc. Soc. London, Ser. B* **119,** 61–84.

Whitlock, J. P., and Simpson, R. T. (1976). Removal of histone H1 exposes a fifty base pair DNA segment between nucleosomes. *Biochemistry* **15,** 3307–3313.

Wilson, E. B. (1896). "The Cell in Development and Inheritance," 1st ed. Macmillan, New York.

Wolfe, S. L. (1969). Molecular organization of chromosomes. *Biol. Basis Med.* **4,** 3–42.
Wolfe, S. L., and Grim, J. N. (1967). The relationship of isolated chromosome fibers to the fibers of the embedded nucleus. *J. Ultrastruct. Res.* **19,** 382–397.
Woodcock, C. L. F. (1973). Ultrastructure of inactive chromatin. *J. Cell Biol.* **59,** 368a.
Woodcock, C. L. F., Safer, J. P., and Stanchfield, J. E. (1976). Structural repeating units in chromatin. I. Evidence for their general occurrence. *Exp. Cell Res.* **97,** 101–110.
Worcel, A., and Benyajati, C. (1977). Higher order coiling of DNA in chromatin. *Cell* **12,** 83–100.
Wray, W. (1973). Isolation of metaphase chromosomes, mitotic apparatus, and nuclei. *Methods Cell Biol.* **6,** 283–306.
Wray, W., and Stubblefield, E. (1970). A new method for the rapid isolation of chromosomes, mitotic apparatus, or nuclei from mammalian fibroblasts at near neutral pH. *Exp. Cell Res.* **59,** 469–478.
Wu, M., Holmes, D. S., and Davidson, N. (1976). The relative positions of sea urchin histone genes on the chimeric plasmids pSp2 and pSp17 as studied by electron microscopy. *Cell* **9,** 163–169.
Wunderlich, F., and Herlan, G. (1977). A reversibly contractile nuclear matrix. Its isolation, structure, and composition. *J. Cell Biol.* **73,** 271–278.
Yasmineh, W. G., and Yunis, J. J. (1970). Localization of mouse satellite DNA in constitutive heterochromatin. *Exp. Cell Res.* **59,** 69–75.
Yunis, J. J. (1976). High resolution of human chromosomes. *Science* **191,** 1268–1270.
Yunis, J. J., and Yasmineh, W. G. (1971). Heterochromatin, satellite DNA, and cell function. *Science* **174,** 1200–1209.
Yunis, J. J., and Yasmineh, W. G. (1972). Model for mammalian constitutive heterochromatin. *Adv. Cell Mol. Biol.* **2,** 1–46.

7

Strandedness of Chromosomes and Segregation of Replication Products

W. J. Peacock

I. INTRODUCTION

A. Current Problems of Chromosome Structure

The present-day student of chromosome structure is concerned primarily with the linear organization of the chromosome. Following the demon-

CELL BIOLOGY, VOL. 2

ISBN 0-12-289502-9

stration that the genome is heterogeneous with regard to DNA sequence representation (Britten and Kohne, 1968), considerable effort has focused on the elucidation of the roles and arrangement of sequences which are repeated once in the genome versus those which are repeated hundreds (middle repetitive) or many thousands (highly repeated sequences) of times. Added impetus to the analysis of primary chromosomal DNA sequence has been given by genetic studies in *Drosophila,* studies which have identified a single complementation unit per band of the polytene chromosome (Judd *et al.,* 1972) even though the amount of DNA in a band, in the unit chromatid, is known to vary between 5 and 100 times the sequence length expected for one coding segment. Interest in secondary structure of chromosomal DNA sequences has been expanding since it was shown that the highly repeated sequences are confined largely to those regions of chromosomes that have been defined cytologically as heterochromatin (Pardue and Gall, 1970; Jones and Robertson, 1970; Peacock *et al.,* 1977). The burgeoning field, however, is that of tertiary structure, in which the relationship between chromosomal DNA and chromosomal proteins, both histones and nonhistones, is being probed from structural and functional standpoints (see Cold Spring Harbor Symposium 1977).

B. Previous Analyses of Chromosome Structure

The research areas mentioned above largely reflect the appearance of new techniques permitting experiments which were not possible earlier. In the first half of this century, when direct cytological observation with the light microscope was the principal means for analyzing chromosome structure, attention consequently was focused on what might be regarded as a quaternary level of chromosome organization. One of the major points of the controversy was that of chromosome strandedness. The genetic processes of mutation and recombination, which were being unraveled in the same time period, certainly sat most easily with the concept of a singleness to the anaphase chromatid; however, cytological observations indicated otherwise and were sufficiently persuasive to warrant a high level of research interest in the problem.

A duality to the anaphase chromatid was cited by Wilson (1925) as being observed as early as 1883 by van Beneden, and other cytologists have frequently claimed to have seen half-chromatids. In many cases, the visible substructure of the chromatid may well have resulted from poor fixation or optical artifact, but observations such as (1) Manton's (1945) photographs of *Todea* chromosomes using ultraviolet wavelengths, (2) Hughes-Schrader's (1940) report that half-chromatids actually proceed independently to the anaphase poles in the scale insect *Llaveilla,* (3) photographs which show relationally coiled half-chromatids separated to the degree normally seen between chromatids of the metaphase chromosome (Peacock, 1965), (4) Bajer's (1965) filmed observations of living anaphase cells of *Haemanthus* showing bipartite chromatids, and (5) Maguire's (1968) Nomarski interference contrast pictures of maize chromosomes, all still attest to a possible complexity in the structure of the anaphase chromatid. Nebel (1939) commented that "the disagreement on the real thread number will probably appear somewhat artificial 20 years hence when knowledge will be available to understand why the same element may change its appearance so confusingly." Almost twice the length of time predicted by Nebel for enlightenment has passed, and it is still not possible to unequivocally explain his conundrum.

The basic difficulty with the above evidence is that it lacked experimental rigor—observation of predictable consequences of altered variables was not obtainable. This same difficulty applied to the application of the more powerful electron microscope to the problem of describing the structure of the anaphase or metaphase chromosome. An experimental analysis was attempted with the study of radiation or chemically induced chromosome aberrations. Certain anaphase bridge aberrations were considered to provide evidence for the breakage and reunion of subchromatids (e.g., Crouse, 1954, 1961; Wilson and Sparrow, 1960), but the added dimension of tracing the consequences of these aberrations in subsequent cell division led to apparently conflicting observations (Ostergren and Wakonig, 1954; Peacock, 1961; Kihlman and Hartley, 1967). The uncertainties of target specificity of the mutagen, of the time of its action in the cell cycle, and of induced delays in cell division are among the factors detracting from the incisiveness of this class of data.

One other line of experiment that indicated the complexity of DNA arrangement within a chromatid was the trypsin digests of *Vicia* metaphase chromosomes by Trosko and Wolff (1965). Although emphasizing the doubleness seen by many cytologists, their experiments took the analysis no further—the critical identification of genetic identity of the half-chromatids again was lacking.

C. The Chromosome as One DNA Molecule

1. *Chromosomal Autoradiography*

The data which convinced this author that the cytological lines of evidence must, despite appearances to the contrary, be compatible with a unineme (one DNA molecule) chromosome structure all stem from the simple directness of experiments in molecular biology. This, in the main, is the account of the patterns of incorporation of isotopically labeled thymidine into chromosomal DNA and the tracing of its subsequent distribution at a cellular and chromosomal level. The technical tool of prime importance has been autoradiography, and it is the analysis of chromosome structure as unfolded by the autoradiographic imaging of DNA-incorporated thymidine that will be dealt with in this chapter. The review is historical and personal and does not claim to be comprehensive. Nevertheless, it traces the beginnings and the major landmarks of the molecular analysis of the eukaryote chromosome.

2. *Viscoelasticity and Genetics*

Before discussing autoradiography, I would like to recount the essentials of a more recent experiment which, in itself, goes far toward establishing that a eukaryote chromosome can be thought of in terms of a single DNA molecule running the entire length of the chromosome. This molecule, of course, is associated with histones and other proteins in particular ways and has a packing ratio of approximately 1000 : 1 into the structure we see as the mitotic metaphase chromosome.

Kavenoff and Zimm (1973) extracted DNA of *Drosophila* mitotic nuclei into a chamber in which they measured the viscoelastic relaxation of the solution of DNA molecules which had been stretched and allowed to recoil. Their technique was peculiarly sensitive to the longest molecules in solution, and they detected molecules of molecular weights similar to those expected for whole *Drosophila* chromosomes as calculated from the uv measurements of Rudkin (1965). However, the strength of the experiment really was established not by this 1 : 1 correspondence of molecular weights—absolute values might be explained away—but by the imposition of genetic rigidity. They were able to use the arsenal of the *Drosophila* chromosome advantageously. When they took cells from *Drosophila* which were known to have most of the X chromosome translocated onto most of chromosome 3, they found the longest molecules now to be 40% longer than the molecules in the wild-type control. This was precisely the result expected on the basis of the well-defined cytogenetic knowledge of this translocation. This and other related experiments established that the chromosomal DNA molecule runs uninterrupted through the centromere or kinetochore region, extending the full length of the chromosome from one telomere to the other.

II. MITOTIC CHROMOSOME DNA DUPLICATION

A. The Basic Observation

The association of biologists and physicists at the Brookhaven National Laboratory and Columbia University in New York in the years following the Watson–Crick DNA structure paper produced a beautifully simple experiment showing the segregation of chromosomal DNA to be semiconservative, a result that would be expected if the chromosome contained only one molecule which replicated in a semiconservative fashion. Taylor *et al.* (1957) realized that the specificity of exogenously supplied tritiated thymidine as a probe for replicating DNA coupled with the low energy of the beta particles emitted from the decaying tritium was a powerful aid in examining events at the molecular level in the light microscope. They chose *Vicia faba,* a plant favored by cytologists because of its large and easily managed chromosomes, and grew its root tips for a period in liquid culture medium containing the radioactive DNA precursor, after which it was transferred to culture medium not containing any tritiated thymidine, where it was allowed to continue growing. Using autoradiography, they then examined the distribution of incorporated radioactivity in the first and second metaphases following the chromosomal duplication which occurred while the root cells were in contact with the tritiated thymidine solution. Colchicine was used to block the first anaphase after the isotope treatment, hence permitting the unequivocal identification of the second metaphase by virtue of the doubled chromosome number. The basic observations and the inferences drawn from them are shown in Figure 1. The finding that all metaphase chromosomes showed both sister chromatids labeled in the first metaphase and only one chromatid labeled in the second metaphase was interpreted as showing that each anaphase chromatid contained two subunits, extending the length of the chromatid, each of which became associated with a new partner subunit during chromosome duplication. This is exactly the pattern expected if the chromatid contained one double helix of DNA. Semiconservative replication of bacterial DNA molecules was demonstrated by Meselson and Stahl (1958) shortly thereafter. However, in the climate of controversy over chromosome structure, it was necessary to have more compelling identification of the chromatid subunit as a polynucleotide chain of a DNA molecule.

B. Dissimilarity of Chromosome Subunits

The first attempt to characterize further the duplication subunits of a chromatid was made by Taylor (1958) using a phenomenon first noted as an exception to the basic pattern of label distribution discussed above. In some chromosomes of the second metaphase following isotope incorporation, the radioactive labeling switched from one chromatid to its sister. This appeared to be a physical reciprocal exchange between homologous regions of sister chromatids (Figure 2, observation panel; Figure 3). In these tetraploid cells, each chromosome of the original diploid cell is represented by two daughter chromosomes, and Taylor reasoned that if

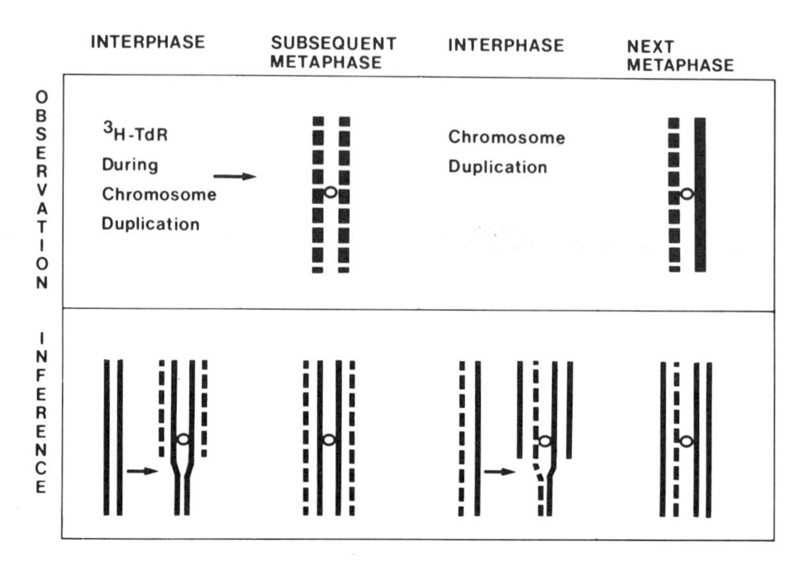

Fig. 1. Chromosomal DNA duplication. The basic experiment by Taylor *et al.* (1957) provided tritiated thymidine (^{3}H-TdR) during one DNA replication period, and the segregation of the isotope then was determined in the next two successive metaphases. Presence of the isotope is indicated by the broken line.

the event which led to the switch of labcl (thc sister chromatid exchange) occurred in the first chromosome duplication cycle, then it would be perpetuated in both daughter chromosomes during the second duplication cycle. In fact, he did find a high frequency of pairwise concordance of label exchange positions in homologous chromosomes. He called these twin-sister chromatid exchanges (top pair of chromosomes, Figure 2). He also observed label switches that were unique in position, occurring in only one homologue (three such exchanges are shown in the bottom pair of chromosomes in Figure 2). The single sister chromatid exchange would be expected if the actual physical exchange between sister chromatids occurred in the second duplication cycle when each chromosome of tetraploid cells would be an independent unit.

Taylor foresaw that if the probability of a sister chromatid exchange event in any one chromosome was considered to be the same in the two duplication cycles involved in the experiment, then the relative frequencies of twin and single exchanges could provide information on the nature of the duplication subunit. If the two subunits were dissimilar and only were able to join with a similar or like subunit in a sister chromatid exchange event, then, since there are twice as many chromosomes in the second duplication cycle, there would be twice as many single sister

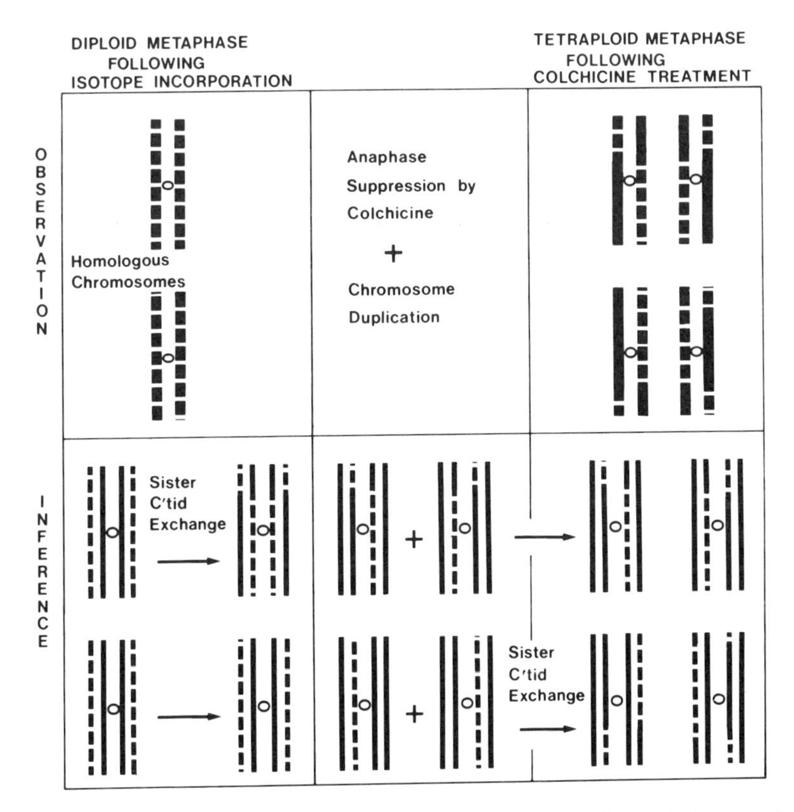

Fig. 2. Twin and single sister chromatid exchange. Pairs of homologues in the tetraploid metaphases (second metaphase postisotope incorporation) sometimes showed label (broken line) switches between sister chromatids in identical positions (top row) and were termed twin exchanges by Taylor (1958). Other exchanges had unique positions among the homologues and were termed single sister chromatid exchange (bottom row).

chromatid exchanges compared to the number of twin exchanges (see inference panel, Figure 2). On the other hand, if subunits were similar to each other and could rejoin randomly, then four possible ways of rejoining could occur in an exchange event. In the first duplication period, one-fourth of the joining events would mirror the event which was obligatory assuming dissimilar subunits—this would lead to a twin exchange. Half of the time, a switch of radioactivity would be detectable in only one of the chromatids, leading ultimately to a single exchange in the tetraploid cell, and the remaining rejoining possibility would not lead to a detectable exchange at all. In the second duplication period, all four modes of rejoining would result in a single exchange. Thus, twin and single exchange frequencies can be expected to provide information on the possible mode

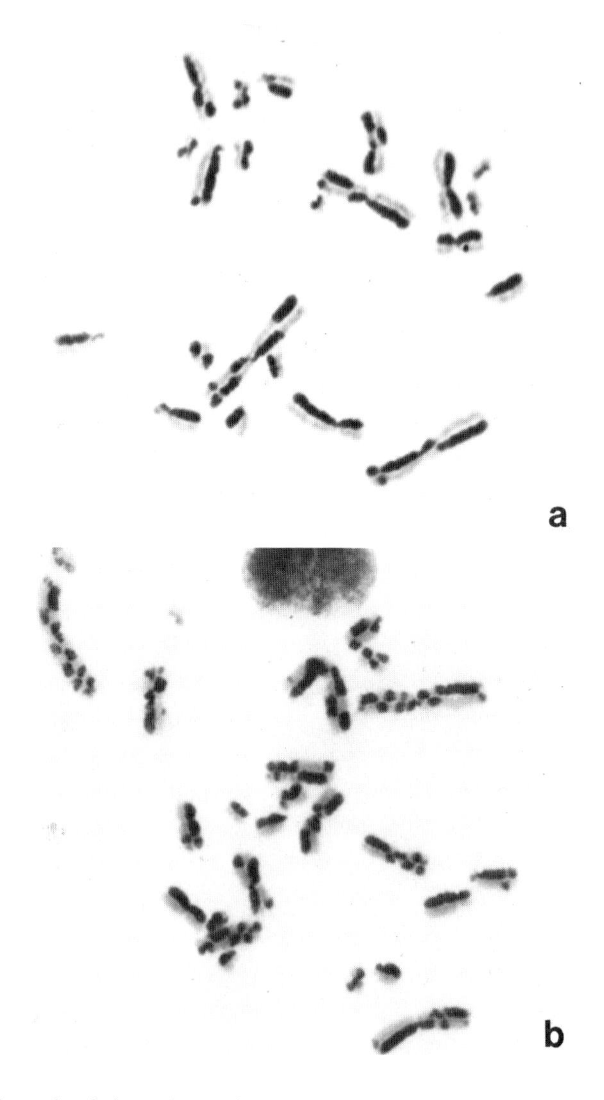

Fig. 3. Example of sister chromatid exchanges in metaphase chromosomes of Chinese hamster cell lines (courtesy of Professor Sheldon Wolff). Bromodeoxyuridine has been provided as a substitute for thymidine during a DNA replication cycle, and as a marker this is equivalent to the use of isotopically labeled thymidine. The cells were stained with a fluorescent dye, exposed to light, and then stained with Giemsa (Wolff and Perry, 1974). (a) A cell from a control culture. (b) A cell from an experiment which included a treatment to increased the frequency of exchange.

of rejoining of the subunits, which in turn reflects the characteristics of the subunits. Dissimilarity of subunits would be indicated by a ratio of two single exchanges to every twin exchange, whereas identity of subunits would be implied if there were ten single exchanges to every twin exchange.

Taylor found 81 twin exchanges and 30 single exchanges in his first experiment with the liliaceous species *Bellevalia romana,* but in later experiments (Taylor, 1959) he found that the colchicine treatment in fact was modifying the probability of exchange, and in his best controlled experiment he obtained 26 singles to 14 twins, arguing in support of restricted rejoining among the subunits of sister chromatids. Probably the best data on the relative frequencies of these classes of sister chromatid exchange were obtained by Herreros and Gianelli (1967) in endotetraploid human lymphocytes (288 singles : 128 twins) and by Geard and Peacock (1969) in *Vicia faba* (e.g., 178 singles : 91 twins). In the latter study, provision was made for the estimation of "false twins" when two bonafide single exchanges were cytologically indistinguishable in position, a feature first pointed out by Heddle (1968). The corrections still left the data strongly consistent with the subunit dissimilarity prediction.

C. Polarity of Chromosome Subunits

Although the categorization of twin and single exchanges was subject to the criticism that the scoring involves cytological subjectivity, these results certainly strengthened the case for the metaphase chromatid as one double helix of DNA. Each subunit replicated and segregated as a unit, and the two subunits of any one chromatid differed from each other in that their ends were differentiated in exchange events between sister chromatids. The final requirement for equivalence of the subunits to the polynucleotide chains of a double helix of DNA was that their dissimilarity in fact stemmed from a molecular polarity difference between them. This was demonstrated simply and unequivocally in an experiment in which the restriction against like subunits rejoining during sister chromatid exchange was shown not to apply when the exchange event involved a reverse reunion (a U rather than an X reunion). Brewen and Peacock (1969a) used X rays to induce, in tritiated thymidine-labeled chromosomes, isochromatid breaks (breaks in identical regions of sister chromatids) and examined that class of chromosome in which the broken ends rejoined so as to form a dicentric chromatid and an acentric fragment (Figure 4). In their experiment, they used colchicine to block the anaphase with its expected bridge formation and looked at the distribution of label in the tetraploid cell which then contained a dicentric chromosome.

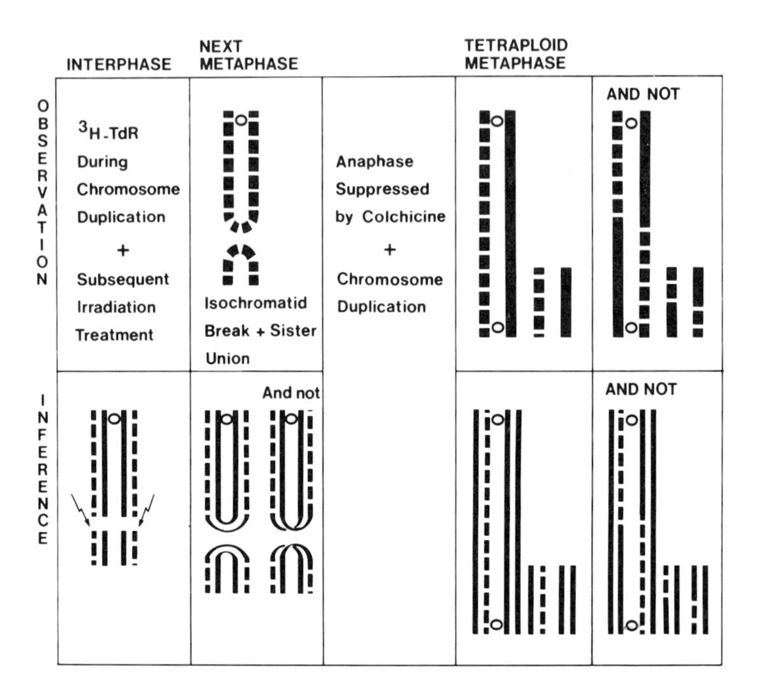

Fig. 4. Polarity of chromosome duplication subunits. Chromosomes which had incorporated tritiated thymidine were irradiated to induce chromosome rearrangements. Some chromosomes were broken in identical regions of sister chromatids (isochromatid break), and the reunion of the broken ends produced a dicentric chromatid and an acentric fragment (sister union). Brewen and Peacock (1969a) found that the label in the intercentromeric region was confined to one chromatid. This showed that the initial reunion always was confined to certain of the subunits (label to label and nonlabel to nonlabel in the particular experiment). This observation, coupled with those on sister chromatid exchanges, showed that the replication subunits had polarity differences as expected for the strands of a DNA double helix.

In 137 such dicentric chromosomes, 104 showed the label to be confined, in the intercentromeric segment, to only one of the sister chromatids. This result conformed to the expectation based on an hypothesis of restricted rejoining of subunits and not to that expected for random rejoining. In the latter case, there would have been a number of dicentrics with label switches in the middle of the intercentromeric region equal to the number with label in only one of the chromatids. The 33 chromosomes which did not show conservation of label to one chromatid in the intercentromeric region had radioactivity patterns all explicable in terms of sister chromatid exchange events superimposed on the basic pattern of label conservation.

This experiment showed that, whereas chromosome subunits certainly were not identical and could not join with each other in a sister chromatid event, they could and did rejoin with each other if the mode of rejoining of broken ends was a reversed exchange. Thus, the property enforcing the dissimilarity of subunits was shown to be directionally dependent—the two subunits thus can be said to have opposite polarity as they are normally present in the chromosome. However, they must be comparable otherwise, since in reversed reunions they can rejoin to form a single continuous subunit fully competent in the subsequent duplication cycle.

D. A Puzzling Exception

The autoradiographic case identifying the chromosome as one DNA molecule was now strong: (1) it had defined two subunits, each of which was conserved during replication, and (2) the two subunits had directional or polarity differences which affected molecular reunion for both X-ray or spontaneous events. Nevertheless, one observation did not agree well with this simple position. Some chromosomes in the second metaphase after labeling had both, rather than one, sister chromatids labeled in homologous positions. This first was reported in *Vicia* by La Cour and Pelc (1958) and analyzed in more detail by Peacock (1963). This phenomenon of isolabeling was seen in other plant cells (e.g., Darlington and Haque, 1969) and animal cells (e.g., Walen, 1965; Deaven and Stubblefield, 1969). Peacock considered four possible origins of isolabeling. Of these, the possibility that overlap occurred between successive replication cycles during the period of isotope availability was dismissed on the basis of the available evidence which suggested rapid depletion of the tritiated thymidine pool (Wimber, 1960). Taylor's (1962) postulate that interchromosomal exchange between homologues was involved also was rejected, since the labeling pattern occurred in experiments in which colchicine was omitted. The remaining two explanations are illustrated in Figure 5. Multiple sister chromatid exchanges could lead to the appearance of isolabeling if the segments between successive exchanges were small enough. Callan (1972) subscribed to this explanation of isolabeling as did Prescott (1970), however, in 1963 I discounted it on the basis of the statistical distribution of observed exchange events. The final possible explanation of isolabeling, given that it was real and not apparent (autoradiographic image artifact), lay in the acceptance of there being more than one DNA molecule per chromosome (bineme structure). A lateral duplicity admittedly was a contrived postulate in that it demanded additional features to account for the more frequent semiconservative duplication and segregation described above; however, this hypothesis of

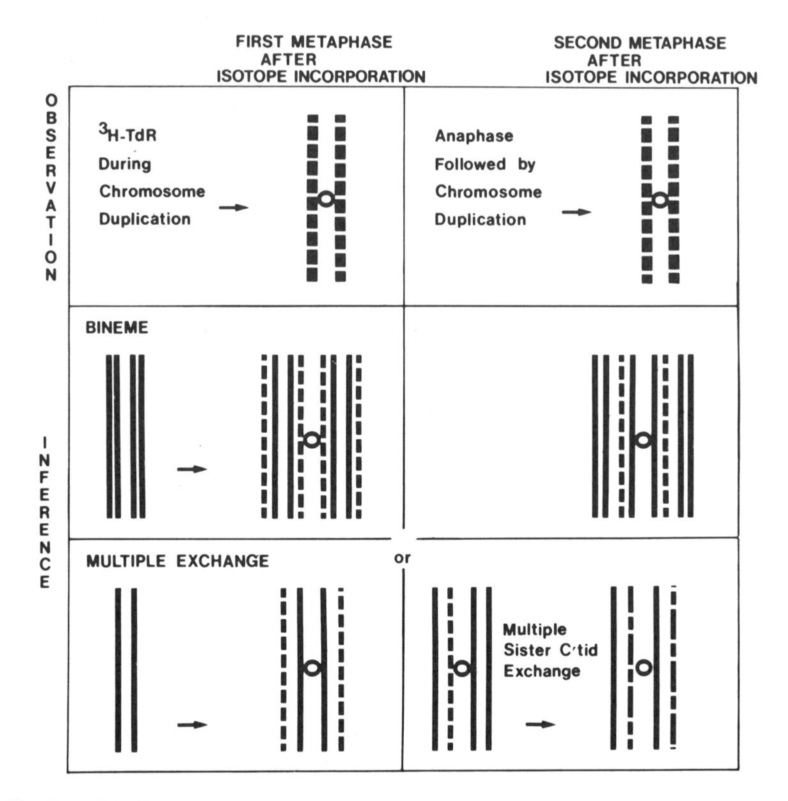

Fig. 5. Isolabeling. In tritium autoradiography experiments, it sometimes was found that both sister chromatids were labeled in the second postisotope incorporation metaphase (contrary to the more usual condition shown in Figure 1). Two of the four possible explanations considered by Peacock (1963) are shown. Subsequent experiments using staining methods, which have greater resolution than the autoradiographic technique, have shown that the multiple exchange explanation is the principal cause of isolabeling.

course was in concordance with the other bodies of cytological evidence that at least pointed to the existence of half-chromatids.

It was not until recently that isolabeling was explained satisfactorily. The essential ingredient was the development of a new method of detecting sister chromatid exchanges (see legend to Figure 3). Giemsa staining of chromatids in which one duplication subunit has bromodeoxyuridine substituted for thymidine is different in appearance from that of chromatids in which both subunits are substituted (Ikushima and Wolff, 1974). The difference is particularly marked if a combination of fluorescent dye and Giemsa staining is used (e.g., Perry and Wolff, 1974). This technique has a greater resolving power than that obtainable with tritium autoradiography, and it soon was shown that chromosomes which were isolabeled by autoradiographic criteria actually had a large number of sister chromatid exchanges (Wolff and Perry, 1974). Two of the chromosomes in Figure 3b could well have appeared to have isolabeled segments in autoradiographic analysis. The frequency distributions of the exchanges that are visible by autoradiography sometimes do fit Poisson expectations as originally found by Peacock (1963), but Geard and Peacock (1969) had noted that the distribution of exchanges did not always fit the expected Poisson. Now it seems probable that the underdispersion they found resulted from the limitations of autoradiographic detection of exchange events. Results obtained with the newer staining methodology are consistent with random distribution of exchanges among cells and chromosomes (Wolff and Perry, 1974; Galloway and Evans, 1975). A less frequent cause of isolabeling also has been demonstrated by the harlequin staining procedure. This, in fact, is the first possibility Peacock considered (1963), that of overlap between successive replication cycles. Crossen *et al.* (1975) and S. Wolff (personal communication) have found "isolabeling" of this type in late replicating regions of chromosomes.

It is clear that the phenomenon of isolabeling which for some years remained the major obstacle in the path of the simplest interpretation of most other autoradiographic data, in terms of one DNA molecule per chromosome, now is explicable in just those terms.

III. SEGREGATION OF CHROMOSOME SUBUNITS

A. Segregation at the Chromosome Level

The experiments described in preceding sections showed that the two subunits of the chromosome segregated one into each daughter chromatid during chromosome duplication. In this regard, sister chromatid exchanges provide only a trivial perturbation to the pattern of semiconservative subunit replication and segregation. Since in most eukaryotes the length of the chromosomal subunit (a single polynucleotide strand of DNA) is between 1 cm and 1 m, the process of subunit segregation must be a highly controlled one. Spectacular evidence that this is the case has been given by analyses of chromosome duplication in endoreduplicated cells. In these tetraploid cells, sister homologues remain in apposition and are termed diplochromosomes. Walen (1965) observed in endotetraploid cells of the marsupial *Potorous* that the two labeled chromatids of diplochromosomes were always the two outermost chromatids (Figure 6). This

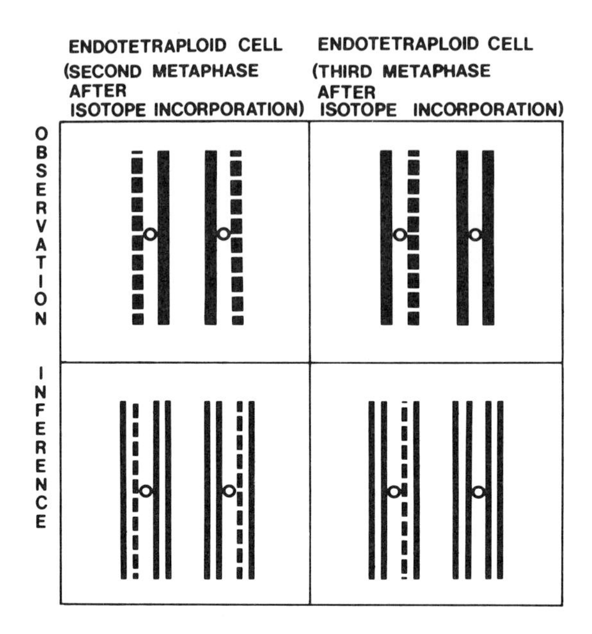

Fig. 6. Chromosome subunit segregation. In diplochromosomes of endotetraploid cells, the labeled chromatids have defined spatial positions (Walen, 1965). The figure shows the outer and inner positions of labeled chromatids in the second and third postisotope incorporation metaphases and details the segregation of the labeled subunit relative to subunits of other replication periods.

analysis was extended to the third postisotope incorporation mitosis in human lymphocytes (Herreros and Gianelli, 1967) and hamster cells (Peacock and Brewen, cited in Peacock, 1971). In diplochromosomes of the third division, the labeled chromatid always has an inside position (Figure 6). These observations can be restated as the rule that newly replicated subunits of sister chromatids are segregated to the outer chromatid of diplochromosomes in the next replication cycle. How is this achieved with DNA molecular weights ranging from 2×10^{10} to 2×10^{12}, with the existence of thousands of replicons in each chromosome, and with a packing ratio of approximately 100 : 1 in interphase chromosomes, and 1000 : 1 in metaphase chromosomes? In both human and hamster material, it has been established that the centromere is not responsible—acentric fragments show precisely the same patterns as centric fragments or whole chromosomes. Apart from this, I know of no other pertinent datum. When somatic cell genetics is developed sufficiently, genetic analysis presumably could provide the key to identification of the principal control processes in duplication subunit segregation.

B. Segregation at the Genome Level

Although mechanistically obscure, the regularity of the pattern of subunit distribution in diplochromosomes evidences a highly controlled cellular process. It was an obvious step to look for similar controls at the whole chromosome complement level. Lark *et al.* (1966) were the first to suggest that all the chromosomal subunits formed in a particular cell cycle subsequently would be cosegregated. This conclusion was based on an observed unequal distribution of radioactivity over sister cells, a situation presumed to have resulted from nonrandom segregation of the chromatids of the chromosomes of the complement. Although these original conclusions were based on animal cells, Lark (1967) extended them to plant cells.

In these studies, the investigators had no unequivocal way of knowing which postisotope replication mitosis the scored nuclei were in, a problem which could have introduced serious errors into the scores. This factor was controlled positively in experiments by Heddle *et al.* (1967) in which metaphase chromosomes were scored; hence, the distribution of label over chromatids was a reliable indicator as to the determination of the cell division. Since it was clear that sister chromatid exchanges would disturb any bimodality in radioactivity distribution, the above study was restricted to the centromere and the immediate pericentromeric chromatin within a one-chromatid width of the actual centromere, and with the additional requirement that label had to be present on both pericentromeric regions for a chromosome to be scored. Only third postisotope replication metaphases were included in the analysis. The results both in root tip cells of *Vicia faba* and tissue culture cells of the marsupial *Potorous tridactylis* demonstrated that segregation of chromatids of the various chromosomes of the complements is random.

Further evidence against any cosegregation of the template strands from a particular mitotic cycle was provided in a definitive set of data for centromere regions of the large and individually distinguishable chromosomes of the swamp wallaby, *Wallabia bicolor* (Geard, 1973). This analysis extended to scores for individual pairs of autosomes, and again the results were consistent with random chromatid distribution. Similar results were obtained for a small group of homologues in human cells (Cuevas-Sosa, 1968).

I think it is clear that in eukaryotes of both the plant and animal kingdoms, there is no convincing evidence that chromosomal DNA subunits of a like age tend to be distributed by a mechanism which would result in a nonrandom distribution of chromatids at mitotic anaphase. This conclusion sharply contrasts, but does not contravene, the demonstration of mechanisms controlling the spatial distribution of replication products at the individual chromosome level.

IV. SEGREGATION OF CHROMOSOME SUBUNITS AT MEIOSIS

If we accept that the mitotic anaphase chromosome, or metaphase chromatid, is one long DNA molecule, we would expect that the chromosomal duplication subunits (one strand of the double helix) would follow a semiconservative segregation in meiosis as well as in mitosis. However, a decade ago it was important to determine whether this was the case. Taylor (1965), using the male meiotic cells of a grasshopper, was the first to show that the semiconservative mode of segregation of chromosome subunits did apply at meiosis. His result was confirmed in spermatogenesis of other grasshoppers (Henderson, 1966; Moens, 1966; Peacock, 1968, 1970) and of a newt (Callan and Taylor, 1968), as well as in microsporogenesis of a plant (Church and Wimber, 1969).

Chromosomes which are labeled in the DNA replication of the last mitotic cycle preceding meiosis show label segregation over sister chromatids in the meiotic divisions (Figure 7). This autoradiographic pattern of chromatid differentiation has been used to analyze the time and mechanism of meiotic recombination. A reciprocal exchange of segments between homologous, nonsister chromatids (one labeled, one unlabeled) would lead to one meiotic chromosome (a dyad) having both chromatids labeled in the region distal to the crossover point, and its sister dyad with a corresponding absence of label in equivalent regions. An exchange between nonsister chromatids which are both labeled, or both unlabeled, would not be visible (Figure 7). The dyad label patterns showing a reductional distribution of label distal to the point of recombination gave the equivalence of a genetic analysis to the cytological examination of meiosis. Taylor's (1965) analysis in *Romalea* showed that at least some recombination involved actual breakage and reunion of existing chromosome segments, and he noted a close relation between exchange and chiasmata frequencies.

Peacock (1970) was able to take this further in *Goniaea* and concluded that all meiotic recombination is the consequence of reciprocal exchange between two nonsister, homologous chromatids and results in the formation of cytologically visible chiasmata. This conclusion was based on analysis of metaphase I bivalents containing a single chiasma (Peacock, 1970, 1971) and on the analysis of label patterns in dyads of the first and second meiotic divisions. The latter analysis was combined with a

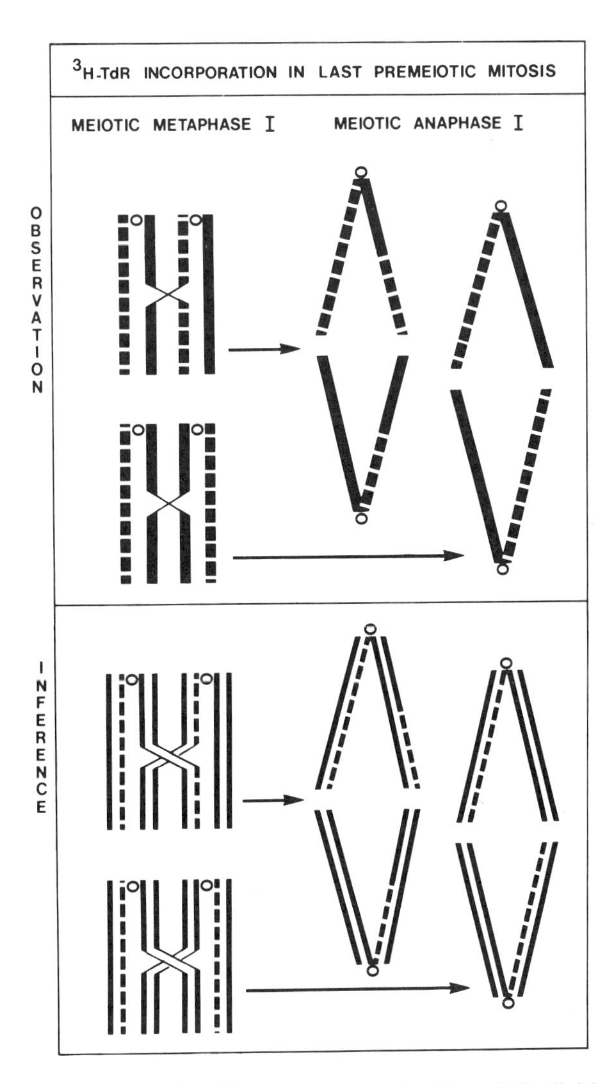

Fig. 7. Meiotic recombination. Chromosomes entering the meiotic divisions with label segregation over sister chromatids have been used to prove that genetic crossing-over produces a biparental recombinant chromosome as a result of reciprocal exchange of nonsister homologous chromatids (Taylor, 1965; Peacock, 1970). Half of the anaphase dyads produced from a bivalent with a single cytological chiasma (one genetic recombination event) have label distributions diagnostic of breakage and reunion of the parental chromatids.

temperature-induced reduction in recombination probability which emphasized the synonymy of the cytological chiasma and the genetic crossover, and which also showed that at least the final molecular events in recombination occurred in pachytene. Label patterns of dyads also were used in two other studies which showed a one-to-one relationship between a chiasma and an exchange (Church and Wimber, 1969; Craig-Cameron and Jones, 1970).

Thus, it was autoradiographic analysis of chromatid segments, appropriately labeled as a consequence of the orderly process of segregation of chromosome duplication subunits, which first showed that, as in bacteriophage (Meselson, 1967), the recombinant chromosome in higher organisms is a biparental structure containing previously duplicated DNA from each of the two parental homologous chromosomes.

V. SISTER CHROMATID EXCHANGE

In the analyses of meiotic label patterns discussed in the preceding section, sister chromatid exchanges similar to those found in mitotic experiments occurred. Taylor (1965) thought that they all might have their origins in the premeiotic mitosis, and certainly Moens (1966) detected exchanges in premeiotic mitosis in *Chorthippus*. I concluded that sister chromatid exchanges were not an integral part of meiotic recombination since they occurred in similar frequencies in bivalents and noncrossover univalents, and also because they occurred at the same frequency per unit length in the X chromosome, which does not have a synaptic partner in *Goniaea,* as in the autosomes. If any of the experimental organisms had contained a ring chromosome, a meiotic sister chromatid exchange would have been detectable as a dicentric ring chromatid at anaphase I. Probable sister chromatid exchanges have been detected in meiosis in ring chromosomes, and, in contrast to my conclusion in *Goniaea,* Schwartz (1954) postulated that they may play a regular role in the mechanism of crossing-over. I think that the autoradiographic analysis makes a clear statement against Schwartz's conclusion.

A human ring chromosome was featured in a series of experiments by Brewen and Peacock (1969b) in which it was shown that tritiated thymidine, an essential component of the autoradiographic system, caused a significant frequency of sister chromatid exchanges. However, they did find a spontaneous level of exchanges (0.12/chromosome/cell cycle). Gibson and Prescott (1972) also found that sister chromatid exchange frequency was dependent on the cellular tritium dose but could not rigorously exclude spontaneous exchange. The newer methods of staining, dependent on bromodeoxyuridine incorporation, also induce sister chromatid exchange (e.g., Wolff and Perry, 1974). Uncertainty about spontaneous versus induced components in sister chromatid exchange frequencies in part has been due to an absence of a dose response in frequency (Marin and Prescott, 1964) due to saturation effects (Wolff, 1964; Gibson and Prescott, 1972).

Ring chromosome data remained the only convincing demonstration that sister chromatid exchanges do occur in the absence of tritium or of other chromosome breakage-inducing agents. That the conversion of monocentric ring chromatids to a dicentric ring involved sister chromatid exchange was proposed by McClintock for her observations in maize (McClintock, 1938, 1941). McClintock's explanation was shown to be correct by the application of the autoradiographic analysis (Brewen and Peacock, 1969b) and is demonstrated most convincingly in a recent work involving harlequin staining (Wolff *et al.*, 1976). In a powerful genetic analysis, Green (1968) has detected sister chromatid exchange events associated with tandem duplications in *Drosophila,* and certainly sister chromatid exchange has become a fashionable mechanism to explain properties of highly repeated chromosomal DNA sequences (e.g., Smith, 1973).

Although ring chromosomes still provide the only cytological means of exchange detection without the use of isotopes or brominated thymidine analogues, credence for spontaneous exchange in rod chromosomes is given in data which show a nonrandom distribution of exchanges within a chromosome. Latt (1974) found a preponderance of exchanges in the interband regions of G-banded human chromosomes, and Carrano and Wolff (1975), exploiting the cytological characteristics of the Indian muntjac, showed that in the X chromosome the heterochromatic segment, the euchromatic segment, and the junction region between those segments all had characteristic exchange potential. High frequencies of exchange in cells from patients with Bloom's syndrome (Chaganti *et al.*, 1974) also point to an underlying mechanism of sister chromatid exchange.

That sister chromatid exchange may play a central role in the dynamics of generation and evolution of arrays of tandemly repeated chromosomal DNA sequences is the only possible role of which this author is aware for its biological significance. These exchanges do present problems in any attachment of significance to them, as well as in their detection, since they represent exchange between genetically identical molecules. Nevertheless, as I have shown in earlier sections of this chapter, sister chromatid exchanges have been particularly valuable in the unraveling of basic chromosome structure in eukaryotes.

VI. CHROMOSOME POLARITY CONTINUITY—A REMINDER OF IGNORANCE

My intent in this limited review has been to trace the demonstration, principally by autoradiographic means, of the eukaryote chromosome as having two DNA subunits. Each of these subunits is a conserved unit of replication, with the consequent result of a semiconservative segregation of duplicated DNA at the chromosomal level. Further, these subunits have been shown to have a directional differentiation or polarity, particularly evidenced in the reunion events following spontaneous or induced interruptions to chromosomal integrity. These data are perfectly compatible with the conclusion based on viscoelasticity (see Section I) that a continuous DNA molecule extends from telomere to telomere in any given chromosome. This is a satisfyingly simple concept of a chromosome, especially since some chromosomes contain three orders of magnitude more DNA than the chromosome of *Escherichia coli!* A difficulty now has surfaced in experiments which were designed to determine whether the polarity of a particular chromosome subunit remains constant

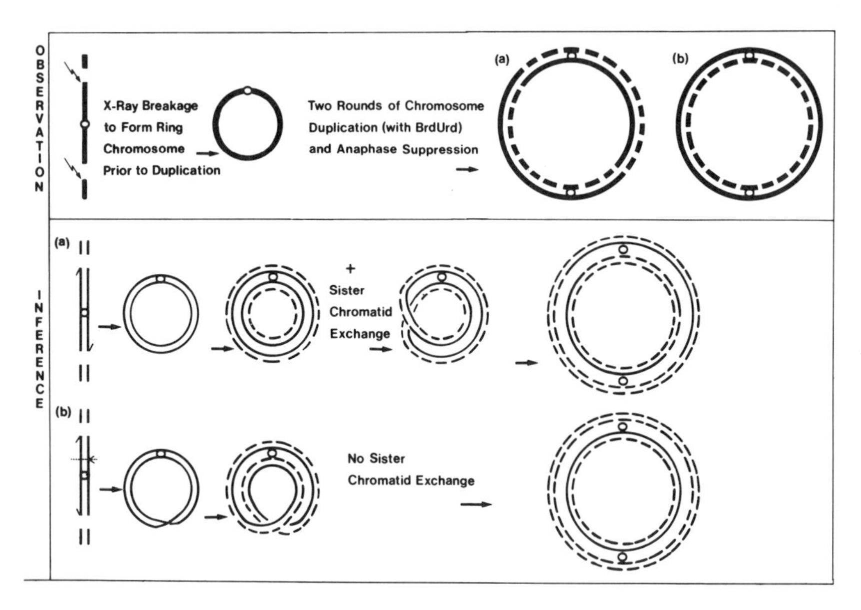

Fig. 8. Continuity of subunit polarity. Irradiation-induced ring chromosomes have provided evidence that there are rare polarity reversals along chromosome subunits (Wolff *et al.*, 1976). The distribution of subunits, marked by the BUdR staining method (Figure 3), in dicentric ring chromosomes of the second postirradiation metaphase, provide staining patterns of sister chromatids which form the pivotal data for this conclusion.

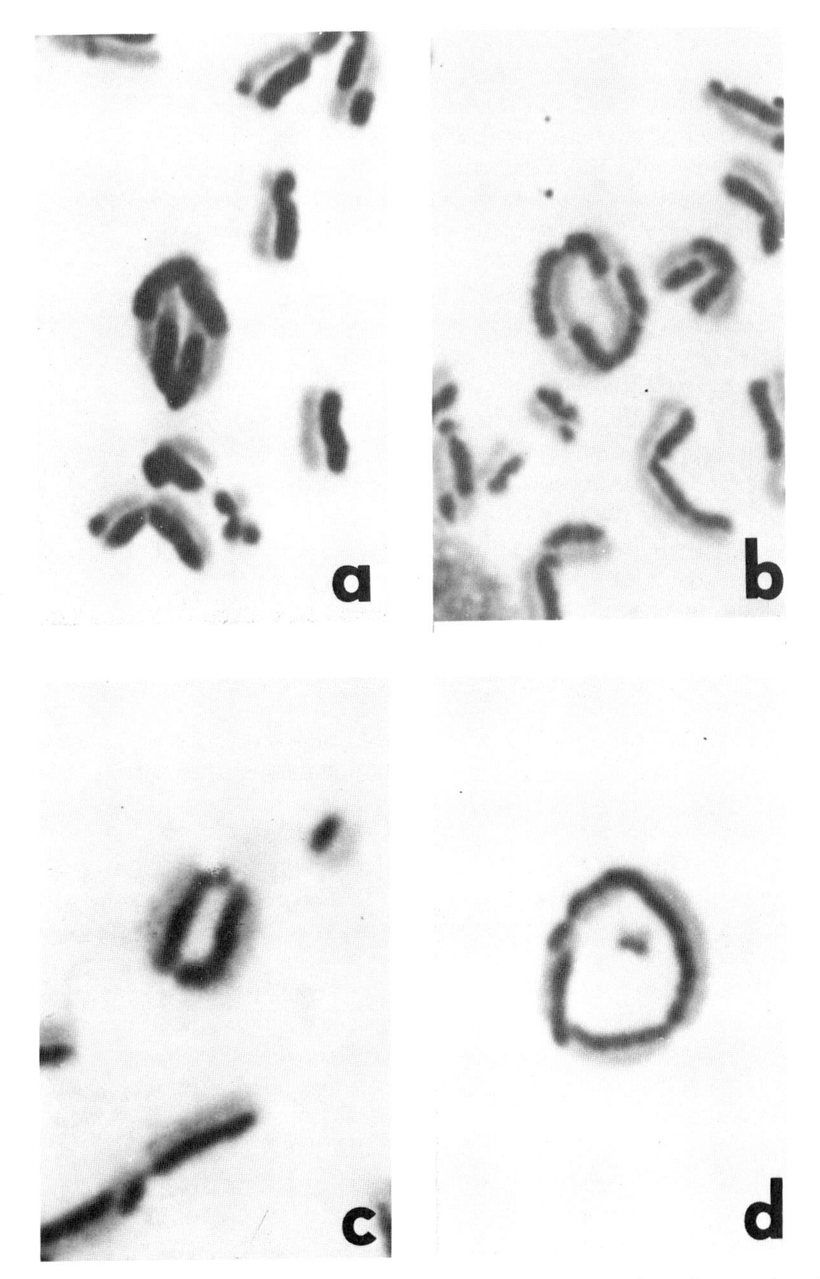

Fig. 9. Example of the BUdR–Giemsa stained ring chromosomes from the experiment outlined in Figure 8. (a,b) Symmetrical dicentric ring chromosomes formed by sister chromatid exchange following breakage and union of segments of like polarity. (c) Dicentric ring chromosome formed by sister chromatid exchange following breakage and reunion of segments of opposing polarity. (d) Asymmetrical dicentric ring chromosome formed by two sister unions.

in direction along the full length of the chromosome. An affirmative answer would be expected on the basis of the straightforward views stated above. However, these experiments, which ask about the type of ring chromosome that is induced by X rays or that could be induced by X rays in prereplication cells, while confirming that chromosomal subunits had different polarities, showed that the polarity of any one subunit could change along the length of the chromatid (Peacock *et al.*, 1973; Wolff *et al.*, 1976). On the basis of continuity of the chromosomal DNA double helix, Wolff *et al.* concluded that 3′–3′/5′–5′ phosphodiester linkages must occur, and they estimated one such bond for every 10^9 normal 3′–5′/3′–5′ phosphodiester linkages in the Chinese hamster genome.

The basis of the analysis was simply that, if two breaks induced in a prereplication chromosome were to rejoin to form a ring chromosome, then if the breaks were in regions of identical polarity, and if the cells were made tetraploid by the addition of colcemid, the tetraploid cell would contain a pair of monocentric rings. If, however, the breaks were in regions of opposing polarity, a single dicentric ring chromosome would be evident in the tetraploid cell. The analysis was somewhat complicated by the fact that sister chromatid exchange can convert a pair of monocentric rings to a dicentric ring (Figure 8), but this complication was overcome by building subunit segregation observations into the design (Figures 8 and 9). The results were straightforward and demanded the conclusion cited above: that polarity reversals occur, albeit in low frequency, along the chromosomal DNA moiety.

While it remains to be seen whether this conclusion is verified by independent lines of experimental inquiry, the data at least serve to remind us that the facile equation of the chromosome to a DNA double helix may be an oversimplification in some respects. We are probably prudent to remain perplexed by observations such as those showing unequal lateral replication of chromosome regions (see discussion on polytene chromosomes in Appels and Peacock, 1978), or those which show that localized sequence regions can be magnified or reduced relative to other parts of a linear chromosome structure (see Tartof, 1973). As was pointed out in the introduction of this chapter, much of the ferment of argument on chromosome strandedness was due to a lack of incisive experimental methodology—we must admit that changes to the simple picture of chromosomal DNA that now applies may result from further increased powers of experimental analysis.

REFERENCES

Appels, R., and Peacock, W. J. (1978). The arrangement and evolution of highly repeated (satellite) DNA sequences with special reference to *Drosophila*. *Int. Rev. Cytol.* (in press).

Bajer, A. (1965). Subchromatid structure of chromosomes in the living state. *Chromosoma* **17,** 291–302.

Brewen, J. G., and Peacock, W. J. (1969a). Restricted rejoining of chromosomal subunits in aberration formation: A test for subunit dissimilarity. *Proc. Natl. Acad. Sci. U.S.A.* **62,** 389–394.

Brewen, J. G., and Peacock, W. J. (1969b). The effect of tritiated thymidine on sister chromatid exchange in a ring chromosome. *Mutat. Res.* **7,** 433–440.

Britten, R. J., and Kohne, D. E. (1968). Repeated sequences in DNA. *Science* **161,** 529–540.

Callan, H. G. (1972). Replication of DNA in the chromosomes of eukaryotes. *Proc. R. Soc. London, Ser. B* **181,** 19–41.

Callan, H. G., and Taylor, J. H. (1968). A radioautographic study of the time course of male meiosis in the newt *Triturus vulgaris*. *J. Cell Sci.* **3,** 615–626.

Carrano, A. V., and Wolff, S. (1975). Distribution of sister chromatid exchanges in the euchromatin and heterochromatin of the Indian Muntjac. *Chromosoma* **53,** 361–369.

Chaganti, R. S. K., Schonberg, S., and German, J. (1974). A many fold increase in sister chromatid exchanges in Bloom's syndrome lymphocytes. *Proc. Natl. Acad. Sci. U.S.A.* **71,** 4508–4512.

Church, K., and Wimber, D. E. (1969). Meiosis in *Ornithogalum virens* (Liliaceae): Meiotic timing and segregation of ^{3}H-thymidine labeled chromosomes. *Can. J. Genet. Cytol.* **11,** 573–581.

Craig-Cameron, T., and Jones, G. H. (1970). The analysis of exchanges in tritium-labelled meiotic chromosomes. 1. *Schistocerca gregaria*. *Heredity* **25,** 223–232.

Crossen, P. E., Pathak, S., and Arrighi, F. E. (1975). A high resolution study of the DNA replication patterns of Chinese hamster chromosomes using sister chromatid differential staining technique. *Chromosoma* **52,** 339–347.

Crouse, H. V. (1954). X ray breakage of lily chromosomes at first meiotic metaphase. *Science* **119,** 485–487.

Crouse, H. V. (1961). Irradiation of condensed meiotic chromosomes in *Lilium longiflorum*. *Chromosoma* **12,** 190–214.

Cuevas-Sosa, A. (1968). Human chromosomology: Segregation of chromatids in diploid cells *in vitro*. *Nature (London)* **218,** 1059–1061.

Darlington, C. D., and Haque, A. (1969). The replication and division of polynemic chromosomes. *Heredity* **24,** 273–280.

Deaven, L. L., and Stubblefield, E. (1969). Segregation of chromosomal DNA in Chinese hamster fibroblasts *in vitro*. *Exp. Cell Res.* **55,** 132–135.

Galloway, S. M., and Evans, H. J. (1975). Sister chromatid exchange in human chromosomes from normal individuals and patients with ataxia telangiectasia. *Cytogenet. Cell Genet.* **15,** 17–29.

Geard, C. R. (1973). Chromatid distribution at mitosis in cultured *Wallabia bicolor* cells. *Chromosoma* **44,** 301–308.

Geard, C. R., and Peacock, W. J. (1969). Sister chromatid exchanges in *Vicia faba*. *Mutat. Res.* **7,** 215–223.

Gibson, D. A., and Prescott, D. M. (1972). Induction of sister chromatid exchanges in chromosomes of rat kangaroo cells by tritium incorporated into DNA. *Exp. Cell Res.* **74,** 397–402.

Green, M. M. (1968). Some genetic properties of intrachromosomal recombination. *Mol. Gen. Genet.* **103,** 209–217.

Heddle, J. A. (1968). The predicted ratios of single to twin sister chromatid exchanges. *Mutat. Res.* **6,** 57–65.

Heddle, J. A., Wolff, S., Whissell, D., and Cleaver, J. (1967). Distribution of chromatids at mitosis. *Science* **158,** 929–931.

Henderson, S. A. (1966). Time of chiasma formation in relation to the time of deoxyribonucleic acid synthesis. *Nature* (*London*) **211,** 1043–1047.

Herreros, B., and Gianelli, F. (1967). Spatial distribution of old and new chromatid subunits and frequency of chromatid exchanges in induced human lymphocyte endoreduplications. *Nature* (*London*) **216,** 286–288.

Hughes-Schrader, S. (1940). The meiotic chromosomes of the male *Llaveiella Taenechina Morrison* (Coccidae) and the question of the tertiary split. *Biol. Bull.* (*Woods Hole, Mass.*) **78,** 312–337.

Ikushima, T., and Wolff, S. (1974). Sister chromatid exchanges induced by light flashes to 5-bromodeoxyuridine- and 5-iododeoxyuridine substituted Chinese hamster chromosomes. *Exp. Cell Res.* **87,** 15–19.

Jones, K. W., and Robertson, F. W. (1970). Localization of reiterated nucleotide sequences in *Drosophila* and mouse by *in situ* hybridisation of complementary RNA. *Chromosoma* **31,** 331–345.

Judd, B. H., Shen, M. W., and Kaufman, T. C. (1972). The anatomy and function of a segment of the X chromosome of *Drosophila melanogaster. Genetics* **71,** 139–156.

Kavenoff, R., and Zimm, B. H. (1973). Chromosome-sized DNA molecules in *Drosophila. Chromosoma* **41,** 1–27.

Kihlman, B. A., and Hartley, B. (1967). "Sub-chromatid" exchanges and the "folded fibre" model of chromosome structure. *Hereditas* **57,** 289–294.

La Cour, L. F., and Pelc, S. R. (1958). Effect of colchicine on the utilization of labelled thymidine during chromosomal reproduction. *Nature* (*London*) **182,** 506–508.

Lark, K. G. (1967). Nonrandom segregation of sister chromatids in *Vicia faba* and *Triticum boeoticum. Proc. Natl. Acad. Sci. U.S.A.* **58,** 352–359.

Lark, K. G., Consigli, R. A., and Minocha, H. C. (1966). Segregation of sister chromatids in mammalian cells. *Science* **154,** 1202–1205.

Latt, S. A. (1974). Localization of sister chromatid exchanges in human chromosomes. *Science* **185,** 74–76.

McClintock, B. (1938). The production of homozygous deficient tissues with mutant characteristics by means of the aberrant mitotic behavior of ring-shaped chromosomes. *Genetics* **23,** 315–376.

McClintock, B. (1941). Spontaneous alterations in chromosome size and form in *Zea mays. Cold Spring Harbor Symp. Quant. Biol.* **9,** 72–81.

Maguire, M. (1968). Nomarski interference contrast resolution of subchromatid structure. *Proc. Natl. Acad. Sci. U.S.A.* **60,** 533–536.

Manton, I. (1945). New evidence on the telophase split in *Todea barbara. Am. J. Bot.* **32,** 342–348.

Marin, G., and Prescott, D. M. (1964). The frequency of sister chromatid exchanges following exposure to varying doses of H^3-thymidine or X-rays. *J. Cell Biol.* **21,** 159–167.

Meselson, M. (1967). The molecular basis of genetic recombination. *In* "Heritage from Mendel" (R. A. Brink, ed.), pp. 81–104. Univ. of Wisconsin Press, Madison.

Meselson, M., and Stahl, F. W. (1958). The replication of DNA in *Escherichia coli. Proc. Natl. Acad. Sci. U.S.A.* **44,** 671–682.

Moens, P. B. (1966). Segregation of tritium-labeled DNA at meiosis in *Chorthippus. Chromosoma* **19,** 277–285.

Nebel, B. R. (1939). Chromosome structure. *Bot. Rev.* **5,** 563–626.

Ostergren, G., and Wakonig, T. (1954). True or apparent subchromatid breakage and the induction of labile states in cytological chromosome loci. *Bot. Not.* **107,** 357–375.

Pardue, M. L., and Gall, J. G. (1970). Chromosomal localization of mouse satellite DNA. *Science* **168,** 1356–1358.

Peacock, W. J. (1961). Subchromatid structure and chromosome duplication in *Vicia faba*. *Nature (London)* **191,** 832–833.

Peacock, W. J. (1963). Chromosome duplication and structure as determined by autoradiography. *Proc. Natl. Acad. Sci. U.S.A.* **49,** 793–801.

Peacock, W. J. (1965). Chromosome replication. *Natl. Cancer Inst., Monogr.* **18,** 101–131.

Peacock, W. J. (1968). Chiasmata and crossing over. *In* "Replication and Recombination of Genetic Material" (W. J. Peacock and R. D. Brock, eds.), pp. 242–252. Aust. Acad. Sci., Canberra.

Peacock, W. J. (1970). Replication, recombination, and chiasmata in *Goniaea Australasiae* (Orthoptera: Acrididae). *Genetics* **65,** 593–617.

Peacock, W. J. (1971). Cytogenetic aspects of the mechanism of recombination in higher organisms. *Stadler Genet. Symp.* **2,** 123–152.

Peacock, W. J., Wolff, S., and Lindsley, D. L. (1973). Continuity of chromosome subunits. *Chromosomes Today* **4,** 85–100.

Peacock, W. J., Appels, R., Dunsmuir, P., Lohe, A. R., and Gerlach, W. L. (1977). Highly repeated DNA sequences: Chromosomal localization and evolutionary conservation. *In* "International Cell Biology" (B. R. Brinkely and R. R. Porter, eds.), pp. 494–506. Rockefeller Univ. Press, New York.

Perry, P., and Wolff, S. (1974). New Giemsa method for the differential staining of sister chromatids. *Nature (London)* **251,** 156–158.

Prescott, D. M. (1970). The structure and replication of eukaryotic chromosomes. *Adv. Cell Biol.* **1,** 57–117.

Rudkin, G. T. (1965). The structure and function of heterochromatin. *Genet. Today, Proc. Int. Congr., 11th, 1963* Vol. 2, pp. 359–374.

Schwartz, D. (1954). Studies on the mechanism of crossing over. *Genetics* **39,** 692–700.

Smith, G. P. (1973). Unequal crossover and the evolution of multigene families. *Cold Spring Harbor Symp. Quant. Biol.* **38,** 507–514.

Tartof, K. D. (1973). Unequal mitotic sister chromatid exchange and disproportionate replication as mechanisms regulating ribosomal RNA gene redundancy. *Cold Spring Harbor Symp. Quant. Biol.* **38,** 491–500.

Taylor, J. H. (1958). Sister chromatid exchanges in tritium-labeled chromosomes. *Genetics* **43,** 515–529.

Taylor, J. H. (1959). Further studies on the mechanism of chromosome duplication. *Proc. Natl. Biophys. Conf.* **1,** 264–274.

Taylor, J. H. (1962). Chromosome reproduction. *Int. Rev. Cytol.* **13,** 39–73.

Taylor, J. H. (1965). Distribution of tritium-labeled DNA among chromosomes during meiosis. I. Spermatogenesis in the grasshopper. *J. Cell Biol.* **25,** No. 2, Part 2, 57–67.

Taylor, J. H., Woods, P. S., and Hughes, W. L. (1957). The organization and duplication of chromosomes as revealed by autoradiographic studies using tritium-labeled thymidine. *Proc. Natl. Acad. Sci. U.S.A.* **43,** 122–128.

Trosko, J. E., and Wolff, S. (1965). Strandedness of *Vicia faba* chromosomes as revealed by enzyme digestion studies. *J. Cell Biol.* **26,** 125–135.

Walen, K. H. (1965). Spatial relationships in the replication of chromosomal DNA. *Genetics* **51,** 915–929.

Wilson, E. B. (1925). "The Cell in Development and Heredity," 3rd ed. Macmillan, New York.

Wilson, G. B., and Sparrow, A. H. (1960). Configurations resulting from iso-chromatid and iso-subchromatid unions after meiotic and mitotic prophase irradiation. *Chromosoma* **11,** 229–244.

Wimber, D. E. (1960). Duration of the nuclear cycle in *Tradescantia paludosa* root tips as measured with H^3-thymidine. *Am. J. Bot.* **47,** 828–834.

Wolff, S. (1964). Are sister chromatid exchanges sister strand crossovers or radiation-induced exchanges? *Mutat. Res.* **1,** 337–343.

Wolff, S., and Perry, P. (1974). Differential Giemsa staining of sister chromatids and the study of sister chromatid exchanges without autoradiography. *Chromosoma* **48,** 341–353.

Wolff, S., Lindsley, D. L., and Peacock, W. J. (1976). Cytological evidence for switches in polarity of chromosomal DNA. *Proc. Natl. Acad. Sci. U.S.A.* **73,** 877–881.

8

Eukaryotic Chromosome Replication and Its Regulation

Roger Hand

I. ORGANIZATION OF THE EUKARYOTIC CHROMOSOME FOR REPLICATION

A. The Replication Unit

Evidence that the eukaryotic chromosome is organized for purposes of replication into multiple discrete units was put forth by Cairns (1966) and Huberman and Riggs (1968). In both studies, the technique of DNA fiber autoradiography was used to investigate DNA replication in mammalian tissue culture cells. With this technique, originated by Cairns (1963) and

CELL BIOLOGY, VOL. 2

ISBN 0-12-289502-9

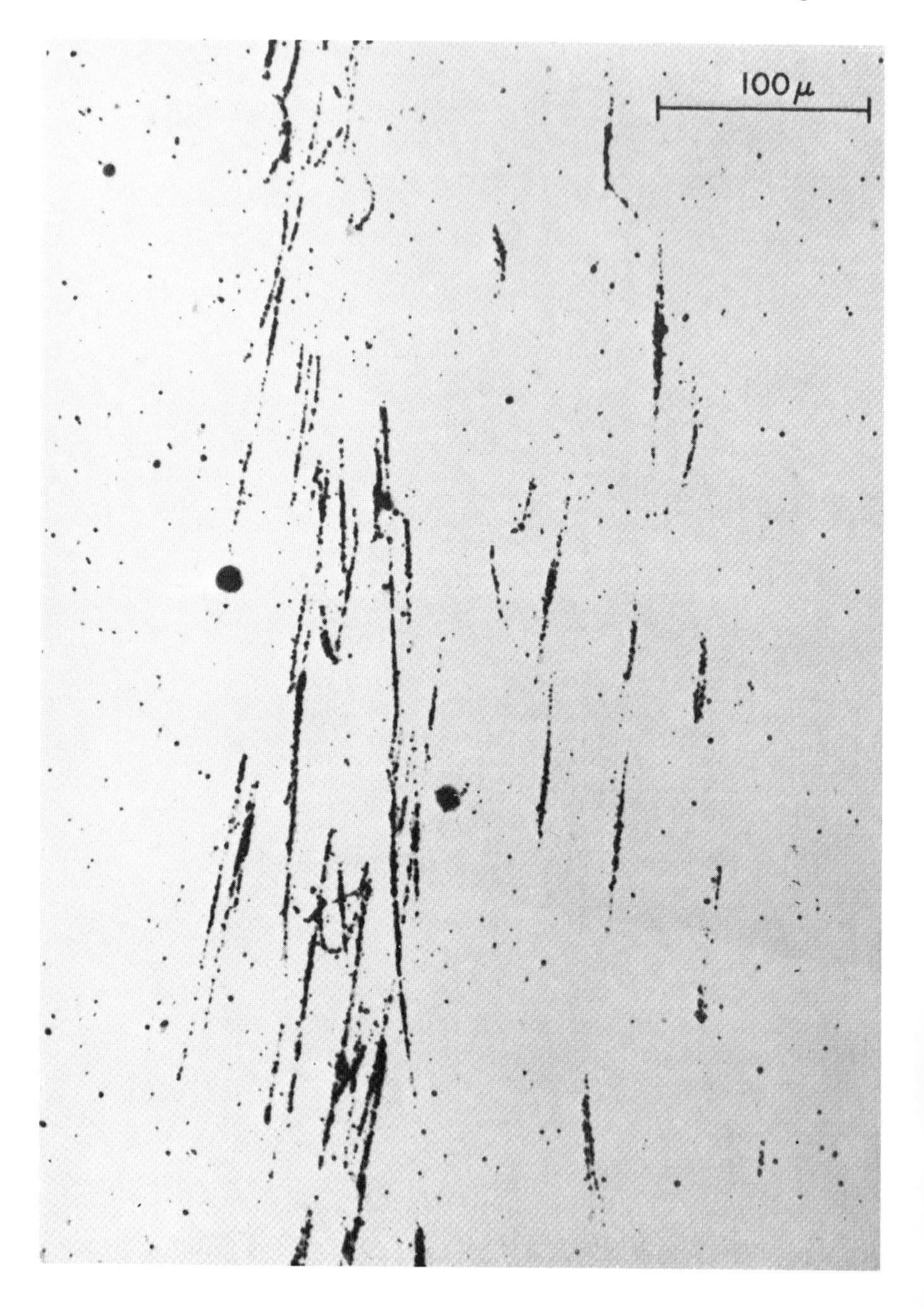
100μ

first applied by him to the study of replication of the bacterial chromosome, DNA is labeled with a suitable radioactive precursor (usually [^{3}H]thymidine) and, after it is released from cells by gentle lysis, the DNA is spread linearly along a microscope slide with a glass rod. After suitable fixation, the preparations are coated with radiation-sensitive emulsion and exposed until arrays of silver grains of sufficient density are produced. The methods of preparation of autoradiographs are described in several publications (Lark *et al.*, 1971; Prescott and Kuempel, 1973). Figure 1, taken from Huberman and Riggs (1968), shows the linear autoradiograms produced when mammalian cellular DNA (from Chinese hamster ovary cells) is labeled with [^{3}H]thymidine and the label chased with nonradioactive thymidine. Replication during the radioactive pulse takes place on many individual units within the chromosomal DNA. The decreasing density of the grains at the ends of the tracks probably results from the decrease in specific activity of the thymidine nucleotide pool following the change from the radioactive to nonradioactive thymidine. The patterns of linear arrays of silver grains with decreasing density at both ends may be taken as evidence that replication initiates at a central origin of each unit, and subsequent chain elongation proceeds bidirectionally. Under the usual conditions of autoradiography, each grain track represents two daughter double helices lying side by side. Occasionally, the grain tracks of the two daughter helices are separated. The ends of the separated tracks meet to form Y-shaped structures, and it may be surmised from this that chain growth occurs via forklike growing points. The newly synthesized chains meet and fuse with chains synthesized on adjacent units, producing the long DNA duplexes characteristic of the eukaryotic chromosome. The diagrammatic representation of this scheme of replication is shown in Figure 2, also taken from Huberman and Riggs (1968). Similar autoradiographic studies have shown that this type of organization for replication occurs in yeast (Petes and Williamson, 1975), insects (Blumenthal *et al.*, 1973), fish (Prescott and Kuempel, 1973), amphibia (Callan, 1972), birds (McFarlane and Callan, 1973), mammals (Huberman and Riggs, 1968; Hand *et al.*, 1971; Hori and Lark, 1973), and humans (Huberman and Riggs, 1968) and probably holds true for all eukaryotes.

Fig. 1. Autoradiograms of DNA from Chinese hamster cells. The cells were treated for 12 hr with fluorodeoxyuridine and then pulse-labeled for 30 min with [^{3}H]thymidine (18 Ci/mmole, 0.5 μg/ml) in the presence of fluorodeoxyuridine. They subsequently were chased with unlabeled thymidine (5 μg/ml). The linear arrays of silver grains indicate individual replication units that initiated during the pulse. Note the tracks of lighter grain density at the ends of most arrays that represent DNA replicated during the chase. Such patterns indicate bidirectional replication. (From Huberman and Riggs, 1968, reproduced with permission.)

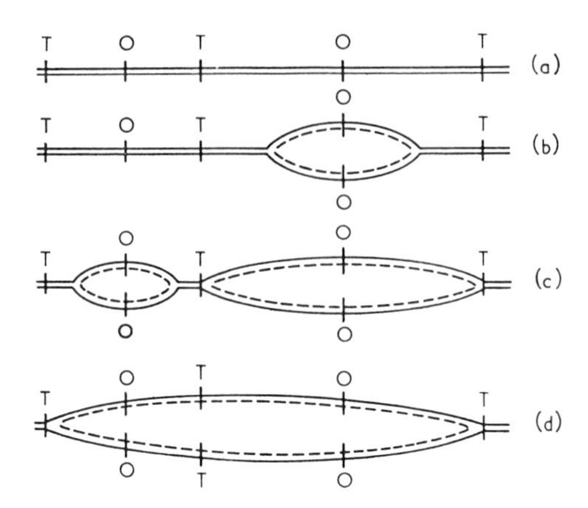

Fig. 2. The bidirectional model for DNA replication. Each pair of horizontal lines represents a portion of a double helix. The solid lines indicate parental chains, while the dashed lines represent daughter chains. Origins (O) and termini (T) are indicated. Replication of two adjacent units is represented. (a) before replication; (b) replication initiated in unit on the right; (c) replication initiated in unit on the left and completed at terminus of unit on the right; (d) replication completed with subsequent fusion of both units at the common terminus. (From Huberman and Riggs, 1968, reproduced with permission.) Since the original publication, the existence of fixed termini has been questioned. Also, it has been shown that initiation on adjacent units is more likely to be synchronous than asynchronous. The remainder of the postulates in the model have been supported experimentally.

The term "replication unit" is used here in preference to the more popular "replicon." The latter term first was applied to prokaryotic replication by Jacob *et al.* (1963), who introduced the concept: "A genetic element such as an episome or a chromosome (of a bacterium or of a phage) constitutes a unit of replication or replicon. Such a unit can only replicate as a whole." Further, they postulated an initiator substance, the replicator, that permits replication to begin at the same point and to proceed linearly until the entire unit is copied. The replicon hypothesis, as stated above, still applies to prokaryotic replication with the only major modification being that some circular prokaryotic chromosomes appear to replicate bidirectionally from the replicator (Bird *et al.*, 1972; Prescott and Kuempel, 1972; Danna and Nathans, 1972; Crawford *et al.*, 1973). The hypothesis was applied intact by a number of investigators to the replication of eukaryotic DNA. Here, as described above, the chromosomes had multiple units of replication, that were termed eukaryotic "replicons" (Taylor, 1963). As can be seen from the definition of Jacob *et al.*, the term "replicon" does not apply strictly to replicating units in eukaryotic chromosomes, since a replicon is a genetic element, and the genetic content of the units in eukaryotic chromosomes is unknown. Huberman and Riggs

(1968) introduced the term "replication section" to refer to any stretch of DNA replicated via a single growing point. Since pairs of replication sections appeared to share a common initiation point from which two growing points proceeded in opposite directions, they proposed the term "replication unit" to mean the basic unit of control in the initiation of replication—presumably an adjacent pair of diverging replication sections. These definitions of replication section and unit have held up well over the past decade as descriptive data on eukaryotic chromosome replication have accumulated, while replicon has been used to mean either section or unit, depending on the author.

B. The Rate of DNA Synthesis

Each cell contains a finite amount of chromosomal DNA that must be replicated completely within the confines of a single synthetic (S) phase of the cell cycle. The rate of synthesis is the amount of DNA made per unit time. The most direct method of measurement would be a biochemical determination of DNA in cells as they progress through the synthetic phase. However, tissue culture cells, the most favored material for the study of eukaryotic DNA replication, can be grown to quantities sufficient for direct biochemical determinations only with great difficulty. Therefore, most estimates of the rate of DNA synthesis have been made with indirect techniques. An indirect method based on the incorporation of exogenous thymidine into mammalian tissue culture cells has been developed by Sieger *et al.* (1974). Here, the synthesis of endogenous thymidylate is blocked, and the rate of incorporation of exogenous thymidine into DNA is determined under conditions in which the specific activity of the thymidine nucleotide pools is the same as that of the exogenous thymidine. Rates of DNA synthesis of slightly more than 0.5 μg/hr per 10^6 cells are obtained, comparing favorably with the gross rate determined from S phase lengths and DNA content.

Saladino and Johnson (1974) have proposed another method based on rates of thymidine incorporation in mammalian cells in tissue culture. The investigators assume that by flooding the cells with thymidine from the external medium, the specific activity of the intracellular thymidine nucleotide pool will be the same as the specific activity of thymidine in the medium, since pool dilution by endogenous thymidylate synthesis will be minimized. Given the external concentrations of thymidine used in these experiments (4×10^{-5} *M*) and the estimated concentrations of TTP in the acid-soluble pools (10^{-4} *M* or greater; Adams, 1969; Skoog and Bjursell, 1974), these assumptions may not be valid. Nevertheless, the estimates of rates of DNA synthesis (1.24×10^{-14} gm/min/cell or 0.75 μg/hr/10^6 cells) are in reasonable agreement with the estimates of Sieger *et al.* (1974).

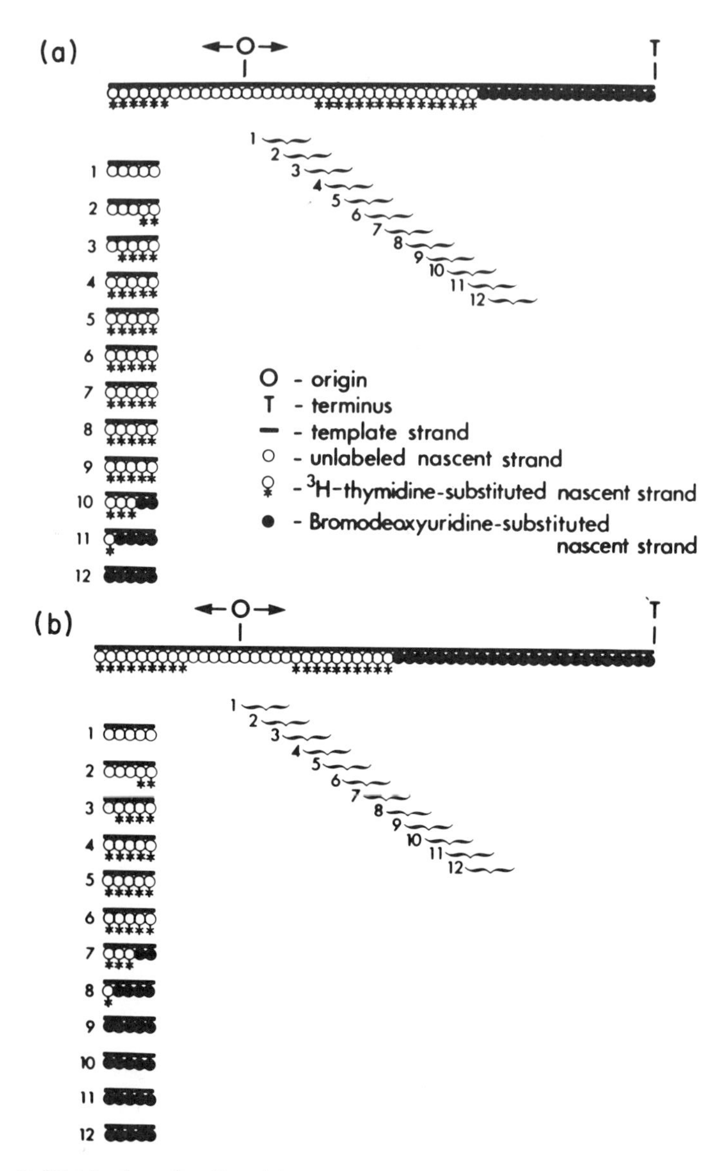

Fig. 3. Distribution of radioactivity and density label following sequential pulses with [^{3}H]thymidine and bromodeoxyuridine. The diagram demonstrates that slower rates of fork progression yield relatively more labeled DNA of greater than normal density. The right-hand section of a replication unit is shown. (a) Replication began at the origin before the [^{3}H]thymidine pulse and continued on the unit through the [^{3}H]thymidine pulse and into the bromodeoxyuridine pulse. If the DNA were extracted and subjected to random shear to the size indicated by the brackets, units with this type of replication would generate the classes of fragments illustrated. The distribution of radioactivity in fragments within and around the radioactive segment is shown in the column below the unit. Classes 2–11 would be detected

The rate of DNA synthesis is not constant through the S phase. Klevecz *et al.* (1975) using an automated method for synchronization of cells by mitotic selection, measured the relative DNA content of individual cells at different points in S by flow microfluorometry and demonstrated variations in the rate of DNA synthesis through the S phase in tissue culture cells derived from Chinese hamsters. Synthesis is very slow at the beginning of S and increases in a saltatory fashion through the S phase. Fifty percent of replication occurs during the last hour of a 5.5-hr S phase in the V79 cell line. The saltatory increases in synthesis correspond well to the three maxima of thymidine incorporation into DNA in S phase diploid cells previously reported (Klevecz and Kapp, 1973; Remington and Klevecz, 1973).

C. Mensuration of Replication

Two parameters of replication are easily measurable, namely, the rate of replication fork progression and the size of replication units. From these, the more refined measures of DNA synthetic rate can be calculated. Before going further, the methods used to measure these merit discussion. There are many methods; however, for the most part, they are based on two techniques, i.e., sedimentation of DNA by ultracentrifugation, and autoradiography of extended fibers of radioactively labeled DNA.

1. Fork Progression

a. Measurements by Sedimentation. Several methods using the ultracentrifuge have been devised. Both rate zonal sedimentation in sucrose density gradients (Lehmann and Ormerod, 1970; Cheevers *et al.*, 1972; Brewer *et al.*, 1974) and equilibrium sedimentation in cesium chloride gradients (Taylor, 1968; Painter and Schaefer, 1969a; Hyodo and Flickenger, 1972) have been used. The method of Painter and Schaefer (1969a) has the advantage of being relatively straightforward but is subject to a few limitations because of mathematical assumptions. It is based on the premise that a short pulse of radioactivity preceding a density label in

---

by analyzing a neutral cesium chloride gradient for radioactivity. Classes 10 and 11 would have buoyant density heavier than normal DNA, and about 10% of the counts would be found on the heavy side of a peak of normal density marker DNA. (b) A similar replication unit with a slower rate of fork progression [10/16 or 62.5% that in (a)]. Sheared DNA from such units would generate classes of fragments indicated in the column below the unit. Classes 2–8 would be detected by analyzing a neutral cesium chloride gradient for radioactivity, and classes 7 and 8 would have greater than normal density. About 16% of the radioactivity would be found on the heavy side of the normal density peak.

DNA will produce little radioactivity associated with the density label if the lengths of radioactive DNA are large compared to the sheared fragments of DNA in a centrifuged specimen. When the lengths of radioactive DNA are small compared to the fragments, there will be a higher frequency of fragments with end-to-end association of the radioactive and density labels. Thus, high rates of fork progression during the radioactive pulse will result in labeled DNA of near normal density, and low rates of fork progression in relatively more labeled DNA of greater than normal density. This is illustrated in Figure 3. Quantitation is achieved by measuring the size of the sheared fragments (B) in daltons, and the fraction of radioactive DNA of greater than normal density (F). The latter may be estimated by comparing the areas under the DNA peaks in experimental and marker normal density DNA centrifuged under the same conditions. The amount of DNA per growing point (L) labeled during the radioactive pulse in daltons is related to B and F by the formula shown in Eq. (1)

$$L = B/2F, \qquad \text{where } B < L \tag{1}$$

The derivation of the formula is straightforward. Let Y = the distance from a transition point of radioactivity and density to the nearest breakpoint in the radioactive segment induced by shear. Since $B < L$, Y is also equal to the length of the radioactively labeled section in the fragment containing both radioactivity and density label generated by a specified pair of breaks. Y varies from 0 to B, and all distances are equally frequent, since shear is random. The average size of radioactively labeled sections of fragments with both radioactive and density label is $\bar{Y}$, which equals $B/2$. For each fragment containing density and radioactive label, there are one or more fragments containing radioactive label alone. The aggregate average length of radioactivity in these non-density-labeled fragments is $L - B/2$. Therefore,

$$F = \frac{B/2}{B/2 + (L - B/2)} = \frac{B}{2L} \tag{2}$$

which rearranges to Eq. (1).

Figure 4, modified from Painter and Schaefer (1969a), shows a hypothetical distribution of density labeled fragments following shear and equilibrium centrifugation. In this case, shear size equals the length of the radioactive segment. The figure points out one of the drawbacks of the method, namely, that the normalization procedure used to facilitate the measurements of the areas under the normal density and heavy DNA peaks underestimates the total amount of DNA of heavy density. There

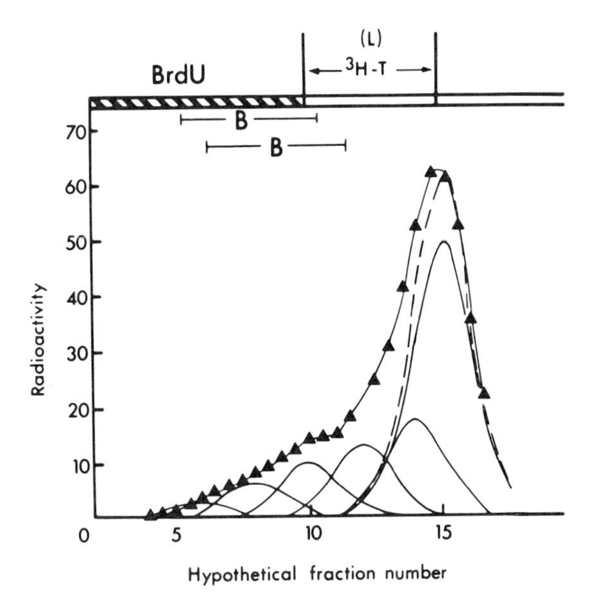

Fig. 4. Hypothetical distribution of ^{3}H counts in a neutral cesium chloride gradient generated by equilibrium centrifugation. Such a distribution could result if DNA were pulse-labeled for 30 min with [^{3}H]thymidine, then labeled for 3 hr with bromodeoxyuridine. Fork movement was from right to left in the section of the unit diagrammed at the top. A length of DNA (L) is labeled with [^{3}H]thymidine. The DNA would be sheared to length B after extraction. Breakpoints would occur randomly along any stretch of DNA containing [^{3}H]thymidine-labeled and bromodeoxyuridine-labeled nucleotides in end-to-end association. Here, the breakpoints are represented as occurring at different points along the DNA molecule at intervals of L/5 with the first breakpoint at L/10. Ten classes of sheared fragment with ^{3}H are created, five containing no bromodeoxyuridine, and five containing bromodeoxyuridine and [^{3}H]thymidine. The five normal density classes would sediment in the distribution indicated by the solid line centered over fraction 15 at the top of the gradient. The five classes with bromodeoxyuridine in association with [^{3}H]thymidine would sediment in the distributions indicated by the overlapping solid lines toward the bottom of the gradient. The filled triangles show the resulting radioactive profile. The dotted line shows the profile of normal density ^{14}C-labeled marker DNA normalized so that the peak fraction has the same amount of radioactivity as the equivalent [^{3}H]fraction. The total number of counts for each curve (solid lines) may be calculated using the arbitrary scale on the ordinate. The actual percentage of counts in heavy DNA is 50; the calculated would be 42%. The difference is accounted for by those bromodeoxyuridine-containing fragments whose density shift is so slight that they sediment largely within the normalized ^{14}C peak (for example, the class distributed about fraction 13). The normalization error in the calculation of rate of fork movement may be minimized by adjusting pulse-labeling conditions so that the fraction of heavy DNA is kept to the minimum necessary to permit accurate calculation. (Modified from Painter and Schaefer, 1969a.)

are other drawbacks as well. Because of mixing of density analogue and radioactive precursor in the pool during the transition from one pulse to another, some DNA will have radioactive label interspersed with density label, rather than association being only end-to-end. The amount of interspersed DNA will vary with pool size. It can be accounted for by shearing an aliquot to size much smaller than B and measuring the fraction of DNA of heavy density. It is assumed that the density-shifted DNA of small size will be composed of fragments in which the radioactive and density labels are interspersed rather than in end-to-end association. In addition, changes in frequency of initiation and termination of replication probably would modify the proportion of density-shifted DNA. The method involves several long centrifugations, usually of replicate samples, and is expensive in both time and material. Its advantages are that it measures rate of fork progression in large populations of cells, that it is objective, and that it lends itself to automation, which results in a substantial saving in effort, if not in time.

b. Measurements by Autoradiography. Autoradiographic methods for measurement of chain growth are direct; they are based simply on the measurements of grain track lengths in fiber autoradiograms from DNA pulse labeled for a known duration of time.

If DNA from a mammalian cell in tissue culture is labeled sequentially with [^{3}H]thymidine of high then low specific activity and fiber autoradiograms are prepared, two types of replication patterns are seen on bidirectional replication units. These are illustrated in Figure 5. Units which began replication before the high specific activity pulse show a pattern with a short clear stretch (presumably indicating DNA replicated before the beginning of the pulse) flanked by two linear grain tracks of heavy density proceeding directly to tracks of lighter density (Figure 5a). Units of this type are prepulse units: the high density tracks represent DNA replicated during the high specific activity pulse, and the low density tracks represent that replicated during the low specific activity pulse. Units which began replication after the high specific activity pulse show a central track of high grain density flanked on either end by tracks of lower grain density (Figure 5b). These are postpulse units.

As pointed out by Huberman and Riggs (1968), the most accurate measure of fork progression probably is derived by measurements of high grain density track lengths in prepulse units. In these units, the beginning of the pulse is marked by the appearance of grains adjacent to the central clear stretch. The sharpness of this transition is achieved by blocking synthesis of TMP by an agent such as fluorodeoxyuridine so that DNA of very high specific activity is synthesized as soon as the pulse begins. The end of the high specific activity pulse is marked by the transition from

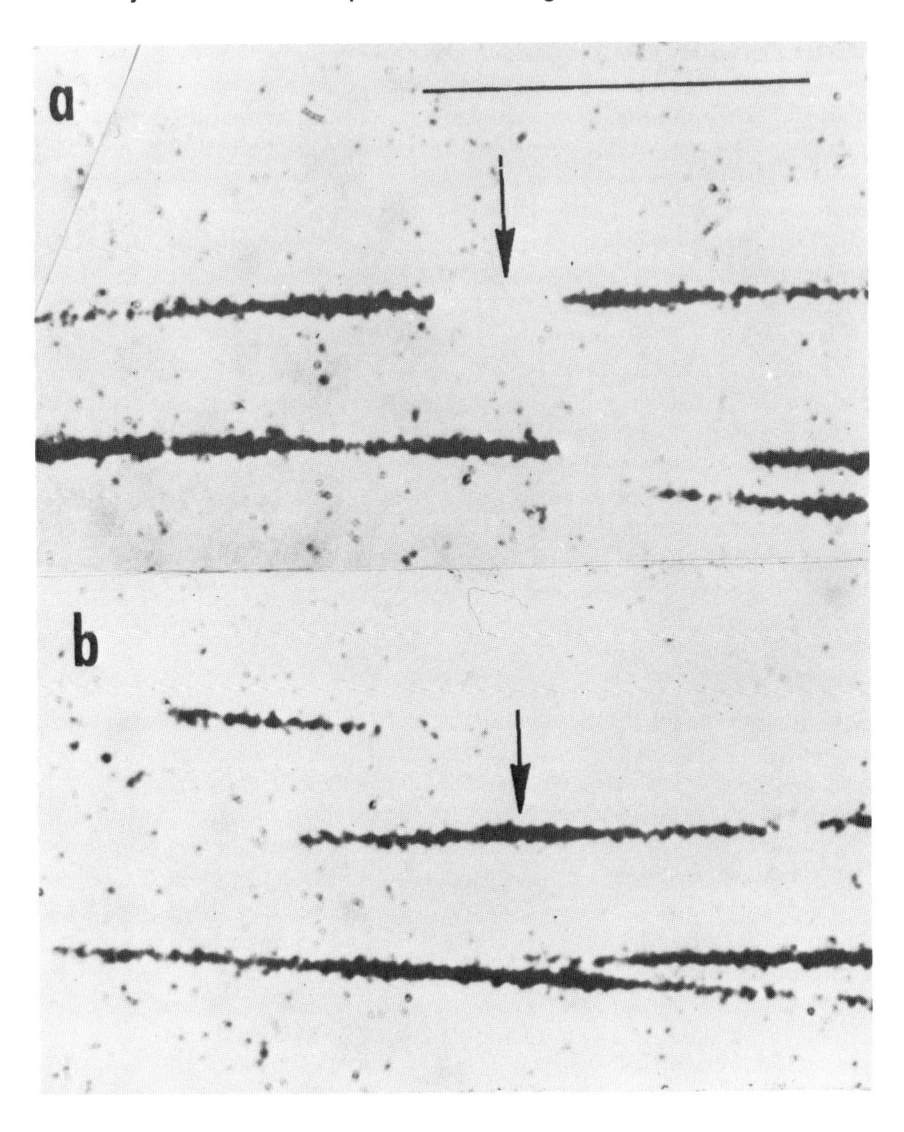

Fig. 5. DNA autoradiograms of individual replication units. Mouse L cells were labeled with [^{3}H]thymidine, 250 μCi/ml, during a 30-min high specific activity pulse (50 Ci/mmole) and then a 30-min low specific activity pulse (5 Ci/mmole). The DNA was prepared for autoradiography. The arrows indicate presumed initiation points. (a) Prepulse initiation unit. (b) Postpulse initiation unit. The bar represents 50 μm; both micrographs are at the same magnification.

heavy to light grain density. With such markers for the beginning and end of the high specific activity pulse, the rate of fork progression or chain growth during that pulse is calculated easily. It is assumed that the DNA molecules are drawn out completely straight by the spreading procedure and that the duplex is not stretched beyond its normal length. Postpulse units cannot be used to measure rate, since the beginning of the high specific activity pulse is not marked. Grain tracks generated by pulses of a single specific activity cannot be used, since prepulse units cannot be differentiated from postpulse units.

The advantages of the method are that it is direct, inexpensive, and technically simple. Its drawbacks are the subjectivity that may result from the observer's selection bias and the difficulty in sampling large populations of cells. To an extent, accuracy is dependent on the enthusiasm the investigator has for sitting at the microscope to measure the tracks.

The agreement between values obtained by sedimentation and autoradiographic methods for DNA from mammalian cells is reasonable [see Table I from Edenberg and Huberman (1975)]. The methods may be viewed as complementary, and both should be used in investigations in which the rate of fork progression is an important parameter.

2. Replication Unit Size

a. Measurements by Sedimentation. Gautschi *et al.* (1973) administered radioactive pulses of 2, 4, and 8 min to mammalian DNA. When the DNA was analyzed by sedimentation in alkaline sucrose density gradients, the radioactivity profiles of the gradients of the three pulses were very similar (Figure 6). They reasoned that this suggested a specific class of nascent DNA molecules reflecting the heterogeneous size distribution of replication units, and that the sizes of the molecules should be half the size of the corresponding replication units. Fifty percent of the labeled DNA in the gradients sedimented between 25 and 60 S. This would correspond to molecules of 5–40 μm and replication unit sizes of 10–80 μm.

As pointed out by Edenberg and Huberman (1975), fused adjacent replication units will sediment as large molecules in an alkaline sucrose gradient. A proportion of the radioactivity should sediment at an S value equivalent to the length of three replication units or more, since the labeled segments at the outboard end of units will join with both flanking units if replication is bidirectional. This would be at least six times the length (approximately 250 μm) of nascent molecules that have not fused with adjacent units. DNA molecules, for the most part, are sheared to sizes less than this by most centrifugation procedures. In the experiment in Figure 6, bulk DNA was sheared to sizes below that of multiple units,

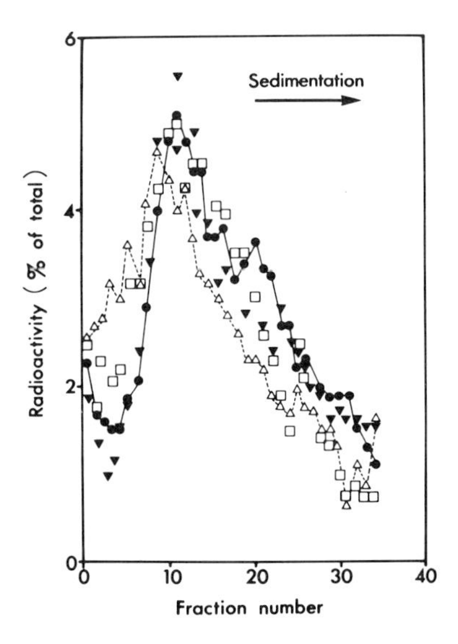

Fig. 6. Sedimentation profiles in alkaline sucrose gradients of pulse-labeled DNA molecules from HeLa cells. Labeling times: 8 min (●, ▼), 4 min (□), and 2 min (△). Bulk DNA labeled continuously for one generation or more would sediment between fraction 30 and the bottom of the gradient. (From Gautschi *et al.*, 1973, reproduced with permission.)

so that the average S value of labeled molecules most probably is not increased significantly by contamination with fused units.

b. Measurements by Autoradiography. A bidirectional replication unit, as defined by the autoradiogram produced by sequential radioactive pulses of differing specific activities, has a central initiation point. The average distance between initiation points on tandemly arranged adjacent units corresponds to the average size of units. Here, the difficulty lies in recognizing widely separated initiation points (and therefore, large units) as being adjacent. This difficulty is compounded when the rate of replication fork progression is slowed by experimental manipulation, since the open spaces between adjacent units are increased. The use of DNA fibers containing multiple units, whose linearity is easy to recognize, minimizes problems of this sort. In mammalian tissue culture cells, good agreement for replication unit size between autoradiographic and sedimentation methods has been realized. Huberman and Riggs (1968) showed that the size of most units was between 15 and 60 μm in Chinese hamster cells. Estimates in other mammalian cells have been of the same order of magnitude (see Section I,E below).

D. Components of the Rate of Replication

The overall rate of replication in a nucleus is determined by two factors, the frequency of initiation of replication and the rate of fork progression. For example, the increase in the rate of DNA synthesis that results from infection of mouse cells with polyomavirus (Dulbecco *et al.*, 1965; Cheevers *et al.*, 1970) is caused both by an increase in fork progression rate and by an increased frequency of initiation (Cheevers *et al.*, 1972). The increase in initiation frequency results from recruitment to replication of initiation sites not ordinarily in operation in the absence of viral infection (Cheevers and Hiscock, 1973). Other factors theoretically could play a role in governing the rate of replication. Premature termination could slow the rate of replication. The nascent strands would not be elongated to their usual size, resulting in unreplicated segments between the ends of the prematurely terminated strands. If the gaps were not filled until sometime later, the result could be a slowing of the overall rate of replication. In such a situation, there also would be an apparent increase in the number of replication units and an apparent decrease in their size.

1. Rate of Replication Fork Progression

Rate of fork progression has been measured in DNA from a wide variety of eukaryotic organisms. In mammalian tissue culture cells, most estimates fall between 0.5 and 2.0 μm/min/fork. In general, estimates obtained by autoradiography (Cairns, 1966; Huberman and Riggs, 1968; Lark *et al.*, 1971; Hand and Tamm, 1972, 1973) are lower than estimates by sedimentation methods by about a factor of 2 (Taylor, 1968; Painter and Shaefer, 1969a; Lehmann and Ormerod, 1970; Gautschi *et al.*, 1973). There is no clear explanation for this discrepancy. In cultures of synchronized cells, the rate of chain growth increases roughly twofold as the cells progress through the S phase (Painter and Schaefer, 1971; Houseman and Huberman, 1975). It has been suggested that this increase might be related to the increased amounts of nucleotide triphosphates available in the DNA precursor pools (Housman and Huberman, 1975), although as yet this has not been demonstrated experimentally.

Rates in nonmammalian tissues have been found to be more variable, ranging from 0.1 to 1.4 μm/min/fork (Callan, 1972; Weintraub and Holtzer, 1972; Hyodo and Flickenger, 1972; McFarlane and Callan, 1973; Blumenthal *et al.*, 1973). However, most of the cell systems have not been maintained in tissue culture (where more uniformity could be expected), and temperatures at which measurements were made have varied from 18° to 37°C. Such variation in rate of fork movement might reflect more accurately the situation under physiological conditions in the whole organism, although it should be mentioned that there has been no systematic study of the effect of temperature on rate of fork progression.

2. *Frequency of Initiation*

Frequency of initiation has not been measured directly. In some studies, investigators have suggested that initiation frequency has changed, because the overall rate of DNA replication has changed with little variation in the rate of replication fork progression. Thus, inhibition of initiation has been proposed as the method by which a number of agents decrease DNA synthesis when reduction in chain growth rate could not account for the decrease (Hand and Tamm, 1972; Gautschi *et al.,* 1973; Hori and Lark, 1973). In other studies, slower rates of overall DNA replication have been correlated with increases in the distance between initiation sites (that is, replication units are larger), while there has been no change in the rate of fork progression (Callan, 1972; Blumenthal *et al.,* 1973; Hand and Tamm, 1974). These results also have been interpreted as showing decreased frequency of initiation.

A method has been devised for estimating the relative frequency of initiation. As indicated in Section I,C, autoradiograms of replicating units show two patterns, prepulse and postpulse. The proportion of each in an autoradiographic preparation is measurable. A putative inhibitor of initiation added at the time of the pulse should decrease the proportion of postpulse patterns in a sample relative to a control. This type of analysis in mammalian tissue culture cells has been used to demonstrate an inhibition of initiation secondary to a decrease in cellular protein synthesis (Hand, 1975b).

An estimate of the number of active replication units and the frequency of initiation in mammalian tissue culture cells may be obtained from the overall rate of replication and the average rate of fork progression. Assume an average rate of DNA replication of 0.75 μg/hr per 10^6 cells (Saladino and Johnson, 1974). This is equivalent to 3.8×10^{-17} moles of DNA nucleotide polymerized/min/cell, if the average molecular weight of a nucleotide is taken as 333. Then $(3.8 \times 10^{-17}) \times (6 \times 10^{23}) = 2.3 \times 10^7$ nucleotides added to DNA/min/cell. If the average rate of replication fork progression is taken as 0.5 μm/min or 3000 nucleotides/min, that is, 1500 base pairs/min, then $(2.3 \times 10^7)/(3.0 \times 10^3) = 7.7 \times 10^3$ replication forks are in operation at any one time in a cell. If it is assumed further that the average distance from origin to terminus is 30 μm, and that each replication fork therefore functions for 1 hr, then it may be estimated there are $7.7 \times 10^3/60$ or about 130 initiation events/min on the average throughout the S phase if one event gives rise to one fork, or 65 events/min if one event gives rise to both forks in a unit.

E. Functional Features of Replication Units

1. *The Size of Replication Units and the Nature of Origins*

Measurements of autoradiograms using the center-to-center distances between initiation sites active during a short pulse have been used most frequently to estimate the size of replication units. In the studies in mammalian cells reporting such measurements (Huberman and Riggs, 1968; Hand and Tamm, 1974; Hand, 1975a), the pulses have been of sufficient duration to give accurate center-to-center distances. It is not likely that unlabeled origins were present in between the labeled origins used for measurements. The range of sizes of units determined from center-to-center distances has been given as 15–60 μm (Huberman and Riggs, 1968). In mouse L cells, 80% of units are 20–90 μm with a mode at a size class 40–50 μm and a mean at roughly 60μm (Hand and Tamm, 1974; Hand, 1975a). The variation in the center-to-center distances is large, and at the class intervals measured, there appears to be no periodicity to the distribution of sizes. It therefore has been proposed that units within the same cell are not of uniform size (Hand and Tamm, 1974).

In amphibian and insect cells, there are changes in the center-to-center distances in cells derived from different tissues of the same organism (Callan, 1972; Blumenthal *et al.*, 1973). In both cases, the increase in distance correlates with an increase in S phase, and it has been suggested that S phase length may be regulated by the frequency of initiations, more frequent initiations resulting in shorter center-to-center distances.

To date, in all eukaryotic cells examined, the chromosomal DNA has been linear rather than circular. Initiation takes place at multiple internal sites along the DNA fiber. In lower eukaryotes, in which the chromosomes are smaller, the electron microscope may be used to visualize the replication bubbles or eye forms generated by these internal origins if the DNA is subjected to some procedure for enrichment of replicating molecules. Figure 7 is an electron micrograph of replicating DNA from yeast. The micrograph illustrates that initiation is internal, and that molecules contain multiple origins.

The center-to-center distances between bubbles in yeast are smaller than those in higher eukaryotes (Newlon *et al.*, 1974). In insect embryonic cells, in which the chromosomes again are small, short center-to-center distances have been observed (Wolstenholme, 1973; Blumenthal *et al.*, 1973; Lee and Pavan, 1974), although as pointed out in Section I,D, the smaller distances are observed in cells with short S phases. The distances are longer in more slowly replicating cells from the same species.

If the center-to-center distances may be varied within cells from the

Fig. 7. Electron micrograph of replicating DNA fiber from yeast. Two bubbles are illustrated indicating internal initiation. Original magnification, 16,300. From Burke (1974), through the courtesy of W. Fangman.

same organism, then it follows that not all origins within a cell need be activated during any given S phase, and the replication units will be correspondingly longer. Although some (<10%) of the origins in mammalian cells appear reproducible from generation to generation (Amaldi *et al.*, 1973), the weight of evidence (Callan, 1972; Blumenthal *et al.*, 1973; Kriegstein and Hogness, 1974; Hand and Tamm, 1974) suggests that in general, origins are not fixed and reproducible.

Callan (1972, 1973a,b) has discussed several theoretical ways in which origin activation might be controlled. There may be qualitatively different sets of origins, each with unique nucleotide sequences, which are recognized by unique sets of enzyme molecules, or a single set of molecules may have varying affinity for different sets of origins. Initiations then would be controlled by variations in the nature and concentration of these molecules. Alternatively, there may exist a single set of origins with individual initiations controlled by variations in the state of compaction of the chromatin. A third possibility raised by Edenberg and Huberman (1975) is that structural features of DNA may specify origins. Palindromes in mammalian cells (Wilson and Thomas, 1974), which may vary with secondary structure of DNA, are obvious candidates for such a role.

Another fundamental question about origins is whether an initiation event gives rise to one or two replication forks. It now is generally accepted that replication in eukaryotes for the most part is bidirectional (Huberman and Riggs, 1968; Amaldi *et al.*, 1972; Weintraub, 1972; Hand and Tamm, 1973; Huberman and Tsai, 1973). However, two initiator sequences or regions might be adjacent at an origin and each might give rise to a single fork. Alternatively, there could be a single initiation site, with the complex of proteins attaching to this site responsible for two forks being generated. In mammalian cells, a small but measurable proportion of replication units shows autoradiograph patterns consistent with unidirectional replication (Hand and Tamm, 1973; Huberman and Tsai, 1973). A unidirectional unit within a group of bidirectional units is illustrated in Figure 8. The proportion increases when DNA synthesis is reduced by inhibition of protein synthesis (Hand, 1975b,c). This may be interpreted as showing that one of the two initiation sites at the origin was blocked, or that the initiation complex was modified so that only one fork was generated.

2. *Replication Forks*

Kriegstein and Hogness (1974) have analyzed the structure of the replication forks from *Drosophila* (fruit fly) embryos using the electron microscope. On more than 60% of the forks, a single-stranded gap on one tine

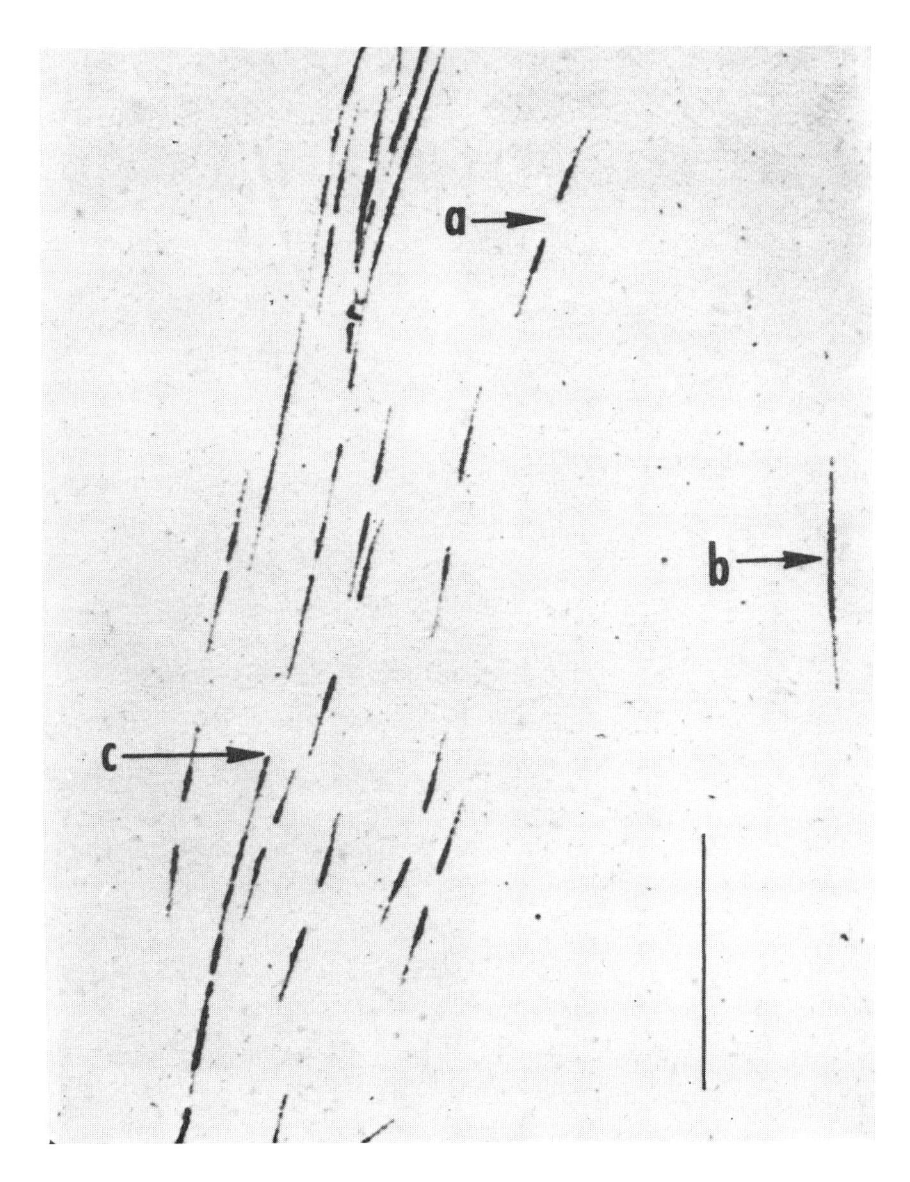

Fig. 8. DNA autoradiograms from mouse L cells. The occasional occurrence of unidirectional replication units is illustrated. Pulse labeling conditions as in Figure 5. (a) A bidirectional prepulse unit that is part of a cluster. (b) A bidirectional postpulse unit. Its cluster is out of the micrograph to the right. (c) A unidirectional unit within a prepulse cluster. Initiation began at the arrow and proceeded downward. The bar represents 100 μm. (Reproduced from Hand, 1975c, *J. Histochem. Cytochem.* **23,** 475–481. Copyright © 1975 by The Histochemical Society.)

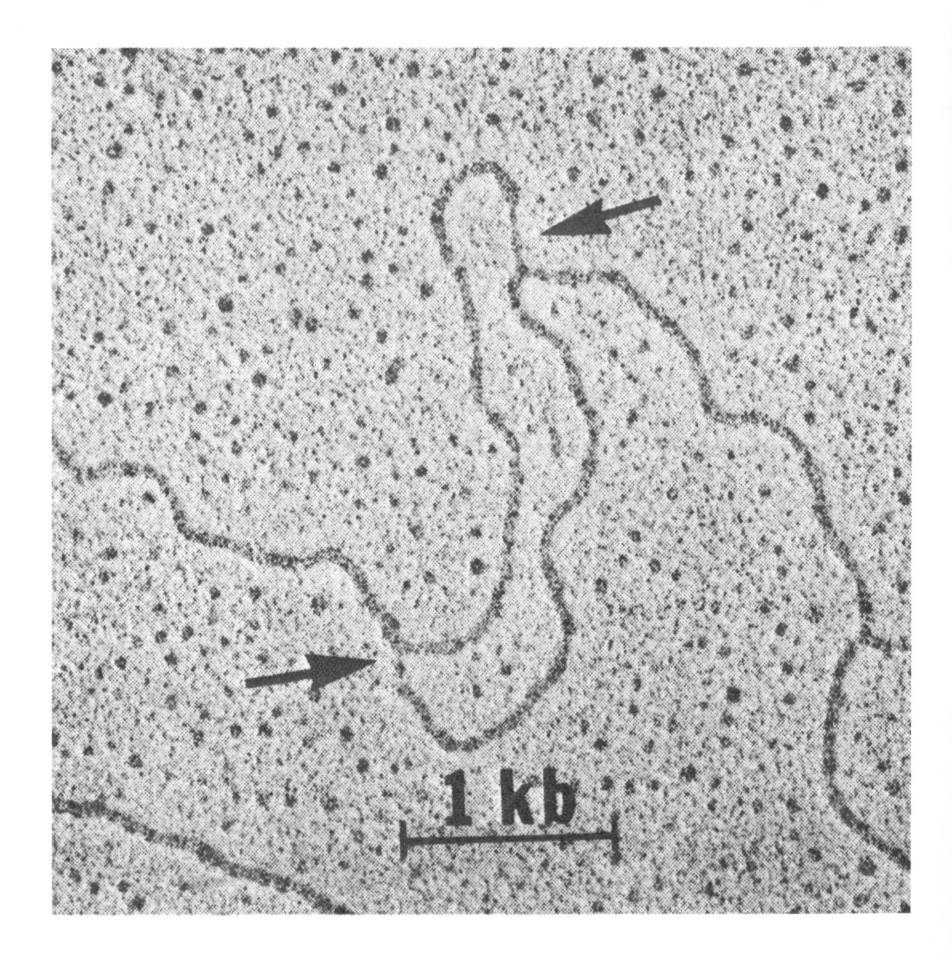

Fig. 9. Electron micrograph of replicating DNA segment from *Drosophila* cleavage nuclei. A single replication bubble is shown. The replication forks show single-stranded gaps (arrows) in a trans configuration. kb = kilobase, the length specified by 1000 bases or base-pairs in single- or double-stranded nucleic acid, respectively. Three kilobases is approximately 1 μm. (Reproduced from Kriegstein and Hogness, 1974.)

was demonstrated directly or the existence of such a gap suggested by the presence of a short-stranded segment (whisker) coming out from the point of confluence of the three branches of the fork.* Bubbles in which both

* Such whiskers probably are strands recently replicated on one parental template while the equivalent section on the other template remained unreplicated. Subsequent to replication, branch migration took place, that is, the newly replicated strand was dissociated from its template and the two parental strands reassociated, resulting in the short strand emanating from the point of confluence of the fork branches.

forks contained one gap invariably had the gaps in trans configuration (Figure 9). The whiskers had 3′-termini as determined by nuclease digestion studies. These characteristics are quite similar to those of forks from the bidirectionally replicating bacteriophages. There are two major differences between eukaryotic and prokaryotic forks. In eukaryotes, the gaps are smaller, and there are fewer forks with two gaps on one tine. The smaller gap size in *Drosophila* forks correlates with the smaller size of Okazaki fragments in this organism. To explain the lower incidence of double-gapped tines, the investigators postulate that, on the gapped helix, the rate at which a strand is elongated on its template is faster than the rate at which the replication fork moves. This implies that polymerization of the strand must stop from time to time during replication to allow the slower process of fork movement (perhaps the unwinding of the duplex) to proceed. The gapped helix probably contains the antiparallel template strand; that is, the strand on which the direction of replication of the nascent strands is opposite that of the replication fork. In prokaryotes, a fast rate of polymerization with subsequent pauses before gap filling results in the frequent presence of two gaps on the antiparallel tine. If, in eukaryotes, the difference between the rate of fork movement and antiparallel chain growth is greater than in prokaryotes, then the gap distal to the fork might be filled more frequently, resulting in the higher proportion of forks with only one gap. Kriegstein and Hogness suggest that the rate of chain growth on the antiparallel tine in *Drosophila* could approach that seen in prokaryotes, while fork progression in the overall direction of replication is at the speed characteristic of eukaryotes. They suggest chromatin proteins as candidates for factors that might be rate limiting for fork progression.

3. *Termini*

There are two alternatives for termination of DNA replication on units in eukaryotic cells. There could be definite structures at fixed sites along the chromosome, or termination could take place whenever two forks from adjacent replication units meet. No positive evidence for the presence of fixed termini has been found (Callan, 1972; McFarlane and Callan, 1973; Hand, 1975a). Blumenthal *et al.* (1973) have pointed out that chromosomal translocations, inversions, and deletions could generate stretches of chromosomes without origins between fixed termini. If fork progression could not proceed beyond the termini, the regions without origins would not be replicated. This would result in gaps that eventually would lead to chromosome breaks, a situation which obviously does not obtain in most cases of these cytogenetic aberrations.

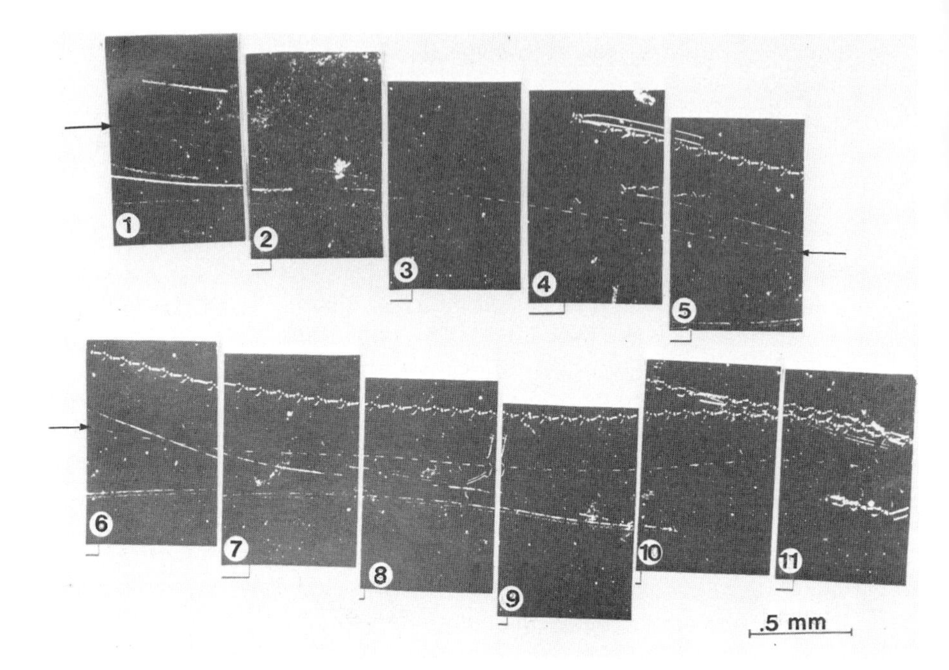

Fig. 10. DNA autoradiogram from mouse L cells labeled for 2 hr with [^{14}C]thymidine (10 μCi/ml, 429 Ci/mole). A single long DNA fiber extends through the 11 dark-field micrographs. The fiber is indicated by the arrows. The extent of overlap between micrographs is shown at the lower left of each frame. From frame 3 to 11, there are 50 labeled segments of 40–50 μm size, none separated by a gap of more than 150 μm. The overall length of the fiber is 5.6 mm. (From Molitor *et al.*, 1974, reproduced with permission.)

4. Clusters

Replication units are activated nonrandomly. Analyses of autoradiographic patterns produced by multiple adjacent units that are simultaneously replicating has led to the concept of clusters (Huberman and Riggs, 1968; Blumenthal *et al.*, 1973; Hori and Lark, 1973; Hand and Tamm, 1974; Hand, 1975a). Several adjacent units are initiated in synchronous fashion (Blumenthal *et al.*, 1973; Hand, 1975a). This synchrony is imperfect, and there may be one or more late replicating units within a cluster (Hori and Lark, 1973). Clusters are present in lower eukaryotes such as yeast (Petes and Williamson, 1975). In the higher eukaryotes, cluster initiation may serve as a method of regulation of replication (Hand, 1975a). Such clusters may correspond to particular labeled regions in autoradiographs of metaphase chromosomes of mammalian cells labeled during short periods of interphase (Taylor, 1960). Estimates of the size of these regions in mammalian chromosomes from data using a fluorescent dye to define replicating DNA (Latt, 1974) are of the order of 100 replication units. Figure 10 shows an autoradiogram prepared from [^{14}C]thymidine-labeled mouse DNA (Molitor *et al.*, 1974) that could represent a cluster of 50 such units. Figure 11 from Kriegstein and Hogness (1974) shows an electron micrograph of a cluster from *Drosophila* DNA containing 23 replicating units over a length of 40 μm. The rate of DNA replication in these cells suggests that if initiation were random, a replication bubble should occur every 3.3 μm rather than one every 1.7 μm, as observed. Thus, the 23 units were activated nonrandomly with some degree of synchrony evident in their initiation.

F. Some Biochemical Events in the Replication Unit

The full details of the biochemistry of DNA replication are covered in Chapter 2, this volume. Here some of the events in eukaryotes will be discussed. The most complete story comes from analyses of replication of two DNA animal viruses, polyoma and SV40. Both have extremely small

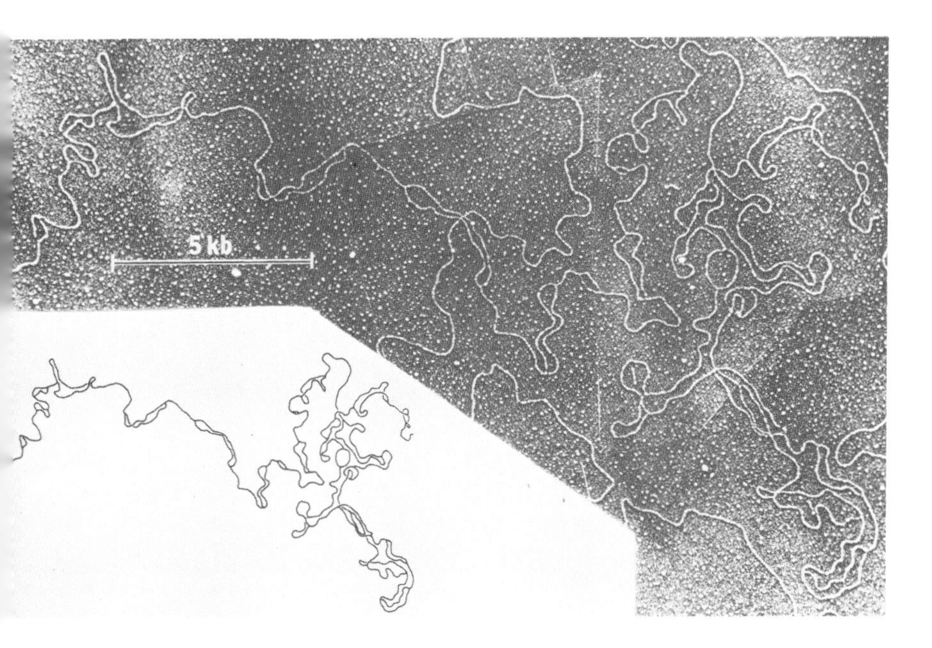

Fig. 11. Replicating DNA segment from *Drosophila* cleavage nuclei. The segment contains 23 replication bubbles in a length of 40 μm. 5 kb = 1.67 μm. (Reproduced from Kriegstein and Hogness, 1974.)

genomes of 3×10^6 daltons, and, so far as is known, their genomes code for only one replication function, the A function, which is involved in initiation. For all subsequent replication functions, the viruses depend on the (mammalian) host cell, and therefore the events in viral replication probably are the same as in mammalian DNA replication.

Initiation occurs at a unique site in each of these viral genomes, and subsequent chain elongation takes place via a discontinuous mechanism (Fareed and Salzman, 1972; Magnusson *et al.*, 1973; Sadoff and Cheevers, 1973) with the formation of short (50–150 bases) DNA strands on the template in a fashion similar to that which occurs in bacteria (Okazaki *et al.*, 1968). There are conflicting data whether Okazaki fragments are formed on one strand (Francke and Hunter, 1974; Francke and Vogt, 1975) or both strands (Fareed and Salzman, 1972; Magnusson *et al.*, 1973). RNA is attached to the 5′-end of these Okazaki fragments (Pigiet *et al.*, 1974) and probably serves as the primer sequence. The RNA piece is a decanucleotide with either adenine or guanine at the 5′-end. Any of the four bases can be at the 3′-end where the RNA is covalently linked to DNA. The Okazaki fragments subsequently are ligated after removal of the RNA primers, and progeny viral DNA is formed (Reichard *et al.*, 1974). Figure 12 is a diagram of these biochemical events as they are thought to occur in polyoma replication.

1. Okazaki Fragments

Okazaki fragments are formed in the course of eukaryotic DNA replication (Schandl and Taylor, 1969; Painter and Schaefer, 1969b; Kidwell and Mueller, 1969; Nuzzo *et al.*, 1970). These are about the same size as the fragments synthesized in the replication of polyoma and SV40. Because of their small size, they are visualized best after extremely short pulses of radioactivity. The labeling of Okazaki fragments from mammalian cells during a short pulse and their subsequent incorporation into bulk DNA during a chase is illustrated in Figure 13. It is not yet clear whether the fragments are synthesized on one or both template strands. Hershey and Taylor (1974) have presented data suggesting that Okazaki fragments are formed on one strand (presumably the one on which overall synthesis is in the 3′ → 5′ direction) and that long DNA strands are synthesized in a continuous manner on the other. Other studies have presented data favoring discontinuous synthesis on both strands (Gautschi and Clarkson, 1975).

In mammalian cells, structures intermediate in size between Okazaki fragments and replication units can be isolated by partial denaturation of replicating DNA (Taylor, 1973; Taylor *et al.*, 1973). These structures sediment at 26 S (equivalent to 6–7 μm by Taylor's estimate) and contain both template and nascent DNA. In addition, single-stranded nascent chains of

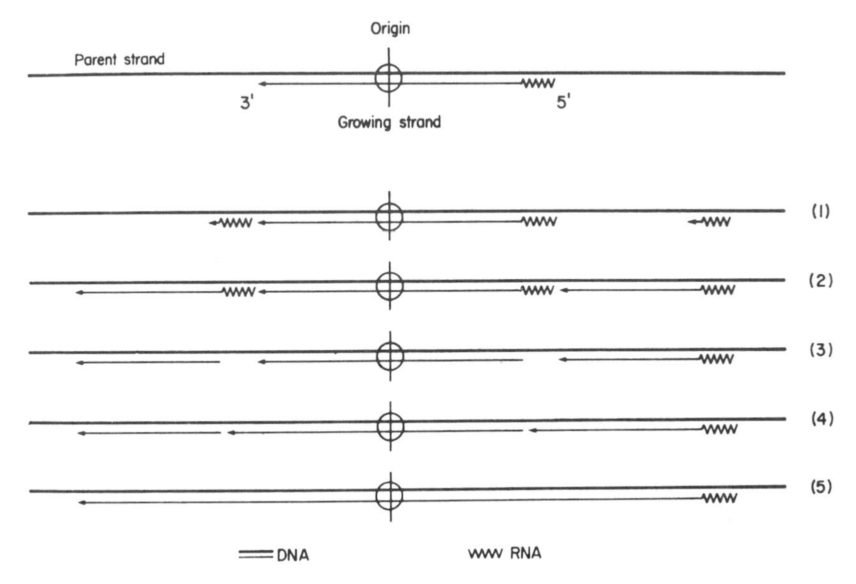

Fig. 12. The model for polyoma DNA chain elongation. Replication is bidirectional and involves initiation and chain growth of 4 S fragments. (1) Initiation by RNA; (2) DNA chain elongation; (3) RNA removal; (4) DNA gap-filling; (5) Ligation. (From Pigiet *et al.*, 1974, reproduced with permission.)

8 S (equivalent to a length of 2 μm) can be identified under these partial denaturing conditions. Taylor proposes that the 26 S segments contain replication forks and that they are released by partial denaturation from bulk DNA because of the presence of nonadjacent single-stranded nicks in front of and behind the moving fork. The 8 S segments represent recently replicated DNA. He suggests that the 26 S structure is the working unit for replication (Figure 14). DNA is replicated in segments of 2 μm, these segments being defined by nicks in the template strand. At some time after the replication fork has passed, the template nicks are sealed and the nascent gaps are filled to produce long continuous duplexes. The nonadjacent nicks could contain the recognition sites for polymerases and other proteins of the replication apparatus. Propagation of DNA chains then occurs by sequential activation of these 2-μm segments after initiation has occurred at one of the junctions via an external control mechanism. In addition, it must be postulated that the replication apparatus temporarily seals the nicks in the template as the fork passes. The data supporting this model have not been confirmed yet by other investigators; nevertheless, it seems possible that a control unit intermediate in size between the Okazaki fragment and the replication unit does exist.

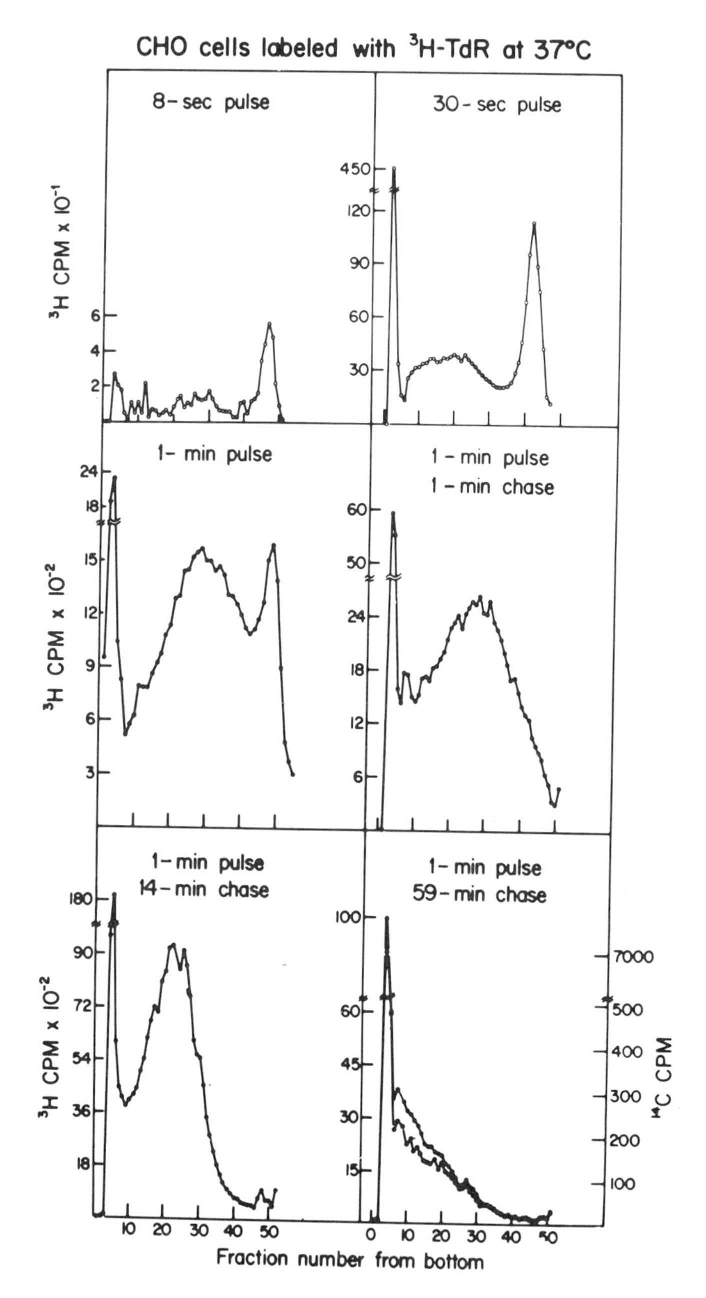
CHO cells labeled with ³H-TdR at 37°C
8-sec pulse
30-sec pulse
1- min pulse
1 - min pulse
1 - min chase
1-min pulse
14-min chase
1-min pulse
59-min chase
³H CPM x 10⁻¹
³H CPM x 10⁻²
³H CPM x 10⁻²
¹⁴C CPM
Fraction number from bottom

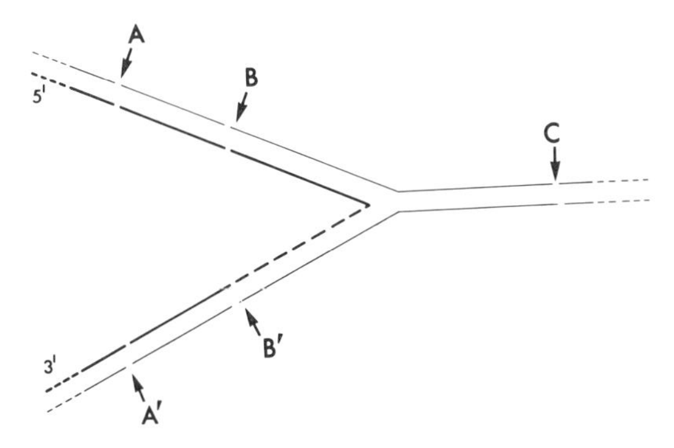

Fig. 14. Mammalian DNA in the region of the replication fork. The thin lines indicate template strand, the thick lines, nascent strands. Nonadjacent single-stranded nicks in the duplex are indicated by arrows. Nascent strands are synthesized as 8 S or 2-μm structures (AB and A′B′) between nicks in the template strands. The 26 S structure (BCB′) can be isolated from replicating DNA after partial denaturation by dissociation of the strands in the regions of the nicks. Discontinuous synthesis via Okazaki fragments is shown only on the antiparallel strand. Modified from Taylor (1973).

DNA intermediate in size between the Okazaki fragment and bulk DNA has been demonstrated by alkaline sucrose gradient sedimentation of pulse-labeled DNA by numerous investigators (Fujiwara, 1972; Goldstein and Rutman, 1973; Gautschi *et al.*, 1973; Seale and Simpson, 1975). Interpretation of these data is complicated by the drawbacks of alkaline sucrose sedimentation. DNA analyzed by this technique is subject to breakage by alkali attack, isotope decay, and shear during cell lysis (McBurney *et al.*, 1972). In addition, denaturation may be incomplete under the usual conditions of sedimentation (Simpson *et al.*, 1973; Jolley and Omerod, 1974).

2. *The RNA Primer*

RNA linked covalently to nascent DNA has been demonstrated in *Physarum* (Waqar and Huberman, 1973, 1975a), Chinese hamster ovary cells (Waquar and Huberman, 1975b), and human lymphocytes (Tseng and Goulian, 1975). These studies have demonstrated this attachment by

---

Fig. 13. Alkaline sucrose gradient profiles of mammalian DNA pulse-labeled with [^{3}H]thymidine. Chinese hamster ovary cells in tissue culture, some of which had been prelabeled overnight with [^{14}C]thymidine, were pulse-labeled or pulse–chase-labeled for various times at 37°C with [^{3}H]thymidine. At the completion of the labeling period, the cells were lysed by treatment with a nonionic detergent, and the resulting nuclei were treated with NaOH. The released DNA then was sedimented in an alkaline sucrose gradient. A marker DNA from P22 phage (2.6×10^{7} daltons) would sediment at about fraction 33. From Horwitz (1975), through the courtesy of Joel Huberman.

means of the transfer of radioactive phosphate from an incorporated deoxynucleotide to an adjacent ribonucleotide. Earlier studies purporting to show covalent attachment of RNA to DNA used equilibrium sedimentation to demonstrate an increase in buoyant density of nascent DNA consistent with its being associated with a small piece of RNA (Fox *et al.*, 1973). The increase in density may have been due to an artifactual noncovalent association between RNA and DNA that occurred during the preparation of DNA for centrifugation. This artifactual association has been demonstrated in mouse Ehrlich ascites tumor cells by Probst *et al.* (1974a). Other investigators have searched for an RNA primer for mammalian DNA replication and have failed to demonstrate it (Probst *et al.*, 1974a; Gautschi and Clarkson, 1975); however, this may be because the attachment is very transient in mammalian cells, and the techniques they used were not sensitive enough to detect it. The concept that DNA replication is RNA-primed is a pleasing one, since all DNA polymerases so far discovered require a primer for *in vitro* activity.

Sheared DNA containing nascent strands has single-stranded regions (Probst *et al.*, 1974b; Kato and Strauss, 1974). These regions may exist at the replication fork. By analogy to the polyoma system (Pigiet *et al.*, 1974), it might be postulated that the single-stranded regions represent the gaps present after removal of the RNA primers [Figure 12 (3)].

3. DNA Polymerases

Three main DNA polymerase activities are present in mammalian cells, and a uniform nomenclature for them has been developed (reviewed by Bollum, 1975; Weissbach, 1975; see also Chapter 2, this volume).

DNA polymerase α is the most abundant. It is a high molecular weight protein (1×10^5–$2. \times 10^5$). Activity is present in nucleus and cytoplasm, although the activity in the latter may represent artifactual leakage resulting from aqueous cell fractionation techniques (Foster and Gurney, 1974). The activity of the α polymerase correlates well with DNA replication, in that its activity is higher in growing than in resting cells, and increases during S phase (Chang *et al.*, 1973; Spadari and Weissbach, 1974a).

DNA β polymerase has a molecular weight of roughly 40,000 (Chang *et al.*, 1973) and is distinct from the α polymerase, although it may aggregate into higher molecular weight material of size similar to the α polymerase (Bollum, 1975). Its activity is predominantly nuclear and does not increase in association with S (Chang *et al.*, 1973; Spadari and Weissbach, 1974a).

The least prominent DNA polymerase is DNA γ polymerase, which is responsible for less than 1% of total cellular DNA polymerase activity (Spadari and Weissbach, 1974a). It has a molecular weight of 110,000 and probably exists in two slightly different forms (Spadari and Weissbach, 1974b). It has a lower K_m for dNTP's than the α and β polymerases, and its activity increases during S phase (Spadari and Weissbach, 1974a).

None of the enzymes has been assigned a specific role in replication. Since activities of α and γ polymerase increase during S phase, they may be involved in chromosome replication, and β polymerase in repair. The low K_m for dNTP's of γ polymerase implicate it in the synthesis of Okazaki fragments. These fragments accumulate in the presence of hydroxyurea, at least in the replication of polyomavirus in mammalian cells (Magnusson, 1973). This drug decreases purine nucleotide pools (Skoog and Nordenskjold, 1971; Adams *et al.*, 1971), and it can be reasoned that γ polymerase is best suited for the continued synthesis of the Okazaki fragments in the presence of the smaller dNTP pools. The α polymerase could be the gap-filling enzyme. However, Spadari and Weissbach (1975) recently have presented data showing that only the α polymerase can extend DNA chains from an RNA primer. β and γ require DNA primers.

G. Specific Classes of Replicating DNA

The various structural classes of chromosomal DNA as defined by renaturation kinctics (Britten and Kohne, 1968) have been studied to see if they replicate at specific times in the S phase. Intermediate repetitious sequences replicate throughout the S phase. Highly repetitious sequences, in general, also replicate throughout the S phase, although in some species (mouse, kangaroo rat), they may be late replicating (Comings and Mattochia, 1970; May and Bello, 1974; Tapiero *et al.*, 1974). The approach using renaturation kinetics supersedes that using buoyant density differences resulting from differing base composition (Tobia *et al.*, 1970; Bostock and Prescott, 1971a,b; Comings, 1972). These earlier studies defined certain satellite classes as replicating throughout or relatively late in S phase. Not all repetitious DNA's sediment as satellite, and sequence complexity seems a more precise way than buoyant density for detailing structural differences in DNA.

Satellite DNA as defined by buoyant density appears to consist of replicating units with the same general characteristics as bulk DNA (Hori and Lark, 1974). In kangaroo rat, replicating DNA with the buoyant density of satellite shows unit size and rate of fork movement comparable to bulk DNA. Figure 15 shows autoradiograms of kangaroo rat main band and satellite DNA.

One well-documented exception to the general scheme of eukaryotic DNA replication outlined above is the amplification of the genes coding for ribosomal RNA which occurs during oogenesis in the toad, *Xenopus*.

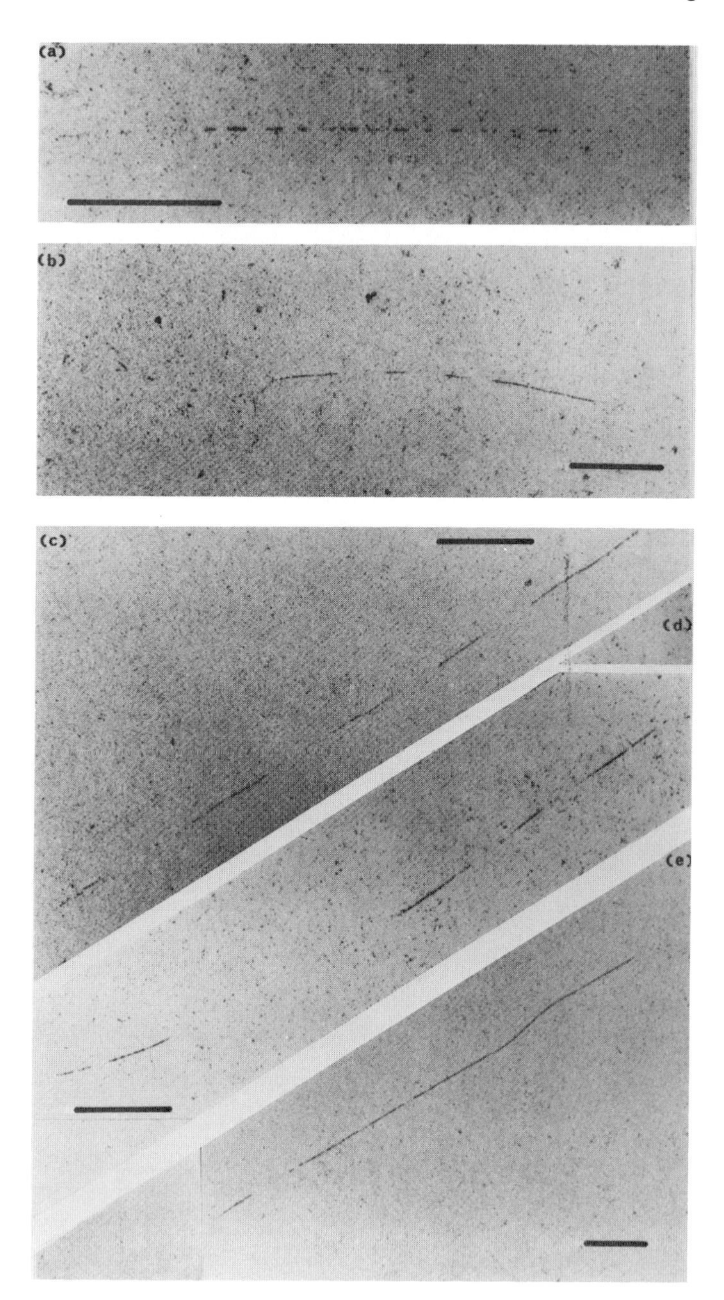
(a)
(b)
(c)
(d)
(e)

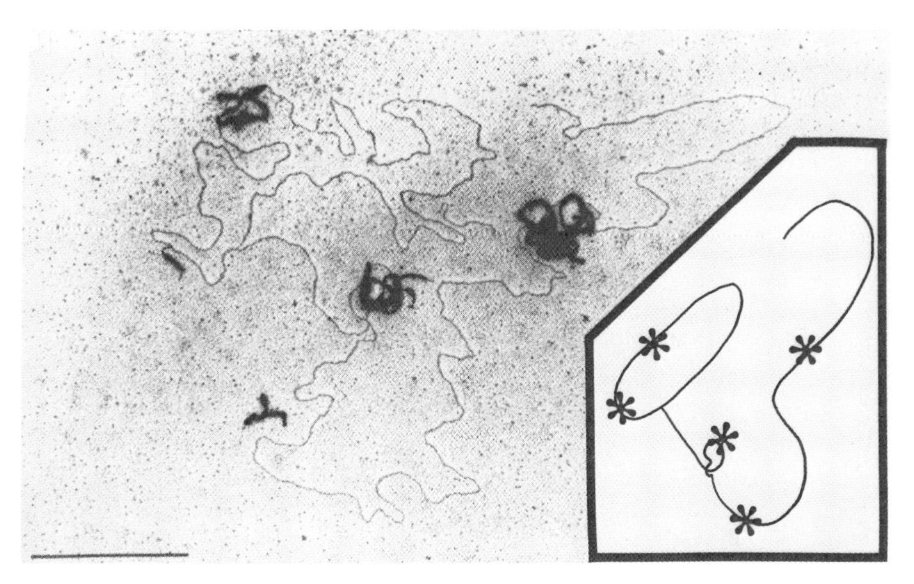

Fig. 16. Electron microscopic autoradiogram of replicating ribosomal DNA from *Xenopus* ovary cells in which the ribosomal DNA is undergoing amplification. A lariat form is shown with radioactivity on the circle (two grains) and the tail (three grains). The length of the circle is equivalent to three ribosomal DNA repeat units. The bar represents 1 μm. (From Rochaix *et al.*, 1974, reproduced with permission.) At the lower right is a diagrammatic representation of the DNA molecule with the contours smoothed out. The five silver grains resulting from radioactive disintegrations are represented by the asterisks.

The approximately 1000-fold increase in gene copies probably takes place via a rolling circle mechanism (Hourcade *et al.*, 1973; Rochaix *et al.*, 1974). These two studies have demonstrated lariat forms by electron microscopy of amplified DNA. These lariat forms contained a pattern of radioactivity consistent with a rolling circle intermediate (Rochaix *et al.*, 1974). Figure 16 shows an electron microscopic autoradiogram of a replicating lariat. Since chromosomal DNA is linear, the problem of how the first template circle is generated in the regions of the ribosomal genes has not been solved, but three alternatives have been proposed: (a) ribosomal genes associated with the chromosomes are contained in circular episomes, (b) the original ribosomal gene is excised from the chromosome,

---

Fig. 15. Autoradiograms from kangaroo rat DNA. Following a 12-hr pretreatment with fluorodeoxyuridine, the cells were labeled for 6 hr with [^{3}H]thymidine (50 Ci/mmole). The DNA was released from lysed cells and centrifuged to equilibrium in a CsCl density gradient. DNA from fractions with the density of satellite (a,b) or main band (c,d,e) DNA was dialyzed and prepared for autoradiography. The bars indicate 100 μm. The longer arrays in main band DNA autoradiograms may result from fusion of adjacent segments. (From Hori and Lark, 1974, reproduced with permission.)

copied into a circular molecule, and the original is reinserted into the chromosome, or (c) copies are made of ribosomal RNA precursors by reverse transcription. The replication of ribosomal DNA from *Xenopus* somatic cells resembles that of bulk chromosomal DNA (Rochaix and Bird, 1975). However, a small proportion of circles still is observed.

II. REGULATION OF DNA REPLICATION

A. Preparation for Entry into S Phase

At the completion of mitosis, cells of some lower eukaryotes (such as certain species of slime molds and amebas) enter directly into DNA synthesis. In almost all cells of higher eukaryotes, a discrete period, the G_1 phase, separates mitosis from the DNA synthetic (S) phase. Nongrowing quiescent cells actually may leave the cycle in G_1 and enter the G_0 phase. Growing cells go through a sequence of biochemical events that ultimately leads to the synthesis of DNA.

The regulation of cell growth and its relation to the cell cycle will be covered in detail in Volume 5 of this Treatise. These subjects have been reviewed extensively (Burger, 1971; Johnson and Rao, 1971; Stein and Baserga, 1972; Clarkson and Baserga, 1974), and are discussed briefly in this section to give an overview of the regulation of DNA replication.

1. Experimental Systems

There are several ways to study the events in G_1 leading to S. One is to investigate the influence of S phase cells or components from cells in S phase on non-S phase cells. The two main approaches here have been nuclear transplantation studies in the appropriate unicellular organisms and cell fusion studies in animal cell tissue cultures. A number of systems have been exploited to study the transition from the quiescent to the growing state in animal cells. In whole organisms, these include the isoproterenol-stimulated rat salivary gland and the rodent liver regenerating after partial hepatectomy. In tissue culture, the properties of serum and density dependence for proliferation of untransformed fibroblasts have been used to study both the biochemical events leading to DNA synthesis and the various factors which stimulate cells to proliferate.

2. Cytoplasmic Control of Nuclear DNA Synthesis

Cell fusion induced by inactivated Sendai virus may be used to study nuclear–cytoplasmic interactions in the control of DNA synthesis. Heterokaryons, multinucleate cells containing nuclei from different cell types, or homokaryons, multinucleate cells containing nuclei from cells of the same type, may be produced.

When heterokaryons are formed between cells capable of synthesizing DNA and cells inactive in DNA synthesis, the inactive nuclei are induced to synthesize DNA (Harris, 1965; Harris *et al.*, 1966). If either parent cell normally is able to synthesize DNA, DNA synthesis will take place in both types of nuclei of the heterokaryon. Heterokaryons may be formed from parent cells of different species. Thus, when HeLa cells of human origin are fused with chick erythrocytes, rabbit macrophages, or rat lymphocytes, under conditions in which the nonhuman cells normally would not synthesize DNA, all are induced to synthesize DNA. The cytoplasmic signals responsible for the control of DNA synthesis cannot be species specific (Harris *et al.*, 1966).

Homokaryons formed from synchronized HeLa cells with parent cells in different phases of the cell cycle also have been used to study nuclear–cytoplasmic interactions (Johnson and Rao, 1971). When S phase cells are fused with G_1 cells, the G_1 nuclei enter S phase sooner than expected (Rao and Johnson, 1970). The greater the ratio of S to G_1 nuclei in the homokaryons, the faster the induction of DNA synthesis. Inducer substances synthesized in S phase cells are capable of shortening or eliminating G_1 in other nuclei brought into the same environment by cell fusion. Repressor substances do not seem to be synthesized by G_1 cells, since the presence of a G_1 nucleus and cytoplasm does not inhibit synthesis in the S phase nucleus in the homokaryon (Johnson and Harris, 1969). There appears to be positive control of DNA synthesis.

Similar conclusions have been reached in nuclear transplantation studies. In eggs from the toad *Xenopus*, transplantation of G_1 nuclei into S phase cytoplasm results in the induction of DNA synthesis (Graham, 1966).

Homokaryons composed of two sister nuclei in the same phase of the cell cycle can be produced by treatment of mitotic cells with cytochalasin B at low doses that inhibit cytokinesis (Fournier and Pardee, 1975). In such homokaryons, the two nuclei are synchronous with respect to DNA synthesis, and the G_1 period is 2 hr shorter than in mononucleate cells. The duration of G_1 does not correlate with either cell size or protein content of the binucleate cells. Thus, substances serving as initiators are probably diffusible through the cytoplasm. The shortened G_1 in the homokaryons argues for some form of nuclear–nuclear cooperativity.

3. *The Requirement for Protein Synthesis*

The factors controlling entry into S very well might be proteins. In animal cells in culture, protein synthesis is necessary in the G_1 period for the commencement of DNA synthesis (Terasima and Yasukawa, 1966). The G_1 events requiring protein synthesis are coupled; that is, if a later G_1 event requiring protein synthesis is inhibited for a sufficient length of time, decay of proteins synthesized at an earlier time occurs, and the entire sequence of events presumably must be repeated after release of the block in order for DNA synthesis to start (Schneiderman *et al.,* 1971). Highfield and Dewey (1972) have presented data suggesting there is a point during the hour before the beginning of S beyond which inhibition of protein synthesis will not delay entry into S. Mammalian cells exposed to an inhibitor of protein synthesis such as cycloheximide beyond this point will enter S and synthesize DNA, albeit at a greatly reduced rate.

In lower eukaryotes, the situation is similar. Protein synthesis is required for entry into S phase in *Tetrahymena,* a ciliated protozoan (Jeffery, 1974). In the slime mold *Physarum,* in which there is no G_1 phase, there is a requirement for protein synthesis in early prophase of mitosis for the nuclei to begin DNA synthesis at the end of telophase (Cummins and Rusch, 1966).

In yeast, in which well-characterized temperature-sensitive cell cycle mutants exist (Hartwell *et al.,* 1973), Hereford and Hartwell (1974) have defined a sequence of four events necessary for the commencement of DNA replication. The two initial steps, the release from the inhibiting action of a pheromone, α factor, and the synthesis of the product of gene *cdc* 28, are interdependent and commit the cell to a mitotic cycle. The third step, the synthesis of the product of gene *cdc* 4, is independent and allows the cell to synthesize other proteins needed for initiation. The fourth step, the synthesis of the product of gene *cdc* 7, also is independent and allows the initiation of DNA synthesis.

B. Regulation of Replication within S Phase

1. Requirement for Protein Synthesis

In prokaryotic DNA replication, initiation at the single replicator signals the start of the DNA synthetic period. New protein synthesis is required for this event, but subsequent chain elongation may proceed in the absence of protein synthesis (Lark, 1969). The requirement for protein synthesis for initiation but not for continuation of DNA synthesis also is found in yeast. Hereford and Hartwell (1973) have shown in the yeast, *Saccharomyces,* that inhibition of protein synthesis at a point 25% into S phase does not stop the completion of a full round of DNA synthesis. Using a yeast strain temperature-sensitive for entry into S phase, these investigators showed that addition of cycloheximide to synchronized cells after they entered S phase produced the same result as a shift to restrictive temperature; that is, in both cases, those cells already synthesizing DNA continued to do so, as demonstrated by similar degrees of incorporation of radioactively labeled DNA precursor during a pulse. The investigators suggest that only initiation events require protein synthesis, and that initiation on all replication units in the yeast genome takes place within the first 25% of S.

In the slime mold, *Physarum*, protein synthesis is required through the S phase (Cummins and Rusch, 1966; Muldoon *et al.*, 1971). The data of Muldoon *et al.* (1971) suggest that the genome of *Physarum* contains ten discrete replicative units. Initiation of replication of these units is staggered through the S phase, and initiation on each of these units requires protein synthesis.

In the protozoan, *Tetrahymena*, inhibition of protein synthesis during S phase in synchronized cells results in an immediate inhibition of DNA synthesis (Jeffery, 1974). In this organism, initiation on individual replication units probably is staggered (Andersen and Zeuthen, 1971). Thus, the reduction in DNA synthesis could reflect inhibition of initiation on late replicating units or a failure to complete synthesis on units which have already initiated (or both processes may be inhibited).

Mueller *et al.* (1962) first noted that protein synthesis was required for the continuation of DNA synthesis in mammalian cells. They synchronized HeLa cells by arresting them at the G_1–S interface with amethopterin and releasing them by addition of thymidine to $10^{-5}M$. Addition of puromycin, an antibiotic inhibiting protein synthesis, before or after reversal of the amethopterin block, resulted in a decrease in the rate of incorporation of [^{14}C]thymidine into DNA. The investigators point out that puromycin does not inhibit DNA synthesis already in progress, but acts to prevent an increase in the rate of DNA synthesis that is observed in the control. Thus, when puromycin was added after the reversal of the block, DNA synthesis continued at near the control rate for some time before inhibition became pronounced. Pulse-labeling experiments in HeLa cells and other cell lines in the presence of a variety of inhibitors of protein synthesis have confirmed the findings that protein synthesis is required for the continuation of ongoing DNA synthesis in mammalian cells (Littlefield and Jacobs, 1965; Young, 1966; Weiss, 1969; Fujiwara, 1972).

Addition of protein synthesis inhibitors to *in vitro* nuclear synthesizing systems does not alter the incorporation of deoxynucleotide triphosphates; however, *in vivo* addition of protein synthesis inhibitors results in decreased DNA synthesis in subsequently isolated nuclei (Hershey *et al.*, 1973). In cells made permeable to triphosphates, protein synthesis also is necessary for the maintenance of DNA synthesis (Seki and Mueller,

1975). Initiation of DNA replication in these *in vitro* systems probably is reduced markedly compared to *in vivo* systems, and the inhibition of DNA synthesis probably reflects decreased daughter chain elongation. The studies demonstrate the requirement for short-lived proteins for the maintenance of ongoing DNA synthesis.

The two component processes of ongoing DNA replication are initiation and chain elongation via fork progression (Section I,D). The requirement for protein synthesis for both these processes in animal cells has been demonstrated.

Weintraub and Holtzer (1972) demonstrated a decrease in the rate of DNA fork progression after inhibition of protein synthesis in embryonic avian erythrocytes. In this system, cycloheximide reduces the incorporation of thymidine into DNA by 50% within 25 sec of addition. Quantitative measurement of the rate of fork movement by equilibrium sedimentation of isotope and density-labeled DNA showed that cycloheximide caused a decrease from 4200 to 2100 base pairs per minute per fork (1.4 to 0.7 μm/min/fork).

Similar results were obtained with a Chinese hamster ovary cell line (Gautschi and Kern, 1973). There is a rapid decrease in the overall rate of DNA replication that seems to be accounted for entirely by a decrease in rate of fork movement. The findings are not unique to this cell line or to the use of cycloheximide as the inhibitor, since similar results are found in HeLa cells treated with puromycin (Gautschi, 1974).

The requirement for protein synthesis in the initiation of DNA synthesis on individual units has not been studied as extensively as its requirement for chain growth. Fujiwara (1972) used sequential labeling with bromodeoxyuridine and [^{3}H]thymidine to study initiation in the absence of protein synthesis. In these experiments, the bulk of DNA labeled during the [^{3}H]thymidine pulse in the presence of cycloheximide had buoyant density heavier than marker DNA, suggesting that this DNA had initiated synthesis before the addition of [^{3}H]thymidine and also before protein synthesis was inhibited. This labeling protocol is simply a reversal of that used by Painter and Schaefer (1969a) to measure fork progression. Thus, density shifts probably reflect changes in both fork progression and initiation. Experiments using equilibrium sedimentation to measure initiation should be interpreted with caution and separate methods must be used to account for changes in other aspects of DNA replication (fork progression, chain termination).

When DNA fiber autoradiography is used to measure fork movement, there is a similar reduction in rate produced by cycloheximide in mammalian cells (Hand and Tamm, 1973). Puromycin inhibits fork progression at comparatively high drug levels in this assay (Hand and Tamm, 1972,

1973). At lower levels of the antibiotic still sufficient to inhibit protein synthesis, there is a decrease in overall DNA synthesis immediately after addition of the inhibitor, but it is not accompanied by a decrease in rate of fork progression (Hand and Tamm, 1972; Hori and Lark, 1973). Both groups of investigators suggested that inhibition of protein synthesis also might reduce initiation of replication.

Fiber autoradiography does permit measurement of the frequency of initiation, since different grain track patterns are produced by replication units depending on whether initiation began before or after the radioactive pulse to the DNA (Section I,C). An inhibitor of initiation added at the time of the pulse will reduce the proportion of replication units with postpulse patterns. By means of this type of analysis, it has been demonstrated that both cycloheximide and puromycin cause a decrease in DNA initiation immediately after addition (Hand, 1975b).

Cycloheximide also alters the synthesis of all components of chromatin (Weintraub, 1973; Seale and Simpson, 1975). Chromatin that is made in the absence of protein synthesis has increased susceptibility to nuclease digestion and lighter buoyant density than control chromatin. These characteristics are similar to those existing in chromatin synthesized during extremely short pulses of [^{3}H]thymidine (30–60 sec). The inhibition of protein synthesis may therefore block the synthesis and possibly the assembly of chromatin proteins, probably histones, onto newly synthesized DNA.

2. *Specific Proteins and Regulation of DNA Synthesis*

A number of different species of proteins have been examined as possible regulators of DNA synthesis. Those which may be involved in the G_1–S transition already have been discussed (Section II,A). The synthesis or enzymatic activity of a number of other proteins increases during the S phase. These will be discussed in this section.

Histone synthesis ordinarily is coupled tightly to DNA replication; both appear to be limited to S phase and inhibition of DNA synthesis inhibits histone synthesis (Robbins and Borun, 1967). There also appear to be both transcriptional and translational controls over histone synthesis that operate in late G_1 or early S and therefore are related to DNA synthesis (Borun *et al.,* 1967, 1975). Models also have been proposed whereby histone may regulate its own synthesis by a negative feedback mechanism and that of DNA by a positive control mechanism (Weintraub, 1973). Although the negative feedback mechanism for control of histone synthesis is a reasonable explanation for a number of well-documented experimental findings (see Weintraub, 1973), there is no convincing evidence that synthesis of histones controls the synthesis of DNA.

Several enzymes involved in the synthesis of DNA precursors and in DNA replication itself have been shown to increase during DNA synthesis. These include thymidine kinase, thymidylate kinase, deoxycytidine kinase (Brent *et al.*, 1965; Littlefield, 1966), deoxycytidylate deaminase (Gelbard *et al.*, 1969), riboside diphosphate reductase (Hwang *et al.*, 1974), DNA polymerase α (Chang *et al.*, 1973), and DNA polymerase γ (Spadari and Weissbach, 1974b). Again, it may be argued that these enzyme activities increase *pari passu* with DNA synthesis as a result of a common regulatory mechanism, rather than any of these activities being regulatory in itself. It is of interest that the synthesis of acidic nuclear proteins is not coupled to DNA synthesis (Stein and Borun, 1972). These proteins have been proposed as regulators of the progression of cells from G_1 into S.

3. Nucleotide Pools and DNA Synthesis

Current techniques for the study of DNA replication depend to a large part on the use of exogenous [^{3}H]thymidine as a specific precursor to DNA. Alterations in the uptake of thymidine or the pool size of its nucleotides can produce apparent changes in rate of DNA synthesis as measured by thymidine incorporation, which may not reflect actual changes in DNA replication (Fuchs and Kohn, 1971; Roller *et al.*, 1974; Grunicke *et al.*, 1975). The subject of nucleotide pools has been reviewed recently by Hauschka (1973).

Endogenous thymidylate (TMP) is formed in the cell from dUMP by thymidylate synthetase. Exogenous thymidine is taken up by mammalian tissue culture cells as they enter S phase (Everhart and Rubin, 1974). Exogenous thymine also can serve as a precursor to DNA in mammalian cells if an exogenous source of deoxyribose is available. The exogenous deoxyribose may be supplied bound to a purine or pyrimidine as part of a deoxynucleoside (Goodman, 1974). The thymine perhaps enters a different triphosphate pool than thymidine, and Goodman postulates the existence of different TTP pools serving as precursors for repair and semiconservative DNA synthesis.

Kuebbing and Werner (1975) has suggested that there might be separate pools derived from endogenously synthesized and exogenously supplied thymidine similar to the situation Werner (1971) has suggested exists in prokaryotes. They demonstrated that the addition of exogenous thymidine to mammalian cells resulted in the immediate cessation of incorporation of endogenous thymidine into DNA. After a 24-hr exposure to exogenous thymidine, the cells adapted by using both endogenous and exogenous thymidine for DNA synthesis. Although these findings are consistent with the existence of two separate pools of TTP, they also are compatible with the existence of a single pool with the utilization of precursor from a section of the pool being determined by the conditions at the site where the DNA is being synthesized.

Thymidine is converted to TMP by thymidine kinase. The activity of thymidine kinase varies through the cell cycle, being lowest in G_1 and highest in S and G_2 (Littlefield, 1966). Its activity also correlates with the rate of cell growth, being relatively high in cells in which both the rate of growth and DNA synthesis are stimulated (Hare, 1970; Nordenskjold *et al.*, 1970). The activity of the enzyme is always in excess of that required to supply thymidine nucleotides for DNA synthesis (Plagemann and Erbe, 1972). TMP is rapidly phosphorylated further to TDP and finally to TTP, in which form it serves as a direct precursor for DNA along with other deoxynucleotide triphosphates. Seventy to 90% of intracellular thymidine nucleotides exist in the form of TTP (Gentry *et al.*, 1965; Adams, 1969).

a. Uptake of Thymidine. This is mediated by a distinct transport site which thymidine may share with deoxyuridine (Plagemann and Erbe, 1974). This suggestion is based on the observations that both substances exhibit similar apparent K_m's and V_{max}'s for transport, that they markedly inhibit the transport of each other, and that the K_i's of other inhibitors for the transport of the two are similar. Fluorodeoxyuridine also may share the same site, since the K_i for its inhibition of thymidine transport is quite low. The observation that thymidine and fluorodeoxyuridine may share the same site is interesting in light of the commonly used laboratory technique of exposing cells to fluorodeoxyuridine to increase the incorporation of thymidine or its analogue bromodeoxyuridine into DNA. Uptake and transport probably occur by facilitated diffusion at low concentrations of thymidine in the external medium and by simple diffusion plus facilitated diffusion at high concentrations (Plagemann and Erbe, 1972). A number of inhibitors of DNA synthesis such as heterologous nucleosides (Steck *et al.*, 1969; Plagemann and Erbe, 1974) and substances which only indirectly affect DNA synthesis, such as puromycin and cycloheximide (Plagemann and Erbe, 1972) and 2-mercapto-1-(β-4-pyridethyl)benzimidazole (Nakata and Bader, 1969), decrease thymidine transport in cells in tissue culture.

b. Thymidine Triphosphate Pool Size. Endogenous intracellular pools of TTP (and other deoxyribonucleotide triphosphates) can be measured specifically by an assay involving the *in vitro* synthesis of DNA by *E. coli* DNA polymerase (Solter and Handschumacher, 1969; Lindberg and Skoog, 1970; Walters *et al.*, 1973). The acid-soluble extracts of small quantities of cells are used to provide limiting amounts of the triphosphate to be measured. These extracts are added to an otherwise complete mixture for the DNA polymerase reaction including a radioactive triphosphate other than the one to be measured. The method has been applied to the study of deoxyribonucleotide pools in synchronized mammalian cells. The total triphosphate concentration in the nucleus in S phase cells is $3 \times 10^{-4} M$. This is three times the cytoplasmic concentration (Skoog and Bjursell, 1974). There are variations in pool size of the individual deoxyribonucleotide triphosphates during the cell cycle, and the pools increase during S and G_2 phases (Skoog *et al.*, 1973; Walters *et al.*, 1973). The endogenous pools of triphosphates are of sufficient size to sustain DNA synthesis for up to 3.8 min (Walters *et al.*, 1973). On the basis of these studies, Housman and Huberman (1975) have suggested that the dNTP pools might be rate limiting for DNA synthesis (at least insofar as that might explain slower rates of DNA chain elongation early in S). The same coordination of DNA synthesis and nucleotide pools size has been observed in *Tetrahymena* (Stocco and Zimmerman, 1974). However, these investigators feel that control is located elsewhere, since the mechanism for production of deoxynucleotides in *Tetrahymena* never is shut off completely, even in the absence of DNA synthesis.

Similarly, Bersier and Braun (1974a) have shown a correlation between an increase in the nucleotide pool size with the onset of DNA synthesis in *Physarum*. When DNA synthesis in this organism was inhibited by cycloheximide, dNTP pools remained the same or increased only slightly (Bersier and Braun, 1974b). Less direct studies on avian (Weintraub and Holtzer, 1972) and mammalian cells (Gautschi *et al.*, 1973) have shown that cycloheximide has either no effect or causes a slight expansion of the dNTP pools. Thus, it appears that inhibition of protein synthesis does not slow DNA synthesis by limiting the size of the precursor pool. Plagemann and Erbe (1972) have demonstrated no immediate decrease in the TTP pool after inhibition of protein synthesis by cycloheximide, and further, the eventual cutoff of thymidine incorporation into the cell probably results from a new equilibrium between synthesis and degradation of TTP that exists when DNA synthesis is stopped by cycloheximide (Plagemann and Erbe, 1974).

III. SUMMARY

Chromosomal DNA replication in eukaryotes is a complex process. During the finite period of an S phase, an exact copy of an extremely large genome is made. There are multiple points before and during synthesis at which the process may be controlled. The first set of regulatory events determine whether the cellular DNA complement will be replicated, that is, whether the cell will enter S phase. Once in S phase, further regulation takes place. Specific sections of individual chromosomes replicate at defined periods of the S phase (Taylor, 1960). Within these sections, there are clusters of replication units that appear to initiate synthesis synchronously. In metaphase chromosomes, in which the replicated DNA is marked by differential fluorescence, these clusters appear to contain about 6 mm of DNA or about 100 units (Latt, 1974). Using fiber autoradiography, clusters of 4 units can be found easily (Hand, 1975b). Larger clusters occasionally can be demonstrated by this technique (Molitor *et al.*, 1974) or by electron microscopy (Kriegstein and Hogness, 1974).

The clusters are made up of the fundamental units for the regulation of initiation, the replication units (Huberman and Riggs, 1968). These units have centrally placed origins, from which replication forks proceed outward in two directions. There are about 30,000 of these units in the DNA of a mammalian cell. Each unit requires a distinct initiation event to start replication. Control of the frequency of these events appears to be the major method of regulation of DNA synthesis within the cell (Callan, 1972; Blumenthal *et al.*, 1973). Chain growth occurs via replication fork movement. There is some evidence to suggest that chain growth may take place by synthesis on multiple subunits within the replication units (Taylor, 1973). These subunits, defined by nicks in the template at intervals of about 2 μm, would be activated in an ordered fashion, beginning with those next to the centrally placed unit origin, and proceeding toward the peripheral termini. Within these subunits, DNA synthesis occurs by a discontinuous mechanism via the formation of Okazaki fragments (Schandl and Taylor, 1969; Painter and Schaefer, 1969b; Kidwell and Mueller, 1969). The synthesis of these fragments is primed by RNA (Waqar and Huberman, 1975b). The biochemical events in the synthesis of the Okazaki fragments in general are similar to the events in prokaryotic organisms (see Chapter 2, this volume).

At all levels, the synthesis of chromosomal DNA is regulated closely, beginning with the first step, the commitment of the cell to DNA synthesis, down to the initiation and elongation of the Okazaki fragment. A complex organization within the genome has developed for purposes of replication, permitting fine control and modulation over each of the synthetic steps.

ACKNOWLEDGMENTS

I would like to thank Walt Fangman, Joel Huberman, and J. Herbert Taylor for providing me with unpublished material to which I otherwise would not have had access. Many of my colleagues provided manuscripts that were in preparation or in press. Critical reading of the manuscript by Igor Tamm, Joel Huberman, and the editors improved its style and content. I especially wish to thank Mademoiselle Marie-Helene Malcangi for editorial assistance and for typing the many drafts of the manuscript.

REFERENCES

Adams, R. L. P. (1969). The effect of endogenous pools of thymidylate on the apparent rate of DNA synthesis. *Exp. Cell Res.* **56,** 55–58.

Adams, R. L. P., Berryman, S., and Thompson, A. (1971). Deoxyribonucleoside triphosphate pools in synchronized and drug-inhibited L929 cells. *Biochim. Biophys. Acta* **240,** 455–462.

Amaldi, F., Carnevali, F., Leoni, L., and Mariotti, D. (1972). Replicon origins in Chinese hamster cell DNA. I. Labeling procedure and preliminary observations. *Exp. Cell Res.* **74,** 367–374.

Amaldi, F., Buongiorno-Nardelli, M., Carnevali, F., Leoni, L., Mariotti, D., and Pomponi, M. (1973). Replicon origins in Chinese hamster cell DNA. II. Reproducibility. *Exp. Cell Res.* **80,** 79–87.

Andersen, H. A., and Zeuthen, E. (1971). DNA replication sequence in *Tetrahymena* is not repeated from generation to generation. *Exp. Cell Res.* **68,** 309–314.

Bersier, D., and Braun, R. (1974a). Pools of deoxyribonucleoside triphosphates in the mitotic cycle of *Physarum. Biochim. Biophys. Acta* **340,** 463–471.

Bersier, D., and Braun, R. (1974b). Effect of cycloheximide on pools of deoxyribonucleoside triphosphates. *Exp. Cell Res.* **84,** 436–440.

Bird, R. W., Louarn, J., Martuscelli, J., and Caro, L. (1972). Origin and sequence of chromosome replication in *Escherichia coli. J. Mol. Biol.* **70,** 549–566.

Blumenthal, A. B., Kriegstein, H. J., and Hogness, D. S. (1973). The units of DNA replication in *Drosophila melanogaster* chromosomes. *Cold Spring Harbor Symp. Quant. Biol.* **38,** 205–223.

Bollum, F. J. (1975). Mammalian DNA polymerases. *Prog. Nucleic Acid Res. Mol. Biol.* **15,** 109–144.

Borun, T. W., Scharff, M. D., and Robbins, E. (1967). Rapidly labeled, polyribosome-associated RNA having the properties of histone messenger. *Proc. Natl. Acad. Sci. U.S.A.* **58,** 1977–1983.

Borun, T. W., Gabrielli, F., Ajird, K., Zweidler, A., and Baglioni, C. (1975). Further evidence of transcriptional and translational control of histone messenger RNA during the HeLa S3 cycle. *Cell* **4,** 59–67.

Bostock, C. J., and Prescott, D. M. (1971a). Buoyant density of DNA synthesized at different stages of the S phase of mouse L cells. *Exp. Cell Res.* **64,** 267–274.

Bostock, C. J., and Prescott, D. M. (1971b). Shift in buoyant density of DNA during the synthetic period and its relation to euchromatin and heterochromatin in mammalian cells. *J. Mol. Biol.* **60,** 151–162.

Brent, T. P., Butler, J. A. V., and Crathorn, A. R. (1965). Variations in phosphokinase activities during the cell cycle in synchronous populations of HeLa cells. *Nature (London)* **207,** 176–177.

Brewer, E. N., Evans, T. E., and Evans, H. H. (1974). Studies of the mechanism of DNA replication in *Physarum polycephalum. J. Mol. Biol.* **90,** 335–342.

Britten, R. J., and Kohne, D. E. (1968). Repeated sequences in DNA. *Science* **161,** 529–540.

Burger, M. M. (1971). Surface changes detected by lectins and implications for growth regulation in normal and in transformed cells. *Biomembranes* **2,** 247–270.

Burke, W. (1974). Ph. D. Thesis, University of Washington, Seattle.

Cairns, J. (1963). The chromosome of *Escherichia coli*. *Cold Spring Harbor Symp. Quant. Biol.* **28,** 43–46.

Cairns, J. (1966). Autoradiography of HeLa cell DNA. *J. Mol. Biol.* **15,** 372–373.

Callan, H. G. (1972). Replication of DNA in the chromosomes of eukaryotes. *Proc. R. Soc. London, Ser. B* **181,** 19–41.

Callan, H. G. (1973a). DNA replication in the chromosomes of eukaryotes. *Cold Spring Harbor Symp. Quant. Biol.* **38,** 195–204.

Callan, H. G. (1973b). Replication of DNA in eukaryotic chromosomes. *Br. Med. Bull.* **29,** 192–195.

Chang, L. M. S., Brown, M., and Bollum, F. J. (1973). Induction of DNA polymerase in mouse L cells. *J. Mol. Biol.* **74,** 1–8.

Cheevers, W. P., and Hiscock, W. B. (1973). DNA synthesis in polyomavirus infection. *J. Mol. Biol.* **78,** 237–245.

Cheevers, W. P., Branton, P. E., and Sheinin, R. (1970). Formation of cellular deoxyribonucleic acid during productive polyomavirus infection. *J. Virol.* **6,** 573–582.

Cheevers, W. P., Kowalski, J., and Yu, K. K.-Y. (1972). Synthesis of high molecular weight cellular DNA in productive polyomavirus infection. *J. Mol. Biol.* **65,** 347–364.

Clarkson, B., and Baserga, R., eds. (1974). "Control of Proliferation in Animal Cells." Cold Spring Harbor Lab., Cold Spring Harbor, New York.

Comings, D. E. (1972). Replicative heterogeneity of mammalian DNA. *Exp. Cell Res.* **71,** 106–112.

Comings, D. E., and Mattocia, E. (1970). Replication of repetitious DNA in the S period. *Proc. Natl. Acad. Sci. U.S.A.* **67,** 448–455.

Crawford, L. V., Syrett, C., and Wilde, A. (1973). The replication of polyoma DNA. *J. Gen. Virol.* **21,** 515–521.

Cummins, J. E., and Rusch, H. P. (1966). Limited DNA synthesis in the absence of protein synthesis in *Physarum polycephalum*. *J. Cell Biol.* **31,** 577–583.

Danna, K. J., and Nathans, D. (1972). Bidirectional replication of simian virus 40 DNA. *Proc. Natl. Acad. Sci. U.S.A.* **69,** 3097–3100.

Dulbecco, R., Hartwell, L. H., and Vogt, M. (1965). Induction of cellular DNA synthesis by polyomavirus. *Proc. Natl. Acad. Sci. U.S.A.* **53,** 403–410.

Edenberg, H. J., and Huberman, J. A. (1975). Eukaryotic chromosome replication. *Annu. Rev. Genet.* **9,** 245–284.

Everhart, L. P., and Rubin, R. W. (1974). Cyclic changes in the cell surface. I. Change in thymidine transport and its inhibition by cytochalasin in Chinese hamster ovary cells. *J. Cell Biol.* **60,** 434–441.

Fareed, G. C., and Salzman, N. P. (1972). Intermediates in SV40 chain growth. *Nature (London), New Biol.* **238,** 274–277.

Fareed, G. C., McKerlie, M. L., and Salzman, N. P. (1973). Characterization of simian virus 40 component II during viral DNA replication. *J. Mol. Biol.* **74,** 95–111.

Foster, D. N., and Gurney, T., Jr. (1974). Sizes of DNA polymerases from nuclei isolated by a nonaqueous method. *J. Cell Biol.* **63,** 103a.

Fournier, R. E., and Pardee, A. B. (1975). Cell cycle studies of mononucleate and cytochalasin B-induced binucleate fibroblasts. *Proc. Natl. Acad. Sci. U.S.A.* **72,** 869–873.

Fox, M., Mendelsohn, J., Barbosa, E., and Goulian, M. (1973). RNA in nascent DNA from cultured human lymphocytes. *Nature (London), New Biol.* **245,** 234–237.

Francke, B., and Hunter, T. (1974). *In vitro* polyoma DNA synthesis: Discontinuous chain growth. *J. Mol. Biol.* **83,** 99–121.

Francke, B., and Vogt, M. (1975). *In vitro* polyoma synthesis: Self-annealing properties of short DNA chains. *Cell* **5,** 205–212.

Fuchs, P., and Kohn, A. (1971). Nature of transient inhibition of deoxyribonucleic acid synthesis in HeLa cells by parainfluenza virus I. (Sendai). *J. Virol.* **5,** 695–700.

Fujiwara, Y. (1972). Effect of cycloheximide on regulatory protein for initiating mammalian DNA replication at the nuclear membrane. *Cancer Res.* **32,** 2089–2095.

Gautschi, J. R. (1974). Effects of puromycin on chain elongation in mammalian cells. *J. Mol. Biol.* **84,** 223–229.

Gautschi, J. R., and Clarkson, J. M. (1975). Discontinuous DNA replication in mouse P815 cells. *Eur. J. Biochem.* **50,** 403–412.

Gautschi, J. R., and Kern, R. M. (1973). DNA replication in mammalian cells in the presence of cycloheximide. *Exp. Cell Res.* **80,** 15–27.

Gautschi, J. R., Kern, R. M., and Painter, R. B. (1973). Modification of replicon operation in HeLa cells by 2,4-dinitrophenol. *J. Mol. Biol.* **80,** 393–403.

Gelbard, A. S., Kim, J. H., and Perez, A. G. (1969). Fluctuation in deoxycytidine monophosphate deaminase activity during the cell cycle in synchronous populations. *Biochim. Biophys. Acta* **182,** 564–566.

Gentry, G. A., Morse, P. A., Jr., Ives, D. H. R., and Potter, V. R. (1965). Pyrimidine metabolism in tissue culture cells derived from rat hepatomas. II. Thymidine uptake in suspension cultures derived from the Novikoff hepatoma. *Cancer Res.* **25,** 509–516.

Goldstein, N. O., and Rutman, R. J. (1973). *In vivo* discontinuous DNA synthesis in Ehrlich ascites tumors. *Nature (London), New Biol.* **244,** 267–269.

Goodman, J. J. (1974). A comparison of the utilization of thymine and thymidine for the synthesis of DNA thymine in Novikoff hepatoma cells. *Exp. Cell Res.* **85,** 415–423.

Graham, C. F. (1966). The regulation of DNA synthesis and mitosis in multinucleate frog eggs. *J. Cell Sci.* **1,** 363–374.

Grunicke, H., Hirsch, F., Wolf, F., Bauer, U., and Kieffer, G. (1975). Selective inhibition of thymidine transport at low doses of the alkylating agent triethylene iminobenzoquinone (Trenimon). *Exp. Cell Res.* **80,** 357–364.

Hand, R. (1975a). Regulation of DNA replication on subchromosomal units of mammalian cells. *J. Cell Biol.* **64,** 89–97.

Hand, R. (1975b). DNA replication in mammalian cells: Altered patterns of initiation during inhibition of protein synthesis. *J. Cell Biol.* **67,** 761–774.

Hand, R. (1975c). DNA fiber autoradiography as a technique for studying the replication of the mammalian chromosome. *J. Histochem. Cytochem.* **23,** 475–481.

Hand, R., and Tamm, I. (1972). Rate of DNA chain growth in mammalian cells infected with cytocidal RNA viruses. *Virology* **47,** 331–337.

Hand, R., and Tamm, I. (1973). DNA replication: Rate and direction of chain growth in mammalian cells. *J. Cell Biol.* **58,** 410–418.

Hand, R., and Tamm, I. (1974). Initiation of DNA replication in mammalian cells and its inhibition by reovirus infection. *J. Mol. Biol.* **82,** 175–183.

Hand, R., Ensminger, W. D., and Tamm, I. (1971). Cellular DNA replication in infections with cytocidal RNA viruses. *Virology* **44,** 527–536.

Hare, J. D. (1970). Quantitative aspects of thymidine uptake into the acid-soluble pool of normal and polyoma-transformed hamster cells. *Cancer Res.* **30,** 684–691.

Harris, H. (1965). Behaviour of differentiated nuclei in heterokaryons of animal cells from different species. *Nature (London)* **206,** 583–588.

Harris, H., Watkins, J. F., Ford, C. E., and Schoeff, G. I. (1966). Artificial heterokaryons of animal cells from different species. *J. Cell Sci.* **1,** 1–30.

Hartwell, L. H., Mortimer, R. K., Culotti, J., and Culotti, M. (1973). Genetic control of the cell division cycle in yeast. V. Genetic analysis of *cdc* mutants. *Genetics* **74,** 267–286.

Hauschka, P. V. (1973). Analysis of nucleotide pools in animal cells. *Methods Cell Biol.* **7,** 362–462.

Hereford, L. M., and Hartwell, L. H. (1973). Role of protein synthesis in the replication of yeast DNA. *Nature (London), New Biol.* **244,** 129–131.

Hereford, L. M., and Hartwell, L. H. (1974). Sequential gene function in the initiation of *Saccharomyces cervisiae* DNA synthesis. *J. Mol. Biol.* **84,** 445–461.

Hershey, H. V., and Taylor, J. H. (1974). DNA replication in eukaryotic nuclei. Evidence suggesting a specific model of replication. *Exp. Cell Res.* **85,** 79–88.

Hershey, H. V., Stieber, J., and Mueller, G. C. (1973). Effect of inhibiting the cellular synthesis of RNA, DNA and protein on DNA replicative activity of isolated S phase nuclei. *Biochim. Biophys. Acta* **312,** 509–517.

Highfield, D. P., and Dewey, W. C. (1972). Inhibition of DNA synthesis in synchronized Chinese hamster cells treated in G_1 or early S phase with cycloheximide or puromycin. *Exp. Cell Res.* **75,** 314–320.

Hori, T-a., and Lark, K. G. (1973). Effect of puromycin on DNA replication in Chinese hamster cells. *J. Mol. Biol.* **77,** 391–404.

Hori, T.-a., and Lark, K. G. (1974). Autoradiographic studies of the replication of satellite DNA in the kangaroo rat. Autoradiographs of satellite DNA. *J. Mol. Biol.* **88,** 221–232.

Horwitz, H. (1975). Ph.D. Thesis, Massachusetts Institute of Technology, Cambridge.

Hourcade, D., Dressler, D., and Wolfson, J. (1973). The nucleolus and the rolling circle. *Cold Spring Harbor Symp. Quant. Biol.* **38,** 537–549.

Housman, D., and Huberman, J. A. (1975). Changes in the rate of DNA replication fork movement during S phase in mammalian cells. *J. Mol. Biol.* **94,** 173–182.

Huberman, J. A., and Riggs, A. D. (1968). On the mechanism of DNA replication in mammalian chromosomes. *J. Mol. Biol.* **32,** 327–341.

Huberman, J. A., and Tsai, A. (1973). The direction of DNA replication in mammalian cells. *J. Mol. Biol.* **75,** 5–12.

Hwang, K. M., Murphee, S. A., Shansky, C. W., and Sartorelli, A. C. (1974). Sequential biochemical events related to DNA replication in the regnerating liver. *Biochim. Biophys. Acta* **366,** 143–148.

Hyodo, M., and Flickenger, R. A. (1972). Replicon growth rates during DNA replication in developing frog embryos. *Biochim. Biophys. Acta* **229,** 24–33.

Jacob, F., Brenner, J., and Cuzin, F. (1963). On the regulation of DNA replication in bacteria. *Cold Spring Harbor Symp. Quant. Biol.* **28,** 329–348.

Jeffrey, W. R. (1974). Macromolecular requirements for the initiation and maintenance of DNA synthesis during the cell cycle of *Tetrahymena pyriformis. J. Cell. Physiol.* **83,** 1–10.

Johnson, R. T., and Harris, H. (1969). DNA synthesis and mitosis in fused cells. HeLa–chick erythrocyte heterokaryons. *J. Cell Sci.* **5,** 625–649.

Johnson, R. T., and Rao, P. N. (1971). Nucleoplasmic interactions in the achievement of nuclear synchrony in DNA synthesis and mitosis in multinucleate cells. *Biol. Rev. Cambridge Philos. Soc.* **46,** 97–155.

Jolley, G. M., and Ormerod, M. G. (1974). The complete separation of complementary strands of high molecular weight DNA in alkali. *Biochim. Biophy. Acta* **353,** 200–214.

Kato, K., and Strauss, B. (1974). Accumulation of an intermediate in DNA synthesis by HEp-2 cells treated with methyl methanesulfonate. *Proc. Natl. Acad. Sci. U.S.A.* **71,** 1969–1973.

Kidwell, W. R., and Mueller, G. C. (1969). The synthesis and assembly of DNA subunits in isolated HeLa cell nuclei. *Biochem. Biophys. Res. Commun.* **36,** 756–763.

Klevecz, R. R., and Kapp, L. N. (1973). Intermittent DNA synthesis and periodic expression of enzyme activity in the cell cycle of WI38. *J. Cell Biol.* **58,** 564–573.

Klevecz, R. R., Keniston, B. A., and Deaven, L. L. (1975). The temporal structure of S phase. *Cell* **5,** 195–204.

Kriegstein, H. J., and Hogness, D. S. (1974). Mechanism of replication in *Drosophila melanogaster* chromosomes: Structure of replication forks and evidence for bidirectionality. *Proc. Natl. Acad. Sci. U.S.A.* **71,** 135–139.

Kuebbing, D., and Werner, R. (1975). A model for compartmentation of *de novo* and salvage thymidine nucleotide pools in mammalian cells. *Proc. Natl. Acad. Sci. U.S.A.* **72,** 3333–3336.

Lark, K. G. (1969). Initiation and control of DNA synthesis. *Annu. Rev. Biochem.* **38,** 569–604.

Lark, K. G., Consigli, R., and Toliver, A. (1971). DNA replication in Chinese hamster cells: Evidence for a single replication fork per replicon. *J. Mol. Biol.* **58,** 873–875.

Latt, S. (1974). Microfluorometric analysis of deoxyribonucleic acid replication kinetics and sister chromatid exchanges in human chromosomes. *J. Histochem. Cytochem.* **22,** 478–491.

Lee, C. S., and Pavan, C. (1974). Replicating DNA molecules from fertilized eggs of *Cochliomyia hominivorax* (Diptera). *Chromosoma* **47,** 429–437.

Lehmann, A. R., and Ormerod, M. G. (1970). The replication of DNA in murine lymphoma cells (L5178Y). 1. Rate of replication. *Biochim. Biophys. Acta* **204,** 128–143.

Lindberg, U., and Skoog, L. (1970). A method for the determination of dATP and dTTP in picomole amounts. *Anal. Biochem.* **34,** 152–160.

Littlefield, J. W. (1966). The periodic synthesis of thymidine kinase in mouse fibroblasts. *Biochim. Biophys. Acta* **114,** 398–403.

Littlefield, J. W., and Jacobs, P. S. (1965). The relation between DNA and protein synthesis in mouse fibroblasts. *Biochim. Biophys. Acta* **108,** 652–658.

McBurney, M. W., Graham, F. L., and Whitmore, G. F. (1972). Sedimentation analysis of DNA from irradiated L cells. *Biophys. J.* **12,** 369–383.

McFarlane, P. W., and Callan, H. G. (1973). DNA replication in the chromosomes of the chicken, *Gallus domesticus*. *J. Cell Sci.* **13,** 821–839.

Magnusson, G. (1973). Hydroxyurea-induced accumulation of short fragments during polyoma DNA replication. I. Characterization of fragments. *J. Virol.* **12,** 600–608.

Magnusson, G., Pigiet, V., Winnacker, E. L., Abrams, R., and Reichard, P. (1973). RNA-linked short fragments during polyoma replication. *Proc. Natl. Acad. Sci. U.S.A.* **70,** 412–415.

May, M. S., and Bello, L. J. (1974). Replication of repeated DNA in human cells. *Exp. Cell Res.* **83,** 79–86.

Molitor, H., Drahovski, D., and Wacker, A. (1974). Structural integrity of chromatid DNA in mouse L cells. *J. Mol. Biol.* **86,** 161–163.

Mueller, G. C., Kajiwara, K., Stubblefield, E., and Rueckert, R. R. (1962). Molecular events in the reproduction of animal cells. I. The effect of puromycin on the duplication of DNA. *Cancer Res.* **22,** 1084–1090.

Muldoon, J. J., Evans, T. E., Nygaard, O. F., and Evans, H. H. (1971). Control of DNA replication by protein synthesis at defined times during the S period in *Physarum polycephalum*. *Biochim. Biophys. Acta* **247,** 310–321.

Nakata, Y., and Bader, J. P. (1969). The uptake of nucleosides by cells in culture. II. Inhibition by 2-mercapto-1-(β-4-pyridethyl)benzimidazodole. *Biochim. Biophys. Acta* **190,** 250–256.

Newlon, C. S., Petes, T. D., Hereford, L. M., and Fangman, W. L. (1974). Replication of yeast chromosomal DNA. *Nature (London)* **247,** 32–35.

Nordenskjold, B. A., Skoog, L., Brown, N. C., and Reichard, P. (1970). Deoxyribonucleotide pools and deoxyribonucleic acid synthesis in cultured mouse embryo cells. *J. Biol. Chem.* **245,** 5360–5368.

Nuzzo, F., Brega, A., and Falaschi, A. (1970). DNA replication in mammalian cells. I. The size of newly synthesized helices. *Proc. Natl. Acad. Sci. U.S.A.* **65,** 1017–1024.

Okazaki, R., Okazaki, T., Sakabe, K., Sugimoto, K., and Sugino, A. (1968). Mechanism of DNA chain growth. I. Possible discontinuity and unusual secondary structure of newly synthesized chains. *Proc. Natl. Acad. Sci. U.S.A.* **59,** 598–605.

Painter, R. B., and Schaefer, A. W. (1969a). Rate of synthesis along replicons of different kinds of mammalian cells. *J. Mol. Biol.* **45,** 467–479.

Painter, R. B., and Schaefer, A. W. (1969b). State of newly synthesized HeLa DNA. *Nature (London)* **221,** 1215–1217.

Painter, R. B., and Schaefer, A. W. (1971). Variation in the rate of DNA chain growth through the S phase in HeLa cells. *J. Mol. Biol.* **58,** 289–295.

Petes, T. D., and Williamson, D. H. (1975). Fiber autoradiography of replicating yeast DNA. *Exp. Cell Res.* **95,** 103–110.

Pigiet, V., Eliasson, R., and Reichard, P. (1974). Replication of polyoma DNA in isolated nuclei. III. The nucleotide sequence and the RNA–DNA junction of the nascent strands. *J. Mol. Biol.* **84,** 197–216.

Plagemann, P. G. W., and Erbe, J. (1972). Thymidine transport by cultured Novikoff hepatoma cells and the uptake by simple diffusion and the relationship to incorporation into deoxyribonucleic acid. *J. Cell Biol.* **55,** 161–178.

Plagemann, P. G. W., and Erbe, J. (1974). The deoxyribonucleoside transport systems of cultured Novikoff rat hepatoma cells. *J. Cell. Physiol.* **83,** 337–344.

Prescott, D., and Kuempel, P. (1972). Bidirectional replication of the chromosome in *Escherichia coli*. *Proc. Natl. Acad. Sci. U.S.A.* **69,** 2842–2845.

Prescott, D., and Kuempel, P. (1973). Autoradiography of individual DNA molecules. *Methods Cell Biol.* **7,** 147–157.

Probst, H., Gentner, P. R., Hofstatter, T., and Jenke, S. (1974a). Newly synthesized mammalian cell DNA: Evidence for effects simulating the presence of RNA in the nascent DNA fraction isolated by nitrocellulose column chromatography. *Biochim. Biophys. Acta* **340,** 361–373.

Probst, H., Hofstatter, T., Jenke, H-S., Gentner, P. R., and Wais, R. (1974b). Newly synthesized mammalian cell DNA, affinity to nitrocellulose and rate of chain growth. *Biochim. Biophys. Acta* **366,** 11–22.

Rao, P. N., and Johnson, R. T. (1970). Mammalian cell fusion: Studies on the regulation of DNA synthesis and mitosis. *Nature (London)* **225,** 159–164.

Reichard, P., Eliasson, R., and Soderman, G. (1974). Initiator RNA in discontinuous polyoma DNA synthesis. *Proc. Natl. Acad. Sci. U.S.A.* **71,** 4901–4905.

Remington, J. A., and Klevecz, R. R. (1973). Families of replicating units in cultured hamster fibroblasts. *Exp. Cell Res.* **76,** 410–418.

Robbins, E., and Borun, T. W. (1967). The cytoplasmic synthesis of histones in HeLa cells and its temporal relationship to DNA replication. *Proc. Natl. Acad. Sci. U.S.A.* **57,** 409–416.

Rochaix, J. D., and Bird, A. P. (1975). Circular ribosomal DNA and ribosomal DNA replication in somatic amphibian cells. *Chromosoma* **52,** 317–328.

Rochaix, J. D., Bird, A., and Bakken, A. (1974). Ribosomal RNA gene amplification by rolling circle. *J. Mol. Biol.* **87,** 473–487.

Roller, B. A., Hirai, K., and Defendi, V. (1974). Effects of cAMP on nucleoside metabolism. I. Effect on thymidine transport and incorporation in monkey cells (CV-I). *J. Cell. Physiol.* **83,** 163–176.

Sadoff, R. B., and Cheevers, W. P. (1973). Evidence for RNA-linked nascent strands in polyoma virus DNA replication. *Biochem. Biophys. Res. Commun.* **53,** 818–823.

Saladino, C. F., and Johnson, H. A. (1974). Rate of DNA synthesis as a function of temperature in cultured hamster fibroblasts (V-79) and HeLa S-3 cells. *Exp. Cell Res.* **85,** 248–254.

Schandl, E. K., and Taylor, J. H. (1969). Early events in the replication and integration of DNA into mammalian chromosomes. *Biochem. Biophys. Res. Commun.* **34,** 291–300.

Schneiderman, M. H., Dewey, W. C., and Highfield, D. P. (1971). Inhibition of DNA synthesis in synchronized Chinese hamster cells treated in G_1 with cycloheximide. *Exp. Cell Res.* **67,** 147–155.

Seale, R. L., and Simpson, R. T. (1975). Effects of cycloheximide on chromatin biosynthesis. *J. Mol. Biol.* **94,** 479–501.

Seki, S., and Mueller, G. C. (1975). A requirement for RNA, protein and DNA synthesis in the establishment of DNA replicase activity in synchronized HeLa cells. *Biochim. Biophys. Acta* **378,** 354–362.

Sieger, M. P., Schaer, J. C., Hirsiger, H., and Schindler, R. (1974). Determination of rates of DNA synthesis in cultured mammalian cell populations. *J. Cell Biol.* **62,** 305–315.

Simpson, J. R., Nagle, W. A., Bick, M. D., and Belli, J. A. (1973). Molecular nature of mammalian cell DNA in alkaline sucrose gradients. *Proc. Natl. Acad. Sci. U.S.A.* **70,** 3660–3664.

Skoog, L., and Bjursell, G. (1974). Nuclear and cytoplasmic pools of deoxyribonucleoside triphosphates in Chinese hamster ovary cells. *J. Biol. Chem.* **249,** 6434–6438.

Skoog, L., and Nordenskjold, B. (1971). Effects of hydroxyurea and 1-beta-D-arabinofuranosyl-cytosine on deoxyribonucleotide pools in mouse embryo cells. *Eur. J. Biochem.* **19,** 81–89.

Skoog, L., Nordenskjold, B. A., and Bjursell, G. (1973). Deoxyribonucleoside–triphosphate pools and DNA synthesis in synchronized hamster cells. *Eur. J. Biochem.* **33,** 428–432.

Solter, A. W., and Handschumacher, R. E. (1969). A rapid quantitative determination of deoxyribonucleoside triphosphates based on the enzymatic synthesis of DNA. *Biochim. Biophys. Acta* **174,** 585–590.

Spadari, S., and Weissbach, A. (1974a). The interrelation between DNA synthesis and various DNA polymerase activities in synchronized HeLa cells. *J. Mol. Biol.* **86,** 11–20.

Spadari, S., and Weissbach, A. (1974b). HeLa cell R-deoxyribonucleic acid polymerases. *J. Biol. Chem.* **249,** 5809–5815.

Spadari, S., and Weissbach, A. (1975). RNA-primed DNA synthesis. Specific catalysis by HeLa cell DNA polymerase α. *Proc. Natl. Acad. Sci. U.S.A.* **72,** 503–507.

Steck, T. I., Nakata, Y., and Bader, J. P. (1969). The uptake of nucleosides by cells in culture. I. Inhibition by heterologous nucleosides. *Biochim. Biophys. Acta* **190,** 237–249.

Stein, G., and Baserga, R. (1972). Nuclear proteins and the cell cycle. *Adv. Cancer Res.* **15,** 287–330.

Stein, G., and Borun, T. (1972). The synthesis of acidic chromosomal proteins during the cell cycle of HeLa S-3 cells. I. The accelerated accumulation of acidic residual nuclear protein before initiation of DNA replication. *J. Cell Biol.* **52,** 292–307.

Stocco, D. M., and Zimmerman, A. M. (1974). Metabolism of acid-soluble nucleotides in starved refed synchronized *Tetrahymena pyriformis*. *Can. J. Biochem.* **52,** 310–318.

Tapiero, H., Shaool, D., Monier, M. P., and Harel, J. (1974). Replication of repetitious DNA in synchronized chick fibroblast cells. *Exp. Cell Res.* **89,** 39–46.

Taylor, J. H. (1960). Asynchronous duplication of chromosomes in cultured cells of Chinese hamster. *J. Biophys. Biochem. Cytol.* **7,** 455–464.

Taylor, J. H. (1963). DNA synthesis in relation to chromosome reproduction and the reunion of breaks. *J. Cell. Comp. Physiol.* **62,** Suppl. 1, 73–86.

Taylor, J. H. (1968). Rates of chain growth and units of replication in DNA of mammalian chromosomes. *J. Mol. Biol.* **31,** 579–594.

Taylor, J. H. (1973). Replication of DNA in mammalian chromosomes: Isolation of replicating segments. *Proc. Natl. Acad. Sci. U.S.A.* **70,** 1083–1087.

Taylor, J. H., Adams, A. G., and Kurek, M. P. (1973). Replication of DNA in mammalian chromosomes. II. Kinetics of ^{3}H-thymidine incorporation and the isolation and partial characterization of labeled subunits at the growing point. *Chromosoma* **41,** 361–384.

Terasima, T., and Yasukawa, M. (1966). Synthesis of G_1 protein preceding DNA synthesis in cultured mammalian cells. *Exp. Cell Res.* **44,** 669–672.

Tobia, A. M., Schildkraut, C. L., and Maio, J. J. (1970). Deoxyribonucleic acid replication in synchronized cultured mammalian cells. *J. Mol. Biol.* **54,** 499–515.

Tseng, B. Y., and Goulian, M. (1975). Evidence for covalent association of RNA with nascent DNA in human lymphocytes. *J. Mol. Biol.* **99,** 339–347.

Walters, R. A., Tobey, R. A., and Ratliff, R. L. (1973). Cell cycle-dependent variation of deoxyribonucleoside triphosphate pools in Chinese hamster cells. *Biochim. Biophys. Acta* **319,** 336–347.

Waqar, M. A., and Huberman, J. A. (1973). Evidence for the attachment of RNA to pulse-labeled DNA in the slime mold, *Physarum polycephalum. Biochem. Biophys. Res. Commun.* **51,** 174–180.

Waqar, M. A., and Huberman, J. A. (1975a). Covalent linkage between RNA and nascent DNA in the slime mold, *Physarum polycephalum. Biochim. Biophys. Acta* **383,** 410–420.

Waqar, M. A., and Huberman, J. A. (1975b). Covalent attachment of RNA to nascent DNA in mammalian cells. *Cell* **6,** 549–555.

Weintraub, H. (1972). Bidirectional initiation of DNA synthesis in developing chick erythroblasts. *Nature (London), New Biol.* **236,** 195–197.

Weintraub, H. (1973). The assembly of newly replicated DNA into chromatin. *Cold Spring Harbor Symp. Quant. Biol.* **38,** 247–256.

Weintraub, H., and Holtzer, H. (1972). Fine control of DNA synthesis in developing chick red blood cells. *J. Mol. Biol.* **66,** 13–35.

Weiss, B. (1969). The dependence of DNA synthesis on protein synthesis in HeLa S3 cells. *J. Cell. Physiol.* **73,** 85–90.

Weissbach, A. (1975). Vertebrate DNA polymerases. *Cell* **5,** 101–108.

Werner, R. (1971). Mechanism of DNA replication. *Nature (London)* **230,** 570–572.

Wilson, D. A., and Thomas, C. A. (1974). Palindromes in chromosomes. *J. Mol. Biol.* **84,** 115–144.

Wolstenholme, D. R. (1973). Replicating DNA molecules from eggs of *Drosophila melanogaster. Chromosoma* **43,** 1–18.

Young, C. W. (1966). Inhibitory effects of acetoxy-cycloheximide, puromycin, and pactomycin upon synthesis of protein and DNA in asynchronous populations of HeLa cells. *Mol. Pharmacol.* **2,** 50–55.

9

DNA Repair and Its Relationship to Mutagenesis, Carcinogenesis, and Cell Death

R. F. Kimball

ISBN 0-12-289502-9

I. INTRODUCTION

Over the past decade, we have come to realize what an impressive array of enzymes cells have for "handling" DNA. Not only are there enzymes for DNA replication, but also enzymes for repair of lesions in DNA, enzymes for recombining DNA molecules, and enzymes for modifying DNA. This chapter will concentrate on the failures of these systems, especially as they are reflected in mutation. It should be remembered, however, that these failures represent only a very small fraction of the operations of the systems. They are of importance simply because very small changes in DNA can have major consequences for the cell, the individual, and even the population of which the individual is a member. Some of the most serious and intractable of human disabilities and disease arise from these failures. Similar rates of error or failure in most other enzyme systems would have negligible consequences that normally would be impossible to detect.

The chapter will begin with summaries of the reviews of biophysical and biochemical studies of DNA repair and of the genes and enzymes involved. The main purpose is to provide essential background for the discussion of biological effects, not a thorough and well-documented description of the molecular processes involved. The remainder of the chapter will be concerned with the relation between DNA repair and several biological endpoints. Most attention will be devoted to mutation, partially because it is of the most interest to the author, partially because the relationship to DNA repair is better developed for this endpoint than for the others. In general, only brief references will be made to evidence about repair processes whose relationships to DNA repair per se are either unformulated or unestablished as yet.

A two-volume work entitled "Molecular Mechanisms for Repair of DNA" (Hanawalt and Setlow, 1975), based on a conference held in 1974, has been published. Anyone interested in the subject would do well to consult these volumes.

II. BIOPHYSICAL AND BIOCHEMICAL STUDIES OF REPAIR

Two major approaches have been applied to the study of the repair of lesions in DNA. One is to look for the disappearance of the lesion with time. Obviously, this depends on the ability to detect and measure the lesion in question. The other is to look for the newly synthesized DNA that frequently, although not always, is the result of the repair process. In this case, the synthesis must be distinguished from normal replicative synthesis either by the fact that it is not semiconservative (repair synthesis) or by the fact that it occurs in the absence of normal replication (unscheduled synthesis).

A. Disappearance of Lesions

The initial lesions introduced into the DNA by an agent can be alterations of bases, losses of bases, cross-links between DNA strands or between DNA and other macromolecules such as proteins, or single- or double-strand breaks. Some of these lesions may result from the primary reaction of the agent with DNA; others may be secondary physical or enzymatic consequences of the primary lesions. In principle, altered bases can be detected by chemical separation procedures without isotopic labeling, but detection can be facilitated greatly if the altered base can be marked by using a radioactively labeled agent (e.g., methyl-labeled ethyl methane sulfonate to detect ethylation of bases) for treatment or by prelabeling specific bases in the DNA (e.g., labeled thymidine for thymine dimers) and separating the altered base from the normal form by chromatography. Cross-links can be detected either because denatured cross-linked DNA renatures more readily than normal DNA or because of the physical association with some other molecule, e.g., protein. Single- and double-strand breaks can be detected by changes in molecular weight as determined by centrifugation in alkaline or neutral sucrose density gradients. In this section, the main emphasis will be upon results with two of the most frequently studied lesions—pyrimidine dimers and single-strand breaks. References to some of the others can be found in Grossman *et al.* (1975) and Hanawalt and Setlow (1975).

B. Photomonomerization of Pyrimidine Dimers

Irradiation with ultraviolet light below about 300 nm produces appreciable numbers of pyrimidine cyclobutane dimers in DNA. These dimers are formed by the saturation of the 5,6-double bonds of two adjacent pyrimidines in the same DNA chain to form a cyclobutane ring. The thymine–thymine dimer has the following chemical structure:

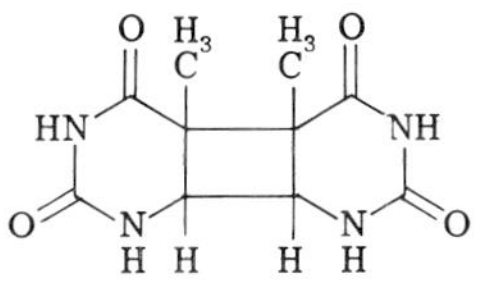

O H_3 H_3 O
C C
HN NH
O N N O
H H H H

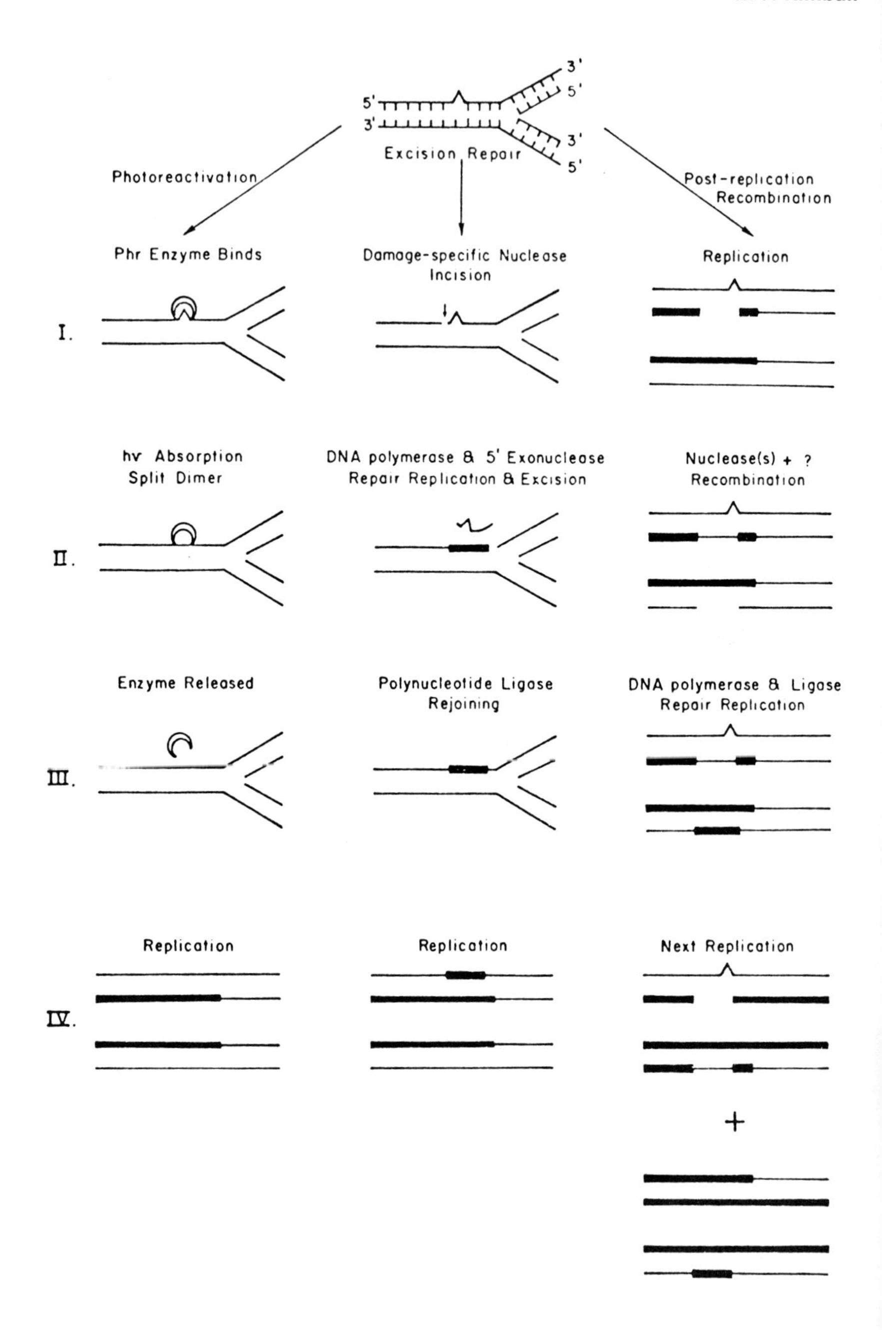
5'
3'
3'
5'
3'
5'
Excision Repair
Photoreactivation
Post-replication Recombination
Phr Enzyme Binds
Damage-specific Nuclease Incision
Replication
I.
hv Absorption Split Dimer
DNA polymerase & 5' Exonuclease Repair Replication & Excision
Nuclease(s) + ? Recombination
II.
Enzyme Released
Polynucleotide Ligase Rejoining
DNA polymerase & Ligase Repair Replication
III.
Replication
Replication
Next Replication
IV.
+

Dimers also can be formed between adjacent cytosines and between a thymine and an adjacent cytosine. Moreover, these dimers appear to be responsible for many, although not all, of the biological consequences of ultraviolet irradiation (J. K. Setlow, 1966). Three major modes for the elimination of such dimers or for repairing the consequences of their presence have been recognized: photoreactivation (photorepair), excision (prereplication) repair, and postreplication repair. These modes are diagrammed in Figure 1.

Enzymatic monomerization of pyrimidine dimers with the aid of energy from long wavelength ultraviolet or short wavelength visible light first was detected by its biological consequences on survival and was called "photoreactivation" (Kelner, 1949; Dulbecco, 1949). Later, it was shown to result from enzymatic action and then was shown to result from monomerization of dimers. The basic reaction scheme is as follows:

$$DNA \xrightarrow{uv} DNA^*$$

$$DNA^* + E \overset{h\nu}{\rightleftharpoons} DNA^*\text{–}E \xrightarrow{h\nu} DNA + E$$

where DNA* is DNA-containing dimers. The photomonomerizing enzyme (E) forms a reversible complex with the dimer-containing DNA which, with the aid of energy from light, can break down with the splitting of the cyclobutane ring to restore the dimerized pyrimidines to their original monomeric form. Jagger (1958), Rupert (1964), J. K. Setlow (1966), and Cook (1970) can be consulted for reviews at various stages of development of the subject. The enzyme involved seems to be nearly ubiquitous in prokaryotic and eukaryotic cells. Although it usually has been reported as absent from placental mammals, Sutherland *et al.* (1973) have reported it to be present in some human cells. As far as is known, the sole function of the enzyme is to monomerize pyrimidine dimers, and there is no evidence that it ever makes errors. For this reason, this form of repair will not be considered in much detail in this chapter. The specificity of the reaction enables photoreactivability to be used as a test for the presence of biologically effective dimers.

Fig. 1. Schematic representation of the principal mechanisms for eliminating or repairing the consequences of pyrimidine dimers induced in DNA by uv. The dimers are represented by an inverted "v" in one of the DNA strands. DNA segments that were present before the dimers were formed are indicated by light lines; DNA segments that were synthesized after the formation of dimers, by heavy lines. As far as is known, photoreactivation is specific for pyrimidine dimers, but the other two processes can repair a number of other lesions as well. (From P. Hanawalt, 1975, reproduced with permission.)

C. Excision Repair

1. Removal of Dimers

Pyrimidine dimers can be removed from DNA by excision (Figure 1) of a section of the strand containing the dimer, followed by resynthesis of the excised section using the other strand as the template [see Cerutti (1974) for a succinct review, Grossman *et al.* (1975) for a review emphasizing *E. coli,* and Painter (1970) and Cleaver (1974) for reviews on mammalian cells]. The original discovery of dimer excision was made nearly simultaneously by Setlow and Carrier (1964) and Boyce and Howard–Flanders (1964). The basic observation was that the dimers disappeared during postirradiation incubation without general breakdown of the DNA. Subsequent work has shown that the following steps are involved: (1) incision of the DNA strand at the dimer by a specific uv-endonuclease, (2) excision of a section of DNA that includes the dimer and may be 20 nucleotides or more long, (3) resynthesis of the excised region, and (4) sealing by DNA ligase of the newly synthesized strand to the preexisting strand. Such ligases have been isolated from a number of different types of cells. They all act in essentially the same way to join a 3′-OH group to a 5′-PO_4 group (for reviews, see Cerutti, 1974; I. R. Lehman, 1974). In *E. coli,* steps (2) and (3) usually are carried out simultaneously by DNA polymerase I, which has both 5′ → 3′ exonucleolytic and polymerizing activity. Under some conditions, however, one or both of the other two DNA polymerases (II and III) may be involved, the former probably with a separate 5′ → 3′ exonuclease. (The DNA polymerases are discussed more fully in Chapter 2 of this volume.) In mammals, separate exonucleolytic and polymerizing enzymes probably are required, since the known DNA polymerases lack exonucleolytic activity. The time for the whole process from initial incision to final ligation usually must be short, since the number of single-strand breaks attributable to excision in progress is only a small fraction of the total dimers excised (R. B. Setlow, 1968).

Ultraviolet-endonucleases from several different sources have been used to detect endonuclease-sensitive sites in uv-irradiated DNA from a variety of cells and to test for the disappearance of such sites with time (e.g., Wilkins, 1973; Paterson *et al.,* 1973; Prakash, 1975). In this case, endonuclease-produced single-strand breaks are used to detect the sites.

2. Repair Replication

In addition to following the disappearance of dimers, two other major methods have been developed to detect excision repair. One method, originally developed by Pettijohn and Hanawalt (1964) for bacteria, is to prelabel the DNA with [^{14}C]thymidine for several cell generations, replace with unlabeled bromodeoxyuridine for a part of a cell generation, irradiate with uv, and put into [^{3}H]bromodeoxyuridine for about 30 min after uv. Equilibrium density centrifugation in $CsCl_2$ then is used to determine whether the ^{3}H-labeled DNA is in the hybrid-density region (semiconservative synthesis) or near the light-density region (repair synthesis). The method can be used with agents other than uv to test whether they induce repair synthesis.

A phenomenon which could complicate the interpretation of results and which, in any case, is of considerable interest in itself, is the induction of multiple new replication forks by X rays (Billen and Cain, 1973). With uv, the replication at preexisting forks may be blocked and may start up again at the initiation point without completion of the previous round (Billen, 1969).

Another method for detecting repair synthesis has been developed more recently (Regan *et al.*, 1971). It consists of prelabeling the DNA with [^{3}H]thymidine, treatment with the agent to be tested, incubating the treated cells with bromodeoxyuridine for a time sufficient to allow repair synthesis to occur, with 313-nm radiation, and measuring the molecular weight of the ^{3}H-prelabeled strands by alkaline sucrose gradient sedimentation. Prelabeled strands that have incorporated bromodeoxyuridine as a result of repair synthesis are sensitive to strand breakage. Consequently, they will sediment less rapidly in the gradient because of the decreased molecular weight. Thus, the decrease in molecular weight is proportional to the amount of repair replication. The method can be made quantitative to estimate both the number and the mean length of the repaired regions. In this way, a number of different agents were classified according to whether they induced "long-patch" or "short-patch" repair. For example, in mammalian cells the repaired region after uv contains about twenty bromouracil residues, whereas after ionizing radiation it contains only one or two. Regan and Setlow (1973) can be consulted for a succinct review of this and other methods of measuring repair replication.

In *E. coli,* uv-irradiation induces both short- and long-patch repair (see Grossman *et al.*, 1975), for review). In this case, short-patch repair means regions about twenty nucleotides long, whereas long-patch repair means regions as much as 2000 nucleotides long. Polymerase I is responsible for short-patch repair, and one or both of the other two DNA polymerases is responsible for long-patch repair. It seems likely that a number of agents, if not all, induce more than one type of repair, but that the ratio between the types depends on both the initial lesion (dimer, cross-link, or strand break) and on the set of enzymes active in the given cell.

3. *Unscheduled DNA Synthesis*

Repair replication can also be detected as so-called unscheduled DNA synthesis in cells that are not replicating their DNA. This method has been used rather widely for studies in mammalian cells, since DNA replication normally is confined to certain stages in the cell cycle. For this purpose, autoradiography is used frequently. In untreated cells, G_1 and G_2 cells are unlabeled, and S cells are labeled heavily by tritiated thymidine. In treated cells, G_1 and G_2 cells are labeled lightly as a result of repair replication, and S cells continue to be labeled heavily. Another procedure is to inhibit semiconservative synthesis by some procedure and to look for incorporation in the treated cells. It generally seems to be true that inhibitors of semiconservative synthesis such as hydroxyurea do not inhibit repair synthesis (see Cleaver, 1974, for review). Painter (1970) and Cleaver (1974) can be consulted for some of the work with unscheduled synthesis.

D. Postreplication Repair

Dimers that remain in the DNA until normal replication occurs cause gaps to be left in the newly synthesized complementary strands. The filling-in and sealing of these gaps are referred to as postreplication repair (Figure 1). The first demonstration of this form of repair was made by Rupp and Howard-Flanders (1968) for an *E. coli* strain unable to excise dimers. The observation was essentially that the newly synthesized strands in the uv-irradiated cells were of lower molecular weight than those in the controls, and that the molecular weight returned to normal after a brief time. The evidence suggested that the gaps were in the new strands opposite the dimers, and that the gaps were eliminated by recombination between the gapped strand and the noncomplementary old strand. The gap left in this latter strand by the recombination process presumably was filled in by repair synthesis using the other new strand as the template. For further review of this process in bacteria, see Howard-Flanders and Rupp (1972).

In mammalian cells, gaps also appear to be left in the newly synthesized strands, and these gaps eventually are closed (see A. R. Lehman, 1974, for review). The usual assumption has been that these gaps are opposite dimers, but some evidence that they may be in some other position has been obtained (Meneghini and Hanawalt, 1976). All attempts to demonstrate sister-strand exchanges during closure have failed. The evidence suggests that the gaps are filled by some form of repair synthesis rather than by exchange. There still are a number of uncertainties about this process, however. One possibility is that there is relatively slow repair synthesis, possibly involving random insertion of bases, to close the gaps directly. There also is some evidence that pyrimidine dimers made by uv in the old strands of the DNA can appear in new strands in mammalian cells (Buhl and Regan, 1973; Meneghini and Hanawalt, 1976) as well as in *E. coli* (Ganesan, 1974). This might happen if recombination repair occurred and involved fairly long recombinant regions that might include several dimers. The result, in this case, would be some randomization of dimers between old and new strands. Thus, despite the negative evidence, recombination repair cannot be considered totally excluded for mammalian cells, at least as a minor contributor.

Neither excision repair nor postreplication repair is confined to pyrimidine dimers. Such processes have been identified after treatment with a variety of agents. For *E. coli,* different endonucleases have been identified with specificity for different kinds of lesions (see Cerutti, 1974, for review). The uv-endonuclease can act on several kinds of lesions, such as cross-links, in addition to dimers. Probably they are lesions that cause an appreciable distortion of the DNA double helix. Another enzyme, endonuclease II, recognizes some of the alterations produced in DNA by alkylating agents and possibly other lesions that produce no more than minor conformational change. Evidence for other endonucleases with still different specificities exists (see Chapters 24 to 26 in Hanawalt and Setlow, 1975).

E. Single-Strand Breaks and Alkali-Labile Sites

A set of lesions that has been much studied includes those that act as strand breaks in alkaline sucrose gradient centrifugation. The basic procedure is simple.

Prelabeled DNA is placed on top of the gradient, often by direct lysis of the cells on the gradient (McGrath and Williams, 1966), and spun, and the average molecular weight is determined from the gradient profile (Figure 2). From this average, the number of breaks per strand is inferred. The method does not distinguish between true breaks present in the DNA before it was put on the gradient and alkali-labile sites that were converted to breaks on the gradient.

The conclusions about molecular weight are fairly straightforward, provided the DNA is not too large. This condition is satisfied relatively easily with viral and bacterial DNA's. It is much more difficult to satisfy with the much larger DNA of mammalian cells, and considerable discussion has been generated about the interpretation of such measurements in this case [see, for example, Cleaver (1974) and several of the papers and accompanying discussions in a symposium, "Molecular and Cellular Repair Processes," edited by Beers *et al.* (1972)].

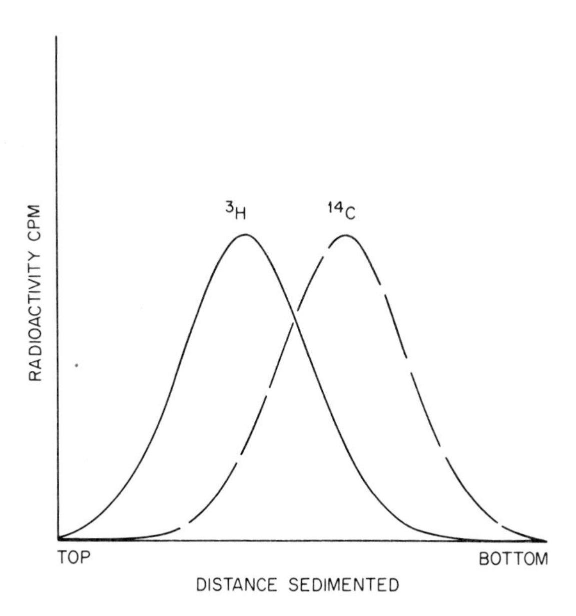

Fig. 2. A diagrammatic example of alkaline sucrose gradient profiles for two DNA's prelabeled respectively with [^{3}H]- and [^{14}C]thymidine. The ^{3}H-labeled DNA was subjected to some procedure, exposure to ionizing radiation, for example, that makes single-strand breaks. Thus, it sediments more slowly on the average than the untreated ^{14}C-labeled DNA. The average molecular weight can be calculated by appropriate weighted averaging over the gradient profile, and the number of strand breaks can be estimated from the ratio of the treated and control molecular weights. Double labeling, as in this example, is useful but not essential since the molecular weight can be determined from the distance sedimented, provided the gradient procedure is kept constant.

Despite such difficulties, a number of studies have been carried out on both bacterial and mammalian cells. In general, the breaks, whatever their origin, tend to disappear as the cells are incubated, showing that such lesions can be repaired. The breaks seen after treatment with a number of agents, including uv, probably are the result of uncompleted excision repair, and their disappearance simply means the completion of this repair. The breaks seen after ionizing radiation may well be produced more directly by ionization or radical formation, and their disappearance represents some form of resealing with or without the insertion of a few bases. It is unlikely that most of these breaks are the result of simple cleavage of the phosphate ester linkages so as to leave the 3′-OH and 5′-PO_4 ends needed for direct ligase action. Thus repair usually must involve at least some enzymatic modification of the initial end groups and quite probably some excision and reinsertion of bases.

Alkylating agents probably produce a number of alkali-labile sites which appear as strand breaks on the gradients. However, there is evidence from lambda phage (Boyce and Farley, 1968) that endonucleolytic cleavage also can occur, and this is confirmed by the identification of enzymatic activities in *E. coli* that can act on apurinic sites, formed as a result of destabilization of the glycosidic linkage, and several different alkylated bases from DNA treated with methyl nitrosourea (see Cerutti, 1974, for review). Breaks also might be produced by destabilization of the phosphate–sugar backbone as a result of direct attack on the phosphate diesters (see Lawley, 1966, for review). The important point for our present purpose is that there are several ways in which single-strand breaks, detectable on alkaline sucrose gradients, can be produced by alkylating agents, but most if not all of these ways appear to produce lesions that are eventually reparable (see, for example, Reiter *et al.*, 1967; Prakash and Strauss, 1970; Kimball *et al.*, 1971b).

III. REPAIR MUTANTS

A. Introduction

The previous section was concerned primarily with the biophysical and biochemical methods by which various repair processes can be identified and measured. Most of the work was carried out with cells without known defects in repair processes. Further dissection of these repair processes and identification of the enzymes involved have depended heavily upon the availability of mutants defective in specific steps in repair. *Escherichia coli* is by far the best-studied species from this point of view, and most of this section will be devoted to it. Repair mutants also are becoming available in a number of prokaryotic and eukaryotic species, including humans. Some attention will be given to the work on humans, but the work with other species will be referred to very briefly, since most of it parallels that in *E. coli*. There are some differences that suggest some evolution of the repair system but, on the whole, the system seems to have been quite conservative in evolution.

Table I lists those mutants of *E. coli* that will be referred to in this chapter. More complete lists with references are given by Clark (1973) and Clark and Ganesan (1975). Grossman *et al.* (1975), Hanawalt and Setlow (1975), and Swenson (1975) can be consulted for further information.

B. Photoreactivation

Mutants of the *phr* locus in *E. coli* cannot be photoreactivated, and presumably this locus is the structural gene for the photomonomerizing

TABLE I

List of Some of the Major Genes in *Escherichia coli* Involved in Repair and Recombination[a]

Functional group	Gene	Function
Excision-repair mutants	*uvrA*	Structural gene for uv-endonuclease, required for initial incision
	uvrB	Same as *uvrA*
	uvrC	Controls some later step in excision repair
Recombination-defective mutants	*recA*	Multiple functions in recombination and repair
	recB	Structural gene for ATP-dependent exonuclease V, involved in both recombination and repair
	recC	Same as *recB*
	recF	Required for alternative pathway in recombination and probably involved in postreplication repair
	sbcB	Structural gene for exonuclease I, suppressor of *recB* and *recC*, mutant allows *recF* recombination pathway to function
Polymerase	*polA*	Structural gene for DNA polymerase I, probably main polymerase for excision repair
	polB	Structural gene for DNA polymerase II
	polC (dnaE)	Structural gene for DNA polymerase III
Others	*exrA (lex)*	Required for uv-induced mutation and for induction of SOS repair
	lig	Structural gene for DNA ligase
	lon	Radiation-induced filament formation
	mfd	Function required for mutation frequency decline
	tif	Thermal induction of phage and of filament formation, SOS repair constitutive at higher temperature
	phr	Structural gene for photoreactivating enzyme

[a] A more complete list, including reference, is given by Clark and Ganesan (1975).

enzyme which monomerizes pyrimidine dimers. This enzyme seems to be nearly universal in both prokaryotes and eukaryotes. Until recently, it was generally accepted that it was missing from placental mammals. However, Sutherland *et al.*(1973) have presented evidence for its occurrence in at least some human cells. Thus, a structural gene for this enzyme exists in a wide variety of organisms, although it may be missing from some.

C. Excision Repair

1. *Escherichia coli*

Both the *uvrA* and *uvrB* mutants (Table I) lack the uv-endonuclease required to make the initial incision at pyrimidine dimers. The respective roles of these two loci in the formation of this enzyme are not yet clear. A mutant at the *uvrC* locus seems to be involved in some later step in excision. One possible explanation is that this locus is involved in an early step that is required to prevent resealing by DNA ligase of the initial incision before any excision or resynthesis can start (see Grossman *et al.*, 1975, for review).

The *polA* locus is the structural gene for DNA polymerase I. It is generally accepted that this is the enzyme that normally is responsible for both excision and resynthesis and that it produces relatively short repair segments (see Grossman *et al.*, 1975). The *polA1* mutant, which is defective in the polymerizing activity, carries out excision repair, but more slowly than normal. The length of the resynthesized region is considerably greater in the absence of this enzymatic activity, and it therefore seems probable that a long-patch type of repair is carried out by DNA polymerase II, DNA polymerase III, or both. These various modes of excision repair are being investigated actively in bacteria treated with toluene to make them permeable to larger, charged molecules. In this way, it is possible to experimentally control the concentration of the cofactors needed by certain repair and replication enzymes and also the availability of the immediate precursors of DNA while leaving more or less intact the structural and enzymatic components needed for replication and repair. Much of this work is reviewed by Grossman *et al.* (1975), and recent papers by Masker and Hanawalt (1974), Waldstein *et al.* (1974), Sharon *et al.* (1975), and Billen and Hellermann (1975) also can be consulted. The general conclusions are that there is more than one form of excision repair.

A locus, *mfd,* whose activity is required for a phenomenon known as mutation frequency decline (see Section IV), is slow to excise pyrimidine dimers (George and Witkin, 1974). The exact role of this locus in excision repair is not established.

The final step in excision repair is the sealing of the newly synthesized section into the old strand. Only a single enzyme, DNA ligase, specified by the structural gene *lig,* appears to be involved. This is an essential enzyme for cell survival, but there exists a viable temperature-sensitive mutant, *lig ts 7,* that has a low but detectable ligase activity at the permissive temperature. It is sensitive to killing by uv and methyl methane sulfonate, presumably because the ligase activity is low (see I. R. Lehman, 1974, for review).

2. *Humans*

The inherited condition, xeroderma pigmentosum, causes high susceptibility to sunlight, with eventual development of skin carcinomas. The cells from many, but not all, such individuals are defective in excision repair. However, several different complementation groups seem to be involved, with different levels of effect on pyrimidine dimer excision (de-Weerd-Kastelen *et al.*, 1972; Cleaver, 1974; Cleaver *et al.*, 1975; Cleaver and Bootsma, 1975). Some even appear normal in dimer excision, but there is some disagreement about whether they are defective in postreplication repair (Buhl *et al.*, 1973; A. R. Lehman *et al.*, 1975). The roles of the several complementation groups that can be responsible for this condition have not been elucidated in any detail.

Evidence has been presented for defective DNA repair in several other inherited conditions in humans, and also in late-passage cultured cells and in nongrowing cells. However, none of these cases is yet well established, and there is conflicting evidence for some. Setlow (1975) can be consulted for a brief critical review of this evidence.

3. *Other Organisms*

Mutants defective in excision repair now have been identified in some viruses and in quite a number of bacteria. No attempt will be made to list all these instances, but it seems clear that the situation is very much the same as in *E. coli,* although there are probably differences in detail. Ultraviolet-sensitive mutants also have been obtained in a number of eukaryotic species, and some of these are probably defective in excision repair. The most extensive set of such mutants in a eukaryote has been obtained for *Saccharomyces cereviseae* (see, for example, Game and Cox, 1971; Resnick and Setlow, 1972; Lemontt, 1973; Lawrence *et al.*, 1974).

D. Postreplication Repair

As stated in the previous section, much if not all postreplication repair in *E. coli* involves recombination between sister strands. It is not surprising, therefore, that mutants affecting recombination are frequently uv-sensitive. Clark (1973) has given a rather complete review of the loci in *E. coli* involved in recombination and also has presented information about their uv sensitivity. The important ones for this chapter are listed in Table I.

The wild-type allele of one of these loci, *recA,* is required for any appreciable amount of recombination to take place, is very uv-sensitive, and is required for several uv-inducible processes such as filament formation (elongation without division), phage induction, and mutation induction. It also is needed to prevent extensive degradation of the DNA by the recBC exonuclease (exonuclease V) after uv irradiation. Thus, the mutant lacking the *recA* function is very uv-sensitive. The wild-type *recA* function is essential for recombinational postreplication repair (Howard-Flanders and Rupp, 1972). It also seems to be essential for the so-called "SOS" repair that will be discussed in the following section. The only repair processes for which this locus does not seem to be essential are photoreactivation and DNA polymerase I-mediated excision repair. At present, however, no enzymatic activity has been associated directly with this locus, and the precise role or roles that it plays in these various processes is uncertain.

Horii and Clark (1973) have developed a scheme for recombination in *E. coli* that involves two alternative pathways, one controlled jointly by the *recB* and *recC* loci and the other controlled by the *recF* locus. The *recB* and *recC* loci are jointly responsible for the production of exonuclease V, but the *recF* gene product has not been identified. Ordinarily, the *recBC* pathway seems to be the major one for recombination, but under some conditions the *recF* pathway can take over. Both pathways also are involved in repair, since mutants in any of these loci are more sensitive to killing by uv than wild type. When both pathways are blocked simultaneously, the sensitivity to radiation is much higher than when either one alone is blocked and only slightly less than that of *recA* mutants. Thus, the two pathways seem to be independently required for repair since blocking both has a greater effect than blocking only one. There is some indication that there may be still a third pathway, since the *recA* mutant is somewhat more sensitive than the doubly blocked mutant. Rothman *et al.* (1975) can be consulted for a more complete discussion of the role of the various *rec* mutants in repair.

Mutants defective in recombination also are defective in postreplication repair in bacterial systems other than *E. coli* (e.g., LeClerc and Setlow, 1972), and it seems probable that this will be a general finding for prokaryotic systems. As discussed in the previous section, postreplication repair in mammals cannot be shown to involve recombination, but since no mutants defective in recombination have been identified, a test of this finding by mutation at present is not possible.

E. "SOS" Repair

Evidence has accumulated that, in addition to the repair processes considered above, all of which appear to be constitutive, there is also a uv-inducible process or processes that Radman (1974) has named "SOS" repair to indicate that it is a system for coping with the emergency produced by large amounts of uv damage. The evidence for such a system is somewhat inferential since neither the enzymes nor the molecular events involved have been identified. The system is induced not only by uv but also by ionizing radiation and by several chemicals. It greatly improves the survival from the lethal damage caused by these agents but is error-prone. The constitutive repair systems for these kinds of damage are less effective in eliminating the lethal effects but appear to be nearly error-free. Thus the SOS system is essential or nearly so for mutation induction by such agents, even though it helps the cells survive. Therefore, it is of particular importance to this discussion. Several loci have been identified that seem to be involved in this form of repair.

Witkin (1967, 1971) demonstrated that the wild-type alleles of the *exr* and *recA* loci were required for mutation induction by uv and advanced the concept of an error-prone repair system. Strains with the mutant alleles are essentially immutable by uv and some other agents. The *lex* locus of the K12 strains probably is identical to the *exr* locus of the B strains. It was found that the wild-type alleles of *exr* and *recA* also were required for certain other uv effects, such as prophage induction, filament formation, reactivation of irradiated phage by an irradiated host (Weigle reactivation), and production of mutants in uv-irradiated phage in an irradiated host. The idea that all these phenomena were due to a uv-induced, error-prone repair system dependent on the wild-type functions of the *recA* and *exr* loci was developed further by Defais *et al.* (1971), Witkin and George (1973), and Radman (1974).

A temperature-sensitive mutant, *tif,* was found (Castellazzi *et al.*, 1972) that mimicked a number of these uv effects when grown at a high temperature. The name comes from the fact that at high temperatures it mimics two effects of uv, namely, induction of prophage and filamentation. Witkin (1974) showed that it increased the induced bacterial mutation frequency at low doses of uv. The interpretation was made that in this mutant at the higher temperature, the SOS system is induced without exposure to uv. Therefore, mutations are produced by the temperature-induced SOS repair system at doses below those needed for full induction by uv.

The original hypothesis was that SOS repair was a form of postreplication repair. However, some of the mutation data to be discussed in Section IV suggest that it may be involved in both excision and postreplication repair. Despite the requirement for the *recA* function, it is unclear whether this form of repair involves recombination, and Radman (1974) has suggested that it may involve gap-filling by some form of repair synthesis.

F. Repair of Single-Strand Breaks and Alkali-Labile Sites

The repair of X-ray-induced strand breaks and alkali-labile sites in prokaryotes, primarily *E. coli,* has been reviewed by Town *et al.* (1973). Estimates of the fraction of apparent strand breaks that really are alkali-labile sites range from 20 to 30%. They distinguish three main types of repair, which they call types I, II, and III. Type I repair apparently is very fast, going to completion in less than 1 min at 0°C. Preliminary evidence suggests that it may involve direct resealing by DNA ligase. It preferentially joins breaks formed under anoxic conditions. The inference is that the breaks produced in the absence of oxygen are more likely to have the 3′-OH and 5′-PO_4 ends needed for ligase action than those produced by radical reactions involving oxygen. About 90% of the remaining breaks, whether produced in the presence or absence of oxygen, also are closed rapidly, 1–2 min at room temperature. This closure requires DNA polymerase I and DNA ligase. Type III repair requires 30–60 min incubation in growth medium, and is absent in *exr* and *recA* mutants. It apparently repairs a small, fixed number of breaks, approximately two per strand, independently of whether oxygen was present during irradiation.

In mammalian cells, several attempts to find type I repair have been unsuccessful (Roots and Smith, 1974; Modig *et al.,* 1974; Palcic and Skarggard, 1975). It would appear that the oxygen effect in mammalian cells is due to a fast chemical reaction (Roots and Smith, 1974), although there also may be some effect on postirradiation repair processes (Modig *et al.,* 1974).

Alkali-labile sites, and possibly single-strand breaks, also are produced by monofunctional alkylating agents. Whatever the nature of the lesions, they disappear with time after treatment. However, considerable breakdown of the DNA also occurs (Kimball *et al.,* 1971b). Prakash and Strauss (1970) report that a *rec* mutant of *Bacillus subtilis* does not carry out repair of this kind of damage as well as does wild type. Howell-Saxton *et al.* (1974) found some repair in the *E. coli recA* and *polA* mutants, but greater breakdown than in wild type. Kimball *et al.* (1971b) found no difference between the wild type and the *recl* mutant of *Haemophilus influenzae* in the ability to perform this kind of repair—despite the fact that this mutant was much more sensitive to killing. The interpretation of the alkaline sucrose gradient results is complicated by the simultaneous repair of some of the DNA and breakdown of some of the rest. In a number of organisms, recombination-defective mutants are more sensitive than wild type to killing by monofunctional alkylating agents, but the relationship between this and the disappearance of detectable lesions in the DNA remains uncertain.

IV. REPAIR AND MUTATION

A. Introduction

In this section, the relationship between repair and mutation induction will be discussed. The view that mutations result from errors in repair is an old one, since a class of mutations (i.e., chromosomal aberrations) long has been thought to result from reunion of broken chromosomes in abnormal arrangements. This hypothesis was developed by Sax (1938, 1941) for *Tradescantia,* and has its origins in the very early days of radiation genetics (see Hollaender, 1954, for a review of early work). More recently, the hypothesis that mutations arise as a result of errors in repair has been developed in a new form by Witkin (1967) for base-pair alterations, especially for *E. coli,* and in this form rapidly has become an important part of the theory of mutation induction.

A rather different hypothesis for the origin of base-pair alterations was proposed by Watson and Crick (1953) as a probable consequence of their double-helix model of DNA. It was that altered bases, in the initial formulation tautomeric forms, could cause the wrong base to be put in during replication. (See also Chapter 1, Volume 1, of this Treatise.) This was the major model for base-pair alterations for a decade or more and was developed in some detail by Freese (e.g., Freese, 1959). Recent thinking has tended to deemphasize the importance of errors in replication relative to errors in repair, at least for mutation induction by uv, ionizing radiation, and some chemicals. Nevertheless, it seems certain that some mutation results from errors in replication, and that mutations arising from errors in repair often do so in association with postreplication repair.

In this section, the error-prone system will be discussed first, followed by a discussion of mutation fixation that presents evidence for errors both in repair and in replication. Finally, a relatively brief discussion will consider some recent thinking about the role of repair in the induction of chromosomal aberrations.

Mutator genes and mutable genes are phenomena of considerable interest from the point of view of the fidelity of repair and replication. However, in the author's view at least, very little has been done so far on the relation to repair processes. Therefore, no attempt has been made to include these phenomena in this section. A fairly recent symposium (Drake, 1973) is a good source of information about them.

B. Error-Prone Repair

Table II shows the uv mutability of several different repair mutants in *E. coli.* Mutants (*uvrA, uvrB*) defective in excision repair but wild type in

TABLE II

Effect of Some *Escherichia coli* Genes on Mutation Induction[a]

Mutant gene	Effect on mutation induction
uvrA, B	Yield of uv-induced mutations considerably increased
recA	No mutations induced by uv, ionizing radiation, or some chemicals
recB, C	Mutations induced by uv
polA	Little or no effect on induced mutation despite sensitivity to killing by uv
exrA, lex	No mutations induced by uv, ionizing radiation, and some chemicals
mfd	Mutation frequency decline during liquid holding does not occur in mutant
tif	Higher than normal mutation induction at low uv doses at high temperature, presumably because SOS repair is constitutive at such temperatures
phr	Ultraviolet-induced mutation not altered by photoreactivating light in mutant

[a] References to action on DNA can be found in Clark and Ganesan (1975).

respect to the *exr* loci are more mutable by uv than wild type (Witkin, 1967), suggesting that pyrimidine dimers are more likely to cause mutations if they remain in the DNA until replication than if they are excised. However, uv induces no mutations in mutants at the *exr* locus in bacteria regardless of whether the strain is excision proficient or deficient (Witkin, 1967). This finding led Witkin (1967) to propose that mutations are produced by a repair system that, although efficient in eliminating lethal damage, operates at the expense of occasional errors leading to mutation. Since uvr^+ *exr* bacteria excise dimers but produce no mutations, she assumed that excision repair was error free, and developed the idea that the error-prone repair system might be recombinational postreplication repair (Witkin, 1969b). Presumably, in *exr* strains, potentially lethal damage was repaired less efficiently than in wild type, but by an error-free process. More recent work has suggested that it is not the main recombinational repair system that makes errors but a special repair system, SOS repair, that is induced by uv (see Section III,D).

Later work (Witkin, 1971) showed that *recA* strains also are immutable, as is the *lex* mutant (Bridges *et al.*, 1968), which seems to be the homologue in K12 strains of the *exr* mutants in the B strains. Mutants of the *recB* and *recC* loci are mutable by uv (Witkin, 1971). Thus, a functioning *recBC* system is not required for error-prone repair. The role, if any, of the *recF* and *sbcB* loci (Section III,C) in mutation induction has not been reported. Thus, it is not clear whether the *recF* recombination and postreplication repair pathway has any relation to error-prone repair.

The *tif* mutant appears to have the error-prone SOS repair system at elevated temperatures without induction by uv (see Section III,D). Witkin (1974) has shown that, at these elevated temperatures, there are more uv-induced mutations at low doses in the *tif* mutant than in wild type. This is interpreted to mean that uv-induction of the repair system is not necessary for mutation production in this strain since, prior to uv, the system is induced by the elevated temperature. Two mutants, *mfd* and *polA1,* which are slow to complete excision, probably for different reasons, appear to induce more error-prone repair at low doses of uv. This suggests that the induction may depend on the rate of excision repair (George and Witkin, 1974, 1975; Witkin and George, 1973).

Confirmation that the error-prone repair system is inducible and can involve postreplication repair comes from the work of Sedgwick (1975) with *E. coli*. He showed that addition of the protein synthesis inhibitor, chloramphenicol, before uv-irradiation prevented a small fraction of postreplication repair and completely eliminated mutation fixation. Chloramphenicol had no effect on postreplication repair in *uvrA exrA* bacteria nor in *uvrA tif* bacteria at 42°C.

It has become increasingly likely that the error-prone pathway does not involve recombination at all. Thus, *exr* and *lex* mutants are no more than slightly deficient in recombination, and the *recA* mutant, although completely deficient in recombination, has so many different effects that there is no assurance that the recombination defect has any bearing on its role in SOS repair. Radman (1974) has suggested that some form of error-prone gap filling might be involved. Moreover, there is fairly strong evidence that *exr*-dependent error-prone repair may occur during the excision repair as well as during postreplication repair. Nishioka and Doudney (1969, 1970) showed that photoreversibility of mutations to streptomycin resistance, as well as of true reversions at the *trp* locus, is lost prior to replication, whereas photoreversibility of mutations to suppressors of *trp* is not lost until replication. If photoreversibility can be taken as an indication of mutation fixation, then fixation in the one case must occur before replication, although in the other it occurred at replication. Since no mutations of any kind are induced in *exr* strains, the *exr*-controlled process must be involved in both kinds of fixation. On the basis of this and other evidence, Witkin and George (1973) have suggested that in excision-proficient strains all dimers are excised from the *str* and *trp* loci prior to replication, and that therefore a fraction of the excision repair events must involve error-prone processes. They postulate that dimer excision is slow in the suppressor (tRNA) loci, possibly because of interference between excision and frequent transcription, and thus dimers remain in these loci to cause postreplication repair. In excisionless strains, mutations at all loci are assumed to arise by error-prone postreplication repair.

A variety of mutagens, but not all, act primarily or entirely through the error-prone *exr*- and *recA*-dependent pathway. Mutagens that act like uv in inducing little if any mutation in *exr* and *recA* strains are gamma radiation and thymine starvation (Bridges *et al.*, 1968), 4-nitroquinoline oxide, X rays, methyl methane sulfonate, and mitomycin C (Kondo *et al.*, 1970; Ishii and Kondo, 1975). Mutagens that induce appreciable numbers of mutations in such strains, although not always as many as in wild type, are nitrosoguanidine, hydroxylamine, and ethyl methane sulfonate (Witkin, 1967; Kondo *et al.*, 1970; Ishii and Kondo, 1975). In this respect, multilocus deletions behave in the same way as base-pair changes (Ishii and Kondo, 1975). Thus, the error-prone repair system controlled by *exr* and *recA* seems capable of causing both base changes within a locus and deletions extending over several loci.

The pathways by which agents that produce mutations in *exr* and *recA* strains act are not clearly established. In some instances, simple base mispairing at replication, resulting from base alterations produced by the agent, may be the mechanism. Kondo and Ichikawa (1973) have suggested that damage to the replicating system might be involved. It is also possible that there are error-prone repair systems other than those dependent on *exr* and *recA*, and some evidence for this will be presented below.

For organisms other than *E. coli*, there is very little information on error-prone repair as a source of mutations, other than that which results in chromosomal aberrations. However, most organisms, including mammalian cells in culture, are mutable by uv and ionizing radiation. Thus, it seems probable that most organisms have a similar error-prone repair system, but whether it is usually constitutive or must be induced by the mutagenic agent is unknown.

A few laboratory "wild-type" strains of bacteria seem to be immutable by uv and presumably lack the error-prone repair system, or at least lack one that can be induced by uv. One such case is the wild-type strain Rd of *Haemophilus influenzae* (R. F. Kimball and J. K. Setlow, unpublished). There is evidence, however, that an error-prone postreplication repair system is involved in mutation induction by nitrosocarbaryl in this strain (Beattie, 1975). More details of the evidence will be given later, but such cases suggest that more than one kind of error-prone repair may exist with specificities for different kinds of initial damage.

A possibility raised by this bacterial work is that different cell types in multicellular organisms might differ with respect to the presence or inducibility of error-prone repair. One might elect to explain in this way the inability of X rays to induce mutations in earlier oocyte stages in the mouse (Russell, 1967). A number of other explanations also might be offered, but this bears upon the importance of considering error-prone repair in discussing the differences between cell systems in response to mutagens.

C. Mutation Fixation

Another approach to determining the relationship between mutation and repair and replication is to determine the time of "fixation" or "stabilization" of mutation relative to cellular events. What we would like to know is the cellular event that converts a premutational lesion to a self-replicating base or base sequence change. Unfortunately, no method to make this determination directly is available, and thus inferences must be made from somewhat indirect evidence. "Fixation" as determined from such evidence might involve the final base or base sequence change itself, or it might involve some intermediate step that makes the mutation appear fixed as far as the test is concerned.

Photoreversibility has been used a number of times to infer the time of fixation of uv-induced mutations. The rationale is that the photoreactivation enzyme is specific for monomerizing pyrimidine dimers, and that photoreversibility of mutation will be lost only when the dimers have been converted to some form that is either the final base change itself or some intermediate form that no longer depends on the presence of a dimer to be converted to final mutation. On this basis, as mentioned above, Nishoioka and Doudney (1969, 1970) concluded that mutations to streptomycin resistance and true reversions at the *trp* locus in *E. coli* were fixed prior to replication, but that mutations to suppressors of the *trp* mutant were fixed at replication. The evidence in this case is quite convincing, but in *Paramecium aurelia* photoreversibility appears to be lost by some mechanism other than the loss of dimers (Kimball, 1969). Thus, it is not clear that this test always will provide clear evidence on the time of fixation.

A number of different agents and conditions, e.g., metabolic inhibitors or nonnutrient conditions, can influence the mutation yield produced by a given dose of mutagen. By giving the treatment at various times after exposure to the mutagen, it can be determined when the treatment ceases to be effective. The simplest hypothesis is that this is the time at which premutational damage has been converted to some irreversible form. However, the mode of action of such treatments is much less specific and much less well understood than is photoreactivation. Thus, it is difficult to ascertain that the effect on mutation yield is not some secondary consequence of the treatment on events that occur some time after the treatment has ceased. Consequently, inferences about the time of fixation made in this way are subject to considerable uncertainty. Earlier work of this sort has been reviewed by Kimball (1966) and Witkin (1969a).

Another procedure for drawing inferences about fixation is to apply the mutagens at various times relative to DNA replication. This method was used with synchronized cells of *Paramecium aurelia* to infer that mutations produced by several different agents (X rays, uv, triethylene-melamine, MNNG) were fixed either solely or at least most efficiently at replication. The basic observation was that the yield of mutations was greatest when the cells were treated in late G_1, appreciably less when the cells were treated earlier in G_1, and very small or undetectable when the cells were treated in G_2. The interpretation was that premutational damage could be repaired until replication, at which time any unrepaired damage had an appreciable chance of being converted to final mutation (see Kimball, 1966, for review). Of course, the latter conversion could be either by direct errors in replication or by errors in postreplication repair.

Another kind of evidence that also points to fixation at replication was that obtained for synchronized *E. coli* treated with *N*-methyl-*N'*-nitro-*N*-nitrosoguanidine (MNNG) (Cerdá-Olmeda *et al.*, 1968). The observation was that mutations at any one locus occurred with maximum frequency in synchronized cultures treated with the mutagen when the locus in question was near the replication fork. It was suggested at the time that the mutagen might act differentially on the special configuration of DNA at the fork. The evidence from *P. aurelia* (Kimball, 1970) that treatment of cells in early G_1 with MNNG induces appreciable numbers of mutations, although fewer than are produced by treating near the G_1/S boundary, appears to be more compatible with an interpretation of both sets of data in terms of repair before replication and fixation at replication. In this case, lesions induced near the replication fork would be more likely to persist to replication than those some distance away. Evidence from *H. influenzae* that fixation of MNNG-induced mutations continues to occur for a large fraction of a cell generation also seems to fit such a hypothesis (Kimball and Setlow, 1974). Evidence also exists in this species that some mutations produced by a related compound, nitrosocarbaryl, are fixed prior to replication (Beattie and Kimball, 1974). Recent evidence by Jiménez-Sánchez and Cerdá-Olmeda (1975) that mutations are induced by MNNG mainly or entirely in *E. coli* in which the progression of the replication fork has been temporarily blocked is more difficult to fit to a hypothesis requiring fixation to be associated at least in part with replication.

The procedure for studying mutation fixation in the bacterium *H. in-*

fluenzae makes use of the transformation system. If the bacteria are treated with MNNG, nitrocarbaryl, or hydrazine, lysed immediately after washing out the mutagen, and the lysates used to transform untreated bacteria, few if any mutations are found in the transformants. However, if the treated bacteria are incubated for approximately one cell generation after washing out the mutagen and then lysed, mutants are found in the transformants at about the frequency expected from the frequency determined in the treated bacteria themselves. For some unknown reason, premutational lesions induced by these mutagens seem to be unable to pass through the transformation process to produce mutations. Only when the lesions are converted to final mutation, or perhaps to some relatively stable intermediate form, can the newly induced mutations be transformed (Kimball and Setlow, 1974; Beattie and Kimball, 1974; Kimball and Hirsch, 1978). By the use of a temperature-sensitive DNA-replication mutant to control the time of replication, it was shown that about two-thirds of the mutations induced by nitrosocarbaryl are fixed at replication and about one-third are fixed in the absence of replication (Beattie and Kimball, 1974). The former class seems to be fixed by an error-prone postreplication repair process requiring the functions controlled by the *rec1* locus; the latter possibly by an error-prone prereplication repair process that involves neither the pyrimidine dimer repair system nor the *rec1* function (Beattie, 1975). (Recent evidence, however, suggests only fixation by replicative error.) Hydrazine-induced mutations appear to be fixed only at replication and probably by base mispairing rather than by error-prone repair (Kimball and Hirsch, 1976).

It seems probable, then, that mutations can arise by error-prone repair either prior to replication (some uv-induced mutations in *E. coli,* some nitrosocarbaryl-induced mutations in *H. influenzae*) or at replication (other uv-induced mutations in *E. coli* and nitrosocarbaryl-induced mutations in *H. influenzae*) or through simple mispairing at replication (hydrazine-induced mutations in *H. influenzae*). Kondo (1973) has suggested classifying DNA alterations into those capable of pairing (and so presumably of mispairing) at replication and those unable to pair at all, thus leaving gaps to be repaired in the newly formed DNA. This scheme takes into account only mutation fixation at replication. It would have to be amended or expanded to take into account mutation fixation that occurs prior to replication—probably as a result of errors in a prereplication repair process.

D. Chromosomal Aberrations and Repair

As mentioned in the introduction to this section, it long has been held that radiation-induced chromosomal aberrations in both plant and animal cells result from breakage followed either by reunion in altered arrangements, i.e., by errors in repair, or by a failure of reunion, i.e., by failure of repair. Reviews of the earlier work may be found in Hollaender (1954). An alternative hypothesis is that all radiation-induced aberrations are the result of exchanges. Revell (1974) can be consulted for a review of this breakage-and-reunion versus exchange controversy. Recent evidence by Wolff and Bodycote (1975), who used a new method of labeling chromatids to look for exchanges, seems to support the original breakage-and-reunion hypothesis for aberrations induced by ionizing radiation. However, a clear decision between these two hypotheses has not been made as yet.

There now seems to be fairly general agreement that the eukaryotic chromosome is a single DNA double helix and that chromosomal aberrations arise directly or indirectly by breakages and exchanges involving this helix. Earlier evidence (e.g., Brewen and Peacock, 1969), namely, that the polarity of the strands involved in aberration production is that expected from such a model, strongly suggested DNA involvement. However, the most convincing evidence that lesions in the DNA are responsible for chromosomal aberrations comes from studies with photoreactivation and with repair-defective mutants. Griggs and Bender (1973) demonstrated that photoreactivation of *Xenopus laevis* cells largely prevented aberration production by uv. Parrington *et al.* (1971) showed that uv induced more chromosomal aberrations in excision-defective (XP) human cells than in normal cells (but see Cleaver, 1974, for some difficulties with this evidence); Sasaki (1973) showed the same result for G_1 cells treated with 4-nitroquinoline oxide, a chemical that is known to induce excisable damage. The possibility remains, nevertheless, that some damage to proteins also is involved in aberration production (see discussion by Cleaver, 1974).

Bender *et al.* (1973a,b, 1974a) have suggested a model for production of chromosomal aberrations by uv, ionizing radiation, and chemicals that requires a single-strand nuclease to convert single-strand nicks or gaps to double-strand breaks and, for some aberrations, postreplication recombination repair. [Although this latter process has not been found in mammals (see Section II,D), it is possible that it exists as a minor part of postreplication repair.] This undoubtedly is the most completely formulated model now available which employs some of the known enzymatic processes for repairing DNA. It appears to fit closely to the data on the frequency of induction of various aberration classes at various stages of the cell cycle by different physical and chemical mutagens. However, critical testing of its assumptions will require mutants defective in some of the processes postulated, and such tests have not yet been done. Indeed, this may prove very difficult, if processes such as recombinational postreplication repair, which are probably at best only minor processes in mammalian cells, must be invoked.

One class of chromosomal alterations, sister chromatid exchanges, may differ in their mode of formation from other types of induced chromosome alterations. These exchanges can be detected by labeling the two sister chromatids differentially either with a radioisotope (e.g., Brewen and Peacock, 1969) or with bromodeoxyuridine (e.g., Wolff *et al.*, 1975). Sister chromatid exchanges can be seen as reciprocal exchanges of labeled and unlabeled material between the two daughter chromatids. On the basis of theoretical considerations, Bender *et al.* (1974b) suggested that such exchanges could arise from recombinational postreplication repair; Wolff *et al.* (1974) presented evidence that the induction of sister-strand exchanges by uv, as observed by Kato (1973), did require an intervening S period to be formed. Wolff *et al.* (1974), however, doubt that they were formed by postreplication recombination repair. The problem with invoking recombinational postreplication repair in mammalian cells is that it has not been demonstrated (see Section II,D), although this does not exclude some repair of this kind as a minor component. Recently, Wolff *et al.* (1975) reported that uv induces about the same amount of sister-strand exchange in cells from XP patients, and in an XP variant defective in postreplication repair as in normal cells. Thus, neither the excision repair system nor the major postreplication repair system appears to have any detectable influence on the production of sister-strand exchanges. On the other hand, the present evidence suggests that a defective excision repair system causes an increase in chromatid-type aberrations (see earlier discussion). If this is so, then it must be concluded that sister-strand exchanges and chromatid-type aberrations are produced by different mechanisms, as far as repair processes are concerned.

Finally, it should be pointed out that small multilocus deletions in *E. coli* seem to be produced by the same error-prone repair system as single locus mutations (Ishii and Kondo, 1975). It is not known whether a similar error-prone repair system, if it exists, is responsible for producing small deletions in eukaryotes, but it is of some interest to note that it was early reported that small deletions induced by X rays in *Drosophila* showed one-hit induction kinetics, whereas larger aberrations appeared to involve a two-hit component as well as a one-hit component) (e.g., Muller, 1941). The usual explanation, that small deletions can be produced by a single ionizing particle while larger ones are much less likely to be, remains very probable. However, the work with bacteria suggests an alternative explanation that should be explored.

V. REPAIR AND CELL DEATH

A. General Background

There seems little question that a major, probably the major, function of DNA repair systems is elimination of potentially lethal damage from the DNA. Indeed, many repair mutants in bacteria were isolated initially on the basis of their sensitivity to killing by radiation or other damaging agents. However, the ways in which damage to DNA can lead to cell death still are understood rather poorly and in many instances may be quite indirect and complex. A brief review will be given of what, to this writer, seem to be some of the major points, but no attempt at a full review or synthesis will be made.

B. Bacteria

Swenson (1975) has reviewed the lethal effects of uv on bacteria in great detail, and his review can be consulted for the detailed information and hypotheses about the lethal effects of this agent. Three main points about the action of this and other agents will be made briefly here. (1) It is highly unlikely that induced lethal mutations ordinarily contribute appreciably to the killing of bacteria even by mutagens. (2) DNA repair systems have major effects on survival after treatment with quite diverse agents. (3) The ways in which unrepaired lesions in the DNA might cause death are numerous.

The evidence against lethal mutations as a major cause of bacterial cell death is that often there is little or no relation between the frequency of mutation and the probability of death. Thus, *E. coli exr* strains are very sensitive to killing but are rendered essentially immutable by uv. Exposure of *H. influenzae* to hydrazine produces a fairly high frequency of mutations with no detectable cell killing, whereas exposure to methyl methane sulfonate (MMS) produces much cell killing with no detectable mutation (Kimball and Hirsch, 1975). Lethal mutations surely must be induced in bacteria, but ordinarily their frequency is too low to have an effect on survival that can be measured by standard methods.

The major function of repair systems seems to be the protection of bacteria against the lethal effects of agents that damage DNA. Thus, excisionless *E. coli* survive much less well than wild type after exposure to uv. Nonetheless, they survive sufficiently well so that some bacteria with dimers in their DNA continue to divide (Bridges and Munson, 1968). Thus, pyrimidine dimers are not unconditionally lethal lesions, although they are more likely to be lethal if not excised. If postreplication repair as well as excision repair is absent (*uvr recA*), then probably even one dimer in the DNA is lethal (e.g., Howard–Flanders, 1968). Undoubtedly this is because the gap left by replication past a dimer must be closed if the cell is to survive.

It would appear probable, however, that the presence of dimers in DNA can lead to cell death in other ways than by causing gaps to be left in newly synthesized DNA. This is a very complex problem. Swenson (1975) has reviewed it in detail and cites twelve possible hypotheses for the lethal action of uv. It seems probable that effects on gene expression, including the induction of certain functions, are important routes through which cell death is brought about.

Some mutagenic agents, alkylating agents for example, would be expected to interact with other cellular components in addition to DNA, and some of these interactions might be cell lethal. Nonetheless, the fact that mutants defective in recombination are especially sensitive to killing by alkylating agents (e.g., Prakash and Strauss, 1970; Kimball *et al.,* 1971a; Kondo *et al.,* 1970) strongly suggests that even with these agents the major cause of lethality is damage to DNA.

The fact that *rec* mutants are sensitive to killing by a wide variety of chemical mutagens has been used as the basis for a rapid assay system (Kada *et al.,* 1972). In this test, the inhibition of growth of wild type and a *rec* mutant by a chemical is compared in a plate assay. Chemicals that inhibit the mutant more than wild type are classified as suspected mutagens.

C. Eukaryotes

The ways in which agents causing damage to DNA might kill a eukaryotic cell are probably even more numerous than the ways in which they might kill a prokaryotic one. A few facts stand out, however. Pyrimidine dimers must be a major cause of killing by uv since increased survival as a result of photoreactivation has been found with a variety of eukaryotic systems (e.g., Cook, 1970). Moreover, cells from humans with xeroderma pigmentosum that are defective in dimer excision are much more sensitive to killing by uv than are cells from normal individuals (Cleaver, 1970).

Cell death caused by ionizing radiation in mammalian and other eukaryotic cells often has been attributed to chromosomal imbalance due to losses of chromosomes or their parts as a result of chromosomal aberrations (e.g., Puck, 1960; Davies and Evans, 1966). There is little doubt that such losses can cause cell death, but there has been considerable discussion about their importance relative to other causes of death. This discussion is beyond the scope of this chapter. Elkind and Whitmore (1967) can be consulted for a review of the state of the subject some years ago. There also has been much evidence from various kinds of dose rate and dose fractionation experiments that some kind of repair of potentially lethal damage occurs, but it seems difficult at present to relate these findings with any certainty to DNA repair. Much of this work also is reviewed by Elkind and Whitmore (1967). Several articles in a symposium on "Molecular and Cellular Repair Processes" (Beers *et al.*, 1972) are also pertinent.

VI. CARCINOGENESIS

There is increasing evidence for a high correlation between the ability of a compound to induce mutations and its ability to induce cancer. Much of the recent evidence has come from extensive tests of known carcinogens and noncarcinogens for mutagenicity in the bacterium *Salmonella typhimurium* (Ames *et al.*, 1973; Ames, 1974). The agreement between the carcinogenicity and mutagenicity results is increased greatly if a liver microsome system is used to "activate" potential carcinogens for use in the *Salmonella* system. This results in the conversion of noncarcinogenic compounds to carcinogens and apparently also of nonmutagens to mutagens.

There still is room for disagreement about the interpretation of such correlations. The simplest interpretation is that the neoplastic transformation is brought about directly by mutation. A number of objections to this view have been raised, however, and other interpretations must be considered as possible or even probable. Many, if not all, mutagens can induce lysogenic viruses, cause filamentation, and induce certain enzymes and enzyme systems in bacteria. Thus, it would appear that alterations in the DNA that can lead to mutation also can cause changes in control and transcriptional events that in turn could have quite a variety of secondary consequences for the cell. A number of such alternatives have been discussed in articles in "Chemical Carcinogenesis," edited by T'so and di Paola (1974); in "Chemical Carcinogenesis Essays" edited by Montesano and Tomatis (1974); and in reviews by Miller and Miller (1971) and by Heidelberger (1975).

Although a high correlation between mutagenicity and carcinogenicity is suggestive evidence that DNA is the target molecule for carcinogenesis, it does not provide irrefutable proof. The main chemical characteristic of most or all chemical carcinogens is that they are electrophiles (relatively electron-deficient) that will attack the nucleophiles (relatively electron-rich) in the cell (Miller and Miller, 1974). Nucleophilic groups exist not only in DNA but also in RNA and protein. Indeed, the idea that proteins were the target molecules for carcinogens was quite popular at one time, although it seems less so now.

The most convincing evidence that alterations in the DNA can cause tumors comes from work on excision-defective cells and on photoreactivation. As mentioned earlier, xeroderma pigmentosum patients are highly susceptible to the induction of skin carcinomas by sunlight. Many of these patients are defective in excision repair, and even those who are not may be defective in postreplication repair. These facts make it appear certain that excisable lesions in the DNA can lead to tumor production.

Fish, unlike mammals, have a readily demonstrable photoreactivation system, presumably specific for the monomerization of pyrimidine dimers. Hart and Setlow (1975) have demonstrated that photoreactivation greatly reduces the frequency of tumors induced by ultraviolet light in these fish, thus providing strong evidence that pyrimidine dimers in the DNA can be an important factor in tumor development.

VII. USE OF REPAIR TO DETECT AND MEASURE DAMAGE TO DNA

A practical problem of increasing importance is the identification of chemicals and other agents that are potential mutagens or carcinogens. It also would be desirable to have a quantitative measure of the effective "dose" to the DNA when such agents are administered by various routes. There are a number of ways in which one might proceed—direct tests for mutagenicity and carcinogenicity, and direct chemical tests for interaction with DNA. One promising possibility is to make use of the cell's own repair system. The rationale for such attempts is that the cell's own repair enzyme systems can recognize a wider variety of alterations in DNA more readily than present biochemical and biophysical techniques, and that the resulting repair can be detected and measured fairly easily and rapidly.

A discussion of some of these possibilities is given by Regan and Setlow (1973). The major techniques for detecting and measuring repair have been given earlier in the chapter (Section II). They are unscheduled DNA synthesis, detection of repair synthesis by density labeling, and detection of repair by incorporation of bromouracil followed by exposure to 313-nm light. All of the methods allow an estimate of the amount of repair synthesis per DNA molecule. The latter two also allow estimations of the length of the repaired region. This is important because, as pointed out in Section II, some agents cause predominantly long-patch, while others cause predominantly short-patch repair. Thus, estimations of the total number of repairable alterations in the DNA from the total amount of repair synthesis must take into account the average length of the patch.

Recently, unscheduled DNA synthesis has been used to estimate the effective dose of ethyl methane sulfonate (EMS) to postmeiotic germ cells in male mice for comparison with the known genetic effects of this agent on these cells (Sega *et al.*, 1974). The method seems quite suitable for developing a quantitative dosimetry for this stage, but is not applicable to late stages in which no repair occurs or to earlier stages in which semiconservative replication is occurring.

Unscheduled DNA synthesis also has been used to test for the potentially active form of carcinogens in mammalian cells (Laishes and Stich, 1973; Stich *et al.*, 1974). The method seems promising for the testing of activating systems.

VIII. INTERRELATIONS

In this chapter, three major ways have been discussed whereby interactions of chemical and physical agents with DNA can be detected and measured. These are (1) direct physical or chemical measurements of the lesions in DNA, (2) measurements of repair replication or unscheduled DNA synthesis, and (3) measurements of mutation. The first gives information about the lesions themselves and to some extent about how the cell handles them. The second shows what the cell's own enzyme systems detect and attempt to repair. The third shows the particular agents to which systems that can make errors during repair or replication are sensitive. Although there is considerable overlap in what these methods detect, it is clear that the overlap is not complete. Thus, there are certainly some alterations in DNA, probably minor alterations in single nucleotides and possibly some strand breaks that are not subject to excision repair even though they may have biological consequences such as mutation or death. Under some circumstances, both excision and postreplication repair can occur so as to produce little if any mutation. The relationships between lesions in the DNA detected by any of these methods and biological endpoints such as cell death and neoplastic transformation are still less certain, although there is convincing evidence that such interrelations exist.

It is possible to make some empirical generalizations, however. Thus, pyrimidine dimers produced by uv are clearly a biologically very important lesion in DNA, while, on the contrary, alkylations of the 7 position of guanine almost certainly are not. There is a high correlation between the ability of an agent to produce mutation in the *Salmonella* test system and its ability to induce cancer. There is some correlation, but a rather poor one, between the ability to induce mutations and the ability to cause cell death. Many other such empirical generalizations could be cited.

There are partial theories attempting to relate physical and chemical effects to biological endpoints (e.g., the induction of mutations by error-prone repair), but a really strong theoretical structure from which to predict the biological consequences of interactions with DNA does not yet exist, and the empirical approach frequently is still the most useful way to proceed. In any case, it seems clear that many biological consequences of exposure to various agents are the consequence of the lesions they produce in DNA. Thus, any method, physical or biological, for detecting such lesions can be useful in identifying agents with potentially serious biological consequences. It still is true, however, that such identifications are tentative; and, in the absence of a strong theoretical structure, the really convincing proof must be a direct test for the biological consequence of interest.

IX. SUMMARY

It has been thoroughly established that quite a number of alterations of DNA produced by interactions with physical or chemical agents can be repaired or eliminated by enzyme systems in the cell. The most completely elucidated system is that in *E. coli,* but similar systems have been found in all cells that have been adequately studied. The impression is that these systems have been relatively conservative during evolution and, although there are a number of differences in detail, the overall features are much alike.

Three main systems have been identified for repairing or eliminating altered bases. These are photoreactivation, prereplication excision repair, and postreplication repair. Photoreactivation appears to be brought about by a single enzyme whose sole known function is to monomerize pyrimidine dimers. Both excision repair and postreplication repair are multistep processes involving several enzymes, and each probably has more than one form—depending on the kind of lesion and on the cell. A number of the enzymes involved in these processes have been identified and purified to various degrees, and some of the steps have been successfully carried out *in vitro*. Mutants defective in some of these processes have been obtained in *E. coli* and, to a lesser extent, in other organisms including humans. In addition to systems that eliminate altered bases, there are systems for repairing strand breaks in DNA.

The relation between mutation induction and repair processes has been studied most thoroughly in *E. coli,* although there is also information from a number of other species. For *E. coli,* it has been fairly well established that mutations produced by uv and ionizing radiations and by certain chemicals result entirely or almost entirely from an induced error-prone repair system that probably operates during both pre- and postreplication repair. Some chemicals, however, produce mutations by processes independent of this error-prone system. These processes probably include simple mispairing at replication without any involvement of repair at all, but they also may include other, as yet not clearly identified, forms of error-prone repair. There is considerable evidence that premutational lesions that persist to replication are more likely to be converted to mutations than are lesions that are repaired before replication. There is good evidence, however, that some conversion to final mutation can occur during prereplication repair processes.

It seems certain that chromosomal aberrations arise by some form of failure of repair (unrejoined breaks), by some form of misrepair (such as rejoining of breaks in abnormal arrangements, induced recombination). That the initial lesions are in DNA, at least in many cases, is shown by studies with photoreactivation and with cells defective for DNA repair. The induction of cytologically observable sister-strand exchanges may involve a different mechanism than that responsible for the induction of true chromosomal aberrations.

There is strong evidence, both for prokaryotes and eukaryotes, linking cell killing to unrepaired but repairable lesions in DNA. However, the processes by which such lesions cause death are understood poorly. It seems unlikely that lethal mutations are an important cause of cell death in prokaryotes, but chromosomal losses might be an important cause in eukaryotes. Effects on transcriptional control certainly can be suspected of being an important contributor to cell death in both groups.

Evidence is increasing that reparable lesions in DNA can cause cancer. Humans with the inherited condition xeroderma pigmentosum, and consequently defective in excision repair, are very prone to skin cancer caused by sunlight. Photoreactivation greatly reduces the frequency of tumors induced in fish by ultraviolet light. Moreover, there is increasing evidence that many, if not all, chemical carcinogens are also mutagens. The mechanisms by which repairable lesions in the DNA can cause a neoplastic transformation are as yet very uncertain. Thus, the interrelationships between lesions in the DNA, repair, mutation, cell death, and cancer still are understood no more than partially, and we still are depend heavily empirical generalizations rather than on firmly based theory for the conclusions we draw. However, interrelations clearly exist and, as empirical knowledge and the understanding of the processes themselves increase, a stronger theoretical basis for predictions from one type of effect to others may be expected to develop.

REFERENCES

Ames, B. N. (1974). A combined bacterial and liver test system for detection and classification of carcinogens and mutagens. *Genetics* **78,** 91–95.

Ames, B. N., Durston, W. E., Yamasaki, E., and Lee, F. D. (1973). Carcinogens are mutagens: A simple test system combining liver homogenates for activation and bacteria for detection. *Proc. Natl. Acad. Sci. U.S.A.* **70,** 2281–2285.

Beattie, K. L. (1975). *N*-nitrosocarbaryl-induced mutagenesis in *Haemophilus influenzae* strains deficient in repair and recombination. *Mutat. Res.* **27,** 201–217.

Beattie, K. L., and Kimball, R. F. (1974). Involvement of DNA replication and repair in mutagenesis of *Haemophilus influenzae* produced by *N*-nitrosocarbaryl. *Mutat. Res.* **24,** 105–115.

Beers, R. F., Jr., Herriott, R. M., and Tilghman, R. C., eds. (1972). "Molecular and Cellular Repair Processes." Johns Hopkins Univ. Press, Baltimore, Maryland.

Bender, M. A., Griggs, H. G., and Walker, P. L. (1973a). Mechanisms of chromosomal aberration production. I. Aberration induction by ultraviolet light. *Mutat. Res.* **20,** 387–402.

Bender, M. A., Griggs, H. G., and Mitchell, J. B. (1973b). Mechanisms of chromosomal aberrations production. II. Aberrations induced by 5-bromodeoxyuridine and visible light. *Mutat. Res.* **20,** 403–416.

Bender, M. A., Griggs, H. G., and Bedford, J. S. (1974a). Mechanisms of chromosomal aberration production. III. Chemicals and ionizing radiation. *Mutat. Res.* **23,** 197–212.

Bender, M. A., Griggs, H. G., and Bedford, J. S. (1974b). Recombinational DNA repair and sister chromatid exchanges. *Mutat. Res.* **24,** 117–123.

Billen, D. (1969). Replication of the bacterial chromosome: Location of new initiation sites after irradiation. *J. Bacteriol.* **97,** 1169–1175.

Billen, D., and Cain, J. (1973). X-ray-induced alterations in marker replication pattern in *Bacillus subtilis* W23. *Radiat. Res.* **56,** 271–281.

Billen, D., and Hellermann, G. R. (1975). Depression by NAD of X-ray-induced repair-type DNA synthesis in toluene-treated *Bacillus subtilis*. *Biochim. Biophys. Acta* **383,** 374–387.

Boyce, R. P., and Farley, J. W. (1968). Production of single-strand breaks in covalent circular λ phage DNA in superinfected lysogens by monoalkylating agents and the joining of broken DNA strands. *Virology* **35,** 601–609.

Boyce, R. P., and Howard-Flanders, P. (1964). Release of ultraviolet light-induced thymine dimers from DNA in *E. coli* K-12. *Proc. Natl. Acad. Sci. U.S.A.* **51,** 293–300.

Brewen, J. G., and Peacock, W. J. (1969). Restricted rejoining of chromosomal subunits in aberration formation: A test for subunit dissimilarity. *Proc. Natl. Acad. Sci. U.S.A.* **62,** 389–394.

Bridges, B. A., and Munson, R. J. (1968). The persistance through several replication cycles of mutation-producing pyrimidine dimers in a strain of *Escherichia coli* deficient in excision repair. *Biochem. Biophys. Res. Commun.* **30,** 620–624.

Bridges, B. A., Law, J., and Munson, R. J. (1968). Mutagenesis in *Escherichia coli*. II. Evidence for a common pathway for mutagenesis by ultraviolet light, ionizing radiation, and thymine deprivation. *Mol. Gen. Genet.* **103,** 266–273.

Buhl, S. N., and Regan, J. D. (1973). Repair endonuclease-sensitive sites in daughter DNA of ultraviolet-irradiated human cells. *Nature (London)* **246,** 484.

Buhl, S. N., Setlow, R. B., and Regan, J. D. (1973). Recovery of the ability to synthesize DNA in segments of normal size at long times after ultraviolet irradiation of human cells. *Biophys. J.* **13,** 1265–1275.

Castellazzi, M., George, J., and Buttin, G. (1972). Prophage induction and cell division in *E. coli*. I. Further characterization of the thermosensitive mutation *tif*-1 whose expression mimics the effects of uv irradiation. *Mol. Gen. Genet.* **119,** 139–152.

Cerdá-Olmeda, E., Hanawalt, P. C., and Guerola, N. (1968). Mutagenesis of the replicating point by nitrosoguanidine map and pattern of replication of the *Escherichia coli* chromosome. *J. Mol. Biol.* **33,** 705–719.

Cerutti, P. (1974). Excision repair of DNA base damage. *Life Sci.* **15,** 1567–1575.

Clark, A. J. (1973). Recombination deficient mutants of *E. coli* and other bacteria. *Annu. Rev. Genet.* **7,** 67–86.

Clark, A. J., and Ganesan, A. (1975). Lists of genes affecting DNA metabolism in *Escherichia coli*. *In* "Molecular Mechanisms for Repair of DNA" (P. C. Hanawalt and R. B. Setlow, eds.), Part B, pp. 431–437. Plenum, New York.

Cleaver, J. E. (1970). DNA repair and radiation sensitivity in human (xeroderma pigmentosum) cells. *Int. J. Radiat. Biol.* **18,** 557–565.

Cleaver, J. E. (1974). Repair processes for photochemical damage in mammalian cells. *Adv. Radiat. Biol.* **4,** 1–75.

Cleaver, J. E., and Bootsma, D. (1975). Xeroderma pigmentosum: Biochemical and genetic characteristics. *Annu. Rev. Genet.* **9,** 19–38.

Cleaver, J. E., Bootsma, D., and Friedberg, E. (1975). Human diseases with genetically altered DNA repair processes. *Genetics* **79,** Suppl., 215–225.

Cook, J. S. (1970). Photoreactivation in animal cells. *Photophysiology* **5,** 191 233.

Davies, D. R., and Evans, H. J. (1966). The role of genetic damage in radiation-induced lethality. *Adv. Radiat. Biol.* **2,** 243–353.

Defais, M., Fauquet, P., Redmon, M., and Errera, M. (1971). Ultraviolet reactivation and ultraviolet mutagenesis of λ in different genetic systems. *Virology* **43,** 495–503.

deWeerd- Kastelen, E. A., Keijzer, W., and Bootsma, D. (1972). Genetic heterogeneity of xeroderma pigmentosum demonstrated by somatic cell hybridization. *Nature (London), New Biol.* **238,** 80–83.

Drake, J. W., ed. (1973). The genetic control of mutation. *Genetics* **73,** Suppl.

Dulbecco, R. (1949). Reactivation of ultraviolet-inactivated bacteriophage by visible light. *Nature (London)* **163,** 949–950.

Elkind, M. M., and Whitmore, G. F. (1967). "The Radiobiology of Cultured Mammalian Cells." Gordon & Breach, New York.

Freese, E. (1959). On the molecular explanation of a spontaneous and induced mutations. *Brookhaven Symp. Biol.* **12,** 63–75.

Game, J. C., and Cox, B. S. (1971). Allelism tests of mutants affecting sensitivity to radiation in yeast and a proposed nomenclature. *Mutat. Res.* **12,** 328–331.

Ganesan, A. K. (1974). Resistance of pyrimidine dimers during postreplication repair in ultraviolet light-irradiated *Escherichia coli* K12. *J. Mol. Biol.* **87,** 103–119.

George, D. L., and Witkin, E. M. (1974). Slow excision repair in an *mfd* mutant of *Escherichia coli* B/r. *Mol. Gen. Genet.* **133,** 283–291.

George, D. L., and Witkin, E. M. (1975). Ultraviolet light-induced responses of an *mfd* mutant of *Escherichia coli* B/r having a slow rate of dimer excision. *Mutat. Res.* **28,** 347–354.

Griggs, H. G., and Bender, M. A. (1973). Photoreactivation of ultraviolet-induced chromosomal aberrations. *Science* **179,** 86–88.

Grossman, L., Braun, A., Feldberg, R., and Mahler, I. (1975). Enzymatic repair of DNA. *Annu. Rev. Biochem.* **44,** 19–43.

Hanawalt, P. C. (1975). Repair models and mechanisms: Overview. *In* "Molecular Mechanisms for Repair of DNA" (P. C. Hanawalt and R. B. Setlow, eds.), pp. 421–429. Plenum, New York.

Hanawalt, P. C., and Setlow, R. B., eds. (1975). "Molecular Mechanisms for Repair of DNA." Plenum, New York.

Hart, R. W., and Setlow, R. B. (1975). Direct evidence that pyrimidine dimers in DNA result in neoplastic transformation. *In* "Molecular Mechanisms for Repair of DNA" (P. C. Hanawalt and R. B. Setlow, eds.), Part B, pp. 719–724. Plenum, New York.

Heidelberger, C. (1975). Carcinogenesis. *Annu. Rev. Biochem.* **44,** 79–121.

Hollaender, A., ed. (1954). "Radiation Biology," Vol. 1. McGraw-Hill, New York.

Horii, Z.-I., and Clark, A. J. (1973). Genetic analysis of the *recF* pathway to genetic recombination in *Escherichia coli* K12: Isolation and characterization of mutants. *J. Mol. Biol.* **80,** 327–344.

Howard-Flanders, P. (1968). Genes that control DNA repair and genetic recombination in *Escherichia coli. Adv. Biol. Med. Phys.* **12,** 299–317.

Howard-Flanders, P., and Rupp, W. D. (1972). Recombinational repair in uv-irradiated *Escherichia coli. In* "Molecular and Cellular Repair Processes" (R. F. Beers, Jr., R. M. Herriott, and R. C. Tilghman, eds.), pp. 212–225. Johns Hopkins Univ. Press, Baltimore, Maryland.

Howell-Saxton, E., Smith, D. C., Zamenhoff, P. J., and Zamenhoff, S. (1974). Effects of growth phase and repair capacity on rejoining of ethyl methane sulfonate-induced DNA breaks in *Escherichia coli. Mutat. Res.* **24,** 227–237.

Ishii, Y., and Kondo, S. (1975). Comparative analysis of deletion and base-change mutabilities of *Escherichia coli* B strains differing in DNA repair capacity (*wild-type, uvrA$^-$, polA$^-$, recA$^-$*) by various mutagens. *Mutat. Res.* **27,** 27–44.

Jagger, J. (1958). Photoreactivation. *Bacteriol. Rev.* **22,** 99–142.

Jiménez-Sanchez, A., and Cerdá-Olmedo, E. (1975). Mutation and DNA replication in *Escherichia coli* treated with low concentrations of *N*-methyl-*N'*-nitro-*N*-nitrosoguanidine. *Mutat. Res.* **28,** 337–345.

Kada, T., Tutikawa, K., and Sadaie, Y. (1972). *In vitro* and host-mediated "rec-assay" procedures for screening chemical mutagens; and phloxine, a mutagenic red dye detected. *Mutat. Res.* **16,** 165–174.

Kato, H. (1973). Induction of sister chromatid exchanges by uv light and its inhibition by caffeine. *Exp. Cell Res.* **82,** 383–390.

Kelner, A. (1949). Effect of visible light on the recovery of *Streptomyces griseus* conidia from ultraviolet injury. *Proc. Natl. Acad. Sci. U.S.A.* **35,** 73–79.

Kimball, R. F. (1966). Repair of premutational damage. *Adv. Radiat. Biol.* **2,** 135–166.

Kimball, R. F. (1969). Studies on mutations induced by ultraviolet radiation in *Paramecium aurelia* with special emphasis on photoreversal. *Mutat. Res.* **8,** 79–89.

Kimball, R. F. (1970). Studies on the mutagenic action of *N*-methyl-*N'*-nitro-*N*-nitrosoguanidine in *Paramecium aurelia* with emphasis on repair processes. *Mutat. Res.* **9,** 261–271.

Kimball, R. F., and Hirsch, B. F. (1975). Tests for the mutagenic action of a number of chemicals on *Haemophilus influenzae* with special emphasis on hydrazine. *Mutat. Res.* **30,** 9–20.

Kimball, R. F., and Hirsch, B. F. (1976). Fixation and loss of hydrazine-induced premutational damage in *Haemophilus influenzae. Mutat. Res.* **36,** 39–48.

Kimball, R. F., and Setlow, J. K. (1974). Mutation fixation in MNNG-treated *Haemophilus influenzae* as determined by transformation. *Mutat. Res.* **22,** 1–14.

Kimball, R. F., Setlow, J. K., and Liu, M. (1971a). The mutagenic and lethal effects of monofunctional methylating agents in strains of *Haemophilus influenzae* defective in repair processes. *Mutat. Res.* **12,** 21–28.

Kimball, R. F., Liu, M., and Setlow, J. K. (1971b). Effects of posttreatment on single-strand breaks in DNA of *Haemophilus influenzae* exposed to nitrosoguanidine and methyl methane sulfonate. *Mutat. Res.* **13,** 289–295.

Kondo, S. (1973). Evidence that mutations are induced by errors in repair and replication. *Genetics* **73,** Suppl., 107–122.

Kondo, S., and Ichikawa, H. (1973). Evidence that pretreatment of *Escherichia coli* cells with *N*-methyl-*N'*-nitro-*N*-nitrosoguanidine enhances mutability of subsequently infecting phage λ. *Mol. Gen. Genet.* **126,** 319–324.

Kondo, S., Ichikawa, H., Iwo, K., and Kato, T. (1970). Base-change mutagenesis and prophage induction in strains of *Escherichia coli* with different DNA repair capacities. *Genetics* **66,** 187–217.

Laishes, B. A., and Stich, H. F. (1973). Repair synthesis and sedimentation analysis of DNA of human cells exposed to dimethylnitrosamine and activated dimethylnitrosamine. *Biochem. Biophys. Res. Commun.* **52,** 827–833.

Lawley, P. D. (1966). Effects of some chemical mutagens on nucleic acids. *Prog. Nucleic Acid Res. Mol. Biol.* **5,** 89–131.

Lawrence, C. W., Stewart, J. W., Sherman, F., and Christensen, R. (1974). Specificity and frequency of ultraviolet-induced reversion of an iso-1-cytochrome c ochre mutant in radiation-sensitive strains of yeast. *J. Mol. Biol.* **85,** 137–162.

LeClerc, J. E., and Setlow, J. K. (1972). Postreplication repair of ultraviolet damage in *Haemophilus influenzae. J. Bacteriol.* **110,** 930–934.

Lehman, A. R. (1974). Postreplication repair of DNA in mammalian cells. *Life Sci.* **15,** 2005–2016.

Lehman, A. R., Kirk-Bell, S., Arlett, C. F., Paterson, M. C., Lohman, P. M., deWeerd-Kastelein, E. A., and Bootsma, D. (1975). Xeroderma pigmentosum cells with normal levels of excision repair have a defect in DNA synthesis after uv-irradiation. *Proc. Natl. Acad. Sci. U.S.A.* **72,** 219–223.

Lehman, I. R. (1974). DNA joining enzymes (ligases). *In* "The Enzymes" (P. D. Boyer, ed.), 3rd ed., Vol. 10, pp. 237–259. Academic Press, New York.

Lemontt, J. F. (1973). Genes controlling ultraviolet mutability in yeast. *Genetics* **73,** Suppl., 153–159.

McGrath, R. A., and Williams, R. W. (1966). Reconstruction *in vivo* of irradiated *Escherichia coli* deoxyribonucleic acid; the rejoining of broken pieces. *Nature (London)* **212,** 534–535.

Masker, W. E., and Hanawalt, P. C. (1974). Nucleoside triphosphate dependence of repair replication in toluenized *Escherichia coli. J. Mol. Biol.* **88,** 13–23.

Meneghini, R., and Hanawalt, P. C. (1976). T4-endonuclease V-sensitive sites in DNA from ultraviolet-irradiated human cells. *Biochim. Biophys. Acta* **425,** 428–437.

Miller, E. C., and Miller, J. A. (1971). The mutagenicity of chemical carcinogens: Correlations, problems, and interpretations. *In* "Chemical Mutagens" (A. Hollaender, ed.), Vol. 1, pp. 83–119. Plenum, New York.

Miller, J. A., and Miller, E. C. (1974). Some current thresholds of research in chemical carcinogens. *In* "Chemical Carcinogenesis" (P. O. P. Ts'o and J. A. di Paolo, eds.), Part A, pp. 61–85. Dekker, New York.

Modig, H. G., Edgren, M., and Révész, L. (1974). Dual effect of oxygen on the induction and repair of single-strand breaks in the DNA of X-irradiated mammalian cells. *Int. J. Radiat. Biol.* **26,** 341–353.

Montesano, R., and Tomatis, L., eds. (1974). "Chemical Carcinogenesis Essays." Int. Agency Res. Cancer, Lyon.

Muller, H. J. (1941). Induced mutations in *Drosophila. Cold Spring Harbor Symp. Quant. Biol.* **9,** 151–167.

Nishioka, H., and Doudney, C. O. (1969). Different modes of loss of photoreversibility of mutation and lethal damage in ultraviolet-light resistant and sensitive bacteria. *Mutat. Res.* **8,** 215–228.

Nishioka, H., and Doudney, C. O. (1970). Different modes of loss of photoreversibility of ultraviolet light-induced true and suppressor mutations to tryptophan independence in an auxotrophic strain of *Escherichia coli. Mutat. Res.* **9,** 349–358.

Painter, R. B. (1970). Repair of DNA in mammalian cells. *Curr. Top. Radiat. Res.* **7,** 45–70.

Palcic, B., and Skarggard, L. D. (1975). Absence of ultrafast processes of repair of single-strand breaks in mammalian DNA. *Int. J. Radiat. Biol.* **27,** 121–133.

Parrington, J. M., Delhanty, J. D. A., and Baden, H. P. (1971). Unscheduled DNA synthesis, uv-induced chromosome aberrations and SV40 transformation in cultured cells from xeroderma pigmentosum. *Ann. Hum. Genet.* **35,** 149–160.

Paterson, M. C., Lohman, P. H. M., and Sluyter, M. L. (1973). Use of a uv endonuclease from *Micrococcus luteus* to monitor the progress of DNA repair in uv-irradiated human cells. *Mutat. Res.* **19,** 245–256.

Pettijohn, D., and Hanawalt, P. (1964). Evidence for repair-replication of ultraviolet damaged DNA in bacteria. *J. Mol. Biol.* **9,** 395–410.

Prakash, L. (1975). Repair of pyrimidine dimers in nuclear and mitochondrial DNA of yeast irradiated with low doses of ultraviolet light. *J. Mol. Biol.* **98,** 781–795.

Prakash, L., and Strauss, B. (1970). Repair of alkylation damage: Stability of methyl groups in *Bacillus subtilis* treated with methyl methane sulfonate. *J. Bacteriol.* **102,** 760–766.

Puck, T. T. (1960). The action of radiation on mammalian cells. *Am. Nat.* **94,** 95–109.

Radman, M. (1974). Phenomenology of an inducible mutagenic repair pathway in *Escherichia coli:* SOS repair hypothesis. *In* "Molecular and Environmental Aspects of Mutagenesis" (L. Prakash, ed.), pp. 128–142. Thomas, Springfield, Illinois.

Regan, J. D., and Setlow, R. B. (1973). Repair of chemical damage to human DNA. *In* "Chemical Mutagens" (A. Hollaender, ed.), Vol. 3, pp. 151–170. Plenum, New York.

Regan, J. D., Setlow, R. B., and Ley, R. D. (1971). Normal and defective repair of damaged DNA in human cells: A sensitive assay utilizing the photolysis of bromodeoxyuridine. *Proc. Natl. Acad. Sci. U.S.A.* **68,** 708–712.

Reiter, H., Strauss, B., Robbins, M., and Marone, R. (1967). Nature of the repair of methyl methane sulfonate-induced damage in *Bacillus subtilis. J. Bacteriol.* **93,** 1056–1062.

Resnick, M. A., and Setlow, J. K. (1972). Repair of pyrimidine dimer damage induced in yeast by ultraviolet light. *J. Bacteriol.* **109,** 979–986.

Revell, S. H. (1974). The breakage-and-reunion theory and the exchange theory for chromosomal aberrations induced by ionizing radiations: A short history. *Adv. Radiat. Biol.* **4,** 367–416.

Roots, R., and Smith, K. C. (1974). On the nature of the oxygen effect on X-ray-induced DNA single-strand breaks in mammalian cells. *Int. J. Radiat. Biol.* **26,** 467–480.

Rothman, R. H., Kato, T., and Clark, A. J. (1975). The beginning of an investigation of the role of *recF* in the pathways of metabolism of ultraviolet-irradiated DNA in *Escherichia coli. In* "Molecular Mechanisms for Repair of DNA" (P. C. Hanawalt and R. B. Setlow, eds.), Part A, pp. 283–291. Plenum, New York.

Rupert, C. S. (1964). Photoreactivation of ultraviolet damage. *Photophysiology* **2,** 283–327.

Rupp, W. D., and Howard-Flanders, P. (1968). Discontinuities in the DNA synthesized in an excision-defective strain of *Escherichia coli* following ultraviolet irradiation. *J. Mol. Biol.* **31,** 291–304.

Russell, W. L. (1967). Repair mechanisms in radiation mutation induction in the mouse. *Brookhaven Symp. Biol.* **20,** 179–189.

Sasaki, M. S. (1973). DNA repair capacity and susceptibility to chromosome breakage in xeroderma pigmentosum cells. *Mutat. Res.* **20,** 291–293.

Sax, K. (1938). Induction by X-rays of chromosome aberrations in *Tradescantia* microspores. *Genetics* **23,** 494–516.

Sax, K. (1941). Types and frequencies of chromosomal aberrations induced by X rays. *Cold Spring Harbor Symp. Quant. Biol.* **9,** 93–101.

Sedgwick, S. G. (1975). Inducible error-prone repair in *Escherichia coli*. *Proc. Natl. Acad. Sci. U.S.A.* **72,** 2753–2757.

Sega, G. A., Cumming, R. B., and Walton, M. F. (1974). Dosimetry studies on ethylation of mouse sperm DNA after *in vivo* exposure to [^{3}H]ethyl methane sulfonate. *Mutat. Res.* **24,** 317–333.

Setlow, J. K. (1966). The molecular basis of biological effects of ultraviolet radiation and photoreactivation. *Curr. Top. Radiat. Res.* **2,** 195–248.

Setlow, R. B. (1968). Steps in repair of DNA: Fact and fancy. *Brookhaven Symp. Biol.* **20,** 1–18.

Setlow, R. B. (1975). Relationship among repair, cancer, and genetic deficiency: Overview. *In* "Molecular Mechanisms for Repair of DNA" (P. C. Hanawalt and R. B. Setlow, eds.), Part B, pp. 711–717. Plenum, New York.

Setlow, R. B., and Carrier, W. L. (1964). The disappearance of thymine dimers from DNA: An error-correcting mechanism. *Proc. Natl. Acad. Sci. U.S.A.* **51,** 226–231.

Sharon, R., Miller, C., and Ben-Ishai, R. (1975). Two modes of excision repair in toluene-treated *Escherichia coli*. *J. Bacteriol.* **123,** 1107–1114.

Stich, H. F., Kieser, D., Laishes, B. A., and San, R. H. C. (1974). The use of DNA repair in the identification of carcinogens, precarcinogens, and target tissue. *Proc. Can. Cancer Res. Conf.* **10,** 83–110.

Sutherland, B. M., Sutherland, J. C., and Chamberlin, M. J. (1973). Deoxyribonucleic acid photoreactivating enzyme from *Escherichia coli*. *J. Biol. Chem.* **248,** 4200–4205.

Swenson, P. A. (1975). Physiological responses of *Escherichia coli* to far-ultraviolet radiation. *Photochem. Photobiol. Rev.* **1,** 269–387.

Town, C. D., Smith, K. C., and Kaplan, H. S. (1973). Repair of X-ray damage to bacterial DNA. *Curr. Top. Radiat. Res. Q.* **8,** 351–399.

Ts'o, P. O. P., and di Paolo, J. A., eds. (1974). "Chemical Carcinogenesis." Dekker, New York.

Waldstein, E. A., Sharon, R., and Ben-Ishai, R. (1974). Role of ATP in excision repair of ultraviolet radiation damage in *Escherichia coli*. *Proc. Natl. Acad. Sci. U.S.A.* **71,** 2651–2654.

Watson, J. D., and Crick, F. J. (1953). The structure of DNA. *Cold Spring Harbor. Symp. Quant. Biol.* **18,** 123–131.

Wilkins, R. J. (1973). Endonuclease-sensitive sites in the DNA of irradiated bacteria: A rapid and sensitive assay. *Biochim. Biophys. Acta* **312,** 33–37.

Witkin, E. M. (1967). Mutation-proof and mutation-prone modes of survival in derivatives of *Escherichia coli* B differing in sensitivity to ultraviolet light. *Brookhaven Symp. Biol.* **20,** 17–55.

Witkin, E. M. (1969a). Ultraviolet-induced mutation and DNA repair. *Annu. Rev. Microbiol.* **23,** 487–514.

Witkin, E. M. (1969b). The role of repair in mutagenesis. *Proc. Int. Congr. Genet., 12th, 1968* Vol. III, pp. 225–245.

Witkin, E. M. (1971). The mutability toward ultraviolet light of recombination-deficient strains of *Escherichia coli. Mutat. Res.* **8,** 9–14.

Witkin, E. M. (1974). Thermal enhancement of ultraviolet mutability in a *tif-1 uvrA* derivative of *Escherichia coli* B/r: Evidence for an inducible error-prone system of DNA repair (SOS repair). *Proc. Natl. Acad. Sci. U.S.A.* **71,** 1930–1934.

Witkin, E. M., and George, D. L. (1973). Ultraviolet mutagenesis in *polA* and *uvrA polA* derivatives of *Escherichia coli* B/r: Evidence for an inducible error-prone repair system. *Genetics* **73,** Suppl., 91–108.

Wolff, S., and Bodycote, J. (1975). The induction of chromatid deletions in accord with the breakage-and-reunion hypothesis. *Mutat. Res.* **29,** 85–91.

Wolff, S., Bodycote, J., and Painter, R. B. (1974). Sister chromatid exchanges induced in Chinese hamster cells by uv irradiation of different stages of the cell cycle: The necessity for cells to pass through S. *Mutat. Res.* **25,** 73–81.

Wolff, S., Bodycote, J., Thomas, G. H., and Cleaver, J. E. (1975). Sister chromatid exchange in xeroderma pigmentosum cells that are defective in DNA excision repair or postreplication repair. *Genetics* **81,** 349–355.

Index

D

R